Integrated Soil Management: Food Supply, Environmental Impacts, and Socioeconomic Functions

Integrated Soil Management: Food Supply, Environmental Impacts, and Socioeconomic Functions

Guest Editors

José L. S. Pereira
Vítor João Pereira Domingues Martinho

MDPI

Basel • Beijing • Wuhan • Barcelona • Belgrade • Novi Sad • Cluj • Manchester

Guest Editors
José L. S. Pereira
School of Agriculture (ESAV) and CERNAS-IPV Research Centre
Polytechnic Institute of Viseu
Viseu
Portugal

Vítor João Pereira Domingues Martinho
School of Agriculture (ESAV) and CERNAS-IPV Research Centre
Polytechnic Institute of Viseu
Viseu
Portugal

Editorial Office
MDPI AG
Grosspeteranlage 5
4052 Basel, Switzerland

This is a reprint of the Special Issue, published open access by the journal *Soil Systems* (ISSN 2571-8789), freely accessible at: https://www.mdpi.com/journal/soilsystems/special_issues/IntegratedSoilManagement_FoodSupply_EnvironmentalImpacts_SocioeconomicFunctions.

For citation purposes, cite each article independently as indicated on the article page online and as indicated below:

Lastname, A.A.; Lastname, B.B. Article Title. *Journal Name* **Year**, *Volume Number*, Page Range.

ISBN 978-3-7258-4649-8 (Hbk)
ISBN 978-3-7258-4650-4 (PDF)
https://doi.org/10.3390/books978-3-7258-4650-4

Cover image courtesy of José Luís da Silva Pereira and Vítor João Pereira Domingues Martinho

Contents

About the Editors

José L. S. Pereira

José L. S. Pereira is Coordinator Professor with Habilitation at the Polytechnic Institute of Viseu, Portugal, and holds a Ph.D. in Agricultural Engineering from the University of Trás-os-Montes and Alto Douro, Portugal. He is Coordinator of the CERNAS research centre management unit at the Polytechnic Institute of Viseu. He publishes regularly in international journals and has coordinated and participated in several national and international R&D projects. He has coordinated several research projects on the control of agricultural pollution to the atmosphere, soil, and water and the evaluation of effluent treatment and valorisation processes.

Vítor João Pereira Domingues Martinho

Vítor João Pereira Domingues Martinho is Coordinator Professor with Habilitation at the Polytechnic Institute of Viseu, Portugal, and holds a Ph.D. in Economics from the University of Coimbra, Portugal. He was President of the Scientific Council, President of the Directive Council, and President of the Agricultural Polytechnic School of Viseu, Portugal, from 2006 to 2012. Moreover, he served as Vice-President of the Scientific Council and the Assembly of Representatives in this school. He also had political roles at the local level and supra municipal level and was the President of the Direction of the Association of Forest Producers of Viseu, Portugal. He was an Erasmus student in the Faculty of Economics from the University of Verona, Italy. Furthermore, he participated in various technical and scientific events/projects nationally and internationally, has published several technical and scientific papers, is referee of various scientific and technical journals, and participates in the evaluation of national and international projects.

soil systems

MDPI

Editorial

Integrated Soil Management: Food Supply, Environmental Impacts, and Socioeconomic Functions

José L. S. Pereira [1,2,*] and Vítor J. P. D. Martinho [1,2,*]

1 Agrarian Higher School of Viseu, Polytechnic Institute of Viseu, Avenida António Almeida Henriques, 3500-631 Viseu, Portugal
2 CERNAS-IPV Research Centre, Polytechnic Institute of Viseu, Avenida José Maria Vale Andrade, 3504-510 Viseu, Portugal
* Correspondence: jlpereira@esav.ipv.pt (J.L.S.P.); vdmartinho@esav.ipv.pt (V.J.P.D.M.)

Soil is a key resource for agricultural production and, consequently, for food supply and sustainable development [1,2]. In fact, the quality of soil impacts the characteristics of the outputs obtained and the income of the farms, with implications for the performance of the agricultural sector and the various associated upstream and downstream activities [3]. Soil and climate conditions are among the first variables considered by agricultural decision-makers when they need to select the most adequate production to draw up agricultural plans [4]. On the other hand, food supply chains and various socioeconomic activities have impacts on soil quality, generating, in some cases, what can be called circular and cumulative processes [5]. In these frameworks, integrated soil management is fundamental to preserving the quality of the soil and its functions for sustainable development and to guaranteeing the safety and security of the food obtained for human health and balanced nutrition [6–8]. For adjusted management, soil legislation and policies play a relevant role, as well as the associated institutions at the national, European and international levels. Nonetheless, the legislation and public policies seem to have been more concerned with the air and water quality than soil health. The outputs from the international scientific community reveal that there is a field to be explored, namely through multidisciplinary approaches, considering smart methodologies and addressing gaps related to specific particularities of the soil management dimensions. These insights are fundamental to supporting the policymakers and decision-makers and give suggestions for future research. This is particularly important when it is still needed to convince the national and European institutions to prioritize specific soil health problems [9].

This Editorial refers to the Special Issue "Integrated Soil Management: Food Supply, Environmental Impacts, and Socioeconomic Functions". The Special Issue highlights bringing a broader perspective on soil management, namely in its relationship with food supply, environmental dimensions, and socioeconomic activities. From a total of twenty-six manuscripts submitted for consideration and peer review, fourteen were accepted for publication and inclusion in this Special Issue (two reviews and twelve articles). The published contributions are listed below followed by a description review to encourage the reader to explore them.

Cárceles Rodríguez et al. (contribution 1) reviewed the impact of conservation agricultural practices on soil health and their role in agricultural sustainability. Their study concludes that the main challenge of conserving and improving soil health is guaranteeing its long-term productivity and environmental sustainability, and to reach this will be vital to develop new tools and methodologies to assess soil quality and health that can be used to evaluate and guide soil management decisions.

https://doi.org/10.3390/soilsystems9020064

Rubiales (contribution 2) critically reviewed the achievements and prospects in broomrape (*Orobanche crenata*) management and resistance breeding to facilitate legume reintroduction into Mediterranean rain-feed farming systems. The author stressed that several strategies have been proposed for crenate broomrape management but considering that temperate legumes in the area are low-input crops, they have been found to be largely uneconomical or hard to grow, leaving the use of resistant cultivars as the most desirable option.

Martinho et al. (contribution 3) investigate the spatial correlations of the soil nutrient balance around the world and analyse how this variable is interrelated with agricultural soil emissions, agricultural output, and food supply. Results highlight that there is space for common strategies worldwide to preserve soil quality, as in some parts of the world the problems are similar. In these frameworks, the international organisations may have a determinant contribution.

Hredoy et al. (contribution 4) investigates the impact of landfill leachate of Amin Bazar landfill on the environmental compartments. The authors found that mitigation measures are critical for preventing soil and water contamination because the leachate from the landfill site has a higher degree of contamination with respect to the analysed parameters, which contribute to the surrounding environmental components including the surface water, groundwater, soil, and plants by polluting them adversely. However, the authors claim that a gap remains in investigating the feasibility of becoming involved in the development of a waste management system as well as ensuring ideal environmental standards for municipal solid waste and balancing environmental quality.

Arrobas et al. (contribution 5) investigates fertilisation programmes oriented towards ecological intensification in European chestnut (*Castanea sativa* Mill.) rainfed orchards and managed with increasingly intensive cropping practices. Results highlight that it seems appropriate to base the annual fertilisation plan on leaf nutrient concentration because these large trees had a poor response to the annual application of fertilisers.

Parveen et al. (contribution 6) investigates the impact of phytohormones, e.g., indole acetic acid and gibberellic acid, on mung bean (*Vigna radiata* L.) yield, seed nutritional profile, and soil N availability in the sub-tropical region of Pakistan. The findings of the study highlight that the combined treatment of the two phytohormones followed by the sole application of each phytohormone were most effective treatments to improve the morpho-physiology and nutrient profile of mung beans; however, the underlying molecular mechanisms need to be explored further.

Nacoon et al. (contribution 7) investigates the effects of different species of arbuscular mycorrhizal fungi (*Claroideoglomus etunicatum; Rhizophagus variabilis; Rhizophagus nov. spec.; Acaulospora longula*) on the growth performance and concentrations of bioactive substances of black rice in a pot experiment under laboratory conditions. Results highlight that *Rhizophagus variabilis* was the best inoculum for increasing grain yield and bioactive compounds.

Cavalaris et al. (contribution 8) investigates the impacts on soil compaction, along with the changes in soil carbon, and aims to identify the optimum tillage schemes that compromise benefits and drawbacks. The findings show that permanent no-tillage was the most effective method for sequestering soil carbon and highlight that carbon credits in carbon farming may be halved if periodic deep tillage operations should be introduced to counteract the consequences of extreme soil compaction.

Sheshnitsan et al. (contribution 9) investigates the level of selenium in soils and its accumulation in plants as well as identifying the factors associated with selenium bioaccumulation in the hydrogeochemical province with high selenium in groundwater. The research results indicated that the absence of antagonistic interactions with heavy metals in the soil–plant system contributes to the enhanced selenium accumulation in plants in the Lower Dniester Valley. Finally, this study shows the complexity of the interactions between selenium and heavy metals in the soil–plant system and their potential impact on agricultural practices. The authors conclude that further studies are required to identify the reasons for the high mobility of selenium and the significant content of its water-soluble forms in soils.

Landi et al. (contribution 10) investigates the potential efficacy of pot cultivation systems using commercial substrates to avoid plant-parasitic nematode infestations and simultaneously increase free-living nematode populations. The findings show that substrates rich in organic matter such as coconut fibre, even though they are unable to prevent accidental introduction during cultivation, could still play an important role in suppressing plant-parasitic nematodes. The main conclusion of this study is that a gap in the knowledge exists and more studies are needed to investigate the mechanisms determining differences in plant-parasitic nematodes between plant species and to explore the effectiveness of combining pot cultivation with other control methods.

Kintl et al. (contribution 11) investigates the potential influence of soil heterogeneity in terms of nutrient contents on differences in the chemical composition of individual parts of chickpea (*Cicer arietinum* L.) plants (stems, leaves, pods and seeds). The authors concluded the heterogeneity of certain elements (nutrients) and soil reactions on the ability of chickpea to uptake and translocate these elements into plant organs and seeds, thus proving that soil heterogeneity strongly affects the overall fitness of chickpea. Moreover, the authors claim that farmers can influence detected plot heterogeneity by taking appropriate measures (mineral and organic fertilisation, liming, etc.) using a system of precision agriculture.

Choudhary et al. (contribution 12) investigates the effects of crop residue, nutrient management, and soil moisture on methane emissions from maize, rice, soybean, and wheat production systems. Results highlight the complexity of methane dynamics and emphasise the importance of integrated crop, nutrient, and soil moisture (irrigation) management strategies that need to be developed to minimise methane emissions from agricultural production systems to mitigate climate change. The authors claim that the findings of this study will help develop more accurate models for mitigating greenhouse gas emissions from agricultural soils.

Wang et al. (contribution 13) investigates the effects of varying the rates of five chicken manure applications on the accumulation and distribution of antibiotic resistance genes across different soil depths using metagenomic sequencing. The authors found that antibiotic resistance genes were predominantly concentrated in the surface soil and exhibited a significant decrease in type and abundance with increased soil depth. This paper offers valuable insights for environmental risk assessments regarding the utilisation of livestock manure resources. Additionally, the authors highlight that it furnishes a scientific foundation for farmland application strategies pertaining to livestock manure.

Buckle et al. (contribution 14) investigates the effects of a selection of management practices and environmental factors on the presence and abundance of arbuscular mycorrhizal fungi in upland Welsh grasslands. The research results suggest that grazing sheep and cattle together had the highest overall influence on arbuscular mycorrhizal fungi abundance compared to grazing sheep or cattle separately. Results highlight that high plant diversity correlated with high arbuscule and vesicle abundance, but conversely, the application of lime reduced vesicle abundance. The authors claim that these findings offer new insights into the effects of management practices on arbuscular mycorrhizal fungi. The main conclusion of this study is that mixing livestock, increasing plant diversity, and reducing lime applications are shown here to improve the abundance of arbuscular mycorrhizal fungi and could, therefore, help to inform sustainable farm management decisions in the future.

The contributions published in this Special Issue offer new insights into further studies on soil quality and agricultural performance, the impact of agricultural activities on soil quality, soil functions, and sustainability, the quantification of heavy metals in soil, and organic agriculture and soil parameters. However, a multidisciplinary approach integrating soil quality with food supply, environmental dimensions, and socioeconomic activities could provide new insights. Another knowledge gap highlights that studies on topics such as soil management, food safety and security, climate-smart agriculture and soil management, agriculture 4.0 and soil characteristics, Industry 4.0 and its impact on soil, and the calculation of potentially toxic elements in soil are welcome.

Author Contributions: J.L.S.P. and V.J.P.D.M. equally contributed. All authors have read and agreed to the published version of the manuscript.

Funding: This work is funded by National Funds through the FCT—Foundation for Science and Technology, I.P., within the scope of the project UIDB/00681 (https://doi.org/10.54499/UIDP/00681/2020). This work was developed under the Science4Policy 2023 (S4P-23): annual science for policy project call, an initiative by PlanAPP—Competence Centre for Planning, Policy and Foresight in Public Administration in partnership with the Foundation for Science and Technology, financed by Portugal's Recovery and Resilience Plan.

Acknowledgments: As Guests Editors of the Special Issue "Integrated Soil Management: Food Supply, Environmental Impacts, and Socioeconomic Functions", we would like to express our deep appreciation to all authors whose valuable work was published under this issue and thus contributed to the success of the edition.

Conflicts of Interest: The authors declare no conflicts of interest.

List of Contributions:

1. Cárceles Rodríguez, B.; Durán-Zuazo, V.H.; Soriano Rodríguez, M.; García-Tejero, I.F.; Gálvez Ruiz, B.; Cuadros Tavira, S. Conservation Agriculture as a Sustainable System for Soil Health: A Review. *Soil Syst.* **2022**, *6*, 87. https://doi.org/10.3390/soilsystems6040087.
2. Rubiales, D. Managing Root Parasitic Weeds to Facilitate Legume Reintroduction into Mediterranean Rain-Fed Farming Systems. *Soil Syst.* **2023**, *7*, 99. https://doi.org/10.3390/soilsystems7040099.
3. Martinho, V.J.P.D.; Pereira, J.L.S.; Gonçalves, J.M. Assessment of the Interrelationships of Soil Nutrient Balances with the Agricultural Soil Emissions and Food Production. *Soil Syst.* **2022**, *6*, 32. https://doi.org/10.3390/soilsystems6020032.
4. Hredoy, R.H.; Siddique, M.A.B.; Akbor, M.A.; Shaikh, M.A.A.; Rahman, M.M. Impacts of Landfill Leachate on the Surrounding Environment: A Case Study on Amin Bazar Landfill, Dhaka (Bangladesh). *Soil Syst.* **2022**, *6*, 90. https://doi.org/10.3390/soilsystems6040090.
5. Arrobas, M.; Silva, J.; Busato, M.R.; Ferreira, A.C.; Raimundo, S.; Pereira, A.; Finatto, T.; de Mello, N.A.; Correia, C.M.; Rodrigues, M.Â. Large Chestnut Trees Did Not Respond to Annual Fertiliser Applications, Requiring a Long-Term Approach to Establishing Effective Fertilisation Plans. *Soil Syst.* **2023**, *7*, 2. https://doi.org/10.3390/soilsystems7010002.
6. Parveen, A.; Aslam, M.M.; Iqbal, R.; Ali, M.; Kamran, M.; Alwahibi, M.S.; Akram, M.; Elshikh, M.S. Effect of Natural Phytohormones on Growth, Nutritional Status, and Yield of Mung Bean (*Vigna radiata* L.) and N Availability in Sandy-Loam Soil of Sub-Tropics. *Soil Syst.* **2023**, *7*, 34. https://doi.org/10.3390/soilsystems7020034.
7. Nacoon, S.; Seemakram, W.; Ekprasert, J.; Theerakulpisut, P.; Sanitchon, J.; Kuyper, T.W.; Boonlue, S. Arbuscular Mycorrhizal Fungi Enhance Growth and Increase Concentrations of Anthocyanin, Phenolic Compounds, and Antioxidant Activity of Black Rice (*Oryza sativa* L.). *Soil Syst.* **2023**, *7*, 44. https://doi.org/10.3390/soilsystems7020044.

8. Cavalaris, C.; Gemtos, T.; Karamoutis, C. Rotational Tillage Practices to Deal with Soil Compaction in Carbon Farming. *Soil Syst.* **2023**, *7*, 90. https://doi.org/10.3390/soilsystems7040090.
9. Sheshnitsan, S.; Golubkina, N.; Sheshnitsan, T.; Murariu, O.C.; Tallarita, A.V.; Caruso, G. Selenium and Heavy Metals in Soil–Plant System in a Hydrogeochemical Province with High Selenium Content in Groundwater: A Case Study of the Lower Dniester Valley. *Soil Syst.* **2024**, *8*, 7. https://doi.org/10.3390/soilsystems8010007.
10. Landi, S.; Carletti, B.; Binazzi, F.; Cacini, S.; Nesi, B.; Resta, E.; Roversi, P.F.; Simoni, S. Impact of Pot Farming on Plant-Parasitic Nematode Control. *Soil Syst.* **2024**, *8*, 60. https://doi.org/10.3390/soilsystems8020060.
11. Kintl, A.; Šmeringai, J.; Lošák, T.; Huňady, I.; Sobotková, J.; Hrušovský, T.; Varga, L.; Vejražka, K.; Elbl, J. The Effect of Soil Heterogeneity on the Content of Macronutrients and Micronutrients in the Chickpea (*Cicer arietinum* L.). *Soil Syst.* **2024**, *8*, 75. https://doi.org/10.3390/soilsystems8030075.
12. Choudhary, R.; Lenka, S.; Yadav, D.K.; Lenka, N.K.; Kanwar, R.S.; Sarkar, A.; Saha, M.; Singh, D.; Adhikari, T. Impact of Crop Residue, Nutrients, and Soil Moisture on Methane Emissions from Soil under Long-Term Conservation Tillage. *Soil Syst.* **2024**, *8*, 88. https://doi.org/10.3390/soilsystems8030088.
13. Wang, Y.; Yang, L.; Liu, W.; Zhuang, J. The Effect of Manure Application Rates on the Vertical Distribution of Antibiotic Resistance Genes in Farmland Soil. *Soil Syst.* **2024**, *8*, 89. https://doi.org/10.3390/soilsystems8030089.
14. Buckle, A.L.; Crotty, F.V.; Staddon, P.L. Mixed Grazing Increases Abundance of Arbuscular Mycorrhizal Fungi in Upland Welsh Grasslands. *Soil Syst.* **2024**, *8*, 94. https://doi.org/10.3390/soilsystems8030094.

References

1. Sahoo, S.; Singha, C.; Govind, A.; Moghimi, A. Review of climate-resilient agriculture for ensuring food security: Sustainability opportunities and challenges of India. *Environ. Sustain. Indic.* **2025**, *25*, 100544. [CrossRef]
2. Ntsomboh-Ntsefong, G.; Mbi, K.T.; Seyum, E.G. Advancements in soil science for sustainable agriculture: Conventional and emerging knowledge and innovations. *Acad. Biol.* **2024**, *2*. [CrossRef]
3. Kumar, K.A.; Jayanthi, J.; Singh, R.D.; Sahu, S.K.; Hasan, A. Exploring soil health and sustainability in the Northwestern Himalayas: Assessing indicators amidst changing land use. *Environ. Earth Sci.* **2025**, *84*, 210. [CrossRef]
4. Chaher, N.E.H.; Nassour, A.; Nelles, M. The (FWE)2 nexus: Bridging food, food waste, water, energy, and ecosystems for circular systems and sustainable development. *Trends Food Sci. Technol.* **2024**, *154*, 104788. [CrossRef]
5. Tovar-Ortiz, S.A.; Rodriguez-Gonzalez, P.T.; Tovar-Gómez, R. Modeling the Impact of Global Warming on Ecosystem Dynamics: A Compartmental Approach to Sustainability. *World* **2024**, *5*, 1077–1100. [CrossRef]
6. Xing, Y.; Wang, X.; Mustafa, A. Exploring the link between soil health and crop productivity. *Ecotoxicol. Environ. Saf.* **2025**, *289*, 117703. [CrossRef] [PubMed]
7. Chen, S.; Ding, Y. Precision Agriculture Current Progress from a Novel Bibliometric Method. *World Food Policy* **2025**, *11*, e7000. [CrossRef]
8. Getahun, S.; Kefale, H.; Gelaye, Y. Application of Precision Agriculture Technologies for Sustainable Crop Production and Environmental Sustainability: A Systematic Review. *Sci. World J.* **2024**, *2024*, 2126734. [CrossRef]
9. Martinho, V.J.P.D.; Ferreira, A.J.D.; Cunha, C.; Pereira, J.L.S.; Carreira, M.D.C.S.; Castanheira, N.L.; Ramos, T.C.B. Soil legislation and policies: Bibliometric analysis, systematic review and quantitative approaches with an emphasis on the specific cases of the European Union and Portugal. *Heliyon* **2024**, *10*, e34307. [CrossRef] [PubMed]

soil systems

MDPI

Review

Conservation Agriculture as a Sustainable System for Soil Health: A Review

Belén Cárceles Rodríguez [1], Víctor Hugo Durán-Zuazo [1,*], Miguel Soriano Rodríguez [2], Iván F. García-Tejero [3], Baltasar Gálvez Ruiz [1] and Simón Cuadros Tavira [4]

1 IFAPA Centro "Camino de Purchil", Camino de Purchil s/n, 18004 Granada, Spain
2 Dpto. Agronomía, Universidad de Almería, Ctra, Sacramento s/n, La Cañada de San Urbano, 04120 Almeria, Spain
3 IFAPA Centro "Las Torres", Carretera Sevilla-Alcalá del Río km 12.2, Alcalá del Río, 41200 Sevilla, Spain
4 Dpto. Ingeniería Forestal, Universidad de Córdoba, Campus de Rabanales, Ctra.Madrid-Cadiz, Km.396, 14071 Cordoba, Spain
* Correspondence: victorh.duran@juntadeandalucia.es; Tel.: +34-900-156-000

Abstract: Soil health is a term used to describe the general state or quality of soil, and in an agroecosystem, soil health can be defined as the ability of the soil to respond to agricultural practices in a way that sustainably supports both agricultural production and the provision of other ecosystem services. Conventional agricultural practices cause deterioration in soil quality, increasing its compaction, water erosion, and salinization and decreasing soil organic matter, nutrient content, and soil biodiversity, which negatively influences the productivity and long-term sustainability of the soil. Currently, there are many evidences throughout the world that demonstrate the capability of conservation agriculture (CA) as a sustainable system to overcome these adverse effects on soil health, to avoid soil degradation and to ensure food security. CA has multiple beneficial effects on the physical, chemical, and biological properties of soil. In addition, CA can reduce the negative impacts of conventional agricultural practices on soil health while conserving the production and provision of soil ecosystem services. Today, agricultural development is facing unprecedented challenges, and CA plays a significant role in the sustainability of intensive agriculture. This review will discuss the impact of conservation agricultural practices on soil health and their role in agricultural sustainability.

Keywords: conservation agriculture; indicators; soil health; soil quality; sustainability

1. Introduction

Soil is the surface material that covers most land, containing inorganic particles and organic matter and supplying structural support to agricultural plants, being thus their source of nutrients and water. Agriculture today faces a double-sided challenge—on the one hand, the urgent need to provide food to a growing population, and on the other hand, to do so in a sustainable way [1], without compromising the provision of ecosystem services by the soil, such as carbon sequestration, nutrient supply, and water cycle regulation.

Sustainable agriculture is a difficult concept to define, since the environmental, social and economic impacts of agriculture are diverse and interact with one another [2]. In general, it can be stated that sustainable crop production systems are those that respect the environment, improve efficiency in the use of resources and promote human well-being [3]. They are those food production practices that integrate ecological, biological, physical and chemical principles, without harming the environment, as opposed to unsuitable agricultural practices [4].

Soil health is the state of the soil in relation to its potential ability to maintain its biological productiveness, strengthen environmental quality, and foster plant and animal health. Sustainable agriculture can be defined as agriculture that can be practiced in a productive and profitable way without affecting the health of the soil [5]. Figure 1 shows the main functions exerted by soil. Today, soil health is threatened all over the world. Some of the main threats to soil are erosion, compaction, salinization, nutrient depletion, pollution, and/or overgrazing [6].

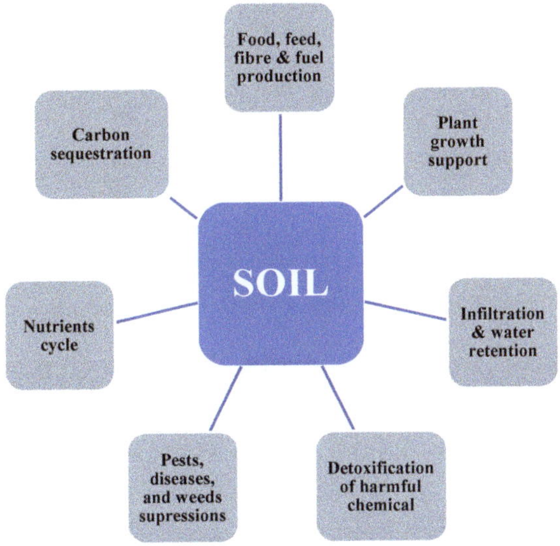

Figure 1. Main functions of soil (Adapted from [7]).

On the other hand, land degradation and deterioration of soil fertility are two of the main causes of the decline in the agricultural productivity of agroecosystems. The intensification of agriculture deteriorates the soil quality, and its negative effects have increased in the past few decades. The aim of conventional agriculture is to produce the highest possible yield of crops by the application of synthetic products, energy inputs, and a number of other industrial products. Biodiversity, soil fertility, and ecosystem health are compromised under conventional systems.

The intensive use of machinery and chemical inputs increases compaction, erosion, and soil salinization and decreases the content of organic matter and soil nutrients, which negatively influences the soil's productivity and long-term sustainability. The degradation of agricultural soil under different cropping systems is a socioeconomic and environmental problem that must be urgently addressed, particularly considering that climate change is expected to have a strong negative impact on food production, as was defined by Smith and Gregory [8]. CA practices are a useful strategy for climate change mitigation and adaptation [9,10]. CA allows slowing down or reducing greenhouse gas emissions and improving carbon sequestration in the soil [11]. The application of CA practices can improve the properties of soil, increasing its resilience to drought, and improving water and nutrient use efficiency. These improvements are essential to maintain the sustainability of agricultural production and mitigate the impacts of climate change on food production [12,13]. To reduce these negative impacts of agricultural systems and guarantee their long-term sustainability, management systems that improve or conserve soil quality are crucial [14]. To this end, agronomic practices of conservation agriculture (CA) are promoted. Figure 2 shows the environmental impacts of conventional agriculture and the benefits of CA on the soil system.

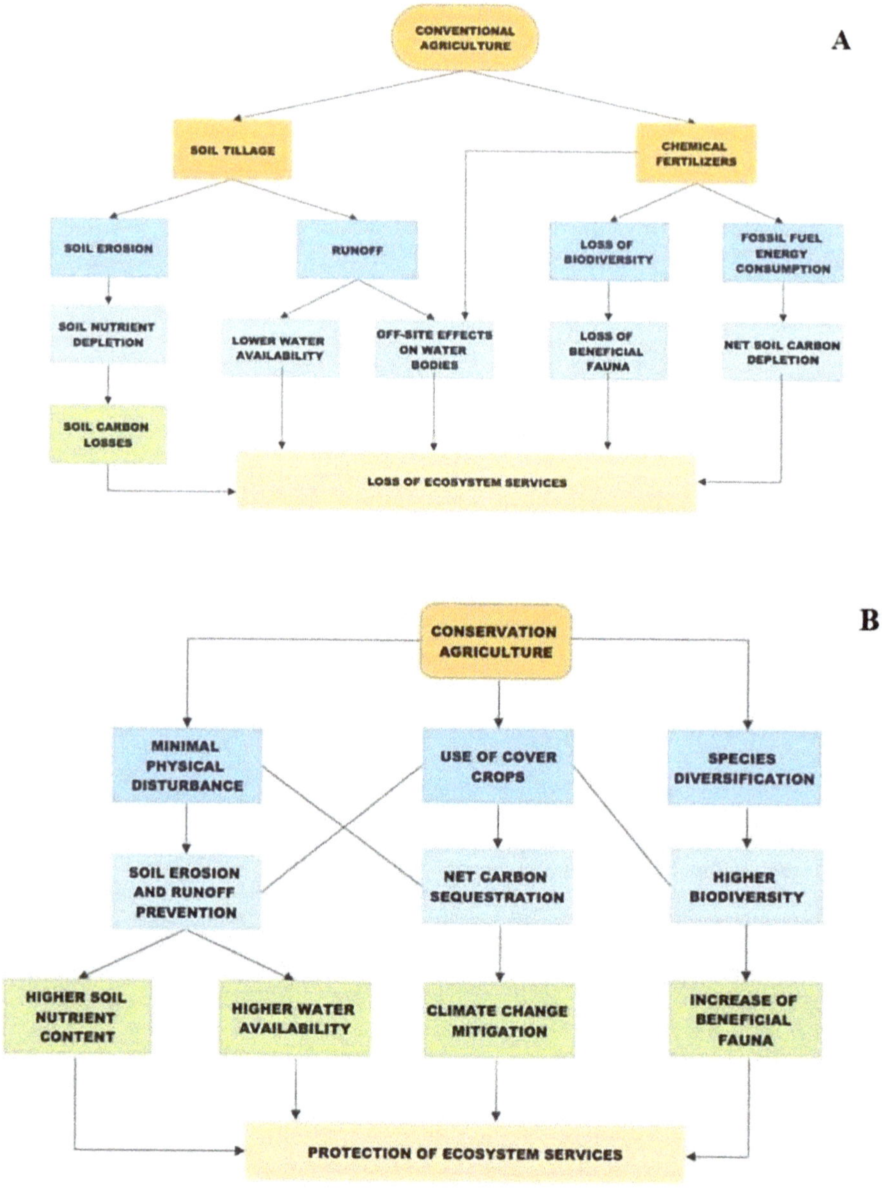

Figure 2. Environmental impacts of conventional agriculture (**A**) and the benefits of conservation agriculture (**B**) on soils.

In this review, we examine and describe advancements in the implementation of conservation agriculture measures as a sustainable system, focusing on their impacts on soil health and its role in supporting the suitable management of land, while fostering food security.

2. Conservation Agriculture

The Food and Agriculture Organization (FAO) defines CA as an agroecosystem management system to ensure food security and improve profits while preserving environmental resources.

Food security, as defined by the United Nations' Committee on World Food Security, means that all people, at all times, have physical, social, and economic access to sufficient, safe, and nutritious food that meets their food preferences and dietary needs for an active and healthy life. Currently almost 800 million people do not have access to enough food, more than 2 billion people experience deficiencies in key micronutrients, and approximately 60% of people in developing countries are food insecure [15]. In addition, it is foreseeable that in the coming decades, the growth of the world population, climate change and environmental impacts will aggravate the problem. The magnitude of the problem globally means that food security is related to all of the Sustainable Development Goals (SDGs) of the United Nations.

Conservation agriculture is an agroecosystem management approach that can be considered as one of the main ways to achieve the sustainability of agriculture, allowing the goal of greater protection while protecting the environment [16]. CA emerged in the 1930s in the USA to combat soil degradation due to water and wind erosion [17]. CA is characterized by the application of three interlinked principles implemented with locally adapted practices, together with other complementary agricultural practices [18]. These three principles are:

(1) Continuous minimum mechanical soil disturbance;
(2) Permanent soil organic cover with crop residues and/or cover crops;
(3) Species diversification through varied crop rotations, sequences, and associations.

The concomitant application of these three individual principles constitutes the classical definition of CA. However, many smallholder farmers cannot apply these three rules at the same time, and CA defined as a fixed package is not often adapted to the particular conditions of small farms. The application separately or in tandem of these components has been shown to have potential benefits, as was reported by many authors [19–21]. However, some of these authors argue that it is necessary to move from the strict definition of CA as a fixed set of three components to talking about conservation practices, which encompass a variety of options for sustainable agricultural intensification [22,23].

CA constitutes the central nucleus of FAO's new sustainable agricultural intensification strategy [24]. According to the FAO, CA is applicable to all "agricultural landscapes and land uses with locally adapted practices", which implies a series of economic, agronomic, and environmental benefits. In this sense, CA is a viable option for the sustainable intensification of agricultural land and obtaining profitable production [25,26].

In 2015/2016, CA was practiced worldwide in 180 M ha (about 12.5% of the total global cropland), an increase of 69% compared to 2008/2009. This growth has been greater in recent years. From 1999 to 2003, the area under CA increased by an average of 8.3 M ha per year [27]. The adoption of CA is not uniform in all regions or among all types of farms. It is generalized in large farms in North America, Australia, and Brazil. In contrast, adoption by smallholder farmers accounts for only 0.3% of the farmland worldwide under CA [28]. Globally, the total CA area is still comparatively small in relation to the total arable land using conventional tillage (CT). As pointed out by Kassam et al. [27], it is expected that large areas of agricultural lands in Asia, Africa, Europe, and Central America will adopt CA in the coming years. The low adoption of CA in developing countries can be attributed in part to the fact that it is a complex system, coupled with insufficient technical knowledge and capacity of farmers. In this context, political and institutional support is essential through incentives for farmers to adopt CA practices and technical support from experts [21].

To increase the implementation of CA techniques and the benefits derived from it, site-specific practices must be designed [22,25,29]. An important constraint is the limited availability in most developing countries of affordable and suitable machinery for no-till seeding, especially for small- and medium-scale farmers [30]. The development and availability of equipment that allows for sufficient germination of crops planted in no-tillage systems, with mulch in the soil, and that can adapt to small- and large-scale farmers should be improved [31]. Therefore, CA is an alternative to enhance productivity and food security, while preserving natural resources and reducing the negative externalities of traditional agricultural practices [32]. Moreover, the CA system can significantly improve the resistance to changing climate conditions in cropping systems [33,34]. In this context, conservation tillage is applied as an alternative to CT in order to alleviate water erosion impacts, reduce production costs, and maintain soil quality [35,36]. The positive effects of minimum tillage on soil quality, environment, and soil water conservation as compared to non-tilled soils in rainfed plantations were highlighted by Jacobs et al. [37] and Busari et al. [38]. Table 1 summarizes the main economic/agronomic and environmental benefits derived from CA practices.

Table 1. Main economic/agronomic and environmental benefits generated by conservation agriculture.

Economic/Agronomic	Environmental
Labor and fuel savings	Lower CO_2 emissions
Cost and time savings	Erosion and surface runoff reductions
Yield gains	Improvement of soil properties
Reduced fertilizer expenditures	Increase in soil biodiversity
Weed control	Increase in microbial activity
Lower irrigation needs	Less pollution of downstream water
Lower risk of pest and disease outbreaks	

Adapted from [31,39].

The cover cropping system as a technique of CA is an essential part of crop rotations in many regions worldwide, dispensing a wide range of benefits and ecosystem services such as N supply and retention [40], weed control [41], soil nematode control [42], water retention [43], and mitigation of nitrate leaching [44]. In addition, in the long term, cover crops can build up soil organic carbon and N [45,46] and lower net N_2O and CO_2 emissions, thus contributing climate change mitigation services [47]. Cover cropping can improve soil organic carbon stocks and potentially promote climate stability and food security, as was reported by Minasny et al. [48]. Similarly, according to Garcia-Tejero et al. [49], who examined Mediterranean rainfed agroecosystems, the use of CA techniques to enhance soil water management and soil carbon storage is vital.

On the other hand, Daryanto et al. [50], in a global quantitative synthesis of ecosystem services from cover crops, reported the suitability of their implementation. Despite the potential benefits of cover crops to improve soil conditions, this measure can add to the complexity of farming operations. According to Clark et al. [51], in the case of hairy vetch (*Vicia villosa* Roth.), which can provide a considerable amount of N demanded by the subsequent crop (maize), a late cover crop harvest is recommended because this allows for higher N accumulation in their biomass and for better synchronization of N release from the decomposing cover crop and maize N uptake [52]. In contrast, the early harvest of the cover crop may be suitable in circumstances where the rainfall amount is low and the depletion of soil moisture reserves by cover crops is a drawback [53].

The CA practices result in soil quality improvement only gradually, and benefits come about only with time. According to Stagnari et al. [54], between 3 to 7 years may be needed for all of the benefits to take hold. Therefore, because long periods are often required before changes in the soil can be detected, studies of CA must be based on long-term research and trials. This transition phase is crucial to ensure the success of the adoption of CA practices. In the initial transition years, problems can arise, such as more difficult weed management [55], lower productivity [56], etc., which can discourage farmers and lead them to abandon these practices.

3. Soil Health

Soil has been receiving increasing political and scientific interest in recent times, given its capability to provide various ecosystem services that contribute to the United Nations Sustainable Development Goals and to the European Union Green Deal [57]. Concepts such as soil health and soil quality are used to refer to this soil capability. The terms soil health and soil quality are often used interchangeably. In fact, the distinction between the two concepts is not clear. According to Laishram et al. [58], soil health refers to a broader concept—the capacity of soil to function as a living system to support plant, animal, and human life. Conversely, soil quality concerns the capacity of a specific kind of soil to sustain a particular use, such as crop production. Bonfante et al. [57] established the following distinction between the two terms: "Soil health is the actual capacity of a particular soil to function, contributing to ecosystem services", while "soil quality is the inherent capacity of a particular soil to function, contributing to ecosystem services". Both concepts, soil health and quality, are used to monitor soil status, analyze the influence of soil management on agricultural sustainability, and direct decision making to avoid degradation [4]. Figure 3 summarizes the management principles and the benefits of soil health.

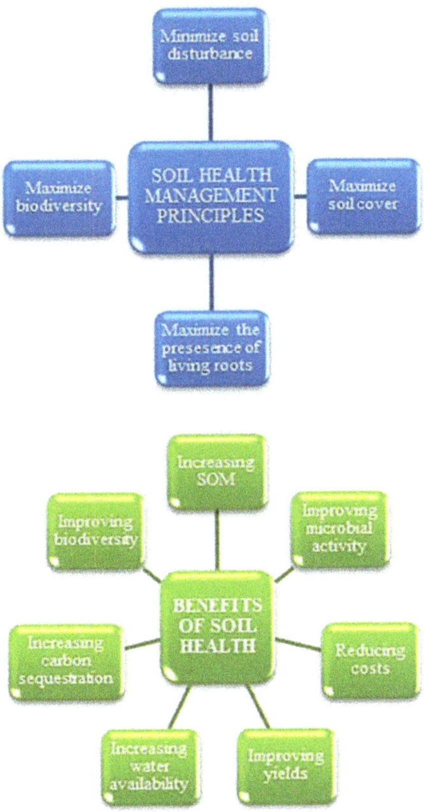

Figure 3. Management principles and benefits of soil health.

Although the concept of soil health emerged in the early 2000s, it is still evolving. It is not an easy concept to define, since soil is an extremely complex ecosystem, as was stated before. There are numerous definitions in the literature. According to Doran and Zeiss [14], soil health is "the capacity of soil to function as a vital living system, within ecosystem and land-use boundaries, to sustain plant and animal productivity, maintain or enhance water and air quality, and promote plant and animal health". The U.S. Department of Agriculture (USDA) [59] defines soil health as "the continued capacity of soil to function as a vital living ecosystem that sustains plants, animals, and humans". Yang et al. [60] defined it as "the capacity of soil to function, within ecosystem boundaries, to sustain crop and animal productivities, maintain or enhance environmental sustainability, and improve human health worldwide".

According to Kibblewhite et al. [5], healthy agricultural soil is "capable of supporting the production of food and fiber, to a level and with a quality sufficient to meet human requirements, together with continued delivery of other ecosystem services that are essential for maintenance of the quality of life for humans and the conservation of biodiversity". According to Wang and Hooks [61], soil health can be defined as having six main characteristics: (i) high biological diversity, (ii) high community stability that can provide resilience and self-recovery to chemical and biological disturbance, (iii) the ability to maintain the integrity of nutrient cycling and energy flow, (iv) the suppression of multiple pests and pathogens, (v) the ability to improve plant health, and (vi) the maintenance of water and air quality.

All of these definitions are conceptual, since they attempt to define what healthy soil is without defining how it is measured. The operational definitions establish a series of key indicators of soil health. It is essential to include indicators of physical, chemical, and biological properties when assessing soil health, as was stated by Bünemann et al. [62]. Ideally, indicators of soil health should be related to relevant soil processes and sensitive to changes in management practices and environmental conditions [60]. There is no universal set of ideal soil characteristics, and their interpretation is always context-dependent [63].

Finally, the concept of soil health can be approached from a "reductionist" or "integrated" point of view. The first is based on estimating the state of the soil using a set of individual indicators of specific soil properties: physical, chemical, and biological. The integrated approach recognizes the complexity of the soil system and the existence of interactions between the different properties and processes of the soil; therefore, soil health is more than simply the sum of a set of specific indicators [5]. According to this integrative approach, the indicators selected to establish soil health must be the result of interactions of the biota with the physicochemical properties of the soil [64]. Thus, healthy soils are crucial for the integrity of agricultural lands to maintain, or recover from perturbations resulting from, agricultural operations, particularly those regarding soil management.

Soil Health Indicators

Knowing and understanding the state of soil health is essential to guarantee the sustainable management of agroecosystems. Soil health is a complex functional concept and cannot be measured directly in the field or laboratory; it can only be inferred indirectly by measuring soil indicators [65]. These indicators are measurable soil parameters that influence soil function and ecosystem services [66].

In general, soil health indicators can be classified as physical, chemical, or biological, although these categories are not always clearly delimited, since there are many soil properties that result from the interaction of multiple processes [67]. Evidently, no single indicator can encompass all processes and parameters of soil health, nor is it feasible (or necessary) to measure all soil attributes. Therefore, it is necessary to select a minimum dataset (MDS) including physical, chemical, and biological parameters of the soil. Establishing a minimal dataset, representative of total data, minimizes costs and efforts in soil health assessment. Table 2 shows an MDS for soil health assessment with the indicators more commonly used.

The desired features of soil health indicators are that they be: (i) easy to measure; (ii) measurable with practical, rapid, and inexpensive measurement methods; (iii) sensitive to variations in management; (iv) relevant to soil ecosystem functions; and (v) informative for management [14,68].

Table 2. Minimum data set (MDS) for soil health assessments.

Key Soil Health Parameters	Reason
BIOLOGICAL	
N mineralization	Capacity of the soil to supply N for crop growth
Microbial biomass	Source and/or drain of C and nutrients
Microbial activity	Related to the availability of nutrients and biogeochemical cycles
Soil respiration	Indicator for biological activity and organic matter
CHEMICAL	
Organic carbon	Important for soil structure and fertility, and water-holding capacity
Bio-available nutrient	Potential of nutrients to support plant development
pH	Availability of nutrients
CEC	Soil's availability to supply plant nutrients
EC	Related to soil structure, infiltration and crop development
Potential pollutants	Potentially harmful for plant growth and plant–soil system health
PHYSICAL	
Penetration resistance	Related to infiltration capacity and erosion and runoff processes
Aggregation	Indicator of soil structure and erosion protection
Infiltration	Indicator for erosion and runoff
Depth to hardpan	Roots growth potential
Texture	Important for soil water and nutrient transfer and retention
Water-holding capacity	Sufficient moisture to support plant growth

CEC, Cation exchange capacity; EC, Electrical conductivity. Compiled by authors from different sources [67,69,70].

Several methods can be used to define an appropriate MDS, including statistical tools (principal component analysis, multiple correlation, etc.), uncertain sets, expert opinion, and farmer/local knowledge [66]. Once the MDS has been established, linear and/or non-linear techniques can be applied to interpret the soil indicators. The non-linear scoring method is more representative of system function than the linear method but is more labor-intensive and requires more knowledge [71]. When individual indicators are scored, they can be integrated into a general index, which can be used to guide management decisions toward promoting the long-term sustainability of the soil resource [72]. These indices have an integrating character, combining multidimensional data on the physical, chemical, and biological properties of soil into a one-dimensional measure of soil health [59]. Many soil health indices can be found in the literature: additives, weighted, decision support system, integrated quality index, Nemoro quality index, etc. [71,73].

The benefits of using these indices are clear—they provide a unique value of soil health, which allows direct comparison between different soils [39]. They are also a decision tool that can help identify the most sustainable management practices [71]. However, they also have drawbacks. For example, the diversity of existing methodologies to build this one-dimensional index means that the resulting value for this index may vary between methods, making it difficult to interpret the results [39]. Furthermore, their use can sometimes give an overly simplified interpretation of the response of the complex agroecosystem to natural or anthropogenic disturbances [60].

4. Impact of Conservation Agriculture on Soil Health

CA measures have been put forward to restore or maintain major soil functions (C cycling and transformation, nutrient cycling, and soil structure maintenance), performing well in terms of crop yield, economic return, greenhouse gas emission mitigation, biodiversity conservation, and soil health improvement. Contrarily, there is an almost general consensus that certain practices of conventional agriculture to increase agricultural production have detrimental effects on the health of the soil. CA is proposed as an alternative to conventional management to ensure sustainability in the provision of ecosystem services through the soil [74], which can improve soil properties and associated processes [13,34].

The total impact of CA systems on soil health varies from location to location and is dependent on site-specific soil and climatic conditions, the amount of time operating under a CA system, features of CA practices (types of cover crops, intensity of the crop rotation, etc.), and the training and experience of farmers [34,70,75].

4.1. Influence on Soil Physical Properties

Traditional agriculture through CT provokes a significant alteration of physical soil properties, such as degradation of the structure, compaction problems, soil bulk density, soil penetration resistance, etc. CA is able to reduce these negative effects of CT. Some of the most important parameters of soil physical health are described in the following sections.

4.1.1. Soil Structure

Soil structure is an important parameter in the sustainability of agroecosystems, due to its role in physical, chemical, and biological dynamics of soil, and determines its resistance to degradation by water erosion. Aggregate stability against different stresses (rainfall, tillage, etc.) is a useful measure to determine soil structural stability.

According to Bronick and Lal [76], soil structure can be significantly modified through management practices. Soil structural development can be enhanced by management systems that reduce soil disturbances, increase organic matter inputs, increase plant cover, and improve soil fertility. In this sense, one of the major negative impacts of conventional long-term tillage is the deterioration of the soil structure due to the reduction in soil organic matter [34].

There is a positive correlation between the mean weight diameter of soil aggregates and total organic carbon content [77,78]. The soil organic matter (SOM) promotes macro-aggregate formation; meanwhile, soil aggregates improve the physical protection of organic matter [79]. Higher aggregate stability under CA is the result of the interaction of various factors: (i) the retention of organic residue on the soil surface protects soil aggregates from raindrop impact and avoids soil compaction [80]; (ii) decomposing organic matter increases the aggregation process [81]; (iii) no soil disturbance increases fungal populations and the persistence of root networks that encourage the stability of the aggregates [82]; and (iv) reducing soil disturbance in CA systems allows the development of a more stable soil structure than in CT systems [83]. Numerous studies have reported an improvement in the stability of soil aggregates due to the application of CA practices [84–86]. In a study in Zambia, CA practices with residue retention and crop rotation showed higher aggregate stability (41–45%) compared with conventional ploughing practices (24%) [87]. This improvement in the stability of the aggregates is a function of the type of soil. Thus, Nyamangara et al. [88] reported a greater increase in the stability of the aggregates due to CA practices in soils high in clay (18.1%) than in soils low in clay (9%), compared to CT. The increase in aggregate stability due to CA practices is greater in the topsoil layer, decreasing with depth. Zhang et al. [89] reported a greater increase in the stability of soil aggregates in the surface layer (0–20 cm) than in the subsurface layer (20–40 cm) in treatments with straw return compared to treatments without straw. A study by Eze et al. [90] with a long-term experiment found that maize-based CA systems result in significant changes to soil hydraulic properties that correlate with improved soil structure. The findings showed increases of 5–15% in total porosity, 0.06–0.22 cm/min in Ksat (saturated hydraulic conductivity), 3–7% in fine pores for water storage, and 3–6% in plant-available water capacity. Furthermore, according to these authors, the maize monocrop under CA practices had an impact on soil hydraulic properties comparable to that of the maize–legume associations.

These improvements in the soil structure, due to CA practices, promote other beneficial effects on the soil, such as higher infiltration rates, greater protection against erosion, increased water-holding capacity, improved habitats to support microbial activity, etc.

4.1.2. Bulk Density

The bulk density is one of the most common physical parameters to assess the impact of tillage and crop residue on agricultural soils, as it is an indicator of the soil's compaction and reflects the soil's ability to function in terms of structural support, water and solute movement, and soil aeration. High bulk densities cause root impedance and lead to poor crop emergence. There is no consensus regarding the effect of CA on soil bulk density, as some studies reported a higher soil bulk density with CA compared to CT [91,92], while others have not found significant differences [86,93] or reported lower soil bulk density in CA in comparison to CT [88,94]. These differences in bulk density in the different trials may be due in part to the typology of the farm. Greater topsoil bulk density recorded in studies on large farms in the USA or Australia can be the result of compaction due to heavy no-till machinery used, but this does not occur in smallholder farms in developing countries, where cultivation is performed manually or with animal draft power [95].

In a global meta-analysis, Li et al. [96] claimed an average increased bulk density of 1.4% in a no-tillage (NT) system with residue retention compared with CT. However, they also concluded that the greatest soil compaction value in conservation tillage practices was below the threshold value that limits plant growth.

According to Mondal et al. [97], no significant differences in bulk density were found in soil depth up to 15 cm after the implementation of CA. However, a greater bulk density was determined in a traditional rice–wheat cropping system than in treatments with CA at soil depth of 15–30 cm. Generally, bulk density was greater for CA than CT for soil depths within the plow layer [13,98]. However, in the top few centimeters in NT, the accumulation of crop residues and soil organic carbon (SOC) on the soil surface led to a lower bulk density [99]. Sometimes, the amount of residue is not enough to limit the increase in bulk density under no-tillage systems. In these cases, the residues can be shredded, thus increasing the covered area and mitigating the hardening of the soil [98].

The effect of conservation tillage systems (minimum/reduced tillage and no tillage) on the apparent density of the soil is not immediate; it is necessary that a few years elapse from the conversion from CT to reduce it [100]. The crop residue incorporation into the soil in conservation tillage plays a pivotal role in decreasing bulk density. In this sense, Nyamadzawo et al. [101] attributed lower bulk density in CA systems to the presence of higher levels of organic matter, which tends to improve soil structure and increase porosity. In contrast, Mondal et al. [102] reported a similar bulk density under CT and NT systems.

According to Islam and Reeder [103], soil bulk density at 0 to 15 and 15 to 30 cm depths under long-term NT decreased significantly compared to CT. At 0 to 15 cm depth, the greatest difference compared to CT occurs with 35 years of continuous zero tillage. The bulk density at depths of 15–30 cm decreased linearly over the years of NT. This decrease in bulk density is associated with an increase in total soil porosity. In a long-term study of maize (*Zea mays* L.) based crop rotations, the bulk density under CA practices (zero tillage and permanent raised beds) was reduced by 4.3–6.9% in soil depths of 0–30 cm compared with CT. In deeper soil layers (30–60 cm), differences between management systems were non-significant [104].

4.1.3. Surface Seal and Soil Crust

Bare soil in conventional systems leads to increased surface seal and crust formation due to the lack of protection against the impact of raindrops. The impact of rainfall causes the breakdown of soil aggregates and the release of finer particles, which are redistributed by the near-surface and fill the most superficial pores. This process causes sealing and surface waterproofing, decreasing water infiltration and, consequently, enhancing the runoff and soil loss [105]. Surface sealing has a negative impact on the physical characteristics of soil, which ultimately affects crop yield [106].

The presence of crop residues in CA practices can help protect the surface of the soil from raindrop impact and prevent surface sealing. In structurally unstable soils or regions where crusting is a serious problem, the maintenance of adequate surface cover is paramount to avoid surface sealing and crust formation [107]. When CA is practiced in the absence of effective soil mulch cover, surface sealing may occur. Usón and Poch [108] showed that reduced tillage did not reduce crust formation in Mediterranean conditions, due to the difficulty of establishing an effective ground cover. In certain circumstances, the quantities of biomass produced and retained in CA systems can be insufficient to avoid soil crusting and compaction [109], but increasing residue above a threshold can have no effect because of sufficient raindrop impact interception [110]. According to Page et al. [111], the surface sealing, due to the inadequate residue cover and the lack of tillage, particularly in drier regions, can be one cause of yield loss in CA systems. In situations where little surface cover from crop residue is available, the creation of surface roughness using strategic tillage is a viable option to break soil crusts, improve water infiltration, and reduce runoff [112].

Thus, a permanent soil surface cover by crop residues significantly reduces surface sealing [113]. Various studies report on the preventive effect against surface sealing in CA exerted by crop residues on the soil surface, protecting the soil from the direct impact of raindrops [114,115]. In this sense, Castellanos-Navarrete et al. [84] reported that in CA systems, soil crusts were not present on the soil surface; however, soil under CT with poor aggregate stability showed soil crust formation.

4.1.4. Soil Compaction

Soil compaction is a form of physical degradation that consists of the densification of the soil, which often results in the destruction of the soil structure; a reduction in biological activity, porosity, and permeability; an increased risk of erosion; a restriction on root development; and, consequently, decreased crop performance. On farmland, the traffic of heavy agricultural machinery is the main cause of soil compaction, and its magnitude increases with the number and intensity of tillage operations and when these are carried out in inappropriate soil moisture conditions. The influence of the machinery is so important that "controlling in-field traffic" is considered a component of CA. Recommended practices include bed planting that reduces compaction by confining traffic to the furrow bottoms [116], or the application of fertilizers at the time of seedbed preparation or seeding to reduce machinery transit [117].

In the long term, tillage promotes soil compaction and the formation of a plough pan in the sub soil. Crop rotation, cover crops, and the addition of crop residues in CA systems can reduce soil compaction. Mondal et al. [118] reported a reduction in the subsurface compaction by CA systems, with a soil penetration resistance significantly less in the 15–30 cm layer under CA. This can have a positive impact on root morphology, which can contribute to increased crop yield. According to Hamza and Anderson [119], increasing the SOM through the retention of crop residues and crop rotations that include plants with deep, strong taproots can delay or prevent soil compaction. The use of root crops in cover crops can significantly reduce soil compaction. In this sense, Islam and Reeder [103] showed that oilseed radish significantly decreased compaction to about 75 cm, with an average improvement effect of about 40% compared with soil between the rows. Chen and Weil [120] reported that the use of cover crops improved maize root penetration in compacted soils and increased the availability of surface soil water. In a study in India, Parihar et al. [104] reported that the CA practices of NT and permanent raised beds reduced the penetration resistance by 15.9 and 30.7%, respectively, compared to CT in maize rotations.

According to Holland [17], there is evidence that the long-term use of conservation tillage can, in certain situations, lead to soil compaction. Similarly, Munkholm et al. [121] concluded that direct drilling provoked the compaction of the arable layer below seeding depth on sandy loam. Thus, the long-term viability of conservation tillage techniques depends on a proper crop rotation [122] and/or the use of strategic or occasional tillage in soils under NT [123,124].

4.1.5. Soil Moisture Content

Water scarcity is one of the greatest challenges facing humanity in the coming decades [125]. CA practices improve soil moisture availability, especially under low-rainfall conditions and could contribute to maintaining crop yield in a changing climate scenario [126]. In this sense, several studies have reported a greater availability of water in CA systems with respect to CT [85,127–129]. Residue retention and cover crops in CA systems improve infiltration [96] and reduce runoff rates [127] and evaporation losses [130,131], as they protect soil from direct contact with solar radiation and act as a barrier to air flow, contributing to higher soil moisture.

No-till practices and residue cover improved soil–water relations in a study in Malawi, with an average increase in soil water content of 22 and 18 mm in NT and CA, respectively, compared to CT [132]. A meta-analysis carried out by Zhao et al. [133] concluded that crop residue retention led to an increase in soil water content by 5.9% compared with crop residue removal. In a rice system study, NT with surface residue and minimum tillage with residue incorporation had higher soil moisture than CT with residue removed [134]. Similarly, Ghosh et al. [127] reported that soil moisture conservation was 108% higher under CA than conventional agriculture plots. Mondal et al. [135] showed that the soil water content was 14% higher in CA relative to CT in the sub-surface layer (15–30 cm), while in other layers, there were no significant differences. A study by Chalise et al. [136] with a corn–soybean (*Glicine max* L.) system highlighted that the use of cover crops with residue returned improved the soil's hydrological properties and increased soil volumetric water content and soil water storage. In maize crops in the sub-humid and semi-arid regions of Kenya, NT with residue retention significantly increased soil water content compared to CT [137]. According to Sindelar et al. [138], residue removal decreased plant-available water by 32% in soil depth of 0 to 5 cm and by 21% in soil depth of 5 to 10 cm. In this context, Li et al. [96] reported that NT with residue retention increased soil-available water capacity by 10.2% compared with NT without residue retention. Similarly, Choudhary et al. [139], in a pearl millet (*Cenchrus americanus* L.)–mustard (*Brassica juncea* L.) rotation system in rainfed semi-arid regions, reported higher soil water content throughout the season in plots with residue retention than in the no-residue plots.

In irrigated plantations, crop residues conserve soil moisture and delay irrigation timing, allowing farmers to save irrigation water. In this sense, Balwinder-Singh et al. [140] found that the use of residue mulch of 8 t ha^{-1} in irrigated wheat led to saving 75 mm of irrigation water. Comparably, Gupta and Sayre [141] reported that NT practices allowed saving between 13 and 21% of irrigation water compared to CT systems. Assefa et al. [142] highlighted that CA practices with a drip irrigation system lessened water needs by about 14–35% for various crops. In irrigated onion and garlic plantations in Ethiopia, CA plots received 49 mm less water than CT treatment [143]. In addition, Jat et al. [144] showed that a CA-based maize–wheat system decreased irrigation water use by 64% compared to conventional management.

Based on field observations, many meta-analysis studies have contrasted the effects of different tillage practices on determining crop production, evapotranspiration, and water-use efficiency (WUE) [122,145–147]. Evidently, CA practices enhance WUE, as the findings by Lu [148] suggested that crop residue return can increase crop yields and WUE. In a study in a semi-arid region of China, Sun et al. [149] stated that conservation tillage significantly enhanced WUE and crop yield with respect to CT. According to Das et al. [150], experimental plots under CA practices had significantly higher WUE and significantly lower water use than CT. That is, the zero tillage with planting on permanent broad beds and residues treatment had higher WUE than the CT. Moreover, zero tillage with planting on permanent broad beds and residues treatment had higher WUE than zero tillage with planting on permanent narrow beds and residues. Thus, CA practices improve water productivity due to their water harvesting and water conservation effects [151].

Although most studies have found positive effects of residue retention on soil water, some negative consequences can also occur in certain environments, such as in rainfed areas. Cover crops in sloping lands with rainfed fruit crops do not result in economic return; however, the environmental return is highly important [152,153]. Cover crops, however, compete for resources (plant nutrients and water) with the trees, which can lead to a decline in productivity [154,155]. In other words, the cover crop benefits are more weather-specific than site-specific because when precipitation is low or not properly distributed, the water reduction after cover crops could have a negative effect on the cash crop growth and yield. In sloping olive orchards, a greater available soil water content was found under a non-tillage system with plant strips (barley and native vegetation) of 4 m width than for a non-tillage system without plant strips, particularly beneath the tree canopies [156]. In addition, Castellini et al. [157] reported the positive influence on soil hydraulic function of minimum tillage compared to non-tilled soil on olive plantations. In this context, Abazi et al. [158], examining rainfed olive orchards, determined that the use of cover crops in a Mediterranean environment has a negative impact on olive transpiration (25% average reduction), although this impact can be attenuated by early-date killing of the cover crop in the middle of March.

Contrarily, in high-rainfall areas, the greater retention of soil moisture under CA can also lead to waterlogging, with associated negative effects on crop growth and yield [91,159,160].

4.1.6. Water Runoff and Soil Loss

Conventional agriculture promotes runoff and soil loss by causing soil compaction, crusting, and surface sealing, and by decreasing porosity. In contrast, CA is associated with a reduction in soil erosion [161] (Figure 4), among other benefits. In particular, in rainfed sloping lands in Mediterranean environments, the crop residue retention and cover crops in CA systems protect the soil surface from raindrop impact and reduce the detachment, displacement, movement, and deposition of soil particles, which causes soil sealing and crust formation [162]. Furthermore, cover crops and their residues slow the velocity of agricultural runoff along the slope, improving infiltration and preventing soil erosion [163].

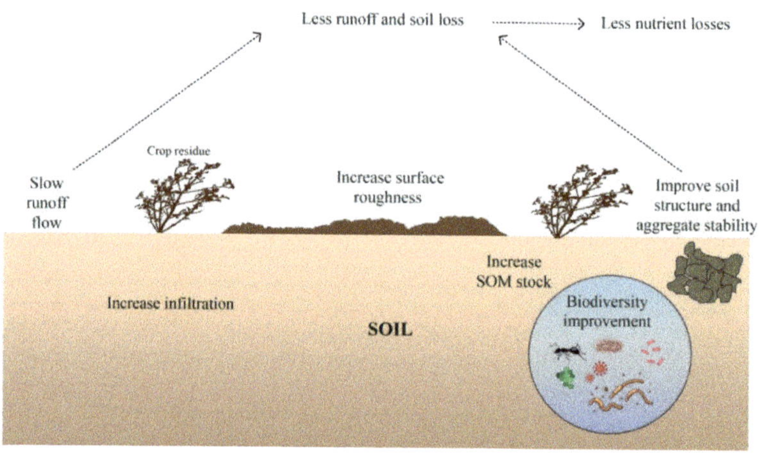

Figure 4. Effect of conservation agriculture on water erosion.

According to Thierfelder and Wall [164], plots with reduced tillage and surface residue retention had less runoff and soil erosion than conventionally tilled plots. Under semiarid rainfed conditions in western India, Kurothe et al. [165] reported that NT reduced runoff by 16.2% and soil loss by 37.2% compared to CT. Panachuki et al. [166] reported a significant reduction in runoff and soil loss in an NT system with soybean residues, compared to an NT system without residues. The retention of residues on the soil surface exerted a greater protective effect than their incorporation into the soil. In an experiment in northern Ethiopia with a wheat (*Triticum* sp.)–teff (*Eragrostis tef*) rotation, after 3 years, soil loss and runoff were significantly lower (5.2 t ha^{-1} and 46.3 mm) in permanent raised beds with 30% standing stubble compared to CT without surface residue (24.2 t ha^{-1} and 98.1 mm) [167]. Ghosh et al. [127] reported that mean runoff coefficients and soil loss with CA plots were ~45% less and ~54% less than conventional agriculture plots, respectively. The efficiency by which surface residues control runoff and soil losses increased with the amount of residue. In this context, Ranaivoson et al. [168] reported that residue levels of 1.5 to 4.5 t dry matter ha^{-1} decreased water runoff by about 50%, and residue amounts of 2 to 4 t dry matter ha^{-1} reduced soil erosion by about 80% compared to bare soil. The amount of residue necessary to reduce runoff and soil loss varies depending on the slope of the field and the intensity or amount of rainfall [169].

According to Du et al. [170], conservation practices decrease surface runoff and erosion, on average, by 67 and 80%, respectively, compared with conventional practices; the use of cover crops is what most reduces erosion and runoff. In northern Ethiopia, permanent raised beds with contour furrows at 60–70 cm intervals significantly reduced runoff and soil loss compared to traditional ploughing, with 255 and 653 m^3 ha^{-1} runoff and 4.7 t ha^{-1} and 19.5 t ha^{-1} soil loss, respectively [171]. In another study in Ethiopia, CA practices also reduced erosion and runoff. CA registered a runoff coefficient of 18.8% and a soil loss of 14.4 t ha^{-1} yr^{-1}, while for plain tillage, these parameters were 30.4% and 35.4 t ha^{-1} yr^{-1}, respectively [172].

Terracing is one of the oldest techniques for the conservation of water and soil in mountainous regions; terraces are built along contour lines to increase the arable surface area. Deng et al. [173] pointed out that these structures provide many ecosystem services, including the control of runoff and sediment by over 41.9 and 52%, respectively, and the improvement of crop yield and soil water content by 44.8 and 12.9%, respectively. In this context, the implementation of cover crops in the taluses of orchard terraces is a key factor for preventing their collapse by water erosion, lessening the runoff, soil loss, and pollution risk in low lands [174,175].

The rainfed plantations in the Mediterranean mountains with traditional practices provoke high soil erosion rates, compromising their long-term sustainability. Francia et al. [176] evaluated erosion rates by the effect of NT, CT, and cover crops in olive (*Olea europea* L.) orchards of 25.6, 5.7, and 2.1 t ha^{-1} yr^{-1}, respectively. Similarly, Gómez et al. [177] determined the soil erosion values for NT, CT, and cover crops as 6.9, 2.9, and 0.8 t ha^{-1} yr^{-1}, respectively. Recently, Cárceles et al. [178] reported that the strategies based on CA proved to be effective. The combination of minimum tillage with plant strips in almond (*Prunus dulcis* L.) and vineyard (*Vitis vinifera* L.) orchards was a more efficient practice in terms of water erosion control than only minimum tillage, averaging declines in soil erosion and runoff rates of 36 and 39%, respectively. Similarly, for olive crops, the association of minimum tillage and plant strips compared to a no-tillage system was able to reduce both soil erosion and runoff rates by 36%. Thus, the implementation of soil management measures based on cover crops is essential for hillslopes and low-fertility soils, encouraging their sustainability.

4.1.7. Soil Temperature

Soil temperature is an important property that affects crop growth and development and impacts numerous soil physical, chemical, and biological processes. Cover crops and retention of residues in CA systems can help moderate and stabilize the fluctuations in soil temperature during the crop growth period as compared to systems with bare soil [34], which can be especially important in regions with large fluctuations in temperatures [179]. The magnitude of variation in soil temperature due to management is higher in the soil top layer, decreasing in the lower layers [180]. Rai et al. [181] reported that the CA practices with mulching were effective for the reduction in soil temperature fluctuations with depth.

Moreover, crop residue retention on the soil surface reflects sunlight and isolates soil from high temperatures and thus reduces evaporative losses of water. The effect of residues on the soil temperature changes depending on the color of the residues. According to Sharratt and Campbell [182], dark residues resulted in higher mid-day temperatures compared to lighter-colored residues. Retention of residues on the soil surface in CA systems decreases daytime soil temperature [183]. Li et al. [184] reported that the crop residue remaining on the soil surface in conservation tillage systems can lessen the soil temperature change because surface residue both increases the reflection of incident solar radiation and acts as an insulating barrier between the soil surface and the warmer or colder atmospheric air above [185]. In this context, lower maximum soil temperature and higher minimum soil temperature in the 0–5 cm surface soil layer were recorded under minimum tillage with mulch treatments, compared to the CT with no-mulch treatment [186]. According to Gupta et al. [187], a zero-tillage system with residue cover had a lower soil temperature than a zero-tillage system without residue and moldboard ploughing. Guzman and Al-Kaisi [188] also reported warmer soil temperatures when crop residues were removed. In the summer season, Oliveira et al. [189] reported that daytime soil temperature in a zero-tillage system with residue retention was 2–8 °C lower than that in the conventional tillage system.

In addition, this lower soil temperature under CA systems in hot regions can help improve plant growth and crop yield [190]. In cooler climates, however, reduced soil temperature from residue cover may be a disadvantage because it can delay seed germination and plant maturity and negatively affect yield [91,191]. In this sense, Chen et al. [192] reported that straw retention decreased soil temperature in spring and delayed the development of winter wheat up to 7 days, on average reducing the final grain yield by 7% compared to systems without straw retention. To address this issue and attempt to adapt this soil management system to temperate zones, the withdrawal of residues from the seed strip has been suggested [191,193].

Tillage operations can also affect soil temperature by changing soil surface microtopography, as inclined ridge surfaces absorbed about 10% more solar radiation than flat surfaces, according to Radke [194]. Additionally, Shen et al. [195] claimed that tillage had significant effects on soil temperature in 10 of 15 weekly periods, with the temperatures of non-tilled soils being 0–1.5 °C lower than those of moldboard plough soils when residue was not returned in the previous autumn. Moreover, the ridge tillage showed no clear advantage over non-tilled soils in increasing soil temperature.

Finally, other studies reported an increase in soil temperature due to stubble retention [196], which helps crops survive during the cold winter and reduces emergence time, improving crop productivity. Kahimba et al. [197] showed that in the Canadian prairies, the presence of a crop cover or perennial vegetation resulted in relatively warmer soil profile temperatures and shallower depth of frozen soil layers. Moreover, according to Al-Darby et al. [198], despite the delay in the growing season due to the lower soil temperature in the CA systems, there was no reduction in dry matter and corn grain yield due to the greater amount of accumulated water.

4.2. *Influence on Soil Chemical Properties*

Agronomical practices may change soil chemical properties and thus fertility. The responses of soil chemical fertility to tillage practices and the magnitude of these changes depend on several factors: soil type, cropping system, climate, fertilizer application, and management practices. Long-term tillage causes severe SOM depletion in agroecosystems and can lead to soil degradation. In contrast, CA practices increase chemical quality by improving the SOC storage and nutrient dynamics. The impacts of CA techniques on some of the most relevant soil chemical properties are presented in the following sections.

4.2.1. Soil Organic Carbon

SOM is a keystone indicator of soil quality because it is linked to other physical, chemical, and biological soil quality indicators [199], playing a crucial role in soil fertility and sustainability, as it increases soil aggregate stability and water retention and provides a reservoir of essential nutrients for crops [200].

In addition, there is currently a growing interest in increasing the stock of SOC in agroecosystems because this can help mitigate climate change. In agricultural practices with high organic inputs, reduced or no tillage and permanent soil cover are capable of increasing SOC stock, acting as a carbon sink and thus mitigating the agricultural impacts on climate change [201,202]. On the other hand, the increase in SOC has positive effects on the quality of the soil, and this can improve the soil resilience, contributing to adaptation to climate change [203].

Soil tillage increases the decomposition rates of SOM, as it implies an alteration of the soil structure and the exposure of the organic matter retained in the micro-aggregates [204]. In a study by Repullo-Ruibérriz de Torres et al. [205], over a 4 year monitoring period on an olive plantation, SOM increased by the effect of different cover crops (*Brachypodium distachyon*, *Eruca vesicaria*, *Sinapis alba*, and native vegetation) between 10.9 and 14.3 Mg ha^{-1} at 0–40 cm soil depth.

The conversion of CT to conservation tillage increases the accumulation of SOC in the soil surface layer. CA increases SOC stock through the reduction in SOC losses by oxidation and erosion, the increase in organic carbon inputs to the soil (plant residues), or a combination of both factors [206,207]. Figure 5 summarizes conservation agriculture practices that may influence SOC stock increases.

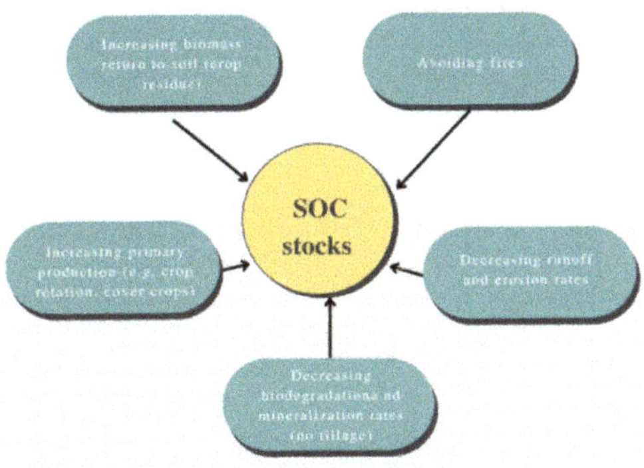

Figure 5. CA practices that increase SOC stock.

Changes in SOC storage with CA practices depend on various factors such as the quantity and quality of plant residues, time period, or edaphoclimatic characteristics [208]. These effects are most evident in the topsoil. In this context, the global analyses by Luo et al. [209] and Mondal et al. [210] indicated that a no-tillage system benefited the storage of SOC only in the upper 10 cm of the soil. Camarotto et al. [211] reported that CA increased the SOC stock in the 0–30 cm layer (0.25 Mg C ha^{-1} yr^{-1}) compared to conventional agriculture. In a maize–mustard rotation, Pooniya et al. [212] reported that CA systems had greater values for SOC than CT at soil depths of 0–0.15 m and 0.15–0.30 m, while at 0.30–0.45 m, there was no difference. Therefore, to obtain a more accurate assessment of CA practices' impact on SOC, the entire plow depth should be sampled [213]. In addition, comparing the results of experiments that compare CA with conventional systems is complicated, since they depend on several factors: depth of the investigated soil, sampling methodologies, duration of the study, edaphoclimatic variability, and crop type [211]. In irrigated almond orchards in Mediterranean semi-arid regions, according to Repullo-Ruibérriz de Torres [214], a crop mixture (65% barley and 35% vetch) and barley cover crops showed higher potential for C sequestration than spontaneous vegetation, augmenting the SOC by more than 1.0 Mg ha^{-1} after two monitoring seasons.

Long-term CA increased SOC content in the 0–5 cm soil layer in an intensive cereal-based cropping system in India [215]. In a study in northern Italy, Perego et al. [216] showed that CA systems in the medium term resulted in significantly higher SOC content and SOC stock than conventional systems. A study in rice (*Oryza sativa*)–wheat cropping systems in a South Asian region showed that the stratification and storage of SOC were higher under CA practices compared to intensive tillage-based conventional agricultural practices [217]. In a meta-analysis to evaluate the effects of minimum tillage and crop residue retention on SOC stock in 0–30 cm soil depths, Li et al. [218] reported that a no-tillage system with residue retention and a reduced tillage system with residue retention increased SOC stock by 13 and 12%, respectively, in comparison to CT. In a rice–wheat system, after 7 years, NT combined with partial residue retention increased SOC stock at 0.6 m depth [219].

4.2.2. Soil pH

The effect of conservation practices on soil pH is generally restricted to the topsoil layers. The effect of crop residues on soil pH depends on the chemical composition of the residues and the properties of the soil [220]. Residues high in ash alkalinity and N, such as some legume residues, will have a greater effect on pH compared to residues with lower content, such as wheat [221]. The initial pH of the soil has a substantial impact on the change in soil pH through the incorporation of crop residues, as it affects the mineralization of N in the residue and the rate of decomposition of organic compounds [222]. Similarly, a long-term study by Muchabi et al. [223] of fields under CA and CT highlighted a significantly higher soil pH (6.18 vs. 5.62), SOC, nodulation, and biological N fixation as a result of CA implementation after 7 years of practice. These findings are comparable with those reported earlier by Duiker and Beagle [224] and Umar et al. [225], who ascribed the upward changes in soil pH to the buffering effect of accumulated organic matter under CA. Recently, Sinha et al. [226] reported that the soil pH generally lowered under zero tillage compared to CT, being the most notable in acidic soil sites, where pH decreased by up to 0.4 units; the lower the initial soil pH, the higher was the decrease in pH under zero tillage.

Several studies have reported an increase in acidity in topsoil layers under reduced tillage treatments in comparison with CT [227,228]. This increase in acidity is attributed to a greater accumulation of soil organic matter on the soil surface in NT, which decomposes and produces acidity. In the deeper layers, there is an increase in pH because the soluble component of the residues moves through the soil profile and contributes to the alkalization of the subsoil layers [228,229]. In acid soils, various authors have reported that CA systems increased soil pH [229,230]. The organic matter that increases with CA practices tends to bring the pH to neutral or slightly acidic by buffering the pH of the soil. A long-term CA experiment carried out by Ligowe et al. [231] registered, on average, 14 and 21% higher pH and SOM, respectively, than the conventional practice, with a positive correlation (74%) between SOM and pH found during the fifth monitoring season.

4.2.3. Cation Exchange Capacity

The cation exchange capacity (CEC) is the ability of a soil to retain and release positive ions due to its content of clays and organic matter, and is considered an indicator of soil fertility. CA practices increase SOM content, and this provokes an increase in CEC [232], as it increases the amount of negative charges [233]. In this context, Ben Moussa-Machraoui et al. [234] reported a positive correlation between SOM and CEC. This increase in CEC driven by improvements in SOM via cover cropping can also lead to an increase in yield stability [235].

According to Sá et al. [233], CEC increased by 0.37 cmolc kg^{-1} for every gram of C per kg of soil. The effects on CEC are generally limited to the topsoil, which is where the SOM content is increased [224]. In this context, Williams et al. [235], in a study in the USA, showed that cover cropping increased SOM compared with no cover crop, implying a rise in CEC. In a tropical soil under no-till farming, CEC increased by 25% in the top soil layer (0–20 cm) with every 1.8 kg m^{-2} of stored organic carbon [236]. After 5 years,

CEC increased in the topsoil when residues were retained compared to soils without residue [237]. Sithole and Magwaza [228], in a long-term study in South Africa, showed that CEC was affected by tillage practices. On average, CT resulted in a significantly lower (71.9 mmolc.kg^{-1}) CEC than rotational tillage (109 mmolc kg^{-1}) and NT (114 mmolc kg^{-1}). A long-term field experiment under rice-based cropping systems showed that the CEC was higher in NT than in CT, amounting to 13.04 and 9.76 cmol (p+) kg^{-1}, respectively [238]. In a tropical rainfed agroecosystem, the adoption of minimum tillage provoked an 11.2% increase in CEC compared with the CT system [239]. Moreover, Mloza-Banda et al. [93] reported a significant increase in CEC after 2 years of conversion to CA (15.24 cmol (+) kg^{-1}) compared to annual ridge tillage (13.38 cmol (+) kg^{-1}). Similarly, Zerihun et al. [240] reported an improvement in CEC with crop rotation and intercropping in CA systems.

Conversely, Fonteyne et al. [241], in a study in Mexico of 20 maize-based trials, did not register differences in CEC between CA and local conventional practices. Comparably, Mrabet et al. [242] did not find significant differences in CEC between CA and CT in a study in Morocco. The lack of difference between the different management systems may be due to the short duration of the studies or due to the influence of local soil conditions.

In other studies, a lower CEC was observed in soils under CA due to a decrease in pH, which resulted in a decrease in pH-dependent cation exchange sites [227,243].

4.2.4. Nutrient Availability

CA practices have a significant impact on nutrient distribution and transformation in soil; thus, they can strongly influence the soil nutrient dynamics [178]. That is, CA systems that cause an increase in organic matter due to the addition of residues can produce a rise in nutrient reserves for plants, registering higher concentrations of nitrogen (N) [244,245], phosphorus (P) [246,247], potassium (K) [228,247], calcium [248], magnesium [249], zinc [250], and manganese [249] in the soil. The nature of crop residues and their management has a significant influence on the plant nutrient availability of soils. For example, in the case of N, the addition of legume residues with a low C/N composition can result in N mineralization, whereas cereal residues with a high C/N composition can temporarily immobilize N during the decomposition process [251,252]. In a review study on the effects of crop residues under CA, Ranaivoson et al. [168] reported, in general, a higher increase in soil mineral N in the case of legume residues than in the case of cereal residues. The availability of nutrients with the retention of residues is also a function of other factors, such as the amount of surface residues or the proportion of soil covered by them [168]. The availability of nutrients in the soil can also be affected by the change in topsoil pH due to CA practices [253].

A greater amount of residues stored in the soil with CA systems does not always lead to a greater availability of nutrients for plants. Soon after CA is implemented, while total stores of N may be higher, the amount of plant-available N may decrease due to lower mineralization rates and higher N immobilization rates [111]; in this case, it is necessary to apply N fertilization to maintain the yield [228].

An NT system with a total absence of soil mixing can lead to the stratification of immobile nutrients such as P and K in the surface layers of soils [254]. In dry areas of Morocco, Mrabet et al. [242] showed that NT caused surface enrichment of P and K compared with CT. This can be a problem, especially in arid regions, as drought conditions can reduce nutrient uptake from the dry soil surface, inaccessible to plant roots [255]. Furthermore, these conditions can increase the risk of N and P losses by surface runoff [256]. Higher moisture content due to CA practices can lead to N losses due to denitrification [257]. Finally, according to Morugán et al. [258], the permanent cover crops in the alleys led to higher increases in SOC and soil N; however, this practice was related to negative effects on available P in the soil. Similarly, Sujatha et al. [259] claimed that the extensive root system of legumes was beneficial for improving their ability to release organic acids from their roots that enhanced K availability in soil. Table 3 shows the implantation effect of

CA practices compared to CT in hillslope farming with rainfed olive orchards in southeast Spain [260].

Table 3. Effect of CA practices on soil physico-chemical parameters in olive orchards throughout 3 year monitoring period (SE Spain).

Soil Management	Year	pH	MCP	BD	SOC	N_T	P	K	CEC
		(H_2O)	(%)	(g cm^{-3})	(g kg^{-1})		(mg kg^{-1})		(cmol (+) kg^{-1})
Minimum tillage and spontaneous vegetation strips	1st	7.5 (±0.1)	11.4 (±4.3)	1.17 (±0.04)	8.4 (±4.8)	0.45 (±0.03)	6.4 (±2.6)	68.7 (±18)	15.8 (±3.0)
	3rd	7.6 (±0.2)	12.6 (±3.6)	1.24 (±0.08)	10.2 (±7.5)	0.68 (0.05)	7.0 (±3.5)	77.7 (±26)	16.7 (±7.8)
Minimum tillage and legume strips	1st	7.5 (±0.2)	10.0 (±3.4)	1.18 (±0.14)	8.0 (±5.7)	0.58 (0.01)	4.6 (±1.7)	84.4 (±14)	10.2 (±4.4)
	3rd	7.7 (±0.5)	11.3 (±3.2)	1.26 (±0.07)	8.9 (±3.4)	0.67 (0.08)	5.2 (±4.2)	94.7 (±22)	14.7 (±7.1)
Conventional tillage	1st	7.5 (±0.1)	11.7 (±2.8)	1.20 (±0.09)	8.3 (±3.4)	0.55 (±0.03)	6.9 (±3.9)	67.5 (±18)	11.8 (±3.5)
	3rd	7.6 (±0.2)	10.1 (±3.1)	1.10 (±0.15)	7.2 (±2.7)	0.48 (±0.05)	7.2 (±2.7)	63.7 (±26)	12.7 (±7.4)

BD, bulk density; MCP, macroporosity; SOC, soil organic carbon; NT, total nitrogen; P, Olsen's extractable phosphorus; K, available potassium; CEC, cation exchange capacity. Values in parentheses are standard deviation.

According to Belay et al. [261], in supplementary irrigation vegetable production systems, CA practices can optimize nutrient use by decreasing nutrient losses through runoff and leaching. In this respect, several studies show that CA practices reduce the loss of nutrients via runoff or nutrients adsorbed in sediments lost by water erosion [176,262–265]. In this context, Jordan et al. [266] registered an 81% decrease in total P loss and a 94% decrease in organic nitrogen with non-inversion tillage compared with plow. In citrus orchards, the straw mulching covering the soil surface reduced runoff and sediment losses and subsequently decreased nutrient losses; the total nitrogen and phosphorus losses were significantly decreased by the straw mulching treatment compared with conventional treatments without mulching [267]. Liu et al. [268], using the Soil and Water Assessment Tool (SWAT), concluded that conservation tillage and contour farming can help reduce runoff by 15.99% and 9.16%, total nitrogen losses by 8.99% and 8%, and total phosphorus losses by 7% and 5%, respectively. In a study by García-Díaz et al. [269], the efficiency of using groundcover in vineyards to reduce mineral N losses via runoff was demonstrated.

As stated by Dinnes et al. [270], the strategies for reducing NO_3 loss through leaching can include CA practices by using cover crops, diversifying crop rotations, and reducing tillage. Cover crops or intercrops with deep-rooted plants reduce nutrient loss, intercepting leached nutrients from the root zone and returning them to the soil surface via mulch or as green manure. Wyland et al. [271] reported a 65–70% reduction in nitrate leaching from cover-cropped plots compared with the fallow control. In a study in Italy, CA practices had lower NO3 concentrations below the maximum rooting zone compared to conventional agricultural practices, thus reducing NO_3 leachate to groundwater [272]. According to Camarotto et al. [245], continuous soil cover and cover crops in CA systems reduced N leaching compared to conventional agriculture.

4.3. Influence on Soil Biological Properties

Soil biota plays a relevant role in soil health and sustainable crop production by supporting important functions such as soil aggregation, soil aeration, nutrient cycling, and bio-control, or the suppression of plant pathogens. Anthropogenic activities and especially intensive agriculture cause a considerable loss of soil biodiversity. Sustainable land uses are linked to the conservation of soil biological diversity [273]. Higher biodiversity means greater resilience to disturbances in the soil system [60]. The response of soil microorganisms and biochemical properties to soil management practices is measured by parameters such as the size and activity of the microbial community and soil enzymatic activities.

4.3.1. Microbial Activity

The soil microbial biomass (SMB) is commonly used to assess soil microbial activity, as this parameter responds quickly to changes in soil management. In this context, Zornoza et al. [274] stated that the quantitative description of the structure and diversity of the microbial community can be used as a tool for the evaluation of soil quality. That is, SMB can be used as an indicator of early changes in cropland management practices [275]. CA creates optimal conditions for microorganisms, with less frequent disturbance of the soil, increased SOM, improved water and thermal conditions, and increased diversity of substrates.

Crop diversification can increase soil microbial diversity and activities because the roots of cover crops release exudates in intercropping systems, contributing to greater microbial biomass [276]. In this context, Lopes and Fernandes [277] registered an increase in microbial biomass C with intercropping compared with monoculture. Singh et al. [278] reported that CA management systems can lead to an improvement in soil biota. Similarly, Wang et al. [279], in a study in drylands of northern China, reported a more diverse soil bacterial community in conservation tillage soils than in CT soils. Moreover, Silva et al. [280] registered a decrease in microbial diversity as tillage practices intensified. Dorr de Cuadros et al. [281] showed that microbial diversity was significantly higher in the NT system at four taxonomic levels (order, family, genus, and species) compared with the CT system. Henneron et al. [282] analyzed the long-term effects of CA on soil biodiversity, finding an improvement in the biomass and biodiversity of microorganisms. Baghel et al. [283], in a rice–wheat cropping system, recorded higher microbial biomass carbon under CA practices compared to CT. In a maize–mustard rotation, the zero-tilled flatbed and permanent bed CA practices improved soil biological properties, with higher SMB-C than CT [212].

Additionally, in a meta-analysis of 96 paired experiments, Li et al. [284] showed that CA practices (NT with residue retention) resulted in higher soil microbial biomass carbon (SMB-C) and nitrogen (SMB-N), and microbial quotient (qMic, Cmic-to-organic C ratio). In a continuous rice–wheat rotation, zero tillage and residue cycling compared to CT and residue removal increased SMB-C by 29 and 56%, respectively, whereas the SMB-N increased by 27 and 84%, respectively [285]. In a pigeon pea (*Cajanus cajan* (L.) Millsp.) and soybean intercropping system, conservation tillage systems recorded significantly higher SMB-C and SMB-N levels than CT without crop residues [286]. Spedding et al. [287] reported higher SMB-C and N levels in plots with residue retention than with residue removal, although the differences were significant only in the 0–10 cm layer. This agrees with Ceja-Navaro et al. [288], who found that in soils under NT with a monoculture of maize and removal of crop residue, microbial diversity was strongly reduced compared to soil under wheat NT where crop residues were retained. According to Legrand et al. [289], soil tillage is the agronomic practice that most influences soil bacterial diversity, with a greater functional and taxonomic diversity of bacteria in agricultural soils with minimal tillage compared to conventional tillage. In this context, Mathew et al. [290] reported a higher microbial biomass at the 0–5 cm depth in a long-term no-tillage system than in a conventional tillage system. According to Lopes and Fernandes [277], the changes in microbial community composition do not coincide with the increased soil physical quality resulting from CA practices, indicating the influence of other factors, such as edaphic or anthropic, on the soil microbial profile.

The crop system also influences microbial diversity. In this respect, Dorr de Cuadros et al. [281] reported greater microbial diversity in soils with a crop system based on cereals without legumes. That is, cereal straw substrates have a higher C:N ratio, which stimulates the microbial community to degrade organic substrate and leads to an increase in the microbial population.

4.3.2. Soil Enzymatic Activities

The microbial enzymatic activities of the soil serve as an indicator of the potential of the soil to decompose organic C and mineralize nutrients (P and N), and thereby nutrients available for plants. Soil enzymatic functions are greatly influenced by the cropping system and the degree of soil disturbance [291].

The main enzymes used to determine soil health are β-glucosidase, N-acetyl-glucosaminidase, and acid phosphatase, which are responsible for mediating C, N, and P cycling in the soil, respectively. According to Bonini-Pires et al. [292], the association of NT and increased crop rotation enhanced enzymatic activity in the soil surface. In a rice–wheat system in India, soil enzyme activities increased (5–18%) under an NT system with residues compared to an NT system without residues and a CT system without residues [293]. The implementation of CA in maize rotations improved soil enzymatic activities [104]. Similarly, Kumar and Babalad [286] registered significantly higher soil urease, dehydrogenase, and total phosphate activities in conservation tillage systems as compared to CT without crop residue. According to Choudhary et al. [285], soil enzyme activities were significantly increased in a conservation agriculture-based maize–wheat system.

In a study by Sharma et al. [294], an NT rice–wheat system with rice residue mulch increased soil dehydrogenase, cellulase, and alkaline phosphatase activities by 23%, 34%, and 14%, respectively, compared to CT. Pooniya et al. [212] reported that CA practices (zero-tilled flatbed and permanent bed) significantly increased dehydrogenase, alkaline phosphatase, and urease activities compared with CT.

The impact of CA practices on soil microbial and enzymatic activities in hillslope farming with rainfed olive orchards compared to CT is shown in Table 4 [259]. Moreover, Kandeler et al. [295] determined that protease and phosphatase activities significantly increased after only 2 years of minimum tillage compared to CT. Similarly, Roldán et al. [296] found that CA techniques based on zero tillage and legume cover remarkably enhanced the soil enzyme activities (dehydrogenase, urease, protease, β-glucosidase, and acid phosphatase). In a study by Pandey et al. [297], the no-till system fostered an improvement in the activities of β-glucosidase as well as microbial biomass carbon and nitrogen compared to CT. Similarly, Sinsabaugh et al. [298] found that minimum tillage promotes β-glucosidase activity due to the augmentation in microbial biomass, more substrate availability, and reduced soil disturbance, as was noted in a CA system compared to CT.

Table 4. Effect of CA practices on soil microbial and enzymatic activities in olive orchards throughout 3 year monitoring period (SE Spain).

Soil Management	Year	MB_N	MB_C	B-GLU	PRO	DHA	PHP
		(mg kg^{-1})		(μg pNP g^{-1} h^{-1})	(μg TRS g^{-1} h^{-1})	(μg TPF g^{-1} h^{-1})	(μg pNP g^{-1} h^{-1})
Minimum tillage and spontaneous vegetation strips	1st	5.8 (±2.2)	3.4 (±1.4)	401 (±1.2)	12.0 (±1.4)	99.20 (±1.9)	131.5 (±11.8)
	3rd	6.9 (±3.4)	3.8 (±1.1)	452 (±2.4)	12.8 (±1.5)	111.8 (±3.4)	139.8 (±22.4)
Minimum tillage and legume strips	1st	5.0 (±1.2)	3.1 (±1.0)	461 (1.9)	11.9 (±0.9)	100.7 (±2.7)	120.4 (±17.1)
	3rd	6.4 (±0.9)	4.2 (±2.4)	483 (±3.5)	12.7 (±1.6)	119.1 (±5.2)	131.4 (±13.7)
Conventional tillage	1st	5.3 (±0.8)	2.0 (±0.8)	131 (±1.2)	11.7 (±1.4)	92.43 (±5.1)	122.0 (±21.5)
	3rd	4.3 (±0.7)	1.3 (±0.9)	196 (±1.8)	12.4 (±1.9)	92.78 (±4.9)	129.6 (±20.9)

β-GLU, β-glucosidase; PRO, protease; DHA, Dehydrogenase; PHP, Phosphatase; MBN, microbial biomass-nitrogen; MBC, microbial biomass-carbon. Values in parentheses are standard deviation.

Ultimately, it is evident that CA practices positively impact soil microorganisms and microbial processes ascribed to changes in the quantity and quality of plant residues that enter the soil, their spatial distribution, changes in the provision of nutrients, and physical alterations. Consequently, the alternative modifications to CT systems, especially those based on methods used in CA, are able to boost important functions for soil health restoration.

4.3.3. Earthworms

Earthworms are one of the most important soil macrofaunal groups and are described as ecosystem engineers because of their effects on soil properties and on the availability of resources for other organisms [299]. They determine the nutrient cycle, microbial activity, the stability of soil aggregates, and the density and distribution of other invertebrates. Soil tillage causes physical damage to earthworms as well as alterations of their habitat, and can vary the community structure and relative abundance of earthworms [300]. The variability in burrowing and feeding behaviors influences the effects that tillage type can have on earthworms [301]. Thus, the species that inhabit the topsoil are most at risk of being adversely affected by plowing [302]. Earthworms have been observed to respond positively to CA practices. Contrarily, a study by Baldivieso-Freitas et al. [303] did not register any positive effects of the combination of CA techniques (reduced tillage by chiseling and green manures) on earthworm populations in a Mediterranean environment. However, organic fertilization showed a more significant role and enhanced their population. Therefore, it is crucial to understand how different factors (soil properties, crop rotations, and climate conditions) interact when designing a sustainable organic system.

According to Van Capelle et al. [304], the increase in earthworm density under no-till systems is due to the interactions of different effects: reduced injuries, decreased exposure to predators at the soil surface, reduced microclimate changes, and increased availability of organic matter. Radford et al. [305] reported that earthworm numbers increased fourfold with a zero-tillage system as compared to CT. Birkás et al. [306], in a study in Hungary, registered significantly more earthworms in soils under a conservation tillage system that included leaving stubble residues on the surface, compared to soils that were deteriorated by tillage pans and left bare without residues. In a study in Zambia, soils under CA practices with residue retention and crop rotation had higher earthworm populations in the top 30 cm than soils under conventionally ploughed practices [87]. Errouissi et al. [307] showed that zero tillage with surface residue increased the populations and diversity of soil invertebrates, including earthworms, compared to CT because of improved soil properties and a lack of soil disturbance. Crop residues retained on the soil surface and minimum soil disturbance improve soil structure, are a food resource, and cool the soil temperature, allowing the number and biomass of earthworms to increase [308]. In a study in central Mexico, Castellanos-Navarrete et al. [84] showed that CA produced an evident increase in the abundance and biomass of earthworms compared to CT. Sharma and Dhaliwal [309], in a study of rice–wheat cropping systems in South Asia, concluded that a zero-tillage system with crop residue retention improved micronutrient contents and provided feeding for soil macrofauna, especially earthworms, as compared to conventional tillage without residue. In a long-term trial in Zambia, Muoni et al. [310] concluded that reduced tillage systems and crop rotations increase biological activity, with the density of termites and earthworms being higher in CA systems than in CT systems. Henneron et al. [282] reported an increase in anecic earthworms in the long term in CA systems. Additionally, Pelosi et al. [302] reported that the decrease in soil tillage intensity led to an increase in functional diversity and an increase in the density of anecic earthworms. Several studies have reported a positive impact of management systems that include diversified crop rotations on earthworm density [311,312].

4.3.4. Soil Respiration

Soil respiration comprises the oxidation of organic matter by microorganisms and rhizosphere respiration [313]. It is a measure of the metabolic activity of the soil microbial community and is considered as the second-largest terrestrial carbon flux worldwide [314]. It is one of the most widely used soil biological indicators in soil quality evaluations [62].

Soil respiration is sensitive to soil disturbances, so it can be used as an indicator to detect soil degradation early [315].

Soil management affects the soil microclimate and biotic factors (soil organic carbon, aboveground biomass, root biomass, and plant residues) that indirectly influence soil respiration [316]. Several studies have reported the effect of conservation agriculture practices on soil microbial respiration [277,317,318], without consistent trends. Some studies did not report significant differences in soil respiration between conventional tillage and conservation agriculture practices [277,319,320]. This may be because tillage seems to affect the temporal distribution more than the total amount of CO_2 emissions from the soil [321]. Therefore, to achieve an accurate assessment of the effects of agricultural practices on soil respiration, it is necessary to design a seasonal sampling [322]. In contrast, other studies recorded significantly higher soil respiration values in CA systems than in CT systems. In a study in Cambodia, Edralin et al. [317] reported higher soil respiration in CA (55.9 ± 4.8 kg CO_2-C ha^{-1} day^{-1}) than in CT (36.2 ± 13.5 kg CO_2-C ha^{-1} day^{-1}). In the long term, NT increased soil respiration compared to CT, by 16, 19 and 26% after 6, 20 and 35 years of implantation [103]. Additionally, a 12 year study showed that, compared to conventional tillage, no-till practices resulted in higher soil microbial respiration [323]. Sapkota et al. [103] reported higher soil respiration in no-tillage systems than in conventional tillage (+44%). In an apricot orchard, cover crops increased soil respiration compared to plots with bare control, herbicide control or mechanical cultivation [324].

According to Williams et al. [325], agricultural practices that imply the greater crop diversity, reduction in mechanical soil disturbance and/or an increase in organic amendment inputs that characterize CA systems improve the microbiological activity of the soil. CA practices increase organic carbon inputs to the soil, for example, through plant residues, improving soil biological activity [326]. In this context, Bera et al. [327] observed a significant and high positive correlation between SOC and basal soil respiration, of 0.84.

5. Conclusions and Future Perspectives

The main challenge of conserving and improving soil health is guaranteeing its long-term productivity and environmental sustainability. As was reviewed, CA systems can be implemented to minimize negative socioeconomic and environmental consequences associated with soil degradation by enhancing soil health and promoting the sustainability and multifunctionality of agroecosystems.

To meet the global challenges of food security and environmental conservation, CA has been identified as one of the technological options for a sustainable intensification of agriculture. CA systems have clear advantages over conventional agricultural systems in improving soil health and the efficient use of natural resources, reducing the environmental impacts of agricultural activities, saving inputs, reducing the cost of production, etc.

Regarding the implementation of CA practices, there are a number of restrictions and challenges that must be addressed in order to increase their adoption on a large scale:

- Unavailability of appropriate equipment and machines, especially for small- and medium-scale farms;
- Use of crop residues for livestock feed and fuel;
- Lack of knowledge about the benefits of CA and how to implement CA;
- Farmer mind-sets that limit the adoption of CA due to traditions or prejudices;
- Lack of technical and financial support from governments, international organizations, and/or extension agencies;
- Technical problems that can arise with the adoption of CA practices such as inadequate weed management, nutrient stratification, lower N availability, development of surface crust, etc., which can translate into a decrease in yield and can motivate farmers to abandon the system.

To overcome these constraints and increase the performance of CA worldwide, it is essential that CA systems be well-adapted to specific agronomic, environmental, social, and economic conditions. Consequently, it is necessary to carry out the following measures, among others:

- Improve the availability of machinery and supplies of plant nutrition;
- Identify and eliminate sociocultural barriers to CA adoption;
- Improve locally adapted management, such as appropriate crop rotations or the frequency and optimal timing of strategic tillage;
- Increase institutional support, research, efficiency of extension services, and information dissemination mechanisms.

Finally, in order to guarantee the long-term productivity and environmental sustainability of agroecosystems, it will be vital to develop new tools and methodologies to assess soil quality and health that can be used to evaluate and guide soil management decisions.

Author Contributions: Conceptualization, B.C.R. and V.H.D.-Z.; writing—review and editing, B.C.R., V.H.D.-Z., M.S.R., I.F.G.-T., B.G.R. and S.C.T. All authors have read and agreed to the published version of the manuscript.

Funding: The first author, Belén Cárceles Rodríguez, has a contract co-financed by the National Institute for Agricultural and Food Research and Technology (FPI-INIA 2018) and the European Social Fund (ESF), "The European Social Fund invests in your future". This publication was sponsored by the following research projects: "Integrated soil and water management in rainfed almond in a context of sustainable agriculture" (RTA2017-00097-00-00) granted by the INIA (National Institute for Agricultural and Food Research and Technology) and the MINECO (Spanish Ministry of Economy and Competitiveness), and partially by the project "Sustainable intensification of almond (Prunus dulcis Mill.) cultivation in a context of climate change (ISACLIMA)" (SG1.SG12020.005, PY20_00999).

Institutional Review Board Statement: Not applicable.

Informed Consent Statement: Not applicable.

Data Availability Statement: Not applicable.

Conflicts of Interest: The authors declare no conflict of interest.

References

1. Foley, J.; Ramankutty, N.; Brauman, K.; Cassidy, E.S.; Gerber, J.S.; Johnston, M.; Mueller, N.D.; O'Connell, C.; Ray, D.K.; West, P.C.; et al. Solutions for a cultivated planet. *Nature* **2011**, *478*, 337–342. [CrossRef] [PubMed]
2. German, R.N.; Thompson, C.E.; Benton, T.G. Relationships among multiple aspects of agriculture's environmental impact and productivity: A meta-analysis to guide sustainable agriculture. *Biol. Rev. Camb. Philos. Soc.* **2017**, *92*, 716–738. [CrossRef] [PubMed]
3. Shah, F.; Wu, W. Soil and Crop Management strategies to ensure higher crop productivity within sustainable environments. *Sustainability* **2019**, *11*, 1485. [CrossRef]
4. Tahat, M.M.; Alananbeh, K.M.; Othman, Y.A.; Leskovar, D.I. Soil health and sustainable agriculture. *Sustainability* **2020**, *12*, 4859. [CrossRef]
5. Kibblewhite, M.G.; Ritz, K.; Swift, M.J. Soil health in agricultural systems. *Philos. Trans. R. Soc. Lond. B Biol. Sci.* **2008**, *363*, 685–701. [CrossRef] [PubMed]
6. FAO & ITPS. *Status of the World's Soil Resources (SWSR): Main Report*; Food and Agriculture Organization of the United Nations and Intergovernmental Technical Panel on Soils: Rome, Italy, 2015; Volume 650. Available online: https://www.fao.org/3/i5199e/i5199e.pdf (accessed on 9 November 2021).
7. Moebius-Clune, B.N.; Moebius-Clune, D.J.; Gugino, B.K.; Idowu, O.J.; Schindelbeck, R.R.; Ristow, A.J.; van Es, H.M.; Thies, J.E.; Shayler, H.A.; McBride, M.B.; et al. *Comprehensive Assessment of Soil Health—The Cornell Framework, 3.2 ed.*; Cornell University: Geneva, NY, USA, 2016.
8. Smith, P.; Gregory, P.J. Climate change and sustainable food production. *Proc. Nutr. Soc.* **2013**, *72*, 21–28. [CrossRef] [PubMed]
9. Choudary, M.; Ghasal, P.C.; Kumar, S.R.P.; Yadav, S.S.; Meena, V.S.; Bisht, J.K. Conservation Agriculture and Climate Change: An Overview. In *Conservation Agriculture*; Bisht, J., Meena, V., Mishra, P., Pattanayak, A., Eds.; Springer: Singapore, 2020. [CrossRef]
10. González-Sánchez, E.J.; Moreno-García, M.; Kassam, A.; Holgado-Cabrera, A.; Triviño-Tarradas, P.; Carbonell-Bojollo, R.; Pisante, M.; Veroz-González, O.; Basch, G. *Conservation Agriculture: Making Climate Change Mitigation and Adaptation Real in Europe*; ECAF: Brussels, Belgium, 2017. [CrossRef]
11. Smith, P.; Olesen, J.E. Synergies between the mitigation of, and adaptation to, climate change in agriculture. *J. Agric. Sci* **2010**, *148*, 543–552. [CrossRef]

12. Lobell, D.B.; Burke, M.B.; Tebaldi, C.; Mastrandrea, M.D.; Falcon, W.P.; Naylor, R.L. Prioritizing climate change adaptation needs for food security in 2030. *Science* **2008**, *319*, 607–610. [CrossRef] [PubMed]
13. Palm, C.; Blanco-Canqui, H.; DeClerck, F.; Gatere, L.; Grace, P. Conservation agriculture and ecosystem services: An overview. *Agric. Ecosyst. Environ.* **2014**, *187*, 87–105. [CrossRef]
14. Doran, J.W.; Zeiss, M.R. Soil health and sustainability: Managing the biotic component of soil quality. *Appl. Soil Ecol.* **2000**, *15*, 3–11. [CrossRef]
15. Pérez-Escamilla, R. Food Security and the 2015–2030 Sustainable Development Goals: From Human to Planetary Health: Perspectives and Opinions. *Curr. Dev. Nutr.* **2017**, *1*, e000513. [CrossRef] [PubMed]
16. Shrestha, J.; Subedi, S.; Timsina, K.; Chaudhary, A.; Kandel, M.; Tripathi, S. Conservation agriculture as an approach towards sustainable crop production: A Review. *Farming Manag.* **2020**, *5*, 7–15. [CrossRef]
17. Holland, J.M. The environmental consequences of adopting conservation tillage in Europe: Reviewing the evidence. *Agric. Ecosyst. Environ.* **2004**, *103*, 1–25. [CrossRef]
18. FAO. Conservation Agriculture. 2015. Available online: https://www.fao.org/conservation-agriculture/en/ (accessed on 19 October 2021).
19. Ikazaki, K.; Nagumo, F.; Simporé, S.; Barro, A. Are all three components of conservation agriculture necessary for soil conservation in the Sudan Savanna? *Soil Sci. Plant Nutr.* **2018**, *64*, 230–237. [CrossRef]
20. Jat, M.L.; Chakraborty, D.; Ladha, J.K.; Rana, D.S.; Gathala, M.K.; McDonald, A.; Gerard, B. Conservation agriculture for sustainable intensification in South Asia. *Nat. Sustain.* **2020**, *3*, 336–343. [CrossRef]
21. Yigezu, Y.A.; El-Shater, T.; Boughlala, M.; Devkota, M.; Mrabet, R.; Moussadek, R. Can an incremental approach be a better option in the dissemination of conservation agriculture? Some socioeconomic justifications from the drylands of Morocco. *Soil Tillage Res.* **2021**, *212*, 105067. [CrossRef]
22. Giller, K.E.; Andersson, J.A.; Corbeels, M.; Kirkegaard, J.; Mortensen, D.; Erenstein, O.; Vanlauwe, B. Beyond conservation agriculture. *Front. Plant Sci.* **2015**, *6*, 870. [CrossRef]
23. Rodenburg, J.; Büchi, L.; Haggar, J. Adoption by adaptation: Moving from Conservation Agriculture to conservation practices. *Int. J. Agric. Sustain.* **2020**, *19*, 437–455. [CrossRef]
24. FAO. *Save and Grow: A Policymaker's Guide to the Sustainable Intensification of Smallholder Crop Production*; Food and Agriculture Organization of the United Nations: Rome, Italy, 2011. Available online: https://www.fao.org/3/i2215e/i2215e.pdf (accessed on 16 November 2021).
25. Das, T.K.; Nath, C.P.; Das, S.; Biswas, S.; Bhattacharyya, R.; Sudhishri, S.; Raj, R.; Singh, B.; Kakralia, S.K.; Rathi, N.; et al. Conservation agriculture in rice-mustard cropping system for five years: Impacts on crop productivity, profitability, water-use efficiency, and soil properties. *Field Crops Res.* **2020**, *250*, 107781. [CrossRef]
26. Jat, H.S.; Choudhary, K.M.; Nandal, D.P.; Yadav, A.K.; Poonia, T.; Singh, Y.; Sharma, P.C.; Jat, M.L. Conservation agriculture-based sustainable intensification of cereal systems leads to energy conservation, higher productivity and farm profitability. *Environ. Manag.* **2020**, *65*, 774–786. [CrossRef]
27. Kassam, A.; Friedrich, T.; Derpsch, R. Global spread of Conservation Agriculture. *Int. J. Environ. Stud.* **2019**, *76*, 29–51. [CrossRef]
28. Derpsch, R.; Friedrich, T.; Kassam, A.; Li, H. Current status of adoption of no-till farming in the world and some of its main benefits. *Int. J. Agric. Biol. Eng.* **2010**, *3*, 1–25. [CrossRef]
29. Lal, R. Sustainable intensification of China's agroecosystems by conservation agriculture. *Int. Soil Water Conserv. Res.* **2018**, *6*, 1–12. [CrossRef]
30. Bhan, S.; Behera, U.K. Conservation agriculture in India—Problems, prospects and policy issues. *Int. Soil Water Conserv. Res.* **2014**, *2*, 1–12. [CrossRef]
31. Hobbs, P.R. Conservation agriculture: What is it and why is it important for future sustainable food production? *J. Agric. Sci.* **2007**, *145*, 127–137. [CrossRef]
32. Sahu, G.; Mohanty, S.; Das, S. Conservation agriculture—A way to improve soil health. *J. Exp. Biol. Agric. Sci.* **2020**, *8*, 355–368. [CrossRef]
33. Gonzalez-Sanchez, E.J.; Veroz, G.O.; Moreno, G.M.; Gomez, A.M.R.; Ordoñez, F.R.; Trivino, T.P.; Kassam, A.; Gil, R.J.A.; Basch, G.; Carbonell, B.R. Climate change adaptability and mitigation with conservation agriculture. In *Food Science, Technology and Nutrition, Rethinking Food and Agriculture*; Woodhead Publishing Series; Kassam, A., Kassam, L., Eds.; Woodhead Publishing: Sawston, UK, 2021; pp. 231–246. [CrossRef]
34. Indoria, A.K.; Rao, C.S.; Sharma, K.L.; Reddy, K.S. Conservation agriculture—A panacea to improve soil physical health. *Curr. Sci.* **2017**, *112*, 52–61. Available online: http://www.jstor.org/stable/24911616 (accessed on 10 November 2022). [CrossRef]
35. Subbulakshmi, S.; Saravanan, N.; Subbian, P. Conventional tillage vs. conservation tillage—A review. *Agric. Rev.* **2009**, *30*, 56–63.
36. Madarász, B.; Juhos, K.; Ruszkiczay, R.Z.; Benke, S.; Jakab, G.; Szalai, Z. Conservation tillage vs. conventional tillage: Long-term effects on yields in continental, sub-humid Central Europe, Hungary. *Int. J. Agric. Sustain.* **2016**, *14*, 408–427. [CrossRef]
37. Jacobs, A.; Helfrich, M.; Hanisch, S.; Quendt, U.; Rauber, R.; Ludwig, B. Effect of conventional and minimum tillage on physical and biochemical stabilization of soil organic matter. *Biol. Fertil. Soils* **2010**, *46*, 671–680. [CrossRef]
38. Busari, A.M.; Kuka, L.S.S.; Amanpreet, K.; Bhatt, R.; Dulazi, A.A. Conservation tillage impacts on soil, crop and the environment. *Int. Soil Water Conserv. Res.* **2015**, *2*, 119–129. [CrossRef]
39. Stevens, A.W. Review: The economics of soil health. *Food Policy* **2018**, *80*, 1–9. [CrossRef]

40. White, C.M.; DuPont, S.T.; Hautau, M.; Hartman, D.; Finney, D.M.; Bradley, B.; LaChance, J.C.; Kaye, J.P. Managing the trade-off between nitrogen supply and retention with cover crop mixtures. *Agric. Ecosyst. Environ.* **2017**, *237*, 121–133. [CrossRef]
41. Schipanski, M.E.; Barbercheck, M.; Douglas, M.R.; Finney, D.M.; Haider, K.; Kaye, J.P.; Kemanian, A.R.; Mortensen, D.A.; Ryan, M.R.; Tooker, J.; et al. A framework for evaluating ecosystem services provided by cover crops in agroecosystems. *Agric. Syst.* **2014**, *125*, 12–22. [CrossRef]
42. Jaffuel, G.; Blanco-Pérez, R.; Büchi, L.; Mäder, P.; Fließbach, A.; Charles, R.; Degen, T.; Turlings, T.C.J.; Campos-Herrera, R. Effects of cover crops on the overwintering success of entomopathogenic nematodes and their antagonists. *Appl. Soil Ecol.* **2017**, *114*, 62–73. [CrossRef]
43. Lyon, D.J.; Nielsen, D.C.; Felter, D.G.; Burgener, P.A. Choice of summer fallow replacement crops impacts subsequent winter wheat. *Agron. J.* **2007**, *99*, 578–584. [CrossRef]
44. Kaspar, T.C.; Jaynes, D.B.; Parkin, T.B.; Moorman, T.B.; Singer, J.W. Effectiveness of oat and rye cover crops in reducing nitrate losses in drainage water. *Agric. Water Manag.* **2012**, *110*, 25–33. [CrossRef]
45. Poeplau, C.; Don, A. Carbon sequestration in agricultural soils via cultivation of cover crops—A meta-analysis. *Agric. Ecosyst. Environ.* **2015**, *200*, 33–41. [CrossRef]
46. Cates, A.M.; Ruark, M.D.; Grandy, A.S.; Jackson, R.D. Small soil C cycle responses to three years of cover crops in maize cropping systems. *Agric. Ecosyst. Environ.* **2019**, *286*, 106649. [CrossRef]
47. Abdalla, M.; Hastings, A.; Cheng, K.; Chadwick, D.; Espenberg, M.; Truu, J.; Rees, R.M.; Smith, P. A critical review of the impacts of cover crops on nitrogen leaching, net greenhouse gas balance and crop productivity. *Glob. Chang. Biol.* **2019**, *25*, 2530–2543. [CrossRef]
48. Minasny, B.; Malone, B.P.; McBratney, A.B.; Angers, D.A.; Arrouays, D.; Chambers, A.; Chaplot, V.; Chen, Z.S.; Cheng, K.; Das, B.S.; et al. Soil carbon 4 per mille. *Geoderma* **2017**, *292*, 59–86. [CrossRef]
49. García-Tejero, I.F.; Carbonell, B.R.; Ordoñez, F.R.; Torres, F.P.; Durán, Z.V.H. Conservation agriculture practices to improve the soil water management and soil carbon storage in Mediterranean rainfed agro-ecosystems. In *Soil Health Restoration and Management*; Meena, R., Ed.; Springer: Singapore, 2020; pp. 203–230. [CrossRef]
50. Daryanto, S.; Fu, B.; Wang, L.; Jacinthe, P.A.; Zhao, W. Quantitative synthesis on the ecosystem services of cover crops. *Earth-Sci. Rev.* **2018**, *185*, 357–373. [CrossRef]
51. Clark, A.J.; Decker, A.M.; Meisinger, J.J.; McIntosh, M.S. Kill date of vetch, rye, and a vetch-rye mixture: I. Cover crop and corn nitrogen. *Agron. J.* **1997**, *89*, 427–434. [CrossRef]
52. Ladan, S.; Jacinthe, P.A. Nitrogen availability and early corn growth on plowed and no till soils amended with different types of cover crops. *J. Soil Sci. Plant Nutr.* **2017**, *1*, 74–90. [CrossRef]
53. Mitchell, J.P.; Shrestha, A.; Irmak, S. Trade-offs between winter cover crop production and soil water depletion in the San Joaquin Valley, California. *J. Soil Water Conserv.* **2015**, *70*, 430–440. [CrossRef]
54. Stagnari, F.; Galieni, A.; Speca, S.; Cafiero, G.; Pisante, M. Effects of straw mulch on growth and yield of durum wheat during transition to conservation agriculture in Mediterranean environment. *Field Crops Res.* **2014**, *167*, 51–63. [CrossRef]
55. Bhullar, M.S.; Pandey, M.; Kumar, S.; Gill, G. Weed management in conservation agriculture in India. *Indian J. Weed Sci.* **2016**, *48*, 1–12. [CrossRef]
56. Farooq, M.; Flower, K.C.; Jabran, K.; Wahid, A.; Siddique, K.H.M. Crop yield and weed management in rainfed conservation agriculture. *Soil Tillage Res.* **2011**, *117*, 172–183. [CrossRef]
57. Bonfante, A.; Basile, A.; Bouma, J. Targeting the soil quality and soil health concepts when aiming for the United Nations Sustainable Development Goals and the EU Green Deal. *Soil* **2020**, *6*, 453–466. [CrossRef]
58. Laishram, J.; Saxena, K.G.; Maikhuri, R.K.; Rao, K.S. Soil quality and soil health: A review. *Int. J. Ecol. Environ. Sci.* **2012**, *38*, 19–37.
59. USDA. Available online: https://www.nrcs.usda.gov/wps/portal/nrcs/main/soils/health/ (accessed on 25 October 2021).
60. Yang, T.; Siddique, K.H.M.; Liu, K. Cropping systems in agriculture and their impact on soil health—A review. *Glob. Ecol. Conserv.* **2020**, *23*, e01118. [CrossRef]
61. Wang, K.H.; Hooks, C.R.R. Chapter 4: Managing soil health and soil health bioindicators through the use of cover crops and other sustainable practices. In *MD Organic Vegetable Growers*; Brust, G.E., Ed.; University of Maryland: College Park, MD, USA, 2011.
62. Bünemann, E.K.; Bongiorno, G.; Bai, Z.; Creamer, R.E.; De Deyn, G.; de Goede, R.; Fleskens, L.; Geissen, V.; Kuyper, T.W.; Mäder, P.; et al. Soil quality—A critical review. *Soil Biol. Biochem.* **2018**, *120*, 105–125. [CrossRef]
63. Fierer, N.; Wood, S.A.; Bueno de Mesquita, C.P. How microbes can, and cannot, be used to assess soil health. *Soil Biol. Biochem.* **2021**, *153*, 108111. [CrossRef]
64. Thoumazeau, A.; Bessou, C.; Renevier, M.S.; Trap, J.; Marichal, R.; Mareschal, L.; Decaëns, T.; Bottinelli, N.; Jaillard, B.; Chevallier, T.; et al. Biofunctool®: A new framework to assess the impact of land management on soil quality: Part A: Concept and validation of the set of indicators. *Ecol. Indic.* **2019**, *97*, 100–110. [CrossRef]
65. Mukherjee, A.; Lal, R. Comparison of soil quality index using three methods. *PLoS ONE* **2014**, *9*, e105981. [CrossRef] [PubMed]
66. Cherubin, M.R.; Karlen, D.L.; Cerri, C.E.P.; Franco, A.L.C.; Tormena, C.A.; Davies, C.A.; Cerri, C.C. Soil quality indexing strategies for evaluating sugarcane expansion in Brazil. *PLoS ONE* **2016**, *11*, e0150860. [CrossRef]
67. Lehmann, J.; Bossio, D.A.; Kögel-Knabner, I.; Rillig, M.C. The concept and future prospects of soil health. *Nat. Rev. Earth Environ.* **2020**, *1*, 544–553. [CrossRef]

68. Rinot, O.; Levy, G.J.; Steinberger, Y.; Svoray, T.; Eshel, G. Soil health assessment: A critical review of current methodologies and a proposed new approach. *Sci. Total Environ.* **2019**, *648*, 1484–1491. [CrossRef]
69. Cardoso, E.J.B.N.; Vasconcellos, R.L.F.; Bini, D.; Miyauchi, M.Y.H.; dos Santos, C.A.; Alves, P.R.L.; de Paula, A.M.; Nakatani, A.S.; Pereira, J.M.; Nogueira, M.A. Soil health: Looking for suitable indicators: What should be considered to assess the effects of use and management on soil health? *Sci. Agric.* **2013**, *70*, 274–289. [CrossRef]
70. Hermans, T.D.G.; Dougill, A.J.; Whitfield, S.; Peacock, C.L.; Eze, S.; Thierfelder, C. Combining local knowledge and soil science for integrated soil health assessments in conservation agriculture systems. *J. Environ. Manag.* **2021**, *286*, 112192. [CrossRef]
71. Andrews, S.S.; Karlen, D.L.; Mitchell, J.P. A comparison of soil quality indexing methods for vegetable production systems in Northern California. *Agric. Ecosyst. Environ.* **2002**, *90*, 25–45. [CrossRef]
72. Morrow, J.G.; Huggins, D.R.; Carpenter-Boggs, L.A.; Reganold, J.P. Evaluating measures to assess soil health in long-term agroecosystem trials. *Soil Sci. Soc. Am. J.* **2016**, *80*, 450–462. [CrossRef]
73. Qi, Y.; Darilek, J.L.; Huang, B.; Zhao, Y.; Sun, W.; Gu, Z. Evaluating soil quality indices in an agricultural region of Jiangsu Province, China. *Geoderma* **2009**, *149*, 325–334. [CrossRef]
74. Caron, P.; Biénabe, E.; Hainzelin, E. Making transition towards ecological intensification of agriculture a reality: The gaps in and the role of scientific knowledge. *Curr. Opin. Environ. Sustain.* **2014**, *8*, 44–52. [CrossRef]
75. Kassam, A.; Derpsch, R.; Friedrich, T. Global achievements in soil and water conservation: The case of conservation agriculture. *Int. Soil Water Conserv. Res.* **2014**, *2*, 5–13. [CrossRef]
76. Bronick, C.J.; Lal, R. Soil structure and management: A review. *Geoderma* **2005**, *124*, 3–22. [CrossRef]
77. Liu, M.; Han, G.; Zhang, Q. Effects of Soil Aggregate Stability on Soil Organic Carbon and Nitrogen under Land Use Change in an Erodible Region in Southwest China. *Int. J. Environ. Res. Public Health* **2019**, *16*, 3809. [CrossRef] [PubMed]
78. Spohn, M.; Giani, L. Impacts of land use change on soil aggregation and aggregate stabilizing compounds as dependent on time. *Soil Biol. Biochem.* **2011**, *43*, 1081–1088. [CrossRef]
79. Six, J.; Bossuyt, H.; Degryze, S.; Denef, K. A history of research on the link between (micro) aggregates, soil biota, and soil organic matter dynamics. *Soil Tillage Res.* **2004**, *79*, 7–31. [CrossRef]
80. Cherubin, M.R.; da Silva Oliveira, D.M.; Feigl, B.J.; Pimentel, L.G.; Lisboa, I.P.; Gmach, M.R.; Varanda, L.L.; Morais, M.C.; Satiro, L.S.; Popin, G.V.; et al. Crop residue harvest for bioenergy production and its implications on soil functioning and plant growth: A review. *Sci. Agric.* **2018**, *75*, 255–272. [CrossRef]
81. Murphy, B.W. Impact of soil organic matter on soil properties—A review with emphasis on Australian soils. *Soil Res.* **2015**, *53*, 605. [CrossRef]
82. Wang, Y.; Xu, J.; Shen, J.H.; Luo, Y.M.; Scheu, S.; Ke, X. Tillage, residue burning and crop rotation alter soil fungal community and water-stable aggregation in arable fields. *Soil Tillage Res.* **2010**, *107*, 71–79. [CrossRef]
83. Azooz, R.H.; Arshad, M.A. Soil infiltration and hydraulic conductivity under long-term no-tillage and conventional tillage systems. *Can. J. Soil Sci.* **1996**, *76*, 143–152. [CrossRef]
84. Castellanos-Navarrete, A.; Rodríguez, A.C.; de Goede, R.G.M.; Kooistra, M.J.; Sayre, K.D.; Brussaard, L.; Pulleman, M.M. Earthworm activity and soil structural changes under conservation agriculture in central Mexico. *Soil Tillage Res.* **2012**, *123*, 61–70. [CrossRef]
85. Govaerts, B.; Sayre, K.D.; Goudeseune, B.; De Corte, P.; Lichter, K.; Dendooven, L.; Deckers, J. Conservation agriculture as a sustainable option for the central Mexican highlands. *Soil Tillage Res.* **2009**, *103*, 222–230. [CrossRef]
86. Sithole, N.J.; Magwaza, L.S.; Thibaud, G.R. Long-term impact of no-till conservation agriculture and N-fertilizer on soil aggregate stability, infiltration and distribution of C in different size fractions. *Soil Tillage Res.* **2019**, *190*, 147–156. [CrossRef]
87. Thierfelder, C.; Wall, P.C. Rotation in conservation agriculture systems of Zambia: Effects on soil quality and water relations. *Exp. Agric.* **2010**, *46*, 309–325. [CrossRef]
88. Nyamangara, J.; Marondedze, A.; Masvaya, E.N.; Mawodza, T.; Nyawasha, R.; Nyengerai, K.; Tirivavi, R.; Nyamugafata, P.; Wuta, M. Influence of basin-based conservation agriculture on selected soil quality parameters under smallholder farming in Zimbabwe. *Soil Use Manag.* **2014**, *30*, 550–559. [CrossRef]
89. Zhang, H.; Niu, L.; Hu, K.; Hao, J.; Li, F.; Gao, Z.; Wang, X. Influence of tillage, straw-returning and mineral fertilization on the stability and associated organic content of soil aggregates in the North China Plain. *Agronomy* **2020**, *10*, 951. [CrossRef]
90. Eze, S.; Dougill, A.J.; Banwart, S.A.; Hermans, T.D.G.; Ligowe, I.S.; Thierfelder, C. Impacts of conservation agriculture on soil structure and hydraulic properties of Malawian agricultural systems. *Soil Tillage Res.* **2020**, *201*, 104639. [CrossRef]
91. Soane, B.D.; Ball, B.C.; Arvidsson, J.; Basch, G.; Moreno, F.; Roger-Estrade, J. No-till in northern, western and south-western Europe: A review of problems and opportunities for crop production and the environment. *Soil Tillage Res.* **2012**, *118*, 66–87. [CrossRef]
92. Somasundaram, J.; Salikram, M.; Sinha, N.K.; Mohanty, M.; Chaudhary, R.S.; Dalal, R.C.; Mitra, R.; Blaise, N.; Coumar, D.; Hati, V.; et al. Conservation agriculture effects on soil properties and crop productivity in a semiarid region of India. *Soil Res.* **2019**, *57*, 187–199. [CrossRef]
93. Mloza-Banda, H.R.; Makwiza, C.N.; Mloza-Banda, M.L. Soil properties after conversion to conservation agriculture from ridge tillage in Southern Malawi. *J. Arid Environ.* **2016**, *127*, 7–16. [CrossRef]
94. Gómez-Muñoz, B.; Jensen, L.S.; Munkholm, L.; Olesen, J.E.; Møller Hansen, E.; Bruun, S. Long-term effect of tillage and straw retention in conservation agriculture systems on soil carbon storage. *Soil Sci. Soc. Am. J.* **2021**, *85*, 1465–1478. [CrossRef]

95. Cheesman, S.; Thierfelder, C.; Eash, N.S.; Kassie, G.T.; Frossard, E. Soil carbon stocks in conservation agriculture systems of Southern Africa. *Soil Tillage Res.* **2016**, *156*, 99–109. [CrossRef]
96. Li, Y.; Li, Z.; Cui, S.; Jagadamma, S.; Zhang, Q.P. Residue retention and minimum tillage improve physical environment of the soil in croplands: A global meta-analysis. *Soil Tillage Res.* **2019**, *194*, 104292. [CrossRef]
97. Mondal, S.; Mishra, J.S.; Poonia, S.P.; Kumar, R.; Dubey, R.; Kumar, S.; Verma, M.; Rao, K.K.; Ahmed, A.; Dwivedi, S.; et al. Can yield, soil C and aggregation be improved under long-term conservation agriculture in the eastern Indo-Gangetic plain of India? *Eur. J. Soil Sci.* **2021**, *72*, 1742–1761. [CrossRef]
98. Laborde, J.P.; Wortmann, C.S.; Blanco-Canqui, H.; McDonald, A.J.; Baigorria, G.A.; Lindquist, J.L. Short-term impacts of conservation agriculture on soil physical properties and productivity in the Midhills of Nepal. *Agron. J.* **2019**, *111*, 2128–2139. [CrossRef]
99. Kay, B.D.; VandenBygaart, A.J. Conservation tillage and depth stratification of porosity and soil organic matter. *Soil Tillage Res.* **2002**, *66*, 107–118. [CrossRef]
100. He, J.; Kuhn, N.J.; Zhang, X.M.; Zhang, X.R.; Li, H.W. Effects of 10 years of conservation tillage on soil properties and productivity in the farming–pastoral ecotone of Inner Mongolia, China. *Soil Use Manag.* **2009**, *25*, 201–209. [CrossRef]
101. Nyamadzawo, G.; Chikowo, R.; Nyamugafata, P.; Giller, K.E. Improved legume tree fallows and tillage effects on structural stability and infiltration rates of a kaolinitic sandy soil from central Zimbabwe. *Soil Tillage Res.* **2007**, *96*, 182–194. [CrossRef]
102. Mondal, S.; Poonia, S.P.; Mishra, J.S.; Bhatt, B.P.; Karnena, K.R.; Saurabh, K.; Rakesh, K.; Chakraborty, D. Short-term (5 years) impact of conservation agriculture on soil physical properties and organic carbon in a rice–wheat rotation in the indo-Gangetic plains of Bihar. *Eur. J. Soil Sci.* **2019**, *71*, 1076–1089. [CrossRef]
103. Islam, R.; Reeder, R. No-till and conservation agriculture in the United States: An example from the David Brandt farm, Carroll, Ohio. *Int. Soil Water Conserv. Res.* **2014**, *2*, 97–107. [CrossRef]
104. Parihar, C.M.; Yadav, M.R.; Jat, S.L.; Singh, A.K.; Kumar, B.; Pradhan, S.; Chakraborty, D.; Jat, M.L.; Jat, R.K.; Saharawat, Y.S.; et al. Long term effect of conservation agriculture in maize rotations on total organic carbon, physical and biological properties of a sandy loam soil in north-western Indo-Gangetic Plains. *Soil Tillage Res.* **2016**, *161*, 116–128. [CrossRef]
105. Gucci, R.; Caruso, G.; Bertolla, C.; Urbani, S.; Taticchi, A.; Esposto, S.; Servili, M.; Sifola, M.I.; Pellegrini, S.; Pagliai, M.; et al. Changes of soil properties and tree performance induced by soil management in a high-density olive orchard. *Eur. J. Agron.* **2012**, *41*, 18–27. [CrossRef]
106. Blanco-Canqui, H.; Lal, R. Crop residue removal impacts on soil productivity and environmental quality. *CRC Crit. Rev. Plant Sci.* **2009**, *28*, 139–163. [CrossRef]
107. Lahmar, R. Adoption of conservation agriculture in Europe: Lessons of the KASSA project. *Land Use Policy* **2010**, *27*, 4–10. [CrossRef]
108. Usón, A.; Poch, R.M. Effects of tillage and management practices on soil crust morphology under a Mediterranean environment. *Soil Tillage Res.* **2000**, *54*, 191–196. [CrossRef]
109. Baudron, F.; Tittonell, P.; Corbeels, M.; Letourmy, P.; Giller, K.E. Comparative performance of conservation agriculture and current smallholder farming practices in semi-arid Zimbabwe. *Field Crops Res.* **2012**, *132*, 117–128. [CrossRef]
110. Baumhardt, R.L.; Lascano, R.J. Rain infiltration as affected by wheat residue amount and distribution in ridged tillage. *Soil Sci. Soc. Am. J.* **1996**, *60*, 1908–1913. [CrossRef]
111. Page, K.L.; Dang, Y.P.; Dalal, R.C. The ability of conservation agriculture to conserve soil organic carbon and the subsequent impact on soil physical, chemical, and biological properties and yield. *Front. Sustain. Food Syst.* **2020**, *4*, 31. [CrossRef]
112. Dang, Y.P.; Seymour, N.P.; Walker, S.R.; Bell, M.J.; Freebairn, D.M. Strategic tillage in no-till farming systems in Australia's northern grains-growing regions: I. Drivers and implementation. *Soil Tillage Res.* **2015**, *152*, 104–114. [CrossRef]
113. Ruan, H.X.; Ahuja, L.R.; Green, T.R.; Benjamın, J.G. Residue cover and surface-sealing effects on infiltration: Numerical simulations for field applications. *Soil Sci. Soc. Am. J.* **2001**, *65*, 853–861. [CrossRef]
114. McGarry, D.; Bridge, B.J.; Radford, B.J. Contrasting soil physical properties after zero and traditional tillage of an alluvial soil in the semi-arid subtropics. *Soil Tillage Res.* **2000**, *53*, 105–115. [CrossRef]
115. Verhulst, N.; Carrillo, G.A.; Moeller, C.; Trethowan, R.; Sayre, K.D.; Govaerts, B. Conservation agriculture for wheat-based cropping systems under gravity irrigation: Increasing resilience through improved soil quality. *Plant Soil* **2011**, *340*, 467–479. [CrossRef]
116. Govaerts, B.; Sayre, K.D.; Deckers, J. Stable high yields with zero tillage and permanent bed planting? *Field Crops Res.* **2005**, *94*, 33–42. [CrossRef]
117. Stagnari, F.; Ramazzotti, S.; Pisante, M. Conservation agriculture: A different approach for crop production through sustainable soil and water management: A Review. In *Organic Farming, Pest Control and Remediation of Soil Pollutants*; Lichtfouse, E., Ed.; Sustainable Agriculture Reviews; Springer: Dordrecht, The Netherlands, 2009; Volume 1, pp. 55–83. [CrossRef]
118. Mondal, S.; Das, T.K.; Thomas, P.; Mishra, A.; Bandyopadhyay, K.; Aggarwal, P.; Chakraborty, D. Effect of conservation agriculture on soil hydro-physical properties, total and particulate organic carbon and root morphology in wheat (*Triticum aestivum*) under rice (*Oryza sativa*)-wheat system. *Indian J. Agric. Sci.* **2019**, *89*, 46–55.
119. Hamza, M.A.; Anderson, W.K. Soil compaction in cropping systems: A review of the nature, causes and possible solutions. *Soil Tillage Res.* **2005**, *82*, 121–145. [CrossRef]

120. Chen, G.; Weil, R.R. Root growth and yield of maize as affected by soil compaction and cover crops. *Soil Tillage Res.* **2011**, *117*, 17–27. [CrossRef]
121. Munkholm, L.J.; Schjønning, P.; Rasmussen, K.J.; Tanderup, K. Spatial and temporal effects of direct drilling on soil structure in the seedling environment. *Soil Tillage Res.* **2003**, *71*, 163–173. [CrossRef]
122. Van den Putte, A.; Govers, G.; Diels, J.; Gillijns, K.; Demuzere, M. Assessing the effect of soil tillage on crop growth: A meta-regression analysis on European crop yields under conservation agriculture. *Eur. J. Agron.* **2010**, *33*, 231–241. [CrossRef]
123. Moreno, F.; Arrúe, J.L.; Cantero-Martínez, C.; López, M.V.; Murillo, J.M.; Sombrero, A.; López-Garrido, R.; Madejón, E.; Moret, D.; Álvaro-Fuentes, J. Conservation agriculture under Mediterranean conditions in Spain. In *Biodiversity, Biofuels, Agroforestry and Conservation Agriculture*; Lichtfouse, E., Ed.; Sustainable Agriculture Reviews; Springer: Dordrecht, The Netherlands, 2010; Volume 5, pp. 175–193. [CrossRef]
124. Wortmann, C.S.; Drijber, R.A.; Franti, T.G. One-time tillage of no-till crop land five years post-tillage. *Agron. J.* **2010**, *102*, 1302–1307. [CrossRef]
125. Gosling, S.N.; Arnell, N.W. A global assessment of the impact of climate change on water scarcity. *Clim. Chang.* **2016**, *134*, 371–385. [CrossRef]
126. Verhulst, N.; Sayre, K.D.; Vargas, M.; Crossa, J.; Deckers, J.; Raes, D.; Govaerts, B. Wheat yield and tillage–straw management system × year interaction explained by climatic co-variables for an irrigated bed planting system in north-western Mexico. *Field Crops Res.* **2011**, *124*, 347–356. [CrossRef]
127. Ghosh, B.N.; Dogra, P.; Sharma, N.K.; Bhattacharyya, R.; Mishra, P.K. Conservation agriculture impact for soil conservation in maize–wheat cropping system in the Indian sub-Himalayas. *Int. Soil Water Conserv. Res.* **2015**, *3*, 112–118. [CrossRef]
128. Sławiński, C.; Cymerman, J.; Witkowska-Walczak, B.; Lamorski, K. Impact of diverse tillage on soil moisture dynamics. *Int. Agrophys.* **2015**, *26*, 301–309. [CrossRef]
129. Thierfelder, C.; Wall, P.C. Investigating conservation agriculture (CA) systems in Zambia and Zimbabwe to mitigate future effects of climate change. *J. Crop. Improv.* **2010**, *24*, 113–121. [CrossRef]
130. Busari, A.M.; Salako, F.K.; Tuniz, C.; Zuppi, G.M.; Stenni, B.; Adetunji, M.T.; Arowolo, T.A. Estimation of soil water evaporative loss after tillage operation using the stable isotope technique. *Int. Agrophys.* **2013**, *27*, 257–264. [CrossRef]
131. Parihar, C.M.; Nayak, H.S.; Rai, V.K.; Jat, S.L.; Parihar, N.; Aggarwal, P.; Mishra, A.K. Soil water dynamics, water productivity and radiation use efficiency of maize under multi-year conservation agriculture during contrasting rainfall events. *Field Crops Res.* **2019**, *241*, 107570. [CrossRef]
132. TerAvest, D.; Carpenter-Boggs, L.; Thierfelder, C.; Reganold, J.P. Crop production and soil water management in conservation agriculture, no-till, and conventional tillage systems in Malawi. *Agric. Ecosyst. Environ.* **2015**, *212*, 285–296. [CrossRef]
133. Zhao, X.; Liu, B.Y.; Liu, S.L.; Qi, J.Y.; Wang, X.; Pu, C.; Li, S.S.; Zhang, X.Z.; Yang, X.G.; Lal, R.; et al. Sustaining crop production in China's cropland by crop residue retention: A meta-analysis. *Land Degrad. Dev.* **2020**, *31*, 694–709. [CrossRef]
134. Ghosh, P.K.; Das, A.; Saha, R.; Kharkrang, E.; Tripathi, A.K.; Munda, G.C.; Ngachan, S.V. Conservation agriculture towards achieving food security in North East India. *Curr. Sci.* **2010**, *99*, 915–922.
135. Mondal, S.; Chakraborty, D.; Das, T.K.; Shrivastava, M.; Mishra, A.K.; Bandyopadhyay, K.K.; Aggarwal, P.; Chaudhari, S.K. Conservation agriculture had a strong impact on the sub-surface soil strength and root growth in wheat after a 7-year transition period. *Soil Tillage Res.* **2019**, *195*, 104385. [CrossRef]
136. Chalise, K.S.; Singh, S.; Wegner, B.R.; Kumar, S.; Pérez, G.J.D.; Osborne, S.L.; Nleya, T.; Guzman, J.; Rohila, J.S. Cover crops and returning residue impact on soil organic carbon, bulk density, penetration resistance, water retention, infiltration, and soybean yield. *Agron. J.* **2018**, *110*, 99–108. [CrossRef]
137. Mutuku, E.A.; Roobroeck, D.; Vanlauwe, B.; Boeckx, P.; Cornelis, W.M. Maize production under combined conservation agriculture and integrated soil fertility management in the sub-humid and semi-arid regions of Kenya. *Field Crops Res.* **2020**, *254*, 107833. [CrossRef]
138. Sindelar, M.; Blanco-Canqui, H.; Jin, V.L.; Ferguson, R.B. Cover crops and corn residue removal: Impacts on soil hydraulic properties and their relationships with carbon. *Soil Sci. Soc. Am. J.* **2019**, *83*, 221–231. [CrossRef]
139. Choudhary, M.; Rana, K.S.; Meena, M.C.; Bana, R.S.; Jakhar, P.; Ghasal, P.C.; Verma, R.K. Changes in physico-chemical and biological properties of soil under conservation agriculture based pearl millet–mustard cropping system in rainfed semi-arid region. *Arch. Agron. Soil Sci.* **2019**, *65*, 911–927. [CrossRef]
140. Singh, B.; Eberbach, P.L.; Humphreys, E.; Kukal, S.S. The effect of rice straw mulch on evapotranspiration, transpiration and soil evaporation of irrigated wheat in Punjab, India. *Field Crops Res.* **2011**, *98*, 1847–1855. [CrossRef]
141. Gupta, R.; Sayre, K. Conservation agriculture in South Asia. *J. Agric. Sci.* **2007**, *145*, 207–214. [CrossRef]
142. Assefa, T.; Jha, M.; Reyes, M.; Worqlul, A.W. Modeling the impacts of conservation agriculture with a drip irrigation system on the hydrology and water management in Sub-Saharan Africa. *Sustainability* **2018**, *10*, 4763. [CrossRef]
143. Belay, S.A.; Schmitter, P.; Worqlul, A.W.; Steenhuis, T.S.; Reyes, M.R.; Tilahun, S.A. Conservation agriculture saves irrigation water in the dry monsoon phase in the Ethiopian Highlands. *Water* **2019**, *11*, 2103. [CrossRef]
144. Jat, H.S.; Kumar, V.; Datta, A.; Choudhary, M.; Singh, Y.; Kakraliya, S.K.; Poonia, T.; McDonald, A.J.; Jat, M.L.; Sharma, P.C. Designing profitable, resource use efficient and environmentally sound cereal based systems for the Western Indo-Gangetic plains. *Sci. Rep.* **2020**, *10*, 19267. [CrossRef] [PubMed]

145. Alvarez, R.; Steinbach, H. A review of the effects of tillage systems on some soil physical properties, water content, nitrate availability and crops yield in the Argentine Pampas. *Soil Tillage Res.* **2009**, *104*, 1–15. [CrossRef]
146. Pittelkow, C.M.; Liang, X.Q.; Linquist, B.A.; van Groenigen, K.J.; Lee, J.; Lundy, M.E.; van Gestel, N.; Six, J.; Venterea, R.T.; van Kessel, C. Productivity limits and potentials of the principles of conservation agriculture. *Nature* **2015**, *517*, 365–368. [CrossRef]
147. Zhao, X.; Liu, S.; Pu, C.; Zhang, X.; Xue, J.; Ren, Y.; Zhao, X.; Chen, F.; Lal, R.; Zhang, H. Crop yields under no-till farming in China: A meta-analysis. *Eur. J. Agron.* **2017**, *84*, 67–75. [CrossRef]
148. Lu, X. A meta-analysis of the effects of crop residue return on crop yields and water use efficiency. *PLoS ONE* **2020**, *15*, e0231740. [CrossRef]
149. Sun, L.; Wang, S.; Zhang, Y.; Li, J.; Wang, X.; Wang, R.; Lyu, W.; Chen, N.; Wang, Q. Conservation agriculture based on crop rotation and tillage in the semi-arid Loess Plateau, China: Effects on crop yield and soil water use. *Agric. Ecosyst. Environ.* **2018**, *251*, 67–77. [CrossRef]
150. Das, T.K.; Bandyopadhyay, K.K.; Bhattacharyya, R.; Sudhishri, S.; Sharma, A.R.; Behera, U.K.; Saharawat, Y.S.; Sahoo, P.K.; Pathak, H.; Vyas, A.K.; et al. Effects of conservation agriculture on crop productivity and water-use efficiency under an irrigated pigeonpea–wheat cropping system in the western Indo-Gangetic Plains. *J. Agric. Sci.* **2016**, *154*, 1327–1342. [CrossRef]
151. Rockström, J.; Kaumbutho, P.; Mwalley, J.; Nzabi, A.W.; Temesgen, M.; Mawenya, L.; Barron, J.; Mutua, J.; Damgaard-Larsen, S. Conservation farming strategies in East and Southern Africa: Yields and rain water productivity from on-farm action research. *Soil Tillage Res.* **2009**, *103*, 23–32. [CrossRef]
152. Milgroom, J.; Soriano, M.A.; Garrido, J.M.; Gómez, J.A.; Fereres, E. The influence of a shift from conventional to organic olive farming on soil management and erosion risk in Southern Spain. *Renew. Agric. Food Syst.* **2007**, *22*, 1–10. [CrossRef]
153. Correia, C.M.; Brito, C.; Sampaio, A.; Dias, A.A.; Bacelar, E.; Gonçalves, B.; Ferreira, H.; Moutinho, P.J.; Rodrigues, M.A. Leguminous cover crops improve the profitability and the sustainability of rainfed olive (*Olea europaea* L.) orchards: From soil biology to physiology of yield determination. *Procedia Environ. Sci.* **2015**, *29*, 282–283. [CrossRef]
154. Arampatzis, G.; Hatzigiannakis, E.; Pisinaras, V.; Kourgialas, N.; Psarras, G.; Kinigopoulou, V.; Panagopoulos, A.; Koubouris, G. Soil water content and olive tree yield responses to soil management, irrigation, and precipitation in a hilly Mediterranean area. *J. Water Clim. Chang.* **2018**, *9*, 672–678. [CrossRef]
155. Krstić, Đ.; Vujić, S.; Jaćimović, G.; D'Ottavio, P.; Radanović, Z.; Erić, P.; Ćupina, B. The effect of cover crops on soil water balance in rain-fed conditions. *Atmosphere* **2018**, *9*, 492. [CrossRef]
156. Durán, Z.V.H.; Rodríguez, P.C.R.; Arroyo, P.L.; Martínez, R.A.; Francia, M.J.R.; Cárceles, R.B. Soil conservation measures in rainfed olive orchards in South-Eastern Spain: Impacts of plant strips on soil water dynamics. *Pedosphere* **2009**, *19*, 453–464. [CrossRef]
157. Castellini, M.; Stellacci, A.M.; Mastrangelo, M.; Caputo, F.; Manici, L.M. Estimating the soil hydraulic functions of some olive orchards: Soil management implications for water saving in soils of Salento peninsula (southern Italy). *Agronomy* **2020**, *10*, 177. [CrossRef]
158. Abazi, U.; Lorite, I.J.; Cárceles, R.B.; Martínez, R.A.; Durán, Z.V.H.; Francia, M.J.R.; Gómez, J.A. WABOL: A conceptual water balance model for analyzing rainfall water use in olive orchards under different soil and cover crop management strategies. *Comput. Electron. Agric.* **2013**, *91*, 35–48. [CrossRef]
159. Rusinamhodzi, L.; Corbeels, M.; van Wijk, M.T.; Rufino, M.C.; Nyamangara, J.; Giller, K.E. A meta-analysis of long-term effects of conservation agriculture on maize grain yield under rain-fed conditions. *Agron. Sustain. Dev.* **2011**, *31*, 657. [CrossRef]
160. Thierfelder, C.; Wall, P.C. Effects of conservation agriculture on soil quality and productivity in contrasting agro-ecological environments of Zimbabwe. *Soil Use Manag.* **2012**, *28*, 209–220. [CrossRef]
161. Montgomery, D.R. Soil erosion and agricultural sustainability. *Proc. Natl. Acad. Sci. USA* **2007**, *104*, 13268–13272. [CrossRef]
162. Cárceles, R.B.; Durán, Z.V.H.; Soriano, R.M.; Cermeño, S.P.; Gálvez, R.B.; Carbonell, B.R.; Ordoñez, F.R.; García, T.I.F. Soil and water conservation measures for Mediterranean fruit crops in rainfed hillslopes. In *Resources Use Efficiency in Agriculture*; Kumar, S., Meena, R.S., Jhariya, M.K., Eds.; Springer: Singapore, 2020; pp. 427–480. [CrossRef]
163. Durán, Z.V.H.; Rodríguez, P.C.R. Soil-erosion and runoff prevention by plant covers. A review. *Agron. Sustain. Dev.* **2008**, *28*, 65–86. [CrossRef]
164. Thierfelder, C.; Wall, P.C. Effects of conservation agriculture techniques on infiltration and soil water content in Zambia and Zimbabwe. *Soil Tillage Res.* **2009**, *105*, 217–227. [CrossRef]
165. Kurothe, R.S.; Kumar, G.; Singh, R.; Singh, H.B.; Tiwari, S.P.; Vishwakarma, A.K.; Sena, D.R.; Pande, V.C. Effect of tillage and cropping systems on runoff, soil loss and crop yields under semiarid rainfed agriculture in India. *Soil Tillage Res.* **2014**, *140*, 126–134. [CrossRef]
166. Panachuki, E.; Bertol, I.; Alves Sobrinho, T.; Sanches de Oliveira, P.T.; Bicca Rodrigues, D.B.B. Soil and water loss and water infiltration in red latosol under different management systems. *Rev. Bras. Cienc. Solo* **2011**, *35*, 1777–1785. [CrossRef]
167. Araya, T.; Cornelis, W.M.; Nyssen, J.; Govaerts, B.; Bauer, H.; Gebreegziabher, T.; Oicha, T.; Raes, D.; Sayre, K.D.; Haile, M.; et al. Effects of conservation agriculture on runoff, soil loss and crop yield under rainfed conditions in Tigray, Northern Ethiopia. *Soil Use Manag.* **2011**, *27*, 404–414. [CrossRef]
168. Ranaivoson, L.; Naudin, K.; Ripoche, A.; Affholder, F.; Rabeharisoa, L.; Corbeels, M. Agro-ecological functions of crop residues under conservation agriculture. A review. *Agron. Sustain. Dev.* **2017**, *37*, 26. [CrossRef]

169. Scopel, E.; Findeling, A.; Chavez Guerra, E.; Corbeels, M. Impact of direct sowing mulch-based cropping systems on soil carbon, soil erosion and maize yield. *Agron. Sustain. Dev.* **2005**, *25*, 425–432. [CrossRef]
170. Du, X.; Jian, J.; Du, C.; Stewart, R.D. Conservation management decreases surface runoff and soil erosion. *Int. Soil Water Conserv. Res.* **2022**, *10*, 188–196. [CrossRef]
171. Gebreegziabher, T.; Nyssen, J.; Govaerts, B.; Getnet, F.; Behailu, M.; Haile, M.; Deckers, J. Contour furrows for in situ soil and water conservation, Tigray, Northern Ethiopia. *Soil Tillage Res.* **2009**, *103*, 257–264. [CrossRef]
172. Lanckriet, S.; Araya, T.; Cornelis, W.; Verfaillie, E.; Poesen, J.; Govaerts, B.; Bauer, H.; Deckers, J.; Haile, M.; Nyssen, J. Impact of conservation agriculture on catchment runoff and soil loss under changing climate conditions in May Zeg-zeg (Ethiopia). *J. Hydrol.* **2012**, *475*, 336–349. [CrossRef]
173. Deng, C.; Zhang, G.; Liu, Y.; Nie, X.; Li, Z.; Liu, J.; Zhu, D. Advantages and disadvantages of terracing: A comprehensive review. *Int. Soil Water Conserv. Res.* **2021**, *9*, 344–359. [CrossRef]
174. Durán, Z.V.H.; Aguilar, R.J.; Martínez, R.A.; Franco, T.D. Impact of erosion in the taluses of subtropical orchard terraces. *Agric. Ecosyst. Environ.* **2005**, *107*, 199–210. [CrossRef]
175. Durán, Z.V.H.; Rodríguez, P.C.R.; Martin, P.F.J.; de Graaff, J.; Francia, M.J.R.; Flanagan, D.C. Environmental impact of introducing plant covers in the taluses of terraces: Implications for mitigating agricultural soil erosion and runoff. *Catena* **2011**, *84*, 79–88. [CrossRef]
176. Francia, M.J.R.; Durán, Z.V.H.; Martínez, R.A. Environmental impact from mountainous olive orchards under different soil-management systems (SE Spain). *Sci. Total Environ.* **2006**, *358*, 46–60. [CrossRef] [PubMed]
177. Gómez, J.A.; Sobrinho, T.A.; Giráldez, J.V.; Fereres, E. Soil management effects on runoff, erosion and soil properties in an olive grove of Southern Spain. *Soil Tillage Res.* **2009**, *102*, 5–13. [CrossRef]
178. Cárceles, B.; Durán, Z.V.H.; Soriano, R.M.; Gálvez, R.B.; García, T.I.F. Soil erosion and the effectiveness of the conservation measures in Mediterranean hillslope farming (SE Spain). *Eurasian Soil Sci.* **2021**, *54*, 792–806. [CrossRef]
179. Blanco-Canqui, H.; Ruis, S.J. No-tillage and soil physical environment. *Geoderma* **2018**, *326*, 164–200. [CrossRef]
180. Sarkar, S.; Paramanick, M.; Goswami, S.B. Soil temperature, water use and yield of yellow sarson (*Brassica napus* L. var. *glauca) in relation to tillage intensity and mulch management under rainfed lowland ecosystem in eastern India. Soil Tillage Res.* **2007**, *93*, 94–101. [CrossRef]
181. Rai, V.; Pramanik, P.; Das, T.K.; Aggarwal, P.; Bhattacharyya, R.; Krishnan, P.; Sehgal, V.K. Modelling soil hydrothermal regimes in pigeon pea under conservation agriculture using Hydrus-2D. *Soil Tillage Res.* **2019**, *190*, 92–108. [CrossRef]
182. Sharratt, B.S.; Campbell, G.S. Radiation balance of a soil-straw surface modified by straw color. *Agron. J.* **1994**, *86*, 200–203. [CrossRef]
183. Verhulst, N.; Govaerts, B.; Verachtert, E.; Castellanos-Navarrete, A.; Mezzalama, M.; Wall, P.; Deckers, J.; Sayre, K.D. Conservation agriculture, improving soil quality for sustainable production systems? In *Advances in Soil Science: Food Security and Soil Quality*; Lal, R., Stewart, B.A., Eds.; CRC Press: Boca Raton, FL, USA, 2010; pp. 137–208.
184. Li, R.; Hou, X.; Jia, Z.; Han, Q.; Ren, X.; Yang, B. Effects on soil temperature, moisture, and corn yield of cultivation with ridge and furrow mulching in the rainfed area of the Loess Plateau, China. *Agric. Water Manag.* **2013**, *116*, 101–109. [CrossRef]
185. Chen, S.Y.; Zhang, X.Y.; Pei, D.; Sun, H.Y. Effects of corn straw mulching on soil temperature and soil evaporation of winter wheat field. *Trans. CSAE* **2005**, *21*, 171–173.
186. Acharya, C.L.; Kapur, O.C.; Dixit, S.P. Moisture conservation for rainfed wheat production with alternative mulches and conservation tillage in the hills of north-west India. *Soil Tillage Res.* **1998**, *46*, 153–163. [CrossRef]
187. Gupta, S.C.; Larson, W.E.; Linden, D.R. Tillage and surface residue effects on soil upper boundary temperatures. *Soil Sci. Soc. Am. J.* **1983**, *47*, 1212–1218. [CrossRef]
188. Guzman, J.G.; Al-Kaisi, M. Residue removal and management practices effects on soil environment and carbon budget. *Soil Sci. Soc. Am. J.* **2014**, *78*, 609–623. [CrossRef]
189. Oliveira, J.; Timm, L.; Tominaga, T.; Cássaro, F.A.M.; Reichardt, K.; Bacchi, O.O.S.; Dourado-Neto, D.; Câmara, G.M. de S. Soil temperature in a sugar-cane crop as a function of the management system. *Plant Soil* **2001**, *230*, 61–66. [CrossRef]
190. Steward, P.R.; Dougill, A.J.; Thierfelder, C.; Pittelkow, C.M.; Stringer, L.C.; Kudzala, M.; Sheckelford, G.E. The adaptive capacity of maize-based conservation agriculture systems to climate stress in tropical and subtropical environments: A meta-regression of yields. *Agric. Ecosyst. Environ.* **2018**, *251*, 194–202. [CrossRef]
191. Kaspar, T.C.; Erbach, D.C.; Cruse, R.M. Corn response to seed-row residue removal. *Soil Sci. Soc. Am. J.* **1990**, *54*, 1112–1117. [CrossRef]
192. Chen, S.Y.; Zhang, X.Y.; Pei, D.; Sun, H.Y.; Chen, S. Effects of straw mulching on soil temperature, evaporation and yield of winter wheat: Field experiments on the North China Plain. *Ann. Appl. Biol.* **2007**, *150*, 261–268. [CrossRef]
193. Fortin, M.C. Soil temperature, soil water, and no-till corn development following in-row residue removal. *Agron. J.* **1993**, *85*, 571–576. [CrossRef]
194. Radke, J.K. Managing early season soil temperatures in the northern corn belt using configured soil surfaces and mulches. *Soil Sci. Soc. Am. J.* **1982**, *46*, 1067–1071. [CrossRef]
195. Shen, Y.; McLaughlin, N.; Zhang, X.; Xu, M.; Liang, A. Effect of tillage and crop residue on soil temperature following planting for a Black soil in Northeast China. *Sci. Rep.* **2018**, *8*, 4500. [CrossRef] [PubMed]

196. Franzluebbers, A.J.; Hons, F.M.; Zuberer, D.A. Tillage-induced seasonal changes in soil physical properties affecting soil CO_2 evolution under intensive cropping. *Soil Tillage Res.* **1995**, *34*, 41–60. [CrossRef]
197. Kahimba, F.; Sri Ranjan, R.; Froese, J.; Entz, M.; Nason, R. Cover crop effects on infiltration, soil temperature, and soil moisture distribution in the Canadian prairies. *Appl. Eng. Agric.* **2008**, *24*, 321–333. [CrossRef]
198. Al-Darby, A.M.; Lowery, B.; Daniel, T.C. Corn leaf water potential and water use efficiency under three conservation tillage systems. *Soil Tillage Res.* **1987**, *9*, 241–254. [CrossRef]
199. Reeves, D.W. The role of soil organic matter in maintaining soil quality in continuous cropping systems. *Soil Tillage Res.* **1997**, *43*, 131–167. [CrossRef]
200. Fageria, N.K. Role of soil organic matter in maintaining sustainability of cropping systems. *Commun. Soil Sci. Plant Anal.* **2012**, *43*, 2063–2113. [CrossRef]
201. Chenu, C.; Angers, D.A.; Barré, P.; Derrien, D.; Arrouays, D.; Balesdent, J. Increasing organic stocks in agricultural soils: Knowledge gaps and potential innovations. *Soil Tillage Res.* **2018**, *188*, 41–52. [CrossRef]
202. Valkama, E.; Kunypiyaeva, G.; Zhapayev, R.; Karabayev, M.; Zhusupbekov, E.; Perego, A.; Schillaci, C.; Sacco, D.; Moretti, B.; Grignani, C.; et al. Can conservation agriculture increase soil carbon sequestration? A modelling approach. *Geoderma* **2020**, *369*, 114298. [CrossRef]
203. Powlson, D.S.; Stirling, C.M.; Thierfelder, K.C.; Rodger, P.; White, R.P.; Jat, M.L. Does conservation agriculture deliver climate change mitigation through soil carbon sequestration in tropical agro-ecosystems? *Agric. Ecosyst. Environ.* **2016**, *220*, 164–174. [CrossRef]
204. Balesdent, J.; Chenu, C.; Balabane, M. Relationship of soil organic matter dynamics to physical protection and tillage. *Soil Tillage Res.* **2000**, *53*, 215–230. [CrossRef]
205. Repullo-Ruibérriz de Torres, M.A.; Carbonell, B.R.M.; Moreno, G.M.; Ordóñez, F.R.; Rodríguez, L.A. Soil organic matter and nutrient improvement through cover crops in a Mediterranean olive orchard. *Soil Tillage Res.* **2021**, *210*, 104977. [CrossRef]
206. Lal, R. Soil carbon sequestration to mitigate climate change. *Geoderma* **2004**, *123*, 1–22. [CrossRef]
207. Six, J.; Ogle, S.M.; Breidt, F.J.; Conant, R.T.; Mosiers, A.R.; Paustian, K. The potential to mitigate global warming with no-tillage management is only realized when practised in the long term. *Glob. Chang. Biol.* **2004**, *10*, 155–160. [CrossRef]
208. González-Sánchez, E.J.; Ordóñez, F.R.; Carbonell, B.R.; Veroz, G.O.; Gil, R.J.A. Meta-analysis on atmospheric carbon capture in Spain through the use of conservation agriculture. *Soil Tillage Res.* **2012**, *122*, 52–60. [CrossRef]
209. Luo, Z.; Wang, E.; Sun, O.J. Can no-tillage stimulate carbon sequestration in agricultural soils? A meta-analysis of paired experiments. *Agric. Ecosyst. Environ.* **2010**, *139*, 224–231. [CrossRef]
210. Mondal, S.; Chakraborty, D.; Bandyopadhyay, K.; Aggarwal, P.; Rana, D.S. A global analysis of the impact of zero-tillage on soil physical condition, organic carbon content, and plant root response. *Land Degrad. Dev.* **2020**, *31*, 557–567. [CrossRef]
211. Camarotto, C.; Piccoli, I.; Dal Ferro, N.; Polese, R.; Chiarini, F.; Furlan, L.; Morari, F. Have we reached the turning point? Looking for evidence of SOC increase under conservation agriculture and cover crop practices. *Eur. J. Soil Sci.* **2020**, *71*, 1050–1063. [CrossRef]
212. Pooniya, V.; Biswakarma, N.; Parihar, C.M.; Swarnalakshmi, K.; Lama, A.; Zhiipao, R.R.; Nath, A.; Pal, M.; Jat, S.L.; Satyanarayana, T.; et al. Six years of conservation agriculture and nutrient management in maize–mustard rotation: Impact on soil properties, system productivity and profitability. *Field Crops Res.* **2021**, *260*, 108002. [CrossRef]
213. VandenBygaart, A.J.; Angers, D.A. Towards accurate measurements of soil organic carbon stock change in agroecosystems. *Can. J. Soil. Sci.* **2006**, *86*, 465–471. [CrossRef]
214. Repullo-Ruibérriz de Torres, M.A.; Moreno, G.M.; Ordóñez, F.R.; Rodríguez, L.A.; Cárceles, R.B.; García, T.I.F.; Durán, Z.V.H.; Carbonell, B.R.M. Cover crop contributions to improve the soil nitrogen and carbon sequestration in almond orchards (SW Spain). *Agronomy* **2021**, *11*, 387. [CrossRef]
215. Roy, D.; Datta, A.; Jat, H.S.; Choudhary, M.; Sharma, P.C.; Singh, P.K.; Jat, M.L. Impact of long term conservation agriculture on soil quality under cereal based systems of North West India. *Geoderma* **2022**, *405*, 115391. [CrossRef]
216. Perego, A.; Rocca, A.; Cattivelli, V.; Tabaglio, V.; Fiorini, A.; Barbieri, S.; Schillaci, C.; Chiodini, M.E.; Brenna, S.; Acutis, M. Agro-environmental aspects of conservation agriculture compared to conventional systems: A 3-year experience on 20 farms in the Po valley (Northern Italy). *Agric. Syst.* **2019**, *168*, 73–87. [CrossRef]
217. Patra, S.; Julich, S.; Feger, K.H.; Jat, M.L.; Sharma, P.C.; Schwärzel, K. Effect of conservation agriculture on stratification of soil organic matter under cereal-based cropping systems. *Arch. Agron. Soil Sci.* **2019**, *65*, 2013–2028. [CrossRef]
218. Li, Y.; Li, Z.; Chang, S.X.; Cui, S.; Jagadamma, S.; Zhang, Q.; Cai, Y. Residue retention promotes soil carbon accumulation in minimum tillage systems: Implications for conservation agriculture. *Sci. Total Environ.* **2020**, *740*, 140147. [CrossRef] [PubMed]
219. Sapkota, T.B.; Jat, R.K.; Singh, R.G.; Jat, M.L.; Stirling, C.M.; Jat, M.K.; Bijarniya, D.; Kumar, M.; Saharawat, Y.S.; Gupta, R.K. Soil organic carbon changes after seven years of conservation agriculture in a rice–wheat system of the eastern Indo-Gangetic Plains. *Soil Use Manag.* **2017**, *33*, 81–89. [CrossRef]
220. Butterly, C.R.; Kaudal, B.B.; Baldock, J.A.; Tang, C. Contribution of soluble and insoluble fractions of agricultural residues to short-term pH changes. *Eur. J. Soil Sci.* **2011**, *62*, 718–727. [CrossRef]
221. Xu, R.K.; Coventry, D.R. Soil pH changes associated with lupin and wheat plant materials incorporated in a red-brown earth soil. *Plant Soil* **2003**, *250*, 113–119. [CrossRef]

222. Xu, J.M.; Tang, C.; Chen, Z.L. The role of plant residues in pH change of acid soils differing in initial pH. *Soil Biol. Biochem.* **2006**, *38*, 709–719. [CrossRef]
223. Muchabi, J.; Lungu, O.I.; Mweetwa, A.M. Conservation agriculture in Zambia: Effects on selected soil properties and biological nitrogen fixation in soya beans (*Glycine max* (L.) Merr). *Sustain. Agric. Res.* **2014**, *3*, 28–36. [CrossRef]
224. Duiker, S.W.; Beegle, D.B. Soil fertility distributions in long-term no-till, chisel/disk and moldboard plow/disk systems. *Soil Tillage Res.* **2006**, *88*, 30–41. [CrossRef]
225. Umar, B.B.; Aune, B.J.; Johnsen, H.F.; Lungu, I.O. Options for improving smallholder conservation agriculture in Zambia. *J. Agric. Sci.* **2011**, *3*, 50–62. [CrossRef]
226. Sinha, A.K.; Ghosh, A.; Dhar, T.; Bhattacharya, P.M.; Mitra, B.; Rakesh, S.; Paneru, P.; Shrestha, S.R.; Manandhar, S.; Beura, S.; et al. Trends in key soil parameters under conservation agriculture-based sustainable intensification farming practices in the Eastern Ganga Alluvial Plains. *Soil Res.* **2019**, *57*, 883–893. [CrossRef]
227. Limousin, G.; Tessier, D. Effects of no-tillage on chemical gradients and topsoil acidification. *Soil Tillage Res.* **2007**, *92*, 167–174. [CrossRef]
228. Sithole, N.J.; Magwaza, L.S. Long-term changes of soil chemical characteristics and maize yield in no-till conservation agriculture in a semi-arid environment of South Africa. *Soil Tillage Res.* **2019**, *194*, 104317. [CrossRef]
229. Butterly, C.R.; Baldock, J.A.; Tang, C. The contribution of crop residues to changes in soil pH under field conditions. *Plant Soil* **2013**, *366*, 185–198. [CrossRef]
230. Husson, O.; Brunet, A.; Babre, D.; Charpentier, H.; Durand, M.; Sarthou, J.P. Conservation agriculture systems alter the electrical characteristics (Eh, pH and EC) of four soil types in France. *Soil Tillage Res.* **2018**, *176*, 57–68. [CrossRef]
231. Ligowe, S.I.; Nalivata, C.P.; Njoloma, J.; Makumba, W.; Thierfelder, C. Medium-term effects of conservation agriculture on soil quality. *Afr. J. Agric. Res.* **2017**, *12*, 2412–2420. [CrossRef]
232. Rashidi, M.; Seilsepour, M. Modeling of soil cation exchange capacity based on soil organic carbon. *ARPN J. Agric. Biol. Sci.* **2008**, *3*, 41–45.
233. Sá, J.C.D.; Cerri, C.C.; Lal, R.; Dick, W.A.; Piccolo, M.D.; Feigl, B.E. Soil organic carbon and fertility interactions affected by a tillage chronosequence in a Brazilian Oxisol. *Soil Tillage Res.* **2009**, *104*, 56–64. [CrossRef]
234. Ben Moussa-Machraoui, S.; Errouissi, F.; Ben-Hammonda, M.; Nouira, S. Comparative effects of conventional and no-tillage management on some soil properties under Mediterranean semi-arid conditions in north western Tunisia. *Soil Tillage Res.* **2010**, *106*, 247–253. [CrossRef]
235. Williams, A.; Jordan, N.R.; Smith, R.G.; Hunter, M.C.; Kammerer, M.; Kane, D.A.; Koide, R.T.; Davis, A.S. A regionally-adapted implementation of conservation agriculture delivers rapid improvements to soil properties associated with crop yield stability. *Sci. Rep.* **2018**, *8*, 8467. [CrossRef] [PubMed]
236. Ramos, F.T.; Dores, E.F.d.C.; Weber, O.L.d.S.; Beber, D.C.; Campelo, J.H., Jr.; Maia, J.C.d.S. Soil organic matter doubles the cation exchange capacity of tropical soil under no-till farming in Brazil. *J. Sci. Food Agric.* **2018**, *98*, 3595–3602. [CrossRef] [PubMed]
237. Govaerts, B.; Sayre, K.D.; Lichter, K.; Dendooven, L.; Deckers, J. Influence of permanent raised bed planting and residue management on physical and chemical soil quality in rain fed maize/wheat systems. *Plant Soil* **2007**, *291*, 39–54. [CrossRef]
238. Kumari, D.; Kumar, S.; Parveen, H.; Pradhan, A.K.; Kumar, S.; Kumari, R. Long-term impact of conservation agriculture on chemical properties of soil. *Int. J. Curr. Microbiol. Appl. Sci.* **2019**, *8*, 2144–2153. [CrossRef]
239. Mohanty, A.; Mishra, K.N.; Roul, P.K.; Dash, S.N.; Panigrahi, K.K. Effects of conservation agriculture production system (CAPS) on soil organic carbon, base exchange characteristics and nutrient distribution in a tropical rainfed agro-ecosystem. *Int. J. Plant Anim. Environ. Sci.* **2015**, *5*, 310–314.
240. Zerihun, A.B.; Tadesse, B.; Shiferaw, T.; Kifle, D. Conservation agriculture: Maize-legume intensification for yield, profitability and soil fertility improvement in maize belt areas of western Ethiopia. *Int. J. Plant Soil Sci.* **2014**, *3*, 969–985. [CrossRef]
241. Fonteyne, S.; Burgueño, J.; Albarrán Contreras, B.A.; Andrio Enríquez, E.; Castillo Villaseñor, L.; Enyanche Velázquez, F.; Escobedo Cruz, H.; Espidio Balbuena, J.; Espinosa Solorio, A.; Garcia Meza, P.; et al. Effects of conservation agriculture on physicochemical soil health in 20 maize-based trials in different agro-ecological regions across Mexico. *Land Degrad. Dev.* **2021**, *32*, 2242–2256. [CrossRef]
242. Mrabet, R.; Moussadek, R.; Fadlaoui, A.; van Ranst, E. Conservation agriculture in dry areas of Morocco. *Field Crops Res.* **2012**, *132*, 84–94. [CrossRef]
243. Thomas, G.A.; Dalal, R.C.; Standley, J. No-till effects on organic matter, pH, cation exchange capacity and nutrient distribution in a Luvisol in the semi-arid subtropics. *Soil Tillage Res.* **2007**, *94*, 295–304. [CrossRef]
244. Alam, M.K.; Bell, R.W.; Haque, M.E.; Islam, M.A.; Kader, M.A. Soil nitrogen storage and availability to crops are increased by conservation agriculture practices in rice–based cropping systems in the Eastern Gangetic Plains. *Field Crops Res.* **2020**, *250*, 107764. [CrossRef]
245. Camarotto, C.; Dal Ferro, N.; Piccoli, I.; Polese, R.; Furlan, L.; Chiarini, F.; Morari, F. Conservation agriculture and cover crop practices to regulate water, carbon and nitrogen cycles in the low-lying Venetian plain. *Catena* **2018**, *167*, 236–249. [CrossRef]
246. Haokip, I.C.; Dwivedi, B.S.; Meena, M.C.; Datta, S.P.; Jat, H.S.; Dey, A.; Tigga, P. Effect of conservation agriculture and nutrient management options on soil phosphorus fractions under maize-wheat cropping system. *J. Indian Soc. Soil Sci.* **2020**, *68*, 45–53. [CrossRef]

247. Jat, H.S.; Datta, A.; Sharma, P.C.; Kumar, V.; Yadav, A.K.; Choudhary, M.; Choudhary, V.; Gathala, M.K.; Sharma, D.K.; Jat, M.L.; et al. Assessing soil properties and nutrient availability under conservation agriculture practices in a reclaimed sodic soil in cereal-based systems of North-West India. *Arch. Agron. Soil Sci.* **2018**, *64*, 531–545. [CrossRef] [PubMed]
248. Chan, K.Y.; Roberts, W.P.; Heenan, D.P. Organic carbon and associated properties of a red earth after 10 years rotation under different stubble and tillage practices. *Aust. J. Soil Res.* **1992**, *30*, 71–83. [CrossRef]
249. Sharma, V.; Irmak, S.; Padhi, J. Effects of cover crops on soil quality: Part II. Soil exchangeable bases (potassium, magnesium, sodium, and calcium), cation exchange capacity, and soil micronutrients (zinc, manganese, iron, copper, and boron). *J. Soil Water Conserv.* **2018**, *73*, 652–668. [CrossRef]
250. Kumar, D.; Kumar, S.; Parveen, H.; Priyanka; Kumar, R.; Kumari, D. Effect of establishment techniques and cropping systems on transformation of zinc in alluvial soil under conservation agriculture. *Int. J. Curr. Microbiol. Appl. Sci.* **2020**, *9*, 2585–2594. [CrossRef]
251. Feng, Y.; Liu, Q.; Tan, C.; Yang, G.; Qin, X.; Xiang, Y. Water and nutrient conservation effects of different tillage treatments in sloping fields. *Arid Land Res. Manag.* **2014**, *28*, 14–24. [CrossRef]
252. Govaerts, B.; Sayre, K.D.; Ceballos, R.J.M.; Luna, G.M.L.; Limon, O.A.; Deckers, L.; Dendooven, L. Conventionally tilled and permanent raised beds with different crop residue management: Effects on soil C and N dynamics. *Plant Soil* **2006**, *280*, 143–155. [CrossRef]
253. Sato, S.; Comerford, N.B. Influence of soil pH on inorganic phosphorus sorption and desorption in a humid Brazilian Ultisol. *Rev. Bras. Cienc. Solo* **2005**, *29*, 685–694. [CrossRef]
254. Deubel, A.; Hofmann, B.; Orzessek, D. Long-term effects of tillage on stratification and plant availability of phosphate and potassium in a loess chernozem. *Soil Tillage Res.* **2011**, *117*, 85–92. [CrossRef]
255. Lupwayi, N.Z.; Clayton, G.W.; O'Donovan, J.T.; Harker, K.N.; Turkington, T.K.; Soon, Y.K. Soil nutrient stratification and uptake by wheat after seven years of conventional and zero tillage in the Northern Grain belt of Canada. *Can. J. Soil Sci.* **2006**, *86*, 767–778. [CrossRef]
256. Obour, A.K.; Holman, J.D.; Simon, L.M.; Schlegel, A.J. Strategic tillage effects on crop yields, soil properties, and weeds in dryland no-tillage systems. *Agronomy* **2021**, *11*, 662. [CrossRef]
257. Hu, Z.H.; Ling, H.; Chen, S.T.; Shen, S.H.; Zhang, H.; Sun, Y.Y. Soil respiration, nitrification, and denitrification in a wheat farmland soil under different managements. *Commun. Soil Sci. Plant Anal.* **2013**, *44*, 3092–3102. [CrossRef]
258. Morugán, C.A.; Linares, P.C.; Gómez, L.M.D.; Faz, A.; Zornoza, R. The impact of intercropping, tillage and fertilizer type on soil and crop yield in fruit orchards under Mediterranean conditions: A meta-analysis of field studies. *Agric. Syst.* **2020**, *178*, 102736. [CrossRef]
259. Sujatha, D.V.; Kavitha, P.; Naidu, M.V.S. Influence of green manure and potassium nutrition on soil potassium fractions and yield of rice crop. *Int. J. Curr. Microbiol. Appl. Sci.* **2017**, *6*, 13–23. [CrossRef]
260. Durán, Z.V.H.; Cárceles, B.; García-Tejero, I.F.; Gálvez, R.B.; Cuadros, T.S. Benefits of organic olive rainfed systems to control soil erosion and runoff and improve soil health restoration. *Agron. Sustain. Dev.* **2020**, *40*, 41. [CrossRef]
261. Belay, S.A.; Assefa, T.T.; Prasad, P.V.V.; Schmitter, P.; Worqlul, A.W.; Steenhuis, T.S.; Reyes, M.R.; Tilahun, S.A. The response of water and nutrient dynamics and of crop yield to conservation agriculture in the Ethiopian highlands. *Sustainability* **2020**, *12*, 5989. [CrossRef]
262. Durán, Z.V.H.; Martínez, R.A.; Aguilar, R.J. Nutrient losses by runoff and sediment from the taluses of orchard terraces. *Water Air Soil Pollut.* **2004**, *153*, 355–373. [CrossRef]
263. Issaka, F.; Zhang, Z.; Zhao, Z.Q.; Asenso, E.; Li, J.H.; Li, Y.T.; Wang, J.J. Sustainable conservation tillage improves soil nutrients and reduces nitrogen and phosphorous losses in maize farmland in southern China. *Sustainability* **2019**, *11*, 2397. [CrossRef]
264. Nummer, A.S.; Qian, S.S.; Harmel, D.R. A meta-analysis on the effect of agricultural conservation practices on nutrient loss. *J. Environ. Qual.* **2018**, *47*, 1172–1178. [CrossRef]
265. Smith, D.R.; Francesconi, W.; Livingston, S.J.; Huang, C. Phosphorus losses from monitored fields with conservation practices in the Lake Erie Basin, USA. *Ambio* **2015**, *44*, 319–331. [CrossRef] [PubMed]
266. Jordan, V.W.; Leake, A.R.; Ogilvy, S.E. Agronomic and environmental implications of soil management practices in integrated farming systems. *Asp. Appl. Biol.* **2000**, *62*, 61–66.
267. Liu, Y.; Tao, Y.; Wan, K.; Zhang, G.; Liu, D.; Xiong, G.Y.; Chen, F. Runoff and nutrient losses in citrus orchards on sloping land subjected to different surface mulching practices in the Danjiangkou Reservoir area of China. *Agric. Water Manag.* **2012**, *110*, 34–40. [CrossRef]
268. Liu, R.; Zhang, P.; Wang, X.; Chen, Y.; Zhenyao, S. Assessment of effects of best management practices on agricultural non-point source pollution in Xiangxi River watershed. *Agric. Water Manag.* **2013**, *117*, 9–18. [CrossRef]
269. García-Díaz, A.; Bienes, R.; Sastre, B.; Novara, A.; Gristina, L.; Cerdà, A. Nitrogen losses in vineyards under different types of soil groundcover. A field runoff simulator approach in central Spain. *Agric. Ecosyst. Environ.* **2017**, *236*, 256–267. [CrossRef]
270. Dinnes, D.L.; Karlen, D.L.; Jaynes, D.B.; Kaspar, T.C.; Hatfield, J.L.; Colvin, T.S.; Cambardella, C.A. Nitrogen management strategies to reduce nitrate leaching in tile-drained Midwestern soils. *Agron. J.* **2002**, *94*, 153–171. [CrossRef]
271. Wyland, L.J.; Jackson, L.E.; Chaney, W.E.; Klonsky, K.; Koike, S.T.; Kimple, B. Winter cover crops in a vegetable cropping system: Impacts on nitrate leaching, soil water, crop yield, pests and management costs. *Agric. Ecosyst. Environ.* **1996**, *59*, 1–17. [CrossRef]

272. Colombani, N.; Mastrocicco, M.; Vincenzi, F.; Castaldelli, G. Modeling soil nitrate accumulation and leaching in conventional and conservation agriculture cropping systems. *Water* **2020**, *12*, 1571. [CrossRef]
273. Thiele-Bruhn, S.; Bloem, J.; de Vries, F.T.; Kalbitz, K.; Wagg, C. Linking soil biodiversity and agricultural soil management. *Curr. Opin. Environ. Sustain.* **2012**, *4*, 523–528. [CrossRef]
274. Zornoza, R.; Guerrero, C.; Mataix Solera, J.; Scow, K.M.; Arcenegui, V.; Mataix-Beneyto. J. Changes in soil microbial community structure following the abandonment of agricultural terraces in mountainous areas of Eastern Spain. *Appl. Soil Ecol.* **2009**, *42*, 315–323. [CrossRef]
275. Kabiri, V.; Raiesi, F.; Ghazavi, M.A. Tillage effects on soil microbial biomass, SOM mineralization and enzyme activity in a semi-arid Calcixerepts. *Agric. Ecosyst. Environ.* **2016**, *232*, 73–84. [CrossRef]
276. Haichar, F.; El, Z.; Santaella, C.; Heulin, T.; Achouak, W. Root exudates mediated interactions belowground. *Soil Biol. Biochem.* **2014**, *77*, 69–80. [CrossRef]
277. Lopes, L.D.; Fernandes, M.F. Changes in microbial community structure and physiological profile in a kaolinitic tropical soil under different conservation agricultural practices. *Appl. Soil Ecol.* **2020**, *152*. [CrossRef]
278. Singh, U.; Choudhary, A.K.; Sharma, S. Comparative performance of conservation agriculture vis-a-vis organic and conventional farming, in enhancing plant attributes and rhizospheric bacterial diversity in *Cajanus cajan*: A field study. *Eur. J. Soil Biol.* **2020**, *99*, 103197. [CrossRef]
279. Wang, Z.; Liu, L.; Chen, Q.; Liao, Y. Conservation tillage increases soil bacterial diversity in the dryland of northern China. *Agron. Sustain. Dev.* **2016**, *36*, 28. [CrossRef]
280. Silva, A.P.; Babujia, L.C.; Matsumoto, L.S.; Guimarães, M.F.; Hungria, M. Bacterial diversity under different tillage and crop rotation systems in an oxisol of Southern Brazil. *Open Agric. J.* **2013**, *7*, 40–47. [CrossRef]
281. Dorr de Quadros, P.; Zhalnina, K.; Davis, R.A.; Fagen, J.R.; Drew, J.; Bayer, C.; Camargo, F.A.O.; Triplett, E.W. The effect of tillage system and crop rotation on soil microbial diversity and composition in a subtropical acrisol. *Diversity* **2021**, *4*, 375. [CrossRef]
282. Henneron, L.; Bernard, L.; Hedde, M.; Pelosi, C.; Villenave, C.; Chenu, C.; Bertrand, M.; Girardin, C.; Blanchart, E. Fourteen years of evidence for positive effects of conservation agriculture and organic farming on soil life. *Agron. Sustain. Dev.* **2015**, *35*, 169–181. [CrossRef]
283. Baghel, J.K.; Das, T.K.; Raj, R.; Sangeeta, P.; Mukherjee, I.; Bisht, M. Effect of conservation agriculture and weed management on weeds, soil microbial activity and wheat (*Triticum aestivum*) productivity under a rice (*Oryza sativa*)-wheat cropping system. *Indian J. Agric. Sci.* **2018**, *88*, 1709–1716.
284. Li, Y.; Chang, S.X.; Tian, L.; Zhang, Q. Conservation agriculture practices increase soil microbial biomass carbon and nitrogen in agricultural soils: A global meta-analysis. *Soil Biol. Biochem.* **2018**, *121*, 50–58. [CrossRef]
285. Choudhary, M.; Datta, A.; Jat, H.S.; Yadav, A.K.; Gathala, M.K.; TeSapkota, T.B.; Das, A.K.; Sharma, P.C.; Jat, M.L.; Singh, R.; et al. Changes in soil biology under conservation agriculture based sustainable intensification of cereal systems in Indo-Gangetic Plains. *Geoderma* **2018**, *313*, 193–204. [CrossRef]
286. Kumar, B.T.N.; Babalad, H.B. Soil organic carbon, carbon sequestration, soil microbial biomass carbon and nitrogen and soil enzymatic activity as influenced by conservation agriculture in pigeonpea and soybean intercropping system. *Int. J. Curr. Microbiol. Appl. Sci.* **2018**, *7*, 323–333. [CrossRef]
287. Spedding, T.A.; Hamel, C.; Mehuys, G.R.; Madramootoo, C.A. Soil microbial dynamics in maize-growing soil under different tillage and residue management systems. *Soil Biol. Biochem.* **2004**, *36*, 499–512. [CrossRef]
288. Ceja-Navarro, J.A.; Rivera, F.N.; Patiño-Zúñiga, L.; Govaerts, B.; Marsch, R.; Vila-Sanjurjo, A.; Dendooven, L. Molecular characterization of soil bacterial communities in contrasting zero tillage systems. *Plant Soil* **2010**, *329*, 127–137. [CrossRef]
289. Legrand, F.; Picot, A.; Cobo, D.J.F.; Carof, M.; Chen, W.; Le Floch, G. Effect of tillage and static abiotic soil properties on microbial diversity. *Appl. Soil Ecol.* **2018**, *132*, 135–145. [CrossRef]
290. Mathew, R.P.; Feng, Y.; Githinji, L.; Ankumah, R.; Balkcom, K.S. Impact of no-tillage and conventional tillage systems on soil microbial communities. *Appl. Environ. Soil Sci.* **2012**, *2012*, 548620. [CrossRef]
291. Habig, J.; Swanepoel, C. Effects of conservation agriculture and fertilization on soil microbial diversity and activity. *Environments* **2015**, *2*, 358–384. [CrossRef]
292. Bonini Pires, C.A.; Amado, T.J.C.; Reimche, G.; Schwalbert, R.; Sarto, M.V.M.; Nicoloso, R.S.; Fiorin, J.E.; Rice, C.W. Diversified crop rotation with no-till changes microbial distribution with depth and enhances activity in a subtropical Oxisol. *Eur. J. Soil Sci.* **2020**, *71*, 1173–1187. [CrossRef]
293. Banerjee, T.; Sharma, S.; Thind, H.S.; Yadvinder, S.; Sidhu, H.S.; Jat, M.L. Soil biochemical changes at different wheat growth stages in response to conservation agriculture practices in a rice-wheat system of north-western India. *Soil Res.* **2017**, *56*, 91–104. [CrossRef]
294. Sharma, S.; Vashisht, M.; Singh, Y.; Thind, H.S. Soil carbon pools and enzyme activities in aggregate size fractions after seven years of conservation agriculture in a rice–wheat system. *Crop. Pasture Sci.* **2019**, *70*, 473–485. [CrossRef]
295. Kandeler, E.; Palli, S.; Stemmer, M.; Gerzabek, M.H. Tillage changes microbial biomass and enzyme activities in particle-size fractions of a Haplic Chernozem. *Soil Biol. Biochem.* **1999**, *31*, 1253–1264. [CrossRef]
296. Roldán, A.; Caravaca, F.; Hernández, M.T.; García, C.; Sánchez, B.C.; Velásquez, M.; Tiscareno, M. No-tillage, crop residue additions and legume cover cropping effects on soil quality characteristics under maize in Patzcuaro watershed (Mexico). *Soil Tillage Res.* **2003**, *72*, 65–73. [CrossRef]

297. Pandey, D.; Agrawal, M.; Bohra, J.S. Effects of conventional tillage and no tillage permutations on extracellular soil enzyme activities and microbial biomass under rice cultivation. *Soil Tillage Res.* **2014**, *136*, 51–60. [CrossRef]
298. Sinsabaugh, R.L.; Lauber, C.L.; Weintraub, M.N.; Ahmed, B.; Allison, S.D.; Crenshaw, C.; Contosta, A.R.; Cusack, D.; Frey, S.; Gallo, M.E. Stoichiometry of soil enzyme activity at global scale. *Ecol. Lett.* **2008**, *11*, 1252–1264. [CrossRef]
299. Kooch, Y.; Jalilvand, H. Earthworms as ecosystem engineers and the most important detritivors in forest soils. *Pak. J. Biol Sci.* **2008**, *11*, 819–825. [CrossRef]
300. Chan, K.Y. An overview of some tillage impacts on earthworm population abundance and diversity—Implications for functioning in soils. *Soil Tillage Res.* **2001**, *57*, 179–191. [CrossRef]
301. Capowiez, Y.; Cadoux, S.; Bouchant, P.; Ruy, S.; Roger, E.J.; Richard, G.; Boizard, H. The effect of tillage type and cropping system on earthworm communities, macroporosity and water infiltration. *Soil Tillage Res.* **2009**, *105*, 209–216. [CrossRef]
302. Pelosi, C.; Pey, B.; Hedde, M.; Caro, G.; Capowiez, Y.; Guernion, M.; Peigné, J.; Piron, D.; Bertrand, M.; Cluzeau, D. Reducing tillage in cultivated fields increases earthworm functional diversity. *Appl. Soil Ecol.* **2014**, *83*, 79–87. [CrossRef]
303. Baldivieso-Freitas, P.; Blanco, M.J.M.; Gutiérrez, L.M.; Peigné, J.; Pérez, F.A.; Trigo, A.D.; Sans, F.X. Earthworm abundance response to conservation agriculture practices in organic arable farming under Mediterranean climate. *Pedobiologia* **2018**, *66*, 58–64. [CrossRef]
304. Van Capelle, C.; Schrader, S.; Brunotte, J. Tillage-induced changes in functional diversity of soil biota—A review with a focus on German data. *Eur. J. Soil Biol.* **2012**, *50*, 165–181. [CrossRef]
305. Radford, B.J.; Key, A.J.; Robertson, L.N.; Thomas, G.A. Conservation tillage increases soil water storage, soil animal populations, grain yield and response to fertilizer in the semi-arid tropics. *Aust. J. Exp. Agric.* **1995**, *35*, 223–232. [CrossRef]
306. Birkás, M.; Jolánkai, M.; Gyuricza, C.; Percze, A. Tillage effects on compaction, earthworms and other soil quality indicators in Hungary. *Soil Tillage Res.* **2004**, *78*, 185–196. [CrossRef]
307. Errouissi, F.; Ben Moussa-Machraoui, S.; Ben-Hammouda, M.; Nouira, S. Soil invertebrates in durum wheat (*Triticum durum* L.) cropping system under Mediterranean semi-arid conditions: A comparison between conventional and no-tillage management. *Soil Tillage Res.* **2011**, *112*, 122–132. [CrossRef]
308. Chan, K.Y.; Heenan, D.P. Earthworm population dynamics under conservation tillage systems in southeastern Australia. *Aust. J. Soil Res.* **2006**, *44*, 425–431. [CrossRef]
309. Sharma, S.; Dhaliwal, S.S. Conservation agriculture based practices enhanced micronutrients transformation in earthworm cast soil under rice-wheat cropping system. *Ecol. Eng.* **2021**, *163*, 106195. [CrossRef]
310. Muoni, T.; Mhlanga, B.; Forkman, J.; Sitali, M.; Thierfelder, C. Tillage and crop rotations enhance populations of earthworms, termites, dung beetles and centipedes: Evidence from a long-term trial in Zambia. *J. Agric. Sci.* **2019**, *157*, 504–514. [CrossRef]
311. Bertrand, M.; Barot, S.; Blouin, M.; Whalen, J.; De Oliveira, T.; Roger, E.J. Earthworm services for cropping systems: A review. *Appl. Soil Ecol.* **2014**, *83*, 79–87. [CrossRef]
312. Schmidt, O.; Clements, R.O.; Donaldson, G. Why do cereal-legume intercrops support large earthworm populations? *Appl. Soil Ecol.* **2003**, *22*, 181–190. [CrossRef]
313. Hanson, P.; Edwards, N.; Garten, C.T.; Andrews. J.A. Separating root and soil microbial contributions to soil respiration: A review of methods and observations. *Biogeochemistry* **2001**, *48*, 115–146. [CrossRef]
314. Bondlamberty, B.; Thomson, A. Temperature-associated increases in the global soil respiration record. *Nature* **2010**, *464*, 579–582. [CrossRef]
315. Askari, M.S.; Holden, N.M. Indices for quantitative evaluation of soil quality under grassland management. *Geoderma* **2014**, *230–231*, 131–142. [CrossRef]
316. Xue, H.; Tang, H. Responses of soil respiration to soil management changes in an agropastoral ecotone in Inner Mongolia, China. *Ecol. Evol.* **2018**, *8*, 220–230. [CrossRef]
317. Edralin, D.I.A.; Sigua, G.C.; Reyes, M. Dynamics of Soil Carbon, Nitrogen and Soil Respiration in Farmer's Field with Conservation Agriculture in Cambodia. *Int. J. Plant Sci.* **2016**, *11*, 1–13. [CrossRef]
318. Shi, X.; Zhang, X.; Yang, X.; Drury, C.F.; McLaughlin, N.B.; Liang, A.; Fan, R.; Jia, S. Contribution of winter soil respiration to annual soil CO_2 emission in a Mollisol under different tillage practices in northeast China. *Glob. Biogeochem. Cycles* **2012**, *26*, GB2007. [CrossRef]
319. Cooper, R.J.; Hama-Aziz, Z.Q.; Hiscock, K.M.; Lovett, A.A.; Vrain, E.; Dugdale, S.J.; Sünnenberg, G.; Dockerty, T.; Hovesen, P.; Noble, L. Conservation tillage and soil health: Lessons from a 5-year UK farm trial (2013–2018). *Soil Tillage Res.* **2002**, *202*, 104648. [CrossRef]
320. Ye, R.; Parajuli, B.; Szogi, A.A.; Sigua, G.C.; Ducey, T.F. Soil health assessment after 40 years of conservation and conventional tillage management in Southeastern Coastal Plain soils. *Soil Sci. Soc. Am. J.* **2021**, *85*, 1214–1225. [CrossRef]
321. Rusu, T.; Bogdan, I.; Marin, D.I.; Moraru, P.I.; Pop, A.I.; Duda, B.M. Effect of conservation agriculture on yield and protecting environmental resources. *Agrolife Sci. J.* **2015**, *4*, 141–145.
322. Gyawali, A.J.; Strickland, M.S.; Thomason, W.; Reiter, M.; Stewart, R. Quantifying short-term responsiveness and consistency of soil health parameters in row crop systems: Part 1: Developing a multivariate approach. *Soil Tillage Res.* **2022**, *219*, 105354. [CrossRef]
323. Nunes, M.R.; van Es, H.M.; Schindelbeck, R.; Ristow, A.J.; Ryan, M. No-till and cropping system diversification improve soil health and crop yield. *Geoderma* **2018**, *328*, 30–43. [CrossRef]

324. Demir, Z.; Tursun, N.; Işik, D. Effects of Different Cover Crops on Soil Quality Parameters and Yield in an Apricot Orchard. *Int J. Agric. Biol.* **2018**, *21*, 399–408. [CrossRef]
325. Williams, H.; Colombi, T.; Keller, T. The influence of soil management on soil health: An on-farm study in southern Sweden. *Geoderma* **2020**, *360*, 114010. [CrossRef]
326. Parihar, C.M.; Singh, A.K.; Jat, S.L.; Dey, A.; Nayak, H.S.; Mandal, B.N.; Saharawat, Y.S.; Jat, M.L.; Yadav, O.P. Soil quality and carbon sequestration under conservation agriculture with balanced nutrition in intensive cereal-based system. *Soil Tillage Res.* **2020**, *202*, 104653. [CrossRef]
327. Bera, T.; Sharma, S.; Thind, H.S.; Sidhu, H.S.; Jat, M.L. Changes in soil biochemical indicators at different wheat growth stages under conservation-based sustainable intensification of rice-wheat system. *J. Integr. Agric.* **2018**, *17*, 1871–1880. [CrossRef]

soil systems

MDPI

Review

Managing Root Parasitic Weeds to Facilitate Legume Reintroduction into Mediterranean Rain-Fed Farming Systems

Diego Rubiales

Institute for Sustainable Agriculture, CSIC, 14080 Córdoba, Spain; diego.rubiales@ias.csic.es

Abstract: Grain and forage legumes are important sources of food and feed, key for sustainable agriculture given the environmental services they provide. However, their cultivation is hampered in the Mediterranean Basin and Near East by the widespread occurrence of the root parasitic weed crenate broomrape (*Orobanche crenata*). Other broomrape species such as *O. minor*, *O. foetida*, and *Phelipanche aegyptica* are also of local importance. As for other parasitic weeds, a number of management strategies have been proposed, but considering that temperate legumes in the area are low-input crops, these strategies are largely uneconomical or hard to achieve, leaving the use of resistant cultivars as the most desirable option. Breeding for broomrape resistance is not an easy task, but significant progress has been achieved by classical breeding and selection and will profit from recent developments in phenomics and genomics. Here, achievements and prospects in broomrape management and resistance breeding are presented and critically discussed.

Keywords: *Orobanche*; crop protection; resistance breeding; faba bean; pea; vetches; grass pea

1. The Key Role of Legumes in Cropping Systems

Legumes are the second most important family of cultivated plants after cereals. They not only play a key role in agri-food systems as sources of food and feed but also provide ecosystem services by improving soil fertility, biodiversity and environmental sustainability [1,2]. The legume-rhizobium association provides a source of renewable nitrogen for agriculture that is estimated to reduce total nitrogen fertilizer consumption in all farming systems by between 24% (grain legumes) and 38% (forage legumes) [3]. Global production of nitrogen fertilizers has increased more than fourfold in the last decades, accounting for more than 60% of all fertilizers used in agriculture, either in the form of ammonium, urea, or nitrate. A large part of these fertilizers is not used by plants but is leached and ends up in aquifers. Reduced use of nitrogen fertilizers in turn reduces fossil energy consumption and greenhouse gas emissions associated with the manufacturing process as well as nitrous oxide emissions from soils [4]. Nitrous oxide is a potent greenhouse gas whose main source is microbial activity in soils and waters enriched in nitrates by the massive application of nitrogen fertilizers. It is estimated that cropping systems that include a legume emit, on average, 18% less nitrous oxide, with this reduction rising to 33% in the case of pastures [3].

However, yields of most temperate legumes are relatively low due to limited investment in breeding compared to other crops. As a result, and despite the above-mentioned ecosystem services they provide, the cultivation of most legumes has declined in Europe since the onset of so-called modern agriculture in the middle of the 20th century; nevertheless, legume cultivation is growing worldwide [5]. It is true that, as a result of changes in eating habits, human consumption of grain legumes has markedly declined in the last five decades, a trend that fortunately is starting to reverse. But this decline in human consumption alone does not explain the reduction in cultivation in traditional legume-producing countries in the Mediterranean Basin and Near East, as production is insufficient to cover the domestic demand, forcing imports of about 60 to 80% of the pulses eaten.

The reduced consumption of legumes paired with an increased consumption of meat has led to an ever-increasing demand for feed legumes, resulting in increasing dependence of imported soybeans [6,7]. It is therefore highly desirable from a nutritional point of view to increase legume consumption, but it is naïve to propose that this is the main measure needed to reintegrate legumes into Mediterranean cropping systems. Promoting legume consumption without acting on the necessary measures to promote local legume cultivation would result in a further increase in imports, as in fact is already the case for all grain legumes, and particularly soybean, whose imports continue to rise. Thus, by not growing legumes locally, we continue to lose the ecosystem services they provide [8]. The solution can only be to develop cropping packages that make the crops profitable to farmers by adjusting cultivation techniques and developing adapted varieties [5,9–11].

2. Broomrape as a Major Constraint in Legume Production

Cultivation of annual grain and forage temperate legumes is strongly hampered in the farming systems of the Mediterranean and Middle East by the widespread occurrence of broomrapes, which cause important yield losses [10]. The most widespread and damaging broomrape is crenate broomrape (*Orobanche crenata* Forsk.), but minor broomrape (*O. minor* Sm), foetida broomrape (*O. foetida* Poir), and Egyptian broomrape (*Pelipanche aegyptiaca* (Pers.) Pomel) can be of importance locally [11–16].

O. crenata is not a new problem in legume farming, having been described by authors from ancient Rome. Little progress has been made in its management since then, and unfortunately, the real situation is that instead of being controlled, it is a problem that is spreading to new areas that were considered free of infestation, even outside the Mediterranean Basin, to the north in Europe, to the south in Africa, and to the east in Asia, representing a situation that could worsen with climate change [16–18].

3. Understanding Broomrape Biologic Features Relevant to Management

The most relevant aspect of broomrape biology is that broomrapes are flowering plants that have evolved to feed on other plants, thereby losing photosynthetic capacity [11]. As flowering plants, broomrapes can be managed from a weed science point of view regarding their reproduction, seed dispersal, and chemical control with herbicides. However, unlike standard weeds, the damage of broomrapes is not due to competition for light and water from the soil but to the direct establishment of a permanent interaction with the roots of the host plants on which they feed and alter their physiology [19]. Broomrapes cause therefore true diseases and can be approached with a plant pathology perspective. The infected plant can defend itself against infection, in a similar way to how it defends itself against infections with any other pathogens like fungi, bacteria, or viruses [20,21]. And therefore, breeders can act by developing varieties that are more resistant to broomrape infection, in a similar way to how we would breed varieties resistant to fungi, bacteria, or viruses.

Several features make broomrape difficult to control [11,22]. One is its wide host range; for instance, the host range of *O. crenata* includes most legumes as well as crops such as carrot, lettuce, geranium, or celery. The host range of *P. aegyptiaca* is particularly wide, including many vegetable crops. Another difficulty is that infection with root parasitic plants occurs underground and is not detected until the broomrape emerges from the soil, by which time most of the damage has been done, and it is too late to attempt any control measures. Additionally, a single plant can produce a large number of seeds that have a great capacity for survival in the soil, germinating only when stimulated by signals emitted by host plants. Broomrape plants produce large numbers of seeds that are dispersed a short distance by the wind; thus, their distribution is typically agglomerated [23]. However, they can be spread over longer distances by manure of animals that feed on them, and above all, they can be spread by human action, either by the movement of contaminated machinery between farms or over even longer distances by the trading of crop seed lots containing soil residues and broomrape seeds [24]. Sanitation measures, disinfecting machinery and sowing seed are therefore essential in preventing expansion to new areas as well as in quarantine measures [25,26]. Diagnosis and quantification in soil or crop seed lots is also essential, with genomic tools being developed for this purpose [27,28].

Once the seeds have entered a farm, they are difficult to eradicate. The seeds have a long viability and germinate only in the presence of a host plant. They first need conditioning, associated with specific humidity and temperature conditions, which mimic the growing conditions of the host plant, thus ensuring that they germinate only when there may be plants available to be infected [29,30]. But this is not sufficient for germination; if the seeds are conditioned but are not stimulated by the proximity of a host plant, they return to their dormant state [31]. Broomrape seeds recognize a series of chemical signals emitted by neighboring host plants. The best-known group of broomrape germination stimulants are strigolactones [32,33], but there are many other metabolites that can induce broomrape germination [34]. For instance, a number of metabolites have been described in root exudates of pea or common vetch that differentially stimulate germination of seeds of different broomrape species, contributing to host specificity [35–37]. Once the seed germinates in the proximity of a root of a host plant, it emits a radicle that must find and anchor itself to the root, or it dies of starvation. Once anchored, it begins to feed on the host plant, developing a stem without functional roots, which eventually emerges to the surface and flowers, producing seeds that fall back to the ground, filling the seed bank, and repeating the cycle when a new host crop is found [11,13].

4. Management Strategies

There have been numerous efforts to develop control measures for broomrape management in legume cropping systems [10,11,13,18,38–40]. Unfortunately, the result has not been satisfactory, and the problem remains unresolved in practice. The approaches followed have ranged from agronomic practices to biological control and have given rise to great scientific discovery; unfortunately, they have had limited commercial application since they have either provided only partial protection or simply are not economically affordable for a low-input field crop, as are most of the legumes that we are dealing with. In practice, the only measures that have had some commercial application for legumes, as well as most field crops, are chemical control strategies with herbicides and genetic resistance [38–40].

The first measure always mentioned is hand weeding, namely, removing emerged plants from the field and destroying them. This is labour demanding and is worthwhile only in cases where infestations are still light. The second most recommended agronomic practice is delaying the sowing date [29,41,42], which may reduce the infection, but which in rain-fed cropping systems in Mediterranean climates is associated with a reduction in productive potential by not taking advantage of winter rains [43]. Other recommended practices are no tilling to reduce the incorporation of seeds into the soil [44] or very deep ploughing to plant the seeds quite deep [45].

Solarization can be very effective [46] and can be economical for cash crops in small areas such as greenhouses or orchards, but it is hardly feasible to solarize large farms with low-input crops. Another alternative is the cultivation of highly susceptible species, namely, "catch crops" that are infected and destroyed before the broomrape produces seeds, either by incorporating them into the soil as green manure or used for silage.

Soil fertilization can contribute to broomrape control as infestations are more severe in poor soils [47]. Under nutrient starvation, particularly P, but also N, strigolactone production by plants is increased to promote mycorrhizal colonisation, which is reduced when plants are fertilized [48]. On the other hand, urea and ammonium can have a toxic effect on the seeds and broomrape plants [49].

Another agronomic practice with potential is intercropping. A similar case with some success in subsistence agricultural systems in sub-Saharan Africa is the control of *Striga hermonthica* on corn or sorghum intercropped with *Desmodium* species, a mixture that was actually explored for the control of cereal stemborer insects, which are repelled by *Desmodium* and attracted to a border crop that is used to remove them from the field [50].

This intercropping was also found useful for controlling *Striga* [51]. This technique, called "pull and push", which is very labor-demanding, has some success on small family farms, but its extrapolation to other agricultural systems is not simple. Thus, it has been shown that a series of species can reduce infection with *O. crenata* in several legumes when they are mixed, such as fenugreek, oats, or berseem clover [52–54]. However, it is necessary to adjust cultivation practices so that they can be adopted by farmers.

The allelopathic effect of a series of crops on broomrape can be exploited not only in mixed crops but also in rotations. Thus, a series of crops have been described that can induce germination of broomrape seeds without being infected, having potential as "trap crops" that reduce the seed bank in the soil [55–58]. The principle is the same as that of "suicidal germination" by applying germination stimulants to the soil [59,60]. The theoretical basis is brilliant, and in both cases, it is based on germinating broomrape seeds that then die when they cannot find roots of a host crop to infect, either with crops that stimulate them but are not infected or by directly applying the germination stimulants to the soil in the absence of susceptible crops. However, in both cases the reduction is not complete, so several crop cycles would be needed for effective control. In the case of the direct application of synthetic germination stimulants to the soil, there is the added difficulty of finding an effective method of incorporation into the soil and of its persistence and cost. And once again, the economic factor must be considered since the measure must not only be effective in reducing the seed bank in the soil but also be economically viable for the farmer to adopt.

Within biological control, there have been efforts to promote the use of various insects, such as *Phytomiza orobanchia*, which is specific to *Orobanche*, whose larvae pupate inside broomrape capsules and destroy a large number of seeds [61]. The reality is that this insect is widely distributed naturally, having been found even in wild populations of many broomrape species, and that even if they destroy a percentage of seeds, there are so many thousands of seeds that a single broomrape plant is capable of producing that the effect of the parasite is minimal in areas with high infestation. After many years of study, there is no conclusive result or commercial application, even with breeding and release of adults. Other types of widely studied biocontrol agents are fungi [62] and bacteria [63] that have shown certain levels of control in pot studies under controlled conditions, but conclusive results from field studies have not yet been reported, highlighting above all the difficulty of finding a viable method of application and persistence [64]. The use of a series of natural metabolites produced by fungi or plants has also been proposed [65,66], which has shown an effect in the laboratory, but the mode of extraction or synthesis of these metabolites as well as their incorporation into the field must be optimized to make them applicable. Additionally, application of the amino acid methionine [67,68] or of growth regulators such as uniconazole [69] has been proposed, but this requires validation under field conditions.

Activation of systemic acquired resistance by various means has been proposed in several legumes. For instance, salicylic acid and benzothiadiazole application activated resistance reducing *O. minor* infection in red clover [70] and *O. crenata* in pea [71] and faba bean [72]. Many other inductors of resistance have been postulated in other pathosystems but not tested so far in legumes against broomrape. Symbionts such as mycorrhizae and rhizobium may also have a protective effect since their colonization affects root exudates or by activating resistance [73–77]. However, the effect although significant is small. Therefore, we have to conclude that although the biological control of broomrape still holds great promise, it has not yet resulted in a commercial application. Alternatively, we can foresee "biocontrol" as using broomrapes for food [78] or in pharmaceutical and cosmetic industries [79].

Since broomrape is a plant, it can be controlled by a number of herbicides [11,80,81]. Chemical disinfection of the soil can be very effective, but like physical disinfection (i.e., solarization), it is recommended only for small areas [81,82]. Also, since most legumes are low-input rain-fed crops and the infection occurs in the roots, the number of herbicides that can be used is reduced, practically excluding contact herbicides that would require irrigation to be incorporated into the roots. This has limited the herbicides used to systemic ones, incorporated on the leaves and translocated to the roots. The most recommended has been glyphosate in faba beans, which even so, finds no problems for its wide adoption by farmers. Glyphosate is also toxic to crops, which is why repeated application of a very low dose is necessary in the initial stages of infection; thus, finding a balance between crop damage and infection control is difficult. This has been even more complicated in other crops such as peas that are more sensitive to glyphosate. Imidazolinones have been proposed, even in seed treatment. However, the control is only partial, and the treatments have to be repeated [13,83]. Understanding the temporal variation in parasitism dynamics to predict broomrape parasitism based on thermal time can help for a more effective chemical control [82,84]. Site-specific broomrape management can benefit from geographical information systems and global positioning systems to delineate the spatial variation in infestation within and between fields [82]. Nanoencapsulation of herbicides has been proposed to improve their effectivity but is still under development [85].

All of this makes the development of resistant varieties the most desirable measure since it would eliminate the need for farmers to use any type of control measure. But, on the one hand, genetic resistance is difficult to identify and requires long improvement processes, and on the other hand, genetic resistance does not usually provide complete protection; therefore, it is advisable to incorporate resistance into integrated management packages, which, in addition to complementing the protection, would prolong the durability of the resistance by keeping the populations of the pathogen low and thus reducing its ability to evolve [13,20,40,86–90].

5. A Focus on Resistance Breeding

5.1. Genetic Basis of Resistance

Monogenic resistance has been identified in sunflower against *Orobanche cumana* [90] and in cowpea against *Striga gesnerioides* [91] but not in any legume crops against any broomrapes. This presents advantages and problems similar to those encountered in genetic improvement for resistance to any kind of disease; although monogenic resistance simplifies breeding progress, new races of the pathogen can emerge that break down these resistances [92]. Although it is easier to manage in breeding, monogenic resistance is not a panacea since the ability of pathogens to evolve into new races that overcome these resistances is well known. The risk of the appearance of new races depends not only on the genetic basis of the resistance (i.e., easier in monogenic resistance than in polygenic resistance) but also on the manner of reproduction and dissemination of pathogens [89,93]. Thus, it is known that the greatest risk occurs in organisms that combine sexual reproduction (new genetic combinations) and asexual reproduction (fixation of these new successful combinations), which can involve several complete cycles of reproduction in a breeding season. Cultivation can play a role in dissemination by influencing aerial dispersal over long distances; rust rot is the typical example where new races can appear in a matter of 2–5 years. In the case of broomrape, the risk is moderate since this plant reproduces sexually, with only one cycle per year, and except for accidental cases of movement of seeds by human action over long distances, the natural dispersal of seeds is a few meters [89]. Thus, in the case of *O. cumana*/sunflower, although new races have appeared, this did not happen as fast as observed in rusts but took several decades [92]. In the case of broomrapes infecting legumes (*O. crenata*, *O. minor*, *O. foetida* and *P. aegyptiaca*), no races have been described so far, and there is not even a consensus that there are *formae speciales*, despite the existence of variability in the pathogen as it is partially allogamous [89,94]. Only some weak levels of host specificity have recently been suggested in *O. crenata* populations growing on lentil [95]. Contrary to *O. crenata*, which has been known to infect legumes over centuries, *O. foetida* seems to be a relatively recent problem. Natural non-weedy populations of *O. foetida* are widespread in the western Mediterranean, infecting wild legume species only, not legume crops. However, only a few decades ago, weedy populations of *O. foetida* on faba bean were reported in the Beja region of Tunisia that became established in that area [96]. It seems that evolution of these populations might have been driven by response to host selection pressures including recognition of root exudates [97–99].

One possible explanation is that since monogenic resistance with complete expression has not been identified and exploited on a large scale but different levels of incomplete resistance has, the pathogen has not suffered this selection pressure, and in any case, since these are generally minority crops that occupy small extensions and are rarely repeated in the farm rotation, although more virulent populations have developed, they have not been established, or at least there is no evidence of establishment [89]. But without a doubt, we cannot rule out that if varieties with complete resistance are developed and become popular by repeatedly growing them in large areas, as has been the case with sunflowers, races that evolve to overcome this resistance will appear.

In the case of legumes, progress in broomrape resistance breeding has been slow, as they are rather minor crops in which relatively little has been invested in the last half century [5,9,100]. Most studies on broomrape resistance in legumes have concluded that there is low heritability and that inheritance is complex, highly influenced by the environment. Mapping studies have been performed in pea and faba bean, identifying a series of quantitative trait loci (QTLs) with a small effect, often not reproducible between years [100–106]. Field screenings are most needed but do not allow dissection of the various resistance mechanisms that might be operative and lack sufficient control of crucial environmental factors and of homogeneity of inoculum in the soil [87]. Attention is needed to improve phenotyping, complementing field screenings with dedicated minirhizotron methods or similar approaches allowing the identification of QTLs involved in specific mechanisms of resistance.

This has made the use of marker-assisted selection difficult. Still, the reality is that classical breeding programs have been successful in developing varieties with certain levels of resistance. Because of this, progress has been slower, but perhaps also because of this, there has not been a high selection pressure on the pathogen and no races of *O. crenata*, *O. minor*, *O. foetida*, or *P. aegyptiaca* have been described so far. Still, valuable sources of resistance have been identified in germplasm of most legume crops, including faba bean [107–112], pea [113–115], lentil [116–118], vetches [119–127], chickpea [128–131], grass pea and related *Lathyrus* species [132–136], or barrel medic [137,138], among other legume species. Some of these sources have been exploited in breeding programs, resulting in the release of resistant cultivars particularly in the case of faba bean [43,108] and pea [139–142].

5.2. Focus on Mechanisms of Resistance Operative

The basis of the durability of resistance is diversity, both at the cropping system level and at the level of genes and operative mechanisms [89,93]. There is consensus on the convenience of avoiding the use of monogenic resistance, recommending the use of polygenic resistance, which is what we only have at the moment in the case of legumes. Therefore, it seems that instead of complaining about the lack of monogenic resistance, we should congratulate ourselves and look for effective ways to accumulate information on the various QTLs or minor genes available, despite the greater complexity of their management in genetic improvement. Even so, given the predictable moderate risk of appearance and establishment of new races described in the previous section, the use of monogenic resistance should not be excluded, as long as the virulence of the parasitic populations is monitored to design strategies for the use of these genes, in space and time, to prolong their durability [93,138]. Another important point to discuss is that it is possibly not only the genetic basis of resistance that matters but also the nature of the resistance mechanism [89,93,143]. Thus, there is a series of monogenic resistances that have proven to be durable [144]. Curiously, these examples have in common that they are not based on a hypersensitive reaction due to cell death of the infected cell, like most of the major genes used in breeding, but rather on prehaustorial mechanisms, making cell penetration difficult. It is therefore highly recommended to explore the existence of resistance mechanisms acting in various phases of the infection process, which can be exploited separately, or preferably, combined in the same genotype [145]. And it is that combining two different resistance mechanisms that provide different barriers could be more effective than combining two genes that regulate the same mechanism.

It is therefore interesting to break down the broomrape infection process to identify the possible mechanisms operating at different stages of the infection process [114,145]. Genetically inherited phenological or root morphological traits might help to prevent infection, reducing the chances of contact in time and space, respectively. For instance, both very early or very late genotypes can escape infection [113,133], either by competition for nutrients of the early formed pods or through the late development of most roots when conditions are less favorable for broomrape establishment.

This would be followed by mechanisms that affect the germination and growth phase of broomrape radicles, either by reducing germination by lower exudation of germination stimulants, or by emitting metabolites with an inhibitory effect. Broomrape seeds germinate only when they recognize chemical signals exuded by the roots, which are thought to be primarily strigolactones [32,33]. There are many strigolactones known from different plants, and it is thought that the differential recognition of one or the other, together with other possible metabolites by different broomrape species, is what determines their host specificity [146–148]. A first working hypothesis would be the selection of genotypes that produce fewer strigolactones, but this would be counterproductive since strigolactones are not only signals that favor mycorrhization but also hormones that regulate the correct branching and architecture of the plant. It would therefore be desirable to have efficient methods to determine the strigolactones required by each broomrape species and to quantify them in the plant, so that we can select genotypes with differential levels of production, namely, to avoid those that stimulate the broomrape species but do produce others that allow a correct mycorrhization and architecture. In the absence of this level of knowledge, it has been possible to empirically identify faba bean genotypes displaying a strong resistance in the field, which was due to non-induction of germination of *O. crenata* seeds [109]. Interestingly, this mechanism was also operative against *O. foetida* and *P. aegyptiaca* and was associated with low production of the two strigolactones studied [148]. However, selected plants showed excellent performance in the field, suggesting that other (unquantified) strigolactones might be produced. The genetic basis of this resistance has not been studied, but there are indications that it could be monogenic. In fact, this mechanism has previously been described as monogenic in sorghum against *S. hermonthica* [149]. Similarly, tomato [150,151] and chickpea [130,152] mutants with reduced induction of broomrape seed germination likely due to reduced exudation of strigolactones have shown to be resistant to broomrape. This mechanism has also been described in other legumes such as pea [113–115], lentil [116,117], vetches [122–125], chickpea [128,129], or barrel medic [137,138], among others. In pea, two QTLs were associated with low induction of *O. crenata* seed germination [105]. A later study postulated monogenic inheritance of the trait [153].

Selection could also be exerted for higher exudation of metabolites inhibiting broomrape seed germination or radicle growth. Thus, a series of metabolites with such an effect have been identified [33], although little is known about the genotypic variability for this trait in legumes. A chemotrophic effect has also been postulated, such that a higher concentration of a series of metabolites could cause a directed growth of the broomrape radicle towards the host root [114] and affect the formation of the haustorium once the root has been contacted [154]. Until now, it was thought that a chemical signal was necessary for the differentiation of the haustorium in *Striga* and other parasitic plants but that this was not necessary in broomrape. However, it has recently been shown that broomrape radicles also respond to certain chemical signals to differentiate a haustorium [155], which in theory would make it possible to design genotypes that do not release these signals. Once a haustorium has differentiated on a host root, penetration is produced by a combination of mechanical pressure and enzymatic activity, so that a vascular interconnection occurs between both plants [156]. From here, broomrape acts as a sink for water and nutrients so that it develops at the expense of the host plant. The plant can develop different types of barriers preventing or delaying the infection. A first barrier can be by reinforcement of the cell walls of the cortex by protein cross-linking or with the deposition of suberin or callose, followed by lignification of endodermal and pericycle cell walls or later by occlusion or sealing of host vessels by gel- or gum-like substances, peroxidase-related lignification, mucilage production, or haustorium disorganization, preventing the parasite's survival [156–159].

5.3. Resistance to Herbicides

As indicated above, broomrape can be managed in legume farming systems with systemic herbicides (i.e., glyphosate, imidazolinones, and sulfonylurea) at low rates with repeated treatments. Control could be improved by enhancing the tolerance of the crops to these herbicides, so higher rates could be applied [160]. Natural variation in herbicide tolerance has been identified in several legumes, including tolerance to imazethapyr and metribuzin in faba bean [161] and lentil [162,163]. Mutation breeding has been effectively exploited to develop herbicide-resistant mutants [164,165], offering scope for improving the chemical control of broomrape by using higher rates of herbicides. An alternative strategy is using transgenic techniques, although engineered legume crops harboring herbicide-resistance transgenes are not yet available for broomrape management [166–168].

5.4. Potential Applications of Biotechnology in Broomrape Resistance Breeding

The basis of any genetic improvement program is genetic diversity on which to act using various tools until obtaining resistant varieties that are also agronomically attractive and of good quality [5,89,99]. If the needed genetic diversity is not available, it can be generated by classical or directed mutagenesis [150–152] or by new biotechnological tools [100,166,167]. But it may be sufficient to explore and exploit the existing natural variability within the crop or related species. Thus, there are a large number of insufficiently characterized collections where we could find the desired characters [145]. A battery of field, pot and minirhizotron screening protocols have been proposed to promote the identification of sources of resistance in most species as described above. An often neglected limitation is the availability of fast but sufficiently reliable screening techniques that allow us to find what we need [87]. Hyperspectral imaging is being adjusted for early detection of broomrape infection to help with precise herbicide application in terms of time and space [169], with continuous attempts also to automate image phenotyping in seedling responses in rhizotrons [170], which is still too laborious and time-consuming. As a result, sources of resistance to broomrape are limited and poorly characterized. In spite of these constraints, pea and faba bean breeding has successfully led to the release of resistant cultivars [139–142]. Adoption of modern technologies rapidly developing in legumes will facilitate breeding. Despite the fact that modern genetics was born with Mendel's genetic studies of the pea and, similarly, the bean played an important role in the onset of cytogenetics, progress in the knowledge of these crops proceeded much more slowly later than in other crops, like cereals. Fortunately, in recent years, we have been experiencing spectacular advances in genomic and phenomic techniques in legumes [171], opening enormous opportunities for their application in breeding. Thus, in only a few years, annotated genomes of peas, faba beans, lentils, and most of the legumes have become available. Even in those species not yet sequenced, the reduction in genotyping costs is facilitating molecular analysis [172–174]. Although similar progress in genome sequencing has not been achieved for broomrape species, spectacular progress is being made in sequence information on other parasitic weeds [175,176] that will help in understanding parasite virulence and host resistance mechanisms. The integration of information obtained from QTL analysis with gene and protein expression analysis in response to broomrape infection [177,178] can shortcut conventional breeding or marker-assisted selection to identify candidate genes that could be used for selective gene silencing (RNAs, siRNA) [179] or DNA base editing (CRISPR/Cas9) to deliver broomrape resistance [180–182]. Although legumes are considered recalcitrant to stable genetic transformation protocols, progress is being achieved [183,184]; therefore, transient transformation or TILLING may be used for the functional characterization of candidate genes.

6. Conclusions

A number of strategies have been proposed for crenate broomrape management, but considering that temperate legumes in the area are low-input crops, they have been found to be largely uneconomical or hard to achieve, leaving the use of resistant cultivars as the most desirable option. Breeding for broomrape resistance is not easy, but significant progress has been achieved by classical breeding and selection and will benefit in the short term from recent developments in phenomics and genomics.

Funding: This research was funded by Spanish Agencia Estatal de Investigación (AEI) projects PID2020-114668RB-100 and PDC2021-120930-I00 financed by MCIN/AEI/10.13039/501100011033.

Institutional Review Board Statement: Not applicable.

Informed Consent Statement: Not applicable.

Data Availability Statement: Not applicable.

Acknowledgments: The author acknowledges the efforts of all colleagues working on broomrape breeding and management at IAS-CSIC and elsewhere.

Conflicts of Interest: The author declares no conflict of interest.

References

1. Ditzler, L.; van Apeldoorn, D.F.; Pellegrini, F.; Antichi, D.; Bàrberi, P.; Rossing, W.A.H. Current research on the ecosystem service potential of legume inclusive cropping systems in Europe. A review. *Agron. Sustain. Dev.* **2021**, *41*, 26. [CrossRef]
2. Iannetta, P.P.M.; Hawes, C.; Begg, G.S.; Maaß, H.; Ntatsi, G.; Savvas, D.; Vasconcelos, M.; Hamann, K.; Williams, M.; Styles, D.; et al. A Multifunctional Solution for Wicked Problems: Value-Chain Wide Facilitation of Legumes Cultivated at Bioregional Scales Is Necessary to Address the Climate-Biodiversity-Nutrition Nexus. *Front. Sustain. Food Syst.* **2021**, *5*, 692137. [CrossRef]
3. Reckling, M.; Bergkvist, G.; Watson, C.A.; Stoddard, F.L.; Zander, P.M.; Walker, R.L.; Pristeri, A.; Toncea, I.; Bachinger, J. Trade-offs between economic and environmental impacts of introducing legumes into cropping systems. *Front. Plant Sci.* **2016**, *7*, 669. [CrossRef] [PubMed]
4. Jensen, E.S.; Peoples, M.B.; Boddey, R.M.; Gresshoff, P.M.; Henrik, H.N.; Alves, B.J.R.; Morrison, M.J. Legumes for mitigation of climate change and the provision of feedstock for biofuels and biorefineries. A review. *Agron. Sustain. Dev.* **2012**, *32*, 329–364. [CrossRef]
5. Cusworth, G.; Garnett, T.; Lorimer, J. Legume dreams: The contested futures of sustainable plant-based food systems in Europe. *Glob. Environ. Change* **2021**, *69*, 102321. [CrossRef] [PubMed]
6. Westhoek, H.J.; Rood, G.A.; van den Berg, M.; Janse, J.H.; Nijdam, D.S.; Reudink, M.A.; Stehfest, E.; Jnase, J. The protein puzzle: The consumption and production of meat, dairy and fish in the European Union. *Eur. J. Nutr. Food Saf.* **2011**, *1*, 123–144.
7. Zander, P.; Amjath-Babu, T.S.; Preissel, S.; Reckling, M.; Bues, A.; Schläfke, N.; Kuhlman, T.; Bachinger, J.; Uthes, S.; Stoddard, F.; et al. Grain legume decline and potential recovery in European agriculture: A review. *Agron. Sustain. Dev.* **2016**, *36*, 26. [CrossRef]
8. Foyer, C.H.; Lam, H.M.; Nguyen, H.T.; Siddique, K.H.M.; Varshney, R.K.; Colmer, T.D.; Cowling, W.; Bramley, H.; Mori, T.A.; Hodgson, J.M.; et al. Neglecting legumes has compromised human health and sustainable food production. *Nat. Plants* **2016**, *2*, 16112. [CrossRef]
9. Rubiales, D. Plant breeding is needed to meet agroecological requirements: Legume crops as case study. *Outlook Agric.* **2023**, *52*, 294–302. [CrossRef]
10. Fernández-Aparicio, M.; Flores, F.; Rubiales, D. The effect of *Orobanche crenata* infection severity in faba bean, field pea and grass pea productivity. *Front. Plant Sci.* **2016**, *7*, 1409. [CrossRef]
11. Fernández-Aparicio, M.; Reboud, X.; Gibot-Leclerc, S. Broomrape Weeds. Underground Mechanisms of Parasitism and Associated Strategies for their Control: A Review. *Front. Plant Sci.* **2016**, *7*, 135. [CrossRef] [PubMed]
12. Fernández-Aparicio, M.; Delavault, P.; Timko, M. Management of infection by parasitic weeds: A review. *Plants* **2020**, *9*, 1184. [CrossRef] [PubMed]
13. Rubiales, D.; Fernández-Aparicio, M. Innovations in parasitic weeds management in legume crops. A review. *Agron. Sustain. Dev.* **2012**, *32*, 433–449. [CrossRef]
14. Parker, C. Parasitic Weeds: A World Challenge. *Weed Sci.* **2012**, *60*, 269–276. [CrossRef]
15. Das, T.K.; Ghosh, S.; Gupta, K.; Suman, S.; Biswaranjan, B.; Rishi, R. The weed *Orobanche*: Species distribution, diversity, biology and management. *J. Res. Weed Sci.* **2020**, *3*, 162–180. [CrossRef]

16. Rubiales, D. Broomrape threat to agriculture. *Outlooks Pest Manag.* **2020**, *31*, 141–144. [CrossRef]
17. Grenz, J.H.; Sauerborn, J. Mechanisms limiting the geographical range of the parasitic weed *Orobanche crenata*. *Agric. Ecosyst. Environ.* **2007**, *122*, 275e281. [CrossRef]
18. Negewo, T.; Ahmed, S.; Tessema, T.; Tana, T. Biological Characteristics, Impacts, and Management of Crenate Broomrape (*Orobanche crenata*) in Faba Bean (*Vicia faba*): A Review. *Front. Agron.* **2022**, *4*, 708187. [CrossRef]
19. Westwood, J.H. The physiology of the established parasite–host association. In *Parasitic Orobanchaceae: Parasitic Mechanisms and Control Strategies*; Joel, D.M., Gressel, J., Musselman, L.J., Eds.; Springer: Berlin/Heidelberg, Germany, 2013; pp. 87–114.
20. Aly, R.; Dubey, N. Weed Management for Parasitic Weeds. In *Recent Advances in Weed Management*; Chauhan, B., Mahajan, G., Eds.; Springer: New York, NY, USA, 2014. [CrossRef]
21. Yoder, J.I.; Scholes, J.D. Host plant resistance to parasitic weeds; recent progress and bottlenecks. *Curr. Opin. Plant Biol.* **2010**, *13*, 478–484. [CrossRef]
22. Cartry, D.; Steinberg, C.; Gibot-Leclerc, S. Main drivers of broomrape regulation. A review. *Agron. Sustain. Dev.* **2021**, *41*, 17. [CrossRef]
23. Oveisi, M.; Yousefi, A.R.; Gonzalez-Andujar, J.L. Spatial distribution and temporal stability of crenate broomrape (*Orobanche crenata* Forsk) in faba bean (*Vicia faba* L.): A long-term study at two localities. *Crop Prot.* **2010**, *29*, 717–720. [CrossRef]
24. Ginman, E.; Prider, J.; Matthews, J.; Virtue, J.; Watling, J. Broomrape dispersal by sheep. *Weed Biol. Manag.* **2015**, *15*, 61–69. [CrossRef]
25. Panetta, F. Evaluating the performance of weed containment programs. *Divers. Distrib.* **2012**, *18*, 1024–1032. [CrossRef]
26. Hosseini, P.; Osipitan, O.; Mesgaran, M. Seed germination responses of broomrape species (*Phelipanche ramosa* and *Phelipanche aegyptiaca*) to various sanitation chemicals. *Weed Technol.* **2022**, *36*, 723–728. [CrossRef]
27. Prider, J.N.; Ophel Keller, K.; McKay, A. Molecular diagnosis of parasite seed banks. In *Parasitic Orobanchaceae: Parasitic Mechanisms and Control Strategies*; Joel, D.M., Gressel, J., Musselman, L.J., Eds.; Springer: Berlin/Heidelberg, Germany, 2013; pp. 357–368.
28. Rolland, M.; Dupuy, A.; Pelleray, A.; Delavault, P. Molecular Identification of Broomrape Species from a Single Seed by High Resolution Melting Analysis. *Front. Plant Sci.* **2016**, *7*, 1838. [CrossRef]
29. Grenz, J.H.; Manschadi, A.M.; Uygurc, F.N.; Sauerborn, J. Effects of environment and sowing date on the competition between faba bean (*Vicia faba*) and the parasitic weed *Orobanche crenata*. *Field Crops Res.* **2005**, *93*, 300–313. [CrossRef]
30. Murdoch, A.J.; Kebreab, A. Germination ecophysiology. In *Parasitic Orobanchaceae*; Joel, D.M., Gressel, J., Musselman, L.J., Eds.; Springer: Berlin/Heidelberg, Germany, 2013; pp. 195–219.
31. Kebreab, E.; Murdoch, A.J. A quantitative model for loss of primary dormancy and induction of secondary dormancy in imbibed seeds of *Orobanche* spp. *J. Exp. Bot.* **1999**, *50*, 211–219. [CrossRef]
32. Brun, G.; Braem, L.; Thoiron, S.; Gevaert, K.; Goormachting, S.; Delavault, P. Seed germination in parasitic plants: What insights can we expect from strigolactone research? *J. Exp. Bot.* **2018**, *69*, 2265–2280. [CrossRef]
33. Xie, X.; Yoneyama, K.; Nomura, T.; Yoneyama, K. Evaluation and Quantification of Natural Strigolactones from Root Exudates. In *Strigolactones: Methods in Molecular Biology*; Prandi, C., Cardinale, F., Eds.; Humana: New York, NY, USA, 2021; Volume 2309. [CrossRef]
34. Cimmino, A.; Masi, M.; Rubiales, D.; Evidente, A.; Fernández-Aparicio, M. Allelopathy for parasitic plant management. *Nat. Prod. Commun.* **2018**, *13*, 289–294. [CrossRef]
35. Evidente, A.; Fernández-Aparicio, M.; Cimmino, A.; Rubiales, D.; Andolfi, A.; Motta, A. Peagol and peagoldione, two new strigolactone-like metabolites isolated from pea root exudates. *Tetrahedron Lett.* **2009**, *50*, 6955–6958. [CrossRef]
36. Evidente, A.; Cimmino, A.; Fernández-Aparicio, M.; Andolfi, A.; Rubiales, D.; Motta, A. Polyphenols, Including the New Peapolyphenols A-C, from Pea Root Exudates Stimulate *Orobanche foetida* Seed Germination. *J. Agric. Food Chem.* **2010**, *58*, 2902–2907. [CrossRef] [PubMed]
37. Evidente, A.; Cimmino, A.; Fernández-Aparicio, M.; Rubiales, D.; Andolfi, A.; Melck, D. Soyasapogenol B and trans-22-dehydrocampesterol from common vetch (*Vicia sativa* L.) root exudates stimulate broomrape seed germination. *Pest Manag. Sci.* **2011**, *67*, 1015–1022. [CrossRef] [PubMed]
38. Nosratti, I.; Mobli, A.; Mohammadi, G.; Yousefi, A.; Sabeti, P.; Chauhan, B. The problem of *Orobanche* spp. and *Phelipanche* spp. and their management in Iran. *Weed Sci.* **2020**, *68*, 555–564. [CrossRef]
39. Fernández-Aparicio, M.; Westwood, J.H.; Rubiales, D. Agronomic, breeding, and biotechnological approaches to parasitic plant management through manipulation of germination stimulant levels in agricultural soils. *Botany* **2011**, *89*, 813–826. [CrossRef]
40. Goldwasser, Y.; Rodenburg, J. Integrated agronomic management of parasitic weeds seed banks. In *Parasitic Orobanchaceae*; Joel, D.M., Gressel, J., Mussleman, L.J., Eds.; Springer: Berlin/Heidelberg, Germany, 2013. [CrossRef]
41. López-Granados, F.; García-Torres, L. Effects of environmental factors on dormancy and germination of crenate broomrape (*Orobanche crenata*). *Weed Sci.* **1996**, *44*, 284–289. [CrossRef]
42. Pérez-de-Luque, A.; Sillero, J.C.; Cubero, J.I.; Rubiales, D. Effect of sowing date and host resistance on the establishment and development of *Orobanche crenata* on faba bean and common vetch. *Weed Res.* **2004**, *44*, 282–288. [CrossRef]
43. Rubiales, D.; Moral, A.; Flores, F. Agronomic Performance of Broomrape Resistant and Susceptible Faba Bean Accession. *Agronomy* **2022**, *12*, 1421. [CrossRef]

44. López-Bellido, R.J.; Benítez-Vega, J.; López-Bellido, L. No-tillage improves broomrape control with glyphosate in faba-bean. *Agron. J.* **2009**, *101*, 1394–1399. [CrossRef]
45. Eizenberg, H.; Lande, T.; Achdari, G.; Roichman, A.; Hershenhorn, J. Effect of Egyptian broomrape (*Orobanche aegyptiaca*) burial depth on parasitism dynamics and chemical control in tomato. *Weed Sci.* **2007**, *51*, 152–156. [CrossRef]
46. Mauro, R.P.; Lo Monaco, A.; Lombardo, S.; Restuccia, A.; Mauromicale, G. Eradication of *Orobanche/Phelipanche* spp. seedbank by soil solarization and organic supplementation. *Sci. Hortic.* **2015**, *193*, 62–68. [CrossRef]
47. Jain, R.; Foy, C.L. Nutrient effects on parasitism and germination of Egyptian broomrape (*Orobanche aegyptiaca*). *Weed Technol.* **1992**, *6*, 269–275. [CrossRef]
48. Yoneyama, K.; Xie, X.; Kim, H.I.; Kisugi, T.; Nomura, T.; Sekimoto, H.; Yokota, T.; Yoneyama, K. How do nitrogen and phosphorus deficiencies affect strigolactone production and exudation? *Planta* **2012**, *235*, 1197–1207. [CrossRef]
49. Westwood, J.H.; Foy, C.L. Influence of nitrogen on germination and early development of broomrape (*Orobanche* spp.). *Weed Sci.* **1999**, *47*, 2–7. [CrossRef]
50. Midega, C.A.O.; Khan, Z.R.; Amudai, D.M.; Pittchar, J.; Pickett, J.A. Integrated management of *Striga hermonthica* and cereal stemborers in finger millet (*Eleusine coracana* (L.) Gaertn.) through intercropping with *Desmodium intortum*. *Int. J. Pest Manag.* **2010**, *56*, 145–151. [CrossRef]
51. Pickett, J.A.; Hamilton, M.L.; Hooper, A.M.; Khan, A.R.; Midega, C.A.O. Companion cropping to manage parasitic plants. *Annu. Rev. Phytopathol.* **2010**, *48*, 161–177. [CrossRef] [PubMed]
52. Fernández-Aparicio, M.; Sillero, J.C.; Rubiales, D. Intercropping with cereals reduces infection by *Orobanche crenata* in legumes. *Crop Prot.* **2007**, *26*, 1166–1172. [CrossRef]
53. Fernández-Aparicio, M.; Emeran, A.A.; Rubiales, D. Control of *Orobanche crenata* in legumes intercropped with fenugreek (*Trigonella foenum-graecum*). *Crop Prot.* **2008**, *27*, 653–659. [CrossRef]
54. Fernández-Aparicio, M.; Emeran, A.A.; Rubiales, D. Inter-cropping with berseem clover (*Trifolium alexandrinum*) reduces infection by *Orobanche crenata* in legumes. *Crop Prot.* **2010**, *29*, 867–871. [CrossRef]
55. Lins, R.D.; Colquhoun, J.B.; Mallory-Smith, C.A. Investigation of wheat as a trap crop for control of *Orobanche minor*. *Weed Res.* **2006**, *46*, 313–318. [CrossRef]
56. Fernández-Aparicio, M.; Flores, F.; Rubiales, D. Recognition of root exudates by seeds of broomrape (*Orobanche* and *Phelipanche*) species. *Ann. Bot.* **2009**, *103*, 423–431. [CrossRef]
57. Chai, M.; Zhu, X.; Cui, H.; Jiang, C.; Zhang, J.; Shi, L. Lily cultivars have allelopathic potential in controlling *Orobanche aegyptiaca* Persoon. *PLoS ONE* **2015**, *10*, e0142811. [CrossRef] [PubMed]
58. Aksoy, E.; Arslan, Z.F.; Tetik, Ö.; Eymirli, S. Using the possibilities of some trap, catch and Brassicaceae crops for controlling crenate broomrape a problem in lentil fields. *Int. J. Plant Prod.* **2016**, *10*, 53–62.
59. Johnson, A.W.; Rosebery, G.; Parker, C. A novel approach to *Striga* and *Orobanche* control using synthetic germination stimulants. *Weed Res.* **1976**, *16*, 223–227. [CrossRef]
60. Mwakaboko, A.S.; Zwanenburg, B. Strigolactone analogs derived from ketones using a working model for germination stimulants as a blueprint. *Plant Cell Physiol.* **2011**, *52*, 699–715. [CrossRef] [PubMed]
61. Klein, O.; Kroschel, J. Biological control of *Orobanche* spp. with *Phytomyza orobanchia*, a review. *Biocontrol* **2002**, *47*, 245–277. [CrossRef]
62. Dor, E.; Hershenhorn, J. The use of several phytopathogenic fungi for broomrape control. *Phytoparasitica* **2003**, *31*, 422.
63. Barghouthi, S.; Salman, M. Bacterial inhibition of *Orobanche aegyptiaca* and *Orobanche cernua* radical elongation. *Biocontrol Sci. Technol.* **2010**, *20*, 423–435. [CrossRef]
64. Watson, A.K. Biocontrol. In *Parasitic Orobanchaceae*; Joel, D.M., Gressel, J., Musselman, L.J., Eds.; Springer: Berlin/Heidelberg, Germany, 2013. [CrossRef]
65. Vurro, M.; Boari, A.; Evidente, A.; Andolfi, A.; Zermane, N. Natural metabolites for parasitic weed management. *Pest Manag. Sci.* **2009**, *65*, 566–571. [CrossRef]
66. Cimmino, A.; Fernández-Aparicio, M.; Andolfi, A.; Basso, S.; Rubiales, D.; Evidente, A. Effect of fungal and plant metabolites on broomrapes (*Orobanche* and *Phelipanche* spp.) seed germination and radicle growth. *J. Agric. Food Chem.* **2014**, *62*, 10485–10492. [CrossRef]
67. Vurro, M.; Boari, A.; Pilgeram, A.L.; Sands, D.C. Exogenous amino acids inhibit seed germination and tubercle formation by *Orobanche ramosa* (broomrape): Potential application for management of parasitic weeds. *Biol. Control* **2006**, *36*, 258–265. [CrossRef]
68. Fernández-Aparicio, M.; Bernard, A.; Falchetto, L.; Marget, P.; Chauvel, B.; Steinberg, C.; Morris, C.E.; Gibot-Leclerc, S.; Boari, A.; Vurro, M.; et al. Investigation of amino acids as herbicides for control of *Orobanche minor* parasitism in red clover. *Front. Plant Sci.* **2017**, *8*, 842. [CrossRef]
69. Joel, D.M. The long-term approach to parasitic weeds control: Manipulation of specific developmental mechanisms of the parasite. *Crop Prot.* **2000**, *19*, 753–758. [CrossRef]
70. Kusumoto, D.; Goldwasser, Y.; Xie, X.; Yoneyama, K.; Takeuchi, Y. Resistance of red clover (*Trifolium pratense*) to the root parasitic plant *Orobanche minor* is activated by salicylate but not by jasmonate. *Ann. Bot.* **2007**, *100*, 537–544. [CrossRef]
71. Pérez-de-Luque, A.; Jorrín, J.V.; Rubiales, D. Crenate broomrape control in pea by foliar application of benzothiadiazole (BTH). *Phytoparasitica* **2004**, *32*, 21–29. [CrossRef]

72. Sillero, J.C.; Rojas-Molina, M.M.; Avila, C.M.; Rubiales, D. Induction of systemic acquired resistance against rust, ascochyta blight and broomrape in faba bean by exogenous application of salicylic acid and benzothiadiazole. *Crop Prot.* **2012**, *34*, 55–69. [CrossRef]
73. Dadon, T.; Nun, N.B.; Mayer, A.M. A factor from *Azospirillum brasilense* inhibits germination and radicle growth of *Orobanche aegyptiaca*. *Isr. J. Plant Sci.* **2004**, *52*, 83–86. [CrossRef]
74. Mabrouk, Y.; Simier, P.; Delavault, P.; Delgrange, S.; Sifi, B.; Zourgui, L.; Belhadj, O. Molecular and biochemical mechanisms of defence induced in pea by *Rhizobium leguminosarum* against *Orobanche crenata*. *Weed Res.* **2007**, *47*, 452–460. [CrossRef]
75. Mishev, K.; Dobrev, P.I.; Lacek, J.; Filepová, R.; Yuperlieva-Mateeva, B.; Kostadinova, A.; Hristeva, T. Hormonomic Changes Driving the Negative Impact of Broomrape on Plant Host Interactions with Arbuscular Mycorrhizal Fungi. *Int. J. Mol. Sci.* **2021**, *22*, 13677. [CrossRef]
76. Fernández-Aparicio, M.; García-Garrido, J.M.; Ocampo, J.A.; Rubiales, D. Colonization of field pea roots by arbuscular mycorrhizal fungi reduces *Orobanche* and *Phelipanche* species seed germination. *Weed Res.* **2010**, *50*, 262–268. [CrossRef]
77. López-Ráez, J.A.; Charnikhova, T.; Fernandez, I.; Bouwmeester, H.; Pozo, M.J. Arbuscular mycorrhizal symbiosis decreases strigolactone production in tomato. *J. Plant Physiol.* **2011**, *168*, 294–297. [CrossRef]
78. Renna, M.; Serio, F.; Santamaria, P. Crenate broomrape (*Orobanche crenata* Forskal): Prospects as a food product for human nutrition. *Genet. Resour. Crop Evol.* **2015**, *62*, 795–802. [CrossRef]
79. Shi, R.; Zhang, C.; Gong, X.; Yang, M.; Ji, M.; Jiang, L.; Leonti, M.; Yao, R.; Li, M. The genus *Orobanche* as food and medicine: An ethnopharmacological review. *J. Ethnopharmacol.* **2020**, *263*, 113154. [CrossRef] [PubMed]
80. García-Torres, L.; López-Granados, F. Control of broomrape (*Orobanche crenata* Forsk.) in broad bean (*Vicia faba* L.) with imidazolinones and other herbicides. *Weed Res.* **1991**, *31*, 227–235. [CrossRef]
81. Eizenberg, H.; Hershenhorn, J.; Ephrath, J.H.; Kanampiu, F. Chemical Control. In *Parasitic Orobanchaceae*; Joel, D., Gressel, J., Musselman, L., Eds.; Springer: Berlin/Heidelberg, Germany, 2013. [CrossRef]
82. Eizenberg, H.; Aly, R.; Cohen, Y. Technologies for Smart Chemical Control of Broomrape (*Orobanche* spp. and *Phelipanche* spp.). *Weed Sci.* **2012**, *60*, 316–323. [CrossRef]
83. Rubiales, D.; Pérez-de-Luque, A.; Cubero, J.I.; Sillero, J.C. Crenate broomrape (*Orobanche crenata*) infection in field pea cultivars. *Crop Prot.* **2003**, *22*, 865–872. [CrossRef]
84. Pérez-de-Luque, A.; Flores, F.; Rubiales, D. Differences in crenate broomrape parasitism dynamics on three legume crops using a thermal time model. *Front. Plant Sci.* **2016**, *7*, 1910. [CrossRef] [PubMed]
85. Pérez-de-Luque, A.; Rubiales, D. Nanotechnology for parasitic plant control. *Pest Manag. Sci.* **2009**, *65*, 540–545. [CrossRef] [PubMed]
86. Scott, D.; Freckleton, R.P. Crop diversification and parasitic weed abundance: A global meta-analysis. *Sci. Rep.* **2022**, *12*, 19413. [CrossRef]
87. Rubiales, D.; Pérez-de-Luque, A.; Sillero, J.C.; Román, B.; Kharrat, M.; Khalil, S.; Joel, D.M.; Riches, C.R. Screening techniques and sources of resistance against parasitic weeds in grain legumes. *Euphytica* **2006**, *147*, 187–199. [CrossRef]
88. Rubiales, D. Legume breeding for broomrape resistance. *Czech J. Genet. Plant Breed.* **2014**, *50*, 144–150. [CrossRef]
89. Rubiales, D. Can we breed for durable resistance to broomrapes? *Phytopathol. Mediterr.* **2018**, *57*, 170–185. [CrossRef]
90. Velasco, L.; Pérez-Vich, B.; Fernández-Martínez, J.M. Research on resistance to sunflower broomrape: An integrated vision. *OCL* **2016**, *23*, D203. [CrossRef]
91. Li, J.; Timko, M.P. Gene-for-gene resistance in Striga-cowpea associations. *Science* **2009**, *325*, 1094. [CrossRef]
92. Molinero-Ruiz, L.; Delavault, P.; Pérez-Vich, B.; Pacureanu-Joita, M.; Bulos, M.; Altieri, E.; Domínguez, J. History of the race structure of *Orobanche cumana* and the breeding of sunflower for resistance to this parasitic weed: A review. *Span. J. Agric. Res.* **2015**, *13*, e10R01. [CrossRef]
93. McDonald, B.A.; Linde, C. The population genetics of plant pathogens and breeding strategies for durable resistance. *Euphytica* **2002**, *124*, 163–180. [CrossRef]
94. Satovic, Z.; Joel, D.M.; Rubiales, D.; Cubero, J.I.; Román, B. Population genetics in weedy species of *Orobanche*. *Australas. Plant Pathol.* **2009**, *38*, 228–234. [CrossRef]
95. Ennami, M.; Briache, F.Z.; Gaboun, F.; Abdelwahd, R.; Ghaouti, L.; Belqadi, L.; Westwood, J.; Mentag, R. Host differentiation and variability of *Orobanche crenata* populations from legume species in Morocco as revealed by cross-infestation and molecular analysis. *Pest Manag. Sci.* **2017**, *73*, 1753–1763. [CrossRef]
96. Kharrat, M.; Halila, M.H.; Linke, K.H.; Haddar, T. First report of *Orobanche foetida* Poiret on faba bean in Tunisia. *FABIS Newsl.* **1992**, *30*, 46–47.
97. Román, B.; Satovic, Z.; Alfaro, C.; Moreno, M.T.; Kharrat, M.; Pérez-de-Luque, A.; Rubiales, D. Host differentiation in *Orobanche foetida* Poir. *Flora* **2007**, *202*, 201–208. [CrossRef]
98. Vaz Patto, M.C.; Díaz-Ruiz, R.; Satovic, Z.; Román, B.; Pujadas-Salvà, A.J.; Rubiales, D. Genetic diversity of Moroccan populations of *Orobanche foetida*: Evolving from parasitising wild hosts to crop plants. *Weed Res.* **2008**, *28*, 179–186. [CrossRef]
99. Belay, G.; Tesfaye, K.; Hamwieh, A.; Ahmed, S.; Dejene, T.; de Oliveira Júnior, J.O.L. Genetic Diversity of *Orobanche crenata* Populations in Ethiopia Using Microsatellite Markers. *Int. J. Genom.* **2020**, *2020*, 3202037. [CrossRef] [PubMed]
100. Rubiales, D.; Barilli, E.; Rispail, N. Breeding for Biotic Stress Resistance in Pea. *Agriculture* **2023**, *13*, 1825. [CrossRef]

101. Román, B.; Torres, A.M.; Rubiales, D.; Cubero, J.I.; Satovic, Z. Mapping of quantitative trait loci controlling broomrape (*Orobanche crenata* Forsk.) resistance in faba bean (*Vicia faba* L.). *Genome* **2002**, *45*, 1057–1063. [CrossRef] [PubMed]
102. Díaz-Ruiz, R.; Torres, A.M.; Satovic, Z.; Gutiérrez, M.V.; Cubero, J.I.; Román, B. Validation of QTLs for *Orobanche crenata* resistance in faba bean (*Vicia faba* L.) across environments and generations. *Theor. Appl. Genet.* **2010**, *120*, 909–919. [CrossRef] [PubMed]
103. Gutiérrez, N.; Palomino, C.; Satovic, Z.; Ruiz-Rodríguez, M.D.; Vitale, S.; Gutiérrez, M.V.; Rubiales, D.; Kharrat, M.; Amri, M.; Emeran, A.; et al. QTLs for *Orobanche* spp. resistance in faba bean: Identification and validation across different environments. *Mol. Breed.* **2013**, *32*, 909–922. [CrossRef]
104. Gutiérrez, N.; Torres, A.M. QTL dissection and mining of candidate genes for *Ascochyta fabae* and *Orobanche crenata* resistance in faba bean (*Vicia faba* L.). *BMC Plant Biol.* **2021**, *21*, 551. [CrossRef]
105. Fondevilla, S.; Fernández-Aparicio, M.; Satovic, Z.; Emeran, A.A.; Torres, A.M.; Moreno, M.T.; Rubiales, D. Identification of quantitative trait loci for specific mechanisms of resistance to *Orobanche crenata* Forsk. in pea (*Pisum sativum* L.). *Mol. Breed.* **2010**, *25*, 259–272. [CrossRef]
106. Delvento, C.; Arcieri, F.; Marcotrigiano, A.R.; Guerriero, M.; Fanelli, V.; Dellino, M.; Curci, P.L.; Bouwmeester, H.; Lotti, C.; Ricciardi, L.; et al. High-density linkage mapping and genetic dissection of resistance to broomrape (*Orobanche crenata* Forsk.) in pea (*Pisum sativum* L.). *Front. Plant Sci.* **2023**, *14*, 1216297. [CrossRef]
107. Abd El-Fatah, B.E.S.; Nassef, D.M.T. Inheritance of faba bean resistance to Broomrape, genetic diversity and QTL mapping analysis. *Mol. Biol. Rep.* **2020**, *47*, 11–32. [CrossRef]
108. Maalouf, F.; Khalil, S.; Ahmed, S.; Akintunde, A.N.; Kharrat, M.; El Shama'a, K.; Hajjar, S.; Malhotra, R.S. Yield stability of faba bean lines under diverse broomrape prone production environments. *Field Crops Res.* **2011**, *124*, 288–294. [CrossRef]
109. Fernández-Aparicio, M.; Moral, A.; Kharrat, M.; Rubiales, D. Resistance against broomrapes (*Orobanche* and *Phelipanche* spp.) in faba bean (*Vicia faba*) based in low induction of broomrape seed germination. *Euphytica* **2012**, *186*, 897–905. [CrossRef]
110. Rubiales, D.; Flores, F.; Emeran, A.A.; Kharrat, M.; Amri, M.; Rojas-Molina, M.M.; Sillero, J.C. Identification and multi-environment validation of resistance against broomrapes (*Orobanche crenata* and *O. foetida*) in faba bean (*Vicia faba*). *Field Crops Res.* **2014**, *166*, 58–65. [CrossRef]
111. Rubiales, D.; Sillero, J.C.; Rojas-Molina, M.M. Characterization resistance mechanisms in faba bean (*Vicia faba*) against broomrape species (*Orobanche* and *Phelipanche* spp.). *Front. Plant Sci.* **2016**, *7*, 1747. [CrossRef]
112. Briache, F.Z.; Ennami, M.; Mbasani-Mansi, J.; Gaboun, F.; Abdelwahd, R.; Fatemi, Z.E.A.; El-Rodeny, W.; Amri, M.; Triqui, Z.E.A.; Mentag, R. Field and controlled conditions screenings of some faba bean (*Vicia faba* L.) genotypes for resistance to the parasitic plant *Orobanche crenata* Forsk. and investigation of involved resistance mechanisms. *J. Plant Dis. Prot.* **2019**, *126*, 211–224. [CrossRef]
113. Rubiales, D.; Moreno, M.T.; Sillero, J.C. Search for resistance to crenate broomrape (*Orobanche crenata*) in pea germplasm. *Genet. Resour. Crop Evol.* **2005**, *52*, 853–861. [CrossRef]
114. Pérez-de-Luque, A.; Jorrín, J.; Cubero, J.I.; Rubiales, D. *Orobanche crenata* resistance and avoidance in pea (*Pisum* spp.) operate at different developmental stages of the parasite. *Weed Res.* **2005**, *45*, 379–387. [CrossRef]
115. Pavan, S.; Schiavulli, A.; Marcotrigiano, A.R.; Bardaro, N.; Bracuto, V.; Ricciardi, F.; Charnikhova, T.; Lotti, C.; Bouwmeester, H.; Ricciardi, L. Characterization of low-strigolactone germplasm in pea (*Pisum sativum* L.) resistant to crenate broomrape (*Orobanche crenata* Forsk.). *Mol. Plant-Microbe Interact.* **2016**, *29*, 743–749. [CrossRef]
116. Fernández-Aparicio, M.; Sillero, J.C.; Pérez-de-Luque, A.; Rubiales, D. Identification of sources of resistance to crenate broomrape (*Orobanche crenata*) in Spanish lentil (*Lens culinaris*) germplasm. *Weed Res.* **2008**, *48*, 85–94. [CrossRef]
117. Fernández-Aparicio, M.; Sillero, J.C.; Rubiales, D. Resistance to broomrape in wild lentils (*Lens* spp.). *Plant Breed.* **2009**, *128*, 266–270. [CrossRef]
118. En-nahli, Y.; Hejjaoui, K.; Mentag, R.; Es-safi, N.E.; Amri, M. Large Field Screening for Resistance to Broomrape (*Orobanche crenata* Forsk.) in a Global Lentil Diversity Panel (GLDP) (*Lens culinaris* Medik.). *Plants* **2023**, *12*, 2064. [CrossRef]
119. Gil, J.; Martín, L.M.; Cubero, J.I. Genetics of resistance in *Vicia sativa* L. to *Orobanche crenata* Forsk. *Plant Breed.* **1987**, *99*, 134–143. [CrossRef]
120. Goldwasser, Y.; Kleifeld, Y.; Plakhine, D.; Rubin, B. Variation in vetch (*Vicia* spp.) response to *Orobanche aegyptiaca*. *Weed Sci.* **1997**, *45*, 756–762. [CrossRef]
121. Goldwasser, Y.; Plakhine, D.; Kleifeld, Y.; Zamski, E.; Rubin, B. The Differential Susceptibility of Vetch (*Vicia* spp.) to *Orobanche aegyptiaca*: Anatomical Studies. *Ann. Bot.* **2000**, *85*, 257–262. [CrossRef]
122. Sillero, J.C.; Moreno, M.T.; Rubiales, D. Sources of resistance to crenate broomrape among species of *Vicia*. *Plant Dis.* **2005**, *89*, 23–27. [CrossRef] [PubMed]
123. Nadal, S.; Cubero, J.I.; Moreno, M.T. Sources of resistance to broomrape (*Orobanche crenata* Forsk.) in narbon vetch. *Plant Breed.* **2007**, *126*, 110–112. [CrossRef]
124. Fernández-Aparicio, M.; Sillero, J.C.; Rubiales, D. Resistance to broomrape species (*Orobanche* spp.) in common vetch (*Vicia sativa* L.). *Crop Prot.* **2008**, *28*, 7–12. [CrossRef]
125. González-Verdejo, C.I.; Fernández-Aparicio, M.; Córdoba, E.M.; López-Ráez, J.A.; Nadal, S. Resistance against *Orobanche crenata* in Bitter Vetch (*Vicia ervilia*) Germplasm Based on Reduced Induction of *Orobanche* Germination. *Plants* **2021**, *10*, 348. [CrossRef]
126. Rubio, J.M.; Rubiales, D. Resistance to rusts and broomrape in one-flowered vetch (*Vicia articulata*). *Euphytica* **2021**, *217*, 9. [CrossRef]

127. González-Verdejo, C.I.; Fernández-Aparicio, M.; Córdoba, E.M.; Nadal, S. Identification of *Vicia ervilia* Germplasm Resistant to *Orobanche crenata*. *Plants* **2020**, *9*, 1568. [CrossRef]
128. Rubiales, D.; Pérez-de-Luque, A.; Joel, D.M.; Alcántara, C.; Sillero, J.C. Characterization of resistance in chickpea to broomrape (*Orobanche crenata*). *Weed Sci.* **2003**, *51*, 702–707. [CrossRef]
129. Rubiales, D.; Alcántara, C.; Sillero, J.C. Variation in resistance to crenate broomrape (*Orobanche crenata*) in species of *Cicer*. *Weed Res.* **2004**, *44*, 27–32. [CrossRef]
130. Brahmi, I.; Mabrouk, Y.; Brun, G.; Delavault, P.; Belhadj, O.; Simier, P. Phenotypical and biochemical characterisation of resistance for parasitic weed (*Orobanche foetida* Poir.) in radiation-mutagenised mutants of chickpea. *Pest Manag. Sci.* **2016**, *72*, 2330–2338. [CrossRef] [PubMed]
131. Rubiales, D.; Alcántara, C.; Pérez-de-Luque, A.; Gil, J.; Sillero, J.C. Infection of chickpea (Cicer arietinum) by crenate broomrape (Orobanche crenata) as influenced by sowing date and weather conditions. *Agronomie* **2003**, *23*, 359–362. [CrossRef]
132. Sillero, J.C.; Cubero, J.I.; Fernández-Aparicio, M.; Rubiales, D. Search for resistance to crenate broomrape (*Orobanche crenata*) in *Lathyrus*. *Lathyrus Lathyrism Newsl.* **2005**, *4*, 7–9.
133. Fernández-Aparicio, M.; Flores, F.; Rubiales, D. Field response of *Lathyrus cicera* germplasm to crenate broomrape (*Orobanche crenata*). *Field Crops Res.* **2009**, *113*, 321–327. [CrossRef]
134. Fernández-Aparicio, M.; Flores, F.; Rubiales, D. Escape and true resistance to crenate broomrape (*Orobanche crenata* Forsk.) in grass pea (*Lathyrus sativus* L.) germplasm. *Field Crops Res.* **2011**, *125*, 92–97. [CrossRef]
135. Fernández-Aparicio, M.; Rubiales, D. Characterisation of resistance to crenate broomrape (*Orobanche crenata* Forsk.) in *Lathyrus cicera* L. *Euphytica* **2010**, *173*, 77–84. [CrossRef]
136. Abdallah, F.; Kumar, S.; Amri, A.; Mentag, R.; Kehel, Z.; Mejri, R.K.; Triqui, Z.E.-A.; Hejjaoui, K.; Baum, M.; Amri, M. Wild *Lathyrus* species as a great source of resistance for introgression into cultivated grass pea (*Lathyrus sativus* L.) against broomrape weeds (*Orobanche crenata* Forsk. and *Orobanche foetida* Poir.). *Crop Sci.* **2021**, *61*, 263–276. [CrossRef]
137. Rodríguez-Conde, M.F.; Moreno, M.T.; Cubero, J.I.; Rubiales, D. Characterization of the *Orobanche—Medicago truncatula* association for studying early stages of the parasite-host interaction. *Weed Res.* **2004**, *44*, 218–223. [CrossRef]
138. Fernández-Aparicio, M.; Pérez-de-Luque, A.; Prats, E.; Rubiales, D. Variability of interactions between barrel medic (*Medicago truncatula*) genotypes and *Orobanche* species. *Ann. Appl. Biol.* **2008**, *153*, 117–126. [CrossRef]
139. Rubiales, D.; Fernández-Aparicio, M.; Pérez-de-Luque, A.; Prats, E.; Castillejo, M.A.; Sillero, J.C.; Rispail, N.; Fondevilla, S. Breeding approaches for crenate broomrape (*Orobanche crenata* Forsk.) management in pea (*Pisum sativum* L.). *Pest Manag. Sci.* **2009**, *65*, 553–559. [CrossRef] [PubMed]
140. Fondevilla, S.; Flores, F.; Emeran, A.A.; Kharrat, M.; Rubiales, D. High productivity of dry pea genotypes resistant to crenate broomrape in Mediterranean environments. *Agron. Sustain. Dev.* **2017**, *37*, 61. [CrossRef]
141. Rubiales, D.; Fondevilla, S.; Fernández-Aparicio, M. Development of pea breeding lines with resistance to *Orobanche crenata* derived from pea landraces and wild *Pisum* spp. *Agronomy* **2021**, *11*, 36. [CrossRef]
142. Rubiales, D.; Osuna-Caballero, S.; González-Bernal, M.J.; Cobos, M.J.; Flores, F. Pea breeding lines adapted to autumn sowings in broomrape prone Mediterranean environments. *Agronomy* **2021**, *11*, 769. [CrossRef]
143. Stam, R.; McDonald, B.A. When resistance gene pyramids are not durable—The role of pathogen diversity. *Mol. Plant Pathol.* **2018**, *19*, 521–524. [CrossRef] [PubMed]
144. Niks, R.E.; Rubiales, D. Potentially durable resistance mechanisms in plants to specialised fungal pathogens. *Euphytica* **2002**, *124*, 201–216. [CrossRef]
145. Rubiales, D. Parasitic plants, wild relatives and the nature of resistance. *New Phytol.* **2003**, *160*, 459–461. [CrossRef]
146. Yoneyama, K.; Brewer, P.B. Strigolactones, how are they synthesized to regulate plant growth and development? *Curr. Opin. Plant Biol.* **2021**, *63*, 102072. [CrossRef]
147. Fernández-Aparicio, M.; Yoneyama, K.; Rubiales, D. The role of strigolactones in host specificity of *Orobanche* and *Phelipanche* seed germination. *Seed Sci. Res.* **2011**, *21*, 55–61. [CrossRef]
148. Fernández-Aparicio, M.; Kisugi, T.; Xie, X.; Rubiales, D.; Yoneyama, K. Low strigolactone root exudation: A novel mechanism of broomrape (*Orobanche* and *Phelipanche* spp.) resistance available for faba bean breeding. *J. Agric. Food Chem.* **2014**, *62*, 7063–7071. [CrossRef]
149. Ejeta, G. Breeding for *Striga* resistance in sorghum: Exploitation of an intricate host-parasite biology. *Crop Sci.* **2007**, *47*, 216–227. [CrossRef]
150. Dor, E.; Alperin, B.; Wininger, S.; Ben-Dor, B.; Somvanshi, V.S.; Koltai, H.; Kapulnik, Y.; Hershenhorn, J. Characterization of a novel tomato mutant resistant to *Orobanche* and *Phelipanche* spp. weedy parasites. *Euphytica* **2010**, *171*, 371–373. [CrossRef]
151. Dor, E.; Yoneyama, K.; Wininger, S.; Kapulnik, Y.; Yoneyama, K.; Koltai, H.; Xie, X.; Hershenhorn, J. Strigolactone deficiency confers resistance in tomato line SL-ORT1 to the parasitic weeds *Phelipanche* and *Orobanche* spp. *Phytopathology* **2011**, *101*, 213–222. [CrossRef] [PubMed]
152. Galili, S.; Hershenhorn, J.; Smirnov, E.; Yoneyama, K.; Xie, X.; Amir-Segev, O.; Bellalou, A.; Dor, E. Characterization of a Chickpea Mutant Resistant to *Phelipanche aegyptiaca* Pers. and *Orobanche crenata* Forsk. *Plants* **2021**, *10*, 2552. [CrossRef] [PubMed]
153. Bardaro, N.; Marcotrigiano, A.R.; Bracuto, V.; Mazzeo, R.; Pavan, S.; Ricciardi, L. Genetic analysis of resistance to *Orobanche crenata* (Forsk.) in a pea (*Pisum sativum* L.) low-strigolactone line. *J. Plant Pathol.* **2016**, *98*, 671–675. [CrossRef]
154. Kokla, A.; Melnyk, C.W. Developing a thief: Haustoria formation in parasitic plants. *Dev. Biol.* **2018**, *442*, 53–59. [CrossRef]

155. Fernández-Aparicio, M.; Masi, M.; Maddau, L.; Cimmino, A.; Evidente, M.; Rubiales, D.; Evidente, A. Induction of haustorium development by sphaeropsidones in radicles of the parasitic weeds *Striga* and *Orobanche*. A structure-activity relationship study. *J. Agric. Food Chem.* **2016**, *64*, 5188–5196. [CrossRef] [PubMed]
156. Pérez-de-Luque, A.; Rubiales, D.; Cubero, J.I.; Press, M.C.; Scholes, J.; Yoneyama, K.; Takeuchi, Y.; Plakhine, D.; Joel, D.M. Interaction between *Orobanche crenata* and its host legumes: Unsuccessful haustorial penetration and necrosis of the developing parasite. *Ann. Bot.* **2005**, *95*, 935–942. [CrossRef]
157. Goldwasser, Y.; Hershenhorn, J.; Plakhine, D.; Kleifeld, Y.; Rubin, B. Biochemical factors involved in vetch resistance to *Orobanche aegyptiaca*. *Physiol. Mol. Plant Pathol.* **1999**, *54*, 87–96. [CrossRef]
158. Pérez-de-Luque, A.; Lozano, M.D.; Moreno, M.T.; Testillano, P.S.; Rubiales, D. Resistance to broomrape (*Orobanche crenata*) in faba bean (*Vicia faba*): Cell wall changes associated with pre-haustorial defensive mechanisms. *Ann. Appl. Biol.* **2007**, *151*, 89–98. [CrossRef]
159. Lozano-Baena, M.D.; Prats, E.; Moreno, M.T.; Rubiales, D.; Pérez-de-Luque, A. *Medicago truncatula* as a model host for legumes-parasitic plants interactions: Two phenotypes of resistance for one defensive mechanism. *Plant Physiol.* **2007**, *145*, 437–449. [CrossRef] [PubMed]
160. Singh, N.P.; Yadav, I.S. *Herbicide Tolerant Food Legume Crops: Possibilities and Prospects, Herbicides—Properties, Synthesis and Control of Weeds*; Hasaneen, M.N., Ed.; InTechOpen Ltd.: London, UK, 2012. [CrossRef]
161. Abou-Khater, L.; Maalouf, F.; Patil, S.B.; Balech, R.; Nacouzi, D.; Rubiales, D.; Kumar, S. Identification of tolerance to metribuzin and imazethapyr herbicides in faba bean. *Crop Sci.* **2021**, *61*, 2593–2611. [CrossRef]
162. Redlick, C.; Syrovy, L.D.; Duddu, H.S.N.; Benaragama, D.; Johnson, E.N.; Willenborg, C.J.; Shirtliffe, S.J. Developing an Integrated Weed Management System for Herbicide-Resistant Weeds Using Lentil (*Lens culinaris*) as a Model Crop. *Weed Sci.* **2017**, *65*, 778–786. [CrossRef]
163. Balech, R.; Maalouf, F.; Patil, S.B.; Hejjaoui, K.; Abou-Khater, L.; Rajendran, K.; Rubiales, D.; Kumar, S. Evaluation of performance and stability of new sources for tolerant to post-emergence herbicides in lentil (*Lens culinaris* ssp. culinaris Medik). *Crop Pasture Sci.* **2022**, *73*, 1264–1278. [CrossRef]
164. Rizwan, M.; Aslam, M.; Asghar, M.J.; Abbas, G.; Shah, T.M.; Shimelis, H. Pre-breeding of lentil (*Lens culinaris* Medik.) for herbicide resistance through seed mutagenesis. *PLoS ONE* **2017**, *12*, e0171846. [CrossRef] [PubMed]
165. McMurray, L.; Preston, C.; Vandenberg, A.; Mao, D.; Oldach, K.; Meier, K.; Paull, J. Development of High Levels of Metribuzin Tolerance in Lentil. *Weed Sci.* **2019**, *67*, 83–90. [CrossRef]
166. Gressel, J. Crops with target-site herbicide resistance for *Orobanche* and *Striga* control. *Pest Manag. Sci.* **2009**, *65*, 560–565. [CrossRef] [PubMed]
167. Gressel, J. Biotechnologies for directly generating crops resistant to parasites. In *Parasitic Orobanchaceae*; Joel, D., Gressel, J., Musselman, L., Eds.; Springer: Berlin/Heidelberg, Germany, 2013.
168. Yoder, J.I.; Gunathilake, P.; Wu, B.; Tomilova, N.; Tomilov, A.A. Engineering host resistance against parasitic weeds with RNA interference. *Pest Manag. Sci.* **2009**, *65*, 460–466. [CrossRef]
169. Atsmon, G.; Nehurai, O.; Kizel, F.; Eizenberg, H.; Lati, R.N. Hyperspectral imaging facilitates early detection of *Orobanche cumana* below-ground parasitism on sunflower under field conditions. *Comput. Electron. Agric.* **2022**, *196*, 106881. [CrossRef]
170. Le Ru, A.; Ibarcq, G.; Boniface, M.C.; Baussart, A.; Muños, S.; Chabaud, M. Image analysis for the automatic phenotyping of *Orobanche cumana* tubercles on sunflower roots. *Plant Methods* **2021**, *17*, 80. [CrossRef]
171. Parihar, A.K.; Kumar, J.; Gupta, D.S.; Lamichaney, A.; Naik, S.J.S.; Singh, A.K.; Dixit, G.P.; Gupta, S.; Toklu, F. Genomics enabled breeding strategies for major biotic stresses in pea (*Pisum sativum* L.). *Front. Plant Sci.* **2022**, *13*, 861191. [CrossRef]
172. Jha, U.C.; Nayyar, H.; Parida, S.K.; Bakır, M.; von Wettberg, E.J.B.; Siddique, K.H.M. Progress of Genomics-Driven approaches for sustaining underutilized legume crops in the post-genomic Era. *Front. Genet.* **2022**, *13*, 831656. [CrossRef] [PubMed]
173. Diakostefani, A.; Velissaris, R.; Cvijanovic, E.; Bulgin, R.; Pantelides, A.; Leitch, I.J.; Mian, S.; Morton, J.A.; Gomez, M.S.; Chapman, M.A. Genome resources for underutilised legume crops: Genome sizes, genome skimming and marker development. *Genet. Resour. Crop Evol.* **2023**. [CrossRef]
174. Kagale, S.; Close, T.J. Legumes: Embracing the genome era. *Legume Sci.* **2021**, *3*, e113. [CrossRef]
175. Westwood, J.H.; Depamphilis, C.W.; Das, M.; Fernández-Aparicio, M.; Honaas, L.A.; Timko, M.P.; Wafula, E.K.; Wickett, N.J.; Yoder, J.I. The Parasitic Plant Genome Project: New Tools for Understanding the Biology of *Orobanche* and *Striga*. *Weed Sci.* **2012**, *60*, 295–300. [CrossRef]
176. Xu, Y.; Zhang, J.; Ma, C.; Lei, Y.; Shen, G.; Jin, J.; Eaton, D.A.R.; Wu, J. Comparative genomics of orobanchaceous species with different parasitic lifestyles reveals the origin and stepwise evolution of plant parasitism. *Mol. Plant* **2022**, *15*, 1384–1399. [CrossRef]
177. Castillejo, M.A.; Fernández-Aparicio, M.; Rubiales, D. Proteomic analysis by two-dimensional differential in gel electrophoresis (2D DIGE) of the early response of *Pisum sativum* to *Orobanche crenata*. *J. Exp. Bot.* **2012**, *63*, 107–119. [CrossRef]
178. Die, J.V.; Román, B.; Nadal, S.; Dita, M.Á.; González-Verdejo, C.I. Expression analysis of *Pisum sativum* putative defence genes during *Orobanche crenata* infection. *Crop Pasture Sci.* **2009**, *60*, 490–498. [CrossRef]
179. Aly, R.; Matzrafi, M.; Bari, V.K. Using biotechnological approaches to develop crop resistance to root parasitic weeds. *Planta* **2021**, *253*, 97. [CrossRef]

180. Bhowmik, P.; Konkin, D.; Polowick, P.; Hodgins, C.L.; Subedi, M.; Xiang, D.; Yu, B.; Patterson, N.; Rajagopalan, N.; Babic, V.; et al. CRISPR/Cas9 gene editing in legume crops: Opportunities and challenges. *Legume Sci.* **2021**, *3*, e96. [CrossRef]
181. Bari, V.K.; Nassar, J.A.; Kheredin, S.M.; Gal-On, A.; Ron, M.; Britt, A.; Steele, D.; Yoder, J.; Aly, R. CRISPR/Cas9-mediated mutagenesis of carotenoid cleavage dioxygenase 8 in tomato provides resistance against the parasitic weed *Phelipanche aegyptiaca*. *Sci. Rep.* **2019**, *9*, 11438. [CrossRef]
182. Li, G.; Liu, R.; Xu, R.; Varshney, R.K.; Ding, H.; Li, M.; Yan, X.; Huang, S.; Li, J.; Wang, D.; et al. Development of an agrobacterium-mediated CRISPR/Cas9 system in pea (*Pisum sativum* L.). *Crop J.* **2023**, *11*, 132–139. [CrossRef]
183. Ludvíková, M.; Griga, M. Pea transformation: History, current status and challenges. *Czech J. Genet. Plant Breed.* **2022**, *58*, 127–161. [CrossRef]
184. Choudhury, A.; Rajam, M.V. Genetic transformation of legumes: An update. *Plant Cell Rep.* **2021**, *40*, 1813–1830. [CrossRef]

soil systems

MDPI

Article

Assessment of the Interrelationships of Soil Nutrient Balances with the Agricultural Soil Emissions and Food Production

Vítor João Pereira Domingues Martinho [1,*], José L. S. Pereira [2,3] and José Manuel Gonçalves [4]

1 Agrarian School (ESAV) and CERNAS-IPV Research Centre, Polytechnic Institute of Viseu (IPV), 3504-510 Viseu, Portugal
2 Agrarian School (ESAV), Polytechnic Institute of Viseu (IPV), Quinta da Alagoa, 3500-606 Viseu, Portugal; jlpereira@esav.ipv.pt
3 Centre for the Research and Technology of Agro-Environmental and Biological Sciences (CITAB), Inov4Agro, University of Trás-os-Montes and Alto Douro, Quinta de Prados, 5000-801 Vila Real, Portugal
4 Agrarian School (ESAC), Polytechnic Institute of Coimbra (IPC), Bencanta, 3045-601 Coimbra, Portugal; jmmg@esac.pt
* Correspondence: vdmartinho@esav.ipv.pt

Abstract: Sustainable and adjusted soil management practices are crucial for soil quality, namely in terms of the nutrient budget. On the other hand, soil characteristics are interlinked with agricultural sustainability and food supply. In other words, soil quality influences agricultural performance and food chains, but it is also impacted by agricultural activities. In this context, this research aims to evaluate the spatial correlations of the soil nutrient balance around the world and analyse how this variable is interrelated with agricultural soil emissions, agricultural output, and food supply. To achieve these goals, data from the FAOSTAT database were considered. This statistical information was analysed with spatial autocorrelation approaches to identify spatial clusters around the world that can be considered as a basis for designing common policies. To perform panel data regressions to identify marginal effects between variables, data were first evaluated using correlation matrices and factor analysis. The results highlight that there is space for common strategies worldwide to preserve soil quality, as in some parts of the world the problems are similar. In these frameworks, the international organizations may have a determinant contribution.

Keywords: spatial autocorrelation; matrices of correlation; factor analysis; panel data regressions

1. Introduction

Information about land characteristics is an important factor in integrated soil management and here, beyond the scientific contributions [1], knowledge of local populations about the soil properties provides relevant contributions [2]. Adjusted management plans may make local needs compatible with soil quality conservation [3], where agricultural practices determine the results obtained [4] in terms of sustainability, along with the farming systems adopted [5] and the crops species [6]. Sustainable practices differ for each agricultural activity and also between countries and regions [7].

Different tillage, fertilisation techniques and rotation approaches are agronomic practices that may make a difference in the quality of the soil [8], in rice-wheat systems for example. Soil conservation techniques are often interrelated with water management approaches [9], because the dynamics of these two resources (soil and water) are mutually dependent [10]. Soil quality also impacts the characteristics of the crops obtained [11] and the health of animal activities [12].

To promote sustainable and best management practices in the agricultural sector, with benefits for the environment and soil quality, farmers need to be supported with technical knowledge, and in these conditions, extension services are crucial [13], as well as training programs to increase the technical skills of stakeholders related with the farming activities.

The agricultural institutions (national and international), organizations (cooperatives and associations, for example) and policies (Common Agricultural Policy in European Union, for instance) are crucial in order to achieve sustainable development goals [14].

The new challenges created by the world population growth and the needs of dealing with the climate change contexts call for alternative ways of better managing the available resources [15], specifically in contexts of agricultural intensification [16] and soil erosion [17]. In these frameworks, the agricultural activities are sources and sinks of greenhouse gases, where the soil carbon sequestration is fundamental for the sustainability [18]. The soil degradation and erosion are threats that particularly concern the several society stakeholders [19].

Considering these issues of the soil management and its interrelationships with the several dimensions of the agricultural sector, this study intends to analyse the soil nutrient balance worldwide through spatial assessments to identify clusters between the countries. These analyses will be a basis for the design of joint policies and combining efforts for together solve common problems related with the soil quality. In addition, this research aims to assess the main interlinkages between the soil nutrient budget and the soil emissions and the food supply.

2. Literature Survey

Agricultural soils are impacted by the agronomic practices adopted by the farmers, where the tillage, for example, has its influence in the physical properties [20] and quality [21]. Hence, this must be considered by several stakeholders, particularly farmers and policymakers. Conventional tillage may reduce the soil organic matter and increase the carbon dioxide (CO_2) emissions [22] by soil respiration. Minimum tillage is suggested to achieve the compromise among the agricultural productions loss and the soil preservation [23]. Specifically, soil erosion is comparable to the water erosion [24]. Soil and water dynamics are correlated [25], wherein the formation of the organic matter is a complex process dependent from diverse drivers [26], such as soil temperature and humidity and carbon/nitrogen ratio of the manures.

Other farming practices have their impacts on the soil characteristics and composition, such as compost or manure application (with benefits for the agricultural activities, but with changes in the microbial community) [27], organic/conventional productions [28], organic/inorganic fertilisers [29], pasture in rotation [30], agrochemicals (affects the bacterial diversity [31], for instance) [32], plastic mulching [33], land use changes [34], straw return [35], soil fumigation [36], harvest practices [37], forest-agriculture conversion [38], conventional practices [39], field fallow [40], polymers use [41] and cover crops [42].

The soil quality is influenced by several factors, some of them from extreme phenomena [43] and the climate changes [44], nonetheless the various dimensions associated with the farming contexts explain a part of the sources of problems that bring degradation of the land, specifically those associated with salinity [45].

Soil is a key factor of production for the agricultural sector [46] and food supply [47], however, it is under pressures by the economic activities [48]. A permanent assessment of the soil quality (mainly the soil physical properties [49]) through new techniques [50], approaches [51] and technologies [52] is crucial for an adjusted soil management [53]. Namely, to maintain the levels of carbon and nitrogen through conservation practices [54] and preserve the human health [55] from toxic contaminants [56], including phthalate esters [57], heavy metals (with impacts on food safety [58]) [59] and copper balance [60]. For these evaluations, the availability of information [61] worldwide [62] is fundamental. The assessment of soil quality is also important to support strategy proposals [63] and characteristics prediction [64] under the global warming challenges [65].

The agricultural soil management is responsible for greenhouse gas emission [66], with several environmental impacts originating in the following gases: nitrous oxide (N_2O) [67] by nitrification and denitrification processes, methane (CH_4) by anaerobic conditions and CO_2 by aerobic or anaerobic environment. These greenhouse gas emissions are particularly influenced by soil type, climate, water management and composition of organic matter [68]. Hence, the agricultural soil management is interrelated with the agricultural practices and environmental impacts [69]. Thus, the interlinkages have impacts, for example, on the ecosystems services [70], soil biodiversity [71] and humus composition [72]. For example, the use of biochar into the soil may be an interesting alternative to reduce the environmental impacts and mitigate the climate change consequences [73]. Additionally, adjusted soil management may prevent soilborne diseases [74] and increase the soil organic carbon [75].

For a sustainable agricultural soil management, the agricultural policies and institutions are called to play relevant roles [76] to promote soil conservation practices [77]. This issue is particularly important in the European Union contexts, under the framework of the Common Agricultural Policy (CAP) [14], and to deal with problematic cases created by the post-Second World War contexts [78]. The public policies are specifically important in the cases where the negative impacts are self-reinforced or have dynamics of rebound effects [79].

3. Materials and Methods

To achieve the objectives proposed and considering the several relationships associated with the soil properties highlighted in the literature review, statistical information for the following variables was obtained from the FAOSTAT [80] database: agricultural soil emissions (CO_2eq, namely N_2O emissions.) in kilotonnes per ha of cropland; average value of food production (constant 2004–2006 I$ (international dollar, an international dollar would buy in a country a comparable amount of products a U.S. dollar would buy in the United States [81])/cap, 3-year average); gross agricultural production value per ha (constant 2014–2016, 1000 I$ per ha of cropland); and cropland nutrient flow per unit area (kg per ha). These variables were selected to represent the characteristics of the soil and their different interlinkages, namely those related with the environment, agricultural production, and food supply. Considering the availability of data for the various variables, it was considered the period 2001–2017. To associate the average valued of food production with the other indicators, the middle year for each group of three years was considered.

These indicators were first analysed through spatial autocorrelation, to identify spatial clusters worldwide, where it may be possible to design common strategies to deal with an integrated agricultural soil management. For the spatial assessments, global and local autocorrelation approaches were considered following GeoDa procedures [82,83]. For the global spatial autocorrelation, the Moran's I statistics were used [84]. The Moran's I statistics range between -1 and 0, for negative spatial autocorrelation (the values of a variable are negatively correlated with the values of the same variable in the neighbour countries), and 0 and 1, for positive autocorrelation. For the local spatial autocorrelation, cluster maps were considered. In these maps, the clusters high-high and low-low highlight positive local spatial autocorrelation for higher and lower values, respectively. The clusters high-low and low-high represent negative spatial autocorrelation. For this spatial analysis, shapefiles from the Eurostat [85] for the world countries were used that were explored through the QGIS software [86].

After this first assessment, the variables were considered to obtain indices for the integrated agricultural soil management through factor analysis [87–91] and to find marginal effects based on panel data regressions [91–93]. To identify the best models for the panel data regressions, the Spearman correlations [94] and the Granger cause statistics [95] were carried out.

4. Spatial Autocorrelation Analysis

The spatial autocorrelation analysis reported in this section was assessed using queen contiguity matrix, for an order of contiguity of 1. Figures 1–4 show the level of global and local spatial autocorrelation and the distribution of values of the several variables considered worldwide.

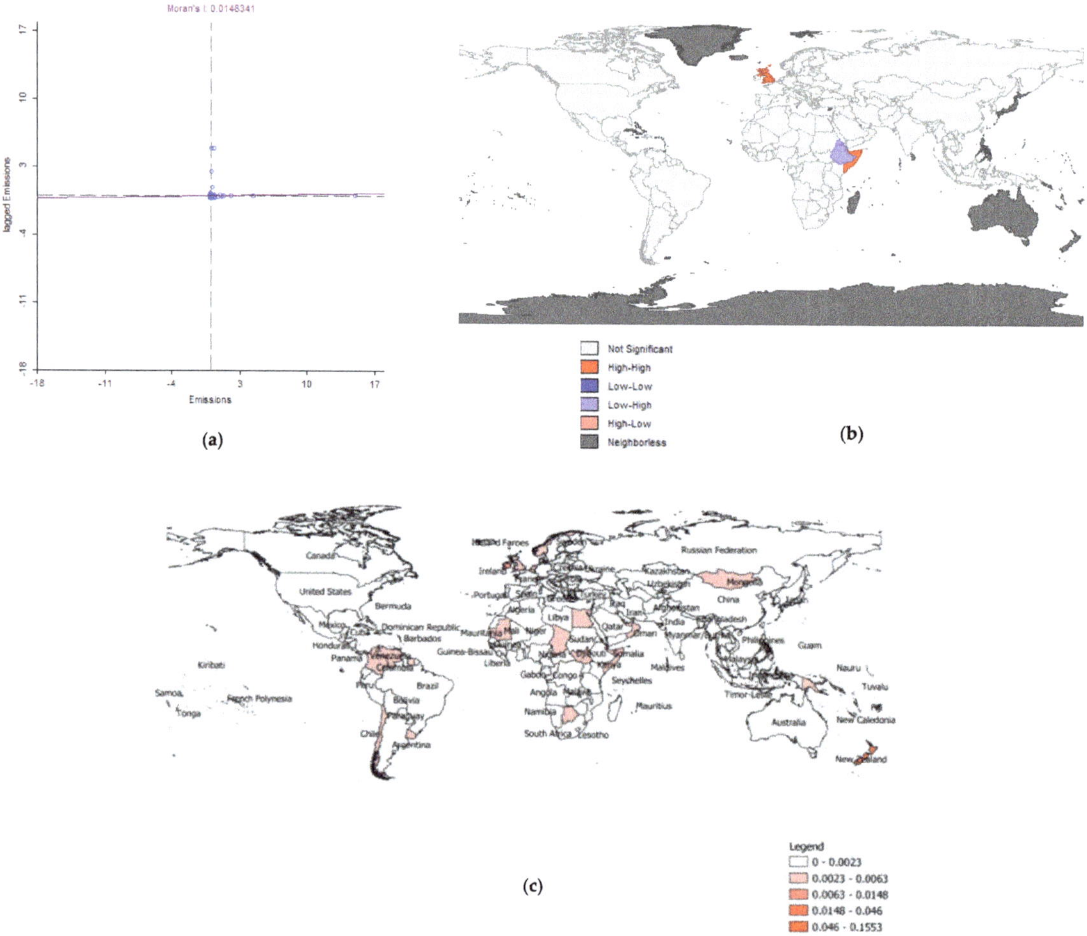

Figure 1. Global and local spatial autocorrelation and worldwide distribution for the agricultural soil emissions (CO_2eq) per ha of cropland, on average over the period 2001–2017 (kilotonnes per ha of cropland); (**a**) Global spatial autocorrelation, (**b**) Local spatial autocorrelation; (**c**) Worldwide distribution.

The global and local spatial autocorrelation was weak for the agricultural soil emissions (CO_2eq) per ha of cropland, and this was a consequence of values relatively low (exception for the case of New Zealand, for example) verified for this variable across the world countries (Figure 1).

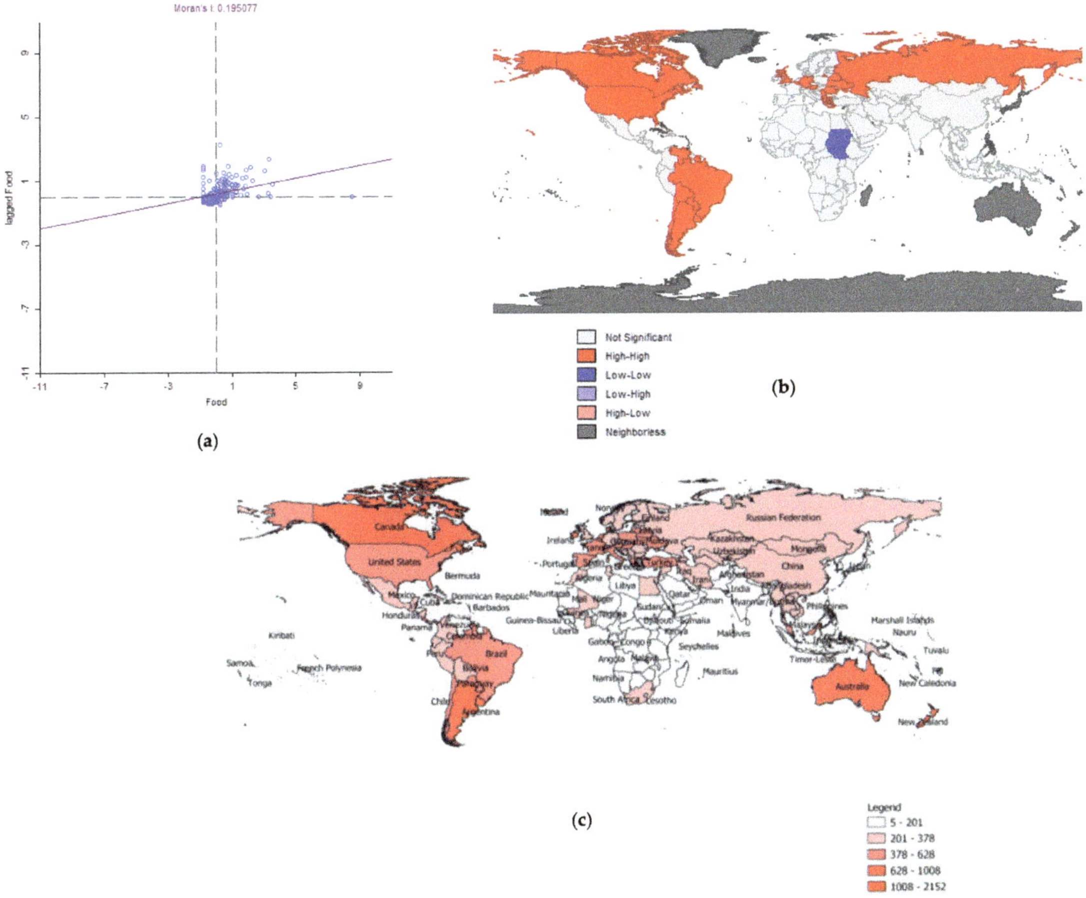

Figure 2. Global and local spatial autocorrelation and worldwide distribution for the average value of food production (constant 2004–2006 I$/cap) (3-year average), on average over the period 2001–2017 (I$ per person); (**a**) Global spatial autocorrelation, (**b**) Local spatial autocorrelation; (**c**) Worldwide distribution.

As can be observed in Figure 2, the scenario was different for the average value of food production, where there are signs of relevant positive global spatial autocorrelation and high-high local spatial autocorrelation in North and South America, Russia, and some European countries. Hence, this means that the strategies developed by the countries inside of each cluster high-high spread among neighbour countries, lead to good findings for future policies.

The gross agricultural production value per ha of cropland was, in general, low worldwide (exception for New Zealand and some European countries, for example), and this explains, at least in part, the reduced level of spatial autocorrelation for this variable (Figure 3). The cropland nutrient flow per unit area had significant signs of positive high-high local spatial autocorrelation in the European countries (Figure 4).

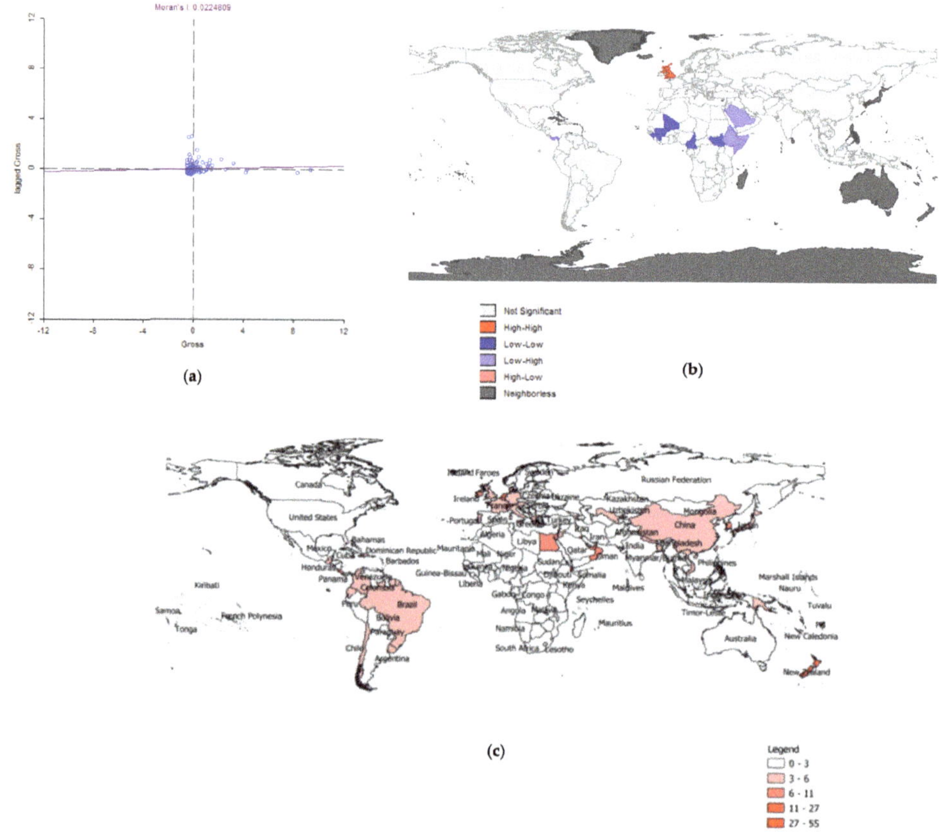

Figure 3. Global and local spatial autocorrelation and worldwide distribution for the gross agricultural production value (constant 2014–2016 thousand I$) per ha of cropland, on average over the period 2001–2017 (thousand I$ per ha of crop land); (**a**) Global spatial autocorrelation, (**b**) Local spatial autocorrelation; (**c**) Worldwide distribution.

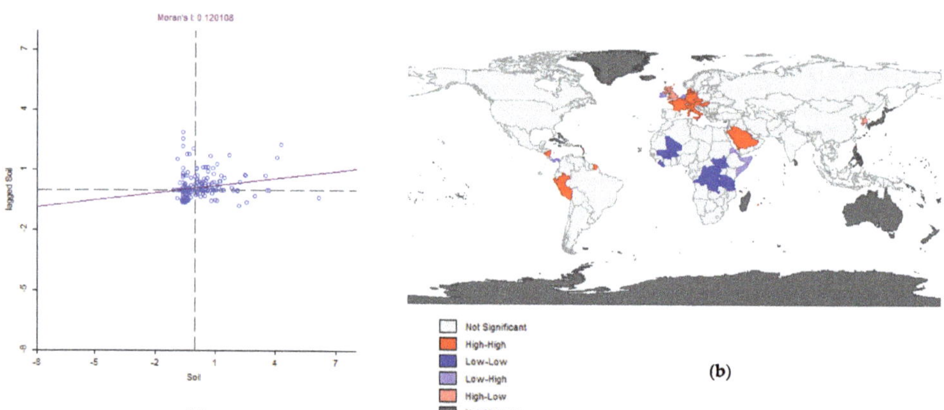

Figure 4. *Cont.*

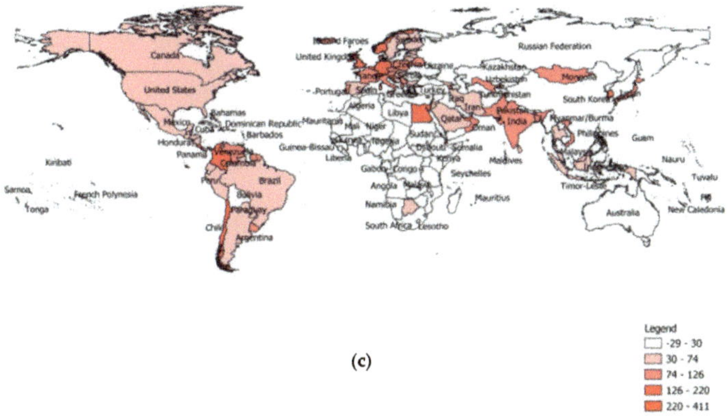

(c)

Figure 4. Global and local spatial autocorrelation and worldwide distribution for the cropland nutrient flow per unit area, on average over the period 2001–2017 (kg per ha); (**a**) Global spatial autocorrelation, (**b**) Local spatial autocorrelation; (**c**) Worldwide distribution.

5. Identifying Indices for an Integrated Agricultural Soil Management

To facilitate the readability of the results presented here, and improve the robustness of the findings, it was obtained a balanced panel data (in which the countries and years with missing values were removed, remaining 183 countries with data for the full period of 2001–2017) and the agricultural soil emissions were converted from kilotonnes per ha into kg per ha and the gross agricultural production from 1000 I$ per ha into I$ per ha.

Table 1 highlights that the stronger correlations are between the agricultural soil emissions per ha, the gross agricultural production per ha and the cropland nutrient flow per ha. There was also strong correlation among the cropland nutrient flow per ha and the gross agricultural production per ha.

Table 1. Spearman's rank correlation matrix for several variables over the period 2001–2017 and across world countries.

	Agricultural Soil Emissions (kg per ha)	Average Food Production (I$ per Person)	Gross Agricultural Production (I$ per ha)	Cropland Nutrient Flow (kg per ha)
Agricultural soil emissions (kg per ha)	1.000			
Average food production (I$ per person)	0.0920 *	1.000		
	(0.000)			
Gross agricultural production (I$ per ha)	0.5996 *	0.2278 *	1.000	
	(0.000)	(0.000)		
Cropland nutrient flow (kg per ha)	0.6691 *	0.2099 *	0.6560 *	1.000
	(0.000)	(0.000)	(0.000)	

Note: *, statistically significant at 1%.

As can be observed in Table 2, it was intended to obtain an integrated agricultural soil management index, through factor analysis, with the most correlated variables. The agricultural soil emissions per ha were not considered in the factor analysis, because it was expected to contribute for the soil sustainability in a different way of the gross agricultural production and the cropland nutrient flow. Hence, the consideration of these three variables (agricultural soil emissions, gross agricultural production and cropland nutrient flow) in the index hampers the interpretation of its results. Thus, the selection of the variables reported in this study considered the objectives proposed (analyse how soil nutrient balance is interrelated with agricultural soil emissions, agricultural output and food supply), nonetheless in future studies could be interesting to benchmark these results with those obtained considering other variables.

Table 2. Factor analysis to obtain an integrated agricultural soil management index over the period 2001–2017 and across world countries.

Method: Principal-Component Factors; Rotation: Orthogonal Varimax (Kaiser Off)				
Factor	Variance	Difference	Proportion	Cumulative
Factor1	1.668		0.834	0.834
Rotated Factor Loadings and Unique Variances				
Variable	Factor1	Uniqueness		
Gross agricultural production (I$ per ha)	0.913	0.166		
Cropland nutrient flow (kg per ha)	0.913	0.166		

Table 3 shows the top 10 countries for the integrated agricultural soil management index and highlights that the countries with higher gross agricultural production per ha, cropland nutrient flow per ha and consequent greater agricultural soil emissions per ha are not the same with better food supply per person.

Table 3. Top 10 countries for the integrated agricultural soil management index, on average over the period 2001–2017.

Countries	Agricultural Soil Emissions (kg per ha)	Average Food Production (I$ per Person)	Gross Agricultural Production (I$ per ha)	Cropland Nutrient Flow (kg per ha)	Index
Belgium	2812	473	10,261	287	3
Malta	2310	175	10,834	260	3
Switzerland	2868	304	9018	251	2
China, Taiwan Province of	1940	209	8754	224	2
Luxembourg	2503	341	4000	298	2
Egypt	3635	228	8658	220	2
United Arab Emirates	4962	100	8615	191	2
Trinidad and Tobago	2945	102	4055	257	2
Republic of Korea	1966	190	8351	173	2
Israel	1897	347	9966	114	1

Note: The country with the highest index is Djibouti, nonetheless because difficulties in validating the data it was not considered in this table.

6. Panel Data Regressions

The Granger causality tests highlight that the cropland nutrient flow per ha impacts the agricultural soil emissions per ha of cropland and the gross agricultural production per ha of cropland. Based on these findings, on the assessments carried out before and on the literature review, the results presented in Tables 4 and 5 were obtained.

Table 4. Panel data regression with the agricultural soil emissions per ha as dependent variable over the period 2001–2017 and across world countries.

Model	Prais-Winsten Regression, Correlated Panels Corrected Standard Errors (PCSEs)
Constant	−34.717 (−0.100) [0.917]
Cropland nutrient flow (kg per ha)	41.279 * (5.910) [0.000]
Pesaran's test of cross sectional independence	3.009 * [0.002]
Modified Wald test for groupwise heteroskedasticity	6.2×10^{10} * [0.000]
Wooldridge test for autocorrelation	1137.221 * [0.000]

Note: *, statistically significant at 1%.

Table 5. Panel data regression with the gross agricultural production per ha as dependent variable over the period 2001–2017 and across world countries.

Model	Prais-Winsten Regression, Correlated Panels Corrected Standard Errors (PCSEs)
Constant	1358.298 * (7.970) [0.000]
Cropland nutrient flow (kg per ha)	25.094 * (7.610) [0.000]
Pesaran's test of cross sectional independence	54.380 * [0.000]
Modified Wald test for groupwise heteroskedasticity	1.2×10^{8} * [0.000]
Wooldridge test for autocorrelation	528.496 * [0.000]

Note: *, statistically significant at 1%.

The results obtained in this study revealed the following statistical problems: cross sectional dependence, heteroscedasticity, and autocorrelation of the data sample. To deal with these frameworks, the Prais–Winsten regression, correlated panels corrected standard errors (PCSEs), following Stata [91] and Torres-Reyna [93] procedures were considered.

These findings reveal that when the cropland nutrient flow increases 1 kg/ha the agricultural soil emissions worldwide increase 41.279 kg/ha and the gross agricultural production increases 25.094 I$ per ha.

These results highlight serious problems of sustainability in the agricultural soil management worldwide because the cropland nutrient flow and the agricultural production are associated with more agricultural soil emissions, but this context is disconnected from the food supply per person.

7. Discussion

This study aimed to analyse the framework of the soil nutrient balances across the world countries and assess their interrelationships with the agricultural soil emissions and the food supply. For that, geographic information system (GIS) approaches were considered, namely, to identify evidence of spatial autocorrelations between the countries for the variables considered. Factor analysis to find indices and panel data regressions to obtain relationships among the variables were also carried out.

The literature review highlighted the impacts on the agricultural soils from the agronomic practices, where the tillage, fertilisation, rotations, and land use changes, for example, have their implications. However, the agricultural soils are also responsible by environmental impacts through the greenhouse gas emissions. Sometimes, these interrelationships create contexts with self-reinforced effects, where the agricultural policies and institutions play a determinant role to reduce the negative externalities.

The spatial autocorrelation analysis shows that the global and local spatial correlations are weak for the agricultural soil emissions, in consequence of relatively values worldwide. For the average value of food production, there are signs positive global and local spatial autocorrelation. These evidences are interesting findings for the several stakeholders, namely for the policymakers, because this means that interventions in countries positively correlated may spread for the neighbours. There are also evidences of positive spatial autocorrelation in some European countries for the cropland nutrient flow per unit area.

A correlation matrix and factor analysis reveal that there are strong correlations between the agricultural soil emissions per ha, gross agricultural production per ha and the cropland nutrient flow per ha. The agricultural soil management was interrelated with agricultural practices and has environmental impacts [66,67,69]. This means that in countries with higher, per unit of area, gross agricultural production, for example, it was expected to find greater agricultural soil emissions and cropland nutrient flow. The regressions with panel data show that there are relevant signs that is the cropland nutrient flow per ha that impacts the agricultural soil emissions per ha and the gross agricultural production per ha. The governments and international organizations may have here important contributions to design policies that encourage adjusted soil management practices that maintain the soil nutrients balances and the agricultural production without compromise the sustainability.

8. Conclusions

There is a great heterogeneity between the countries across the world; however, the clusters found from the spatial autocorrelation analysis, for the food supply and soil nutrient balances, may be relevant findings to support common strategies that promote more sustainable practices. This is particularly important when there are relevant signs that the soil nutrient balances impact the farming production and the agricultural soil emissions. In fact, when the cropland nutrient flow increases 1 kg/ha, the agricultural soil emissions rise 41.279 kg/ha and the gross agricultural production rises 25.094 I$ per ha.

In terms of practical implications, the results obtained in this research highlight that the agricultural soil management is determinant to promote a soil nutrient balance able to maintain or increase the agricultural production to achieve the world demand for food and mitigate the agricultural soil emissions. In these contexts, it is suggested, in terms of policy recommendation, that the public, private, national, and international institutions design policies that mitigate the environmental impacts from the cropland nutrient flow.

For future research, the weak correlation between the food supply per capita, the agricultural production per ha and the soil nutrient flow per ha deserve special attention. In fact, despite the environmental impacts found for the agricultural production, this is not compensated by good indicators for food supply per capita.

Author Contributions: Conceptualization, V.J.P.D.M. and J.L.S.P.; methodology, V.J.P.D.M.; software, V.J.P.D.M.; validation, V.J.P.D.M., J.L.S.P. and J.M.G.; formal analysis, V.J.P.D.M., J.L.S.P. and J.M.G.; investigation, V.J.P.D.M., J.L.S.P. and J.M.G.; data curation, V.J.P.D.M.; writing—original draft preparation, V.J.P.D.M.; writing—review and editing, V.J.P.D.M., J.L.S.P. and J.M.G. All authors have read and agreed to the published version of the manuscript.

Funding: This work is funded by National Funds through the FCT—Foundation for Science and Technology, I.P., under the projects UIDB/00681/2020 and UIDB/04033/2020.

Institutional Review Board Statement: Not applicable.

Informed Consent Statement: Not applicable.

Data Availability Statement: The data that support the findings of this study are available from the corresponding author upon reasonable request.

Acknowledgments: Furthermore, we would like to thank the CERNAS Research Centre and the Polytechnic Institute of Viseu for their support.

Conflicts of Interest: The authors declare no conflict of interest.

References

1. Kone, B.; Diatta, S.; Sylvester, O.; Yoro, G.; Mameri, C.; Desire, D.D.; Ayemou, A. Estimating the inherent fertility of ferralsol using color. *Can. J. Soil Sci.* **2009**, *89*, 331–342. [CrossRef]
2. Barrios, E.; Trejo, M.T. Implications of Local Soil Knowledge for Integrated Soil Management in Latin America. *Geoderma* **2003**, *111*, 217–231. [CrossRef]
3. Cardelus, C.L.; Mekonnen, A.B.; Jensen, K.H.; Woods, C.L.; Baez, M.C.; Montufar, M.; Bazany, K.; Tsegay, B.A.; Scull, P.R.; Peck, W.H. Edge Effects and Human Disturbance Influence Soil Physical and Chemical Properties in Sacred Church Forests in Ethiopia. *Plant Soil* **2020**, *453*, 329–342. [CrossRef]
4. Carsky, R.J.; Oyewole, B.; Tian, G. Integrated Soil Management for the Savanna Zone of W. Africa: Legume Rotation and Fertilizer, N. *Nutr. Cycl. Agroecosyst.* **1999**, *55*, 95–105. [CrossRef]
5. Fujisao, K.; Khanthavong, P.; Oudthachit, S.; Matsumoto, N.; Homma, K.; Asai, H.; Shiraiwa, T. Impacts of the Continuous Maize Cultivation on Soil Properties in Sainyabuli Province, Laos. *Sci. Rep.* **2020**, *10*, 11231. [CrossRef]
6. Kintomo, A.A.; Akintoye, H.A.; Alasiri, K.O. Role of Legume Fallow in Intensified Vegetable-Based Systems. *Commun. Soil Sci. Plant Anal.* **2008**, *39*, 1261–1268. [CrossRef]
7. Santos, M.; Galindro, A.; Santos, C.; Marta-Costa, A.; Martinho, V. Sustainability Evolution of North and Alentejo Vineyard Regions. *Rev. Port. De Estud. Reg.* **2019**, *50*, 49–63.
8. Zhao, Z.; Gao, S.; Lu, C.; Li, X.; Li, F.; Wang, T. Effects of Different Tillage and Fertilization Management Practices on Soil Organic Carbon and Aggregates under the Rice-Wheat Rotation System. *Soil Tillage Res.* **2021**, *212*, 105071. [CrossRef]
9. Shaheen, A.; Naeem, M.A.; Jilani, G.; Shafiq, M. Integrated Soil Management in Eroded Land Augments the Crop Yield and Water-Use Efficiency. *Acta Agric. Scand. Sect. B Soil Plant Sci.* **2010**, *60*, 274–282. [CrossRef]
10. Zhao, Z.; Zheng, W.; Ma, Y.; Wang, X.; Li, Z.; Zhai, B.; Wang, Z. Responses of Soil Water, Nitrate and Yield of Apple Orchard to Integrated Soil Management in Loess Plateau, China. *Agric. Water Manag.* **2020**, *240*, 106325. [CrossRef]
11. Srivastava, A.K.; Singh, S. Citrus Decline: Soil Fertility and Plant Nutrition. *J. Plant Nutr.* **2009**, *32*, 197–245. [CrossRef]
12. Scamell, J.M. Healthy Land for Healthy Cattle. *Cattle Pract.* **2006**, *14*, 143–152.
13. Jayne, T.S.; Snapp, S.; Place, F.; Sitko, N. Sustainable Agricultural Intensification in an Era of Rural Transformation in Africa. *Glob. Food Secur. Agric. Policy* **2019**, *20*, 105–113. [CrossRef]
14. Martinho, V.J.P.D. Output Impacts of the Single Payment Scheme in Portugal: A Regression with Spatial Effects. *Outlook Agric.* **2015**, *44*, 109–118. [CrossRef]
15. Killham, K. Integrated Soil Management—Moving towards Globally Sustainable Agriculture. *J. Agric. Sci.* **2011**, *149*, 29–36. [CrossRef]
16. Takoutsing, B.; Weber, J.; Aynekulu, E.; Rodriguez Martin, J.A.; Shepherd, K.; Sila, A.; Tchoundjeu, Z.; Diby, L. Assessment of Soil Health Indicators for Sustainable Production of Maize in Smallholder Farming Systems in the Highlands of Cameroon. *Geoderma* **2016**, *276*, 64–73. [CrossRef]
17. Vanwalleghem, T.; Gomez, J.A.; Infante Amate, J.; Gonzalez de Molina, M.; Vanderlinden, K.; Guzman, G.; Laguna, A.; Giraldez, J.V. Impact of Historical Land Use and Soil Management Change on Soil Erosion and Agricultural Sustainability during the Anthropocene. *Anthropocene* **2017**, *17*, 13–29. [CrossRef]
18. Komatsuzaki, M.; Ohta, H. Soil Management Practices for Sustainable Agro-Ecosystems. *Sustain. Sci.* **2007**, *2*, 103–120. [CrossRef]
19. Xue, J.; Lyu, D.; Wang, D.; Wang, Y.; Yin, D.; Zhao, Z.; Mu, Z. Assessment of Soil Erosion Dynamics Using the GIS-Based RUSLE Model: A Case Study of Wangjiagou Watershed from the Three Gorges Reservoir Region, Southwestern China. *Water* **2018**, *10*, 1817. [CrossRef]

20. Alonso, A.; Froidevaux, M.; Javaux, M.; Laloy, E.; Mattern, S.; Roisin, C.; Vanclooster, M.; Bielders, C. A Hybrid Method for Characterizing Tillage-Induced Soil Physical Quality at the Profile Scale with Fine Spatial Details. *Soil Tillage Res.* **2022**, *216*, 105236. [CrossRef]
21. Weninger, T.; Kreiselmeier, J.; Chandrasekhar, P.; Julich, S.; Feger, K.-H.; Schwaerzel, K.; Bodner, G.; Schwen, A. Effects of Tillage Intensity on Pore System and Physical Quality of Silt-Textured Soils Detected by Multiple Methods. *Soil Res.* **2019**, *57*, 703–711. [CrossRef]
22. Fernandez-Romero, M.L.; Parras-Alcantara, L.; Lozano-Garcia, B.; Clark, J.M.; Collins, C.D. Soil Quality Assessment Based on Carbon Stratification Index in Different Olive Grove Management Practices in Mediterranean Areas. *Catena* **2016**, *137*, 449–458. [CrossRef]
23. Sun, H.Y.; Koai, P.; Gerl, G.; Schro, R.; Joergensen, R.G.; Munch, J.C. Water-Extractable Organic Matter and Its Fluorescence Fractions in Response to Minimum Tillage and Organic Farming in a Cambisol. *Chem. Biol. Technol. Agric.* **2017**, *4*, 15. [CrossRef]
24. Fiener, P.; Wilken, F.; Aldana-Jague, E.; Deumlich, D.; Gomez, J.A.; Guzman, G.; Hardy, R.A.; Quinton, J.N.; Sommer, M.; Van Oost, K.; et al. Uncertainties in Assessing Tillage Erosion—How Appropriate Are Our Measuring Techniques? *Geomorphology* **2018**, *304*, 214–225. [CrossRef]
25. Urbanek, E.; Hallett, P.; Feeney, D.; Horn, R. Water Repellency and Distribution of Hydrophilic and Hydrophobic Compounds in Soil Aggregates from Different Tillage Systems. *Geoderma* **2007**, *140*, 147–155. [CrossRef]
26. Schulten, H.; Leinweber, P. Influence of the Mineral Matrix on the Formation and Molecular Composition of Soil Organic-Matter in a Long-Term, Agricultural Experiment. *Biogeochemistry* **1993**, *22*, 1–22. [CrossRef]
27. Gravuer, K.; Scow, K.M. Invader-Resident Relatedness and Soil Management History Shape Patterns of Invasion of Compost Microbial Populations into Agricultural Soils. *Appl. Soil Ecol.* **2021**, *158*, 103795. [CrossRef]
28. Guemene, D.; Germain, K.; Aubert, C.; Bouvarel, I.; Cabaret, J.; Chapuis, H.; Corson, M.; Jondreville, C.; Juin, H.; Lessire, M.; et al. Organic poultry production in France: Status, bottlenecks, advantages and perspectives. *Prod. Anim.* **2009**, *22*, 161–178.
29. Hernandez, T.; Berlanga, J.G.; Tormos, I.; Garcia, C. Organic versus Inorganic Fertilizers: Response of Soil Properties and Crop Yield. *AIMS Geosci.* **2021**, *7*, 415–439. [CrossRef]
30. Le Guillou, C.; Prevost-Boure, N.C.; Karimi, B.; Akkal-Corfini, N.; Dequiedt, S.; Nowak, V.; Terrat, S.; Menasseri-Aubry, S.; Viaud, V.; Maron, P.-A.; et al. Tillage Intensity and Pasture in Rotation Effectively Shape Soil Microbial Communities at a Landscape Scale. *MicrobiologyOpen* **2019**, *8*, e676. [CrossRef]
31. Tan, H.; Barret, M.; Mooij, M.J.; Rice, O.; Morrissey, J.P.; Dobson, A.; Griffiths, B.; O'Gara, F. Long-Term Phosphorus Fertilisation Increased the Diversity of the Total Bacterial Community and the PhoD Phosphorus Mineraliser Group in Pasture Soils. *Biol. Fertil. Soils* **2013**, *49*, 661–672. [CrossRef]
32. Meena, R.S.; Kumar, S.; Datta, R.; Lal, R.; Vijayakumar, V.; Brtnicky, M.; Sharma, M.P.; Yadav, G.S.; Jhariya, M.K.; Jangir, C.K.; et al. Impact of Agrochemicals on Soil Microbiota and Management: A Review. *Land* **2020**, *9*, 34. [CrossRef]
33. Mo, F.; Yu, K.-L.; Crowther, T.W.; Wang, J.-Y.; Zhao, H.; Xiong, Y.-C.; Liao, Y.-C. How Plastic Mulching Affects Net Primary Productivity, Soil C Fluxes and Organic Carbon Balance in Dry Agroecosystems in China. *J. Clean Prod.* **2020**, *263*, 121470. [CrossRef]
34. Navarrete, A.A.; Kuramae, E.E.; de Hollander, M.; Pijl, A.S.; van Veen, J.A.; Tsai, S.M. Acidobacterial Community Responses to Agricultural Management of Soybean in Amazon Forest Soils. *FEMS Microbiol. Ecol.* **2013**, *83*, 607–621. [CrossRef]
35. Shan, A.; Pan, J.; Kang, K.J.; Pan, M.; Wang, G.; Wang, M.; He, Z.; Yang, X. Effects of Straw Return with N Fertilizer Reduction on Crop Yield, Plant Diseases and Pests and Potential Heavy Metal Risk in a Chinese Rice Paddy: A Field Study of 2 Consecutive Wheat-Rice Cycles. *Environ. Pollut.* **2021**, *288*, 117741. [CrossRef]
36. Thuerig, B.; Fliessbach, A.; Berger, N.; Fuchs, J.G.; Kraus, N.; Mahlberg, N.; Nietlispach, B.; Tamm, L. Re-Establishment of Suppressiveness to Soil- and Air-Borne Diseases by Re-Inoculation of Soil Microbial Communities. *Soil Biol. Biochem.* **2009**, *41*, 2153–2161. [CrossRef]
37. Val-Moraes, S.P.; de Macedo, H.S.; Kishi, L.T.; Pereira, R.M.; Navarrete, A.A.; Mendes, L.W.; de Figueiredo, E.B.; La Scala, N.; Tsai, S.M.; de Macedo Lemos, E.G.; et al. Liming in the Sugarcane Burnt System and the Green Harvest Practice Affect Soil Bacterial Community in Northeastern So Paulo, Brazil. *Antonie Van Leeuwenhoek* **2016**, *109*, 1643–1654. [CrossRef]
38. Valadares-Pereira, A. de A.; Alves Martins Oliveira, E.C.; Navarrete, A.A.; de Oliveira, W.P.; Tsai, S.M.; Peluzio, J.M.; de Morais, P.B. Fungal Community Structure as an Indicator of Soil Agricultural Management Effects in the Cerrado. *Rev. Bras. Cienc. Solo* **2017**, *41*, e0160489. [CrossRef]
39. Stahl, P.D.; Parkin, T.B.; Christensen, M. Fungal Presence in Paired Cultivated and Uncultivated Soils in Central Iowa, USA. *Biol. Fertil. Soils* **1999**, *29*, 92–97. [CrossRef]
40. Wipf, H.M.-L.; Xu, L.; Gao, C.; Spinner, H.B.; Taylor, J.; Lemaux, P.; Mitchell, J.; Coleman-Derr, D. Agricultural Soil Management Practices Differentially Shape the Bacterial and Fungal Microbiomes of Sorghum Bicolor. *Appl. Environ. Microbiol.* **2021**, *87*, e02345-20. [CrossRef]
41. Yakupoglu, T.; Rodrigo-Comino, J.; Cerda, A. Potential Benefits of Polymers in Soil Erosion Control for Agronomical Plans: A Laboratory Experiment. *Agronomy* **2019**, *9*, 276. [CrossRef]
42. Wang, Y.; Liu, L.; Tian, Y.; Wu, X.; Yang, J.; Luo, Y.; Li, H.; Awasthi, M.K.; Zhao, Z. Temporal and Spatial Variation of Soil Microorganisms and Nutrient under White Clover Cover. *Soil Tillage Res.* **2020**, *202*, 104666. [CrossRef]

43. Garger, E.K.; Paretzke, H.G.; Tschiersch, J. Measurement of Resuspended Aerosol in the Chernobyl Area Part III. Size Distribution and Dry Deposition Velocity of Radioactive Particles during Anthropogenic Enhanced Resuspension. *Radiat. Environ. Biophys.* **1998**, *37*, 201–208. [CrossRef]
44. Holman, I.P.; Hess, T.M.; Rose, S.C. A Broad-Scale Assessment of the Effect of Improved Soil Management on Catchment Baseflow Index. *Hydrol. Process.* **2011**, *25*, 2563–2572. [CrossRef]
45. Endo, T.; Sadahiro, Y.; Haruta, T.; Kitamura, Y.; Li, Z.; Li, P.; Honna, T. Soil Salinization Related to Soil Morphological and Physicochemical Characteristics in the Luohui Irrigation Scheme, China. *Arid Land Res. Manag.* **2012**, *26*, 122–136. [CrossRef]
46. Koch, A.; Chappell, A.; Eyres, M.; Scott, E. Monitor Soil Degradation or Triage for Soil Security? An Australian Challenge. *Sustainability* **2015**, *7*, 4870–4892. [CrossRef]
47. Tiefenbacher, A.; Sanden, T.; Haslmayr, H.-P.; Miloczki, J.; Wenzel, W.; Spiegel, H. Optimizing Carbon Sequestration in Croplands: A Synthesis. *Agronomy* **2021**, *11*, 882. [CrossRef]
48. Aznar-Sanchez, J.A.; Velasco-Munoz, J.F.; Lopez-Felices, B.; del Moral-Torres, F. Barriers and Facilitators for Adopting Sustainable Soil Management Practices in Mediterranean Olive Groves. *Agronomy* **2020**, *10*, 506. [CrossRef]
49. Weninger, T.; Kamptner, E.; Dostal, T.; Spiegel, A.; Strauss, P. Detection of Physical Hazards in Soil Profiles Using Quantitative Soil Physical Quality Assessment in the Pannonian Basin, Eastern Austria. *Int. Agrophys.* **2020**, *34*, 463–471. [CrossRef]
50. Lutz, F.; Herzfeld, T.; Heinke, J.; Rolinski, S.; Schaphoff, S.; von Bloh, W.; Stoorvogel, J.J.; Mueller, C. Simulating the Effect of Tillage Practices with the Global Ecosystem Model LPJmL (Version 5.0-Tillage). *Geosci. Model Dev.* **2019**, *12*, 2419–2440. [CrossRef]
51. Schillaci, C.; Perego, A.; Valkama, E.; Marker, M.; Saia, S.; Veronesi, F.; Lipani, A.; Lombardo, L.; Tadiello, T.; Gamper, H.A.; et al. New Pedotransfer Approaches to Predict Soil Bulk Density Using WoSIS Soil Data and Environmental Covariates in Mediterranean Agro-Ecosystems. *Sci. Total Environ.* **2021**, *780*, 146609. [CrossRef]
52. Papadopoulos, A.; Papadopoulos, F.; Tziachris, P.; Metaxa, I.; Iatrou, M. Site Specific Agricultural Soil Management with the Use of New Technologies. *Glob. Nest. J.* **2014**, *16*, 59–67.
53. Boafo, D.K.; Kraisornpornson, B.; Panphon, S.; Owusu, B.E.; Amaniampong, P.N. Effect of Organic Soil Amendments on Soil Quality in Oil Palm Production. *Appl. Soil Ecol.* **2020**, *147*, 103358. [CrossRef]
54. Castro, J.; Fernandez-Ondono, E.; Rodriguez, C.; Lallena, A.M.; Sierra, M.; Aguilar, J. Effects of Different Olive-Grove Management Systems on the Organic Carbon and Nitrogen Content of the Soil in Jaen (Spain). *Soil Tillage Res.* **2008**, *98*, 56–67. [CrossRef]
55. Frkova, Z.; Vystavna, Y.; Koubova, A.; Kotas, P.; Grabicova, K.; Grabic, R.; Kodesova, R.; Chronakova, A. Microbial Responses to Selected Pharmaceuticals in Agricultural Soils: Microcosm Study on the Roles of Soil, Treatment and Time. *Soil Biol. Biochem.* **2020**, *149*, 107924. [CrossRef]
56. Ma, T.; Zhou, W.; Chen, L.; Christie, P.; Luo, Y.; Wu, P. Phthalate Ester Contamination in Intensively Managed Greenhouse Facilities and the Assessment of Carcinogenic and Non-Carcinogenic Risk: A Regional Study. *Int. J. Environ. Res. Public Health* **2019**, *16*, 2818. [CrossRef]
57. Niu, L.; Xu, Y.; Xu, C.; Yun, L.; Liu, W. Status of Phthalate Esters Contamination in Agricultural Soils across China and Associated Health Risks. *Environ. Pollut.* **2014**, *195*, 16–23. [CrossRef]
58. Zhang, X.; Zhong, T.; Liu, L.; Ouyang, X. Impact of Soil Heavy Metal Pollution on Food Safety in China. *PLoS ONE* **2015**, *10*, e0135182. [CrossRef]
59. Niu, L.; Yang, F.; Xu, C.; Yang, H.; Liu, W. Status of Metal Accumulation in Farmland Soils across China: From Distribution to Risk Assessment. *Environ. Pollut.* **2013**, *176*, 55–62. [CrossRef]
60. Rehman, M.; Liu, L.; Wang, Q.; Saleem, M.H.; Bashir, S.; Ullah, S.; Peng, D. Copper Environmental Toxicology, Recent Advances, and Future Outlook: A Review. *Environ. Sci. Pollut. Res.* **2019**, *26*, 18003–18016. [CrossRef]
61. Porwollik, V.; Rolinski, S.; Heinke, J.; Mueller, C. Generating a Rule-Based Global Gridded Tillage Dataset. *Earth Syst. Sci. Data* **2019**, *11*, 823–843. [CrossRef]
62. Paustian, K.; Collier, S.; Baldock, J.; Burgess, R.; Creque, J.; DeLonge, M.; Dungait, J.; Ellert, B.; Frank, S.; Goddard, T.; et al. Quantifying Carbon for Agricultural Soil Management: From the Current Status toward a Global Soil Information System. *Carbon Manag.* **2019**, *10*, 567–587. [CrossRef]
63. Kebonye, N.M.; Eze, P.N.; Agyeman, P.C.; John, K.; Ahado, S.K. Efficiency of the T-Distribution Stochastic Neighbor Embedding Technique for Detailed Visualization and Modeling Interactions between Agricultural Soil Quality Indicators. *Biosyst. Eng.* **2021**, *210*, 282–298. [CrossRef]
64. Keshavarzi, A.; Tuffour, H.O.; Oppong, J.C.; Zeraatpisheh, M.; Kumar, V. Dealing with Soil Organic Carbon Modeling: Some Insights from an Agro-Ecosystem in Northeast Iran. *Earth Sci. Inform.* **2021**, *14*, 1833–1845. [CrossRef]
65. Smith, J.; Smith, P.; Wattenbach, M.; Gottschalk, P.; Romanenkov, V.A.; Shevtsova, L.K.; Sirotenko, O.D.; Rukhovich, D.I.; Koroleva, P.V.; Romanenko, I.A.; et al. Projected Changes in the Organic Carbon Stocks of Cropland Mineral Soils of European Russia and the Ukraine, 1990–2070. *Glob. Chang. Biol.* **2007**, *13*, 342–356. [CrossRef]
66. Eranki, P.L.; Devkota, J.; Landis, A.E. Carbon Footprint of Corn-Soy-Oats Rotations in the US Midwest Using Data from Real Biological Farm Management Practices. *J. Clean Prod.* **2019**, *210*, 170–180. [CrossRef]
67. Mummey, D.L.; Smith, J.L.; Bluhm, G. Assessment of Alternative Soil Management Practices on N_2O Emissions from US Agriculture. *Agric. Ecosyst. Environ.* **1998**, *70*, 79–87. [CrossRef]
68. Oertel, C.; Matschullat, J.; Zurba, K.; Zimmermann, F.; Erasmi, S. Greenhouse Gas Emissions from Soils—A Review. *Geochemistry* **2016**, *76*, 327–352. [CrossRef]

69. Hansen, E.M.O.; Hauggaard-Nielsen, H.; Justes, E.; Ambus, P.; Mikkelsen, T.N. The Influence of Grain Legume and Tillage Strategies on CO_2 and N_2O Gas Exchange under Varied Environmental Conditions. *Agriculture* **2021**, *11*, 464. [CrossRef]
70. Paul, C.; Kuhn, K.; Steinhoff-Knopp, B.; Weisshuhn, P.; Helming, K. Towards a Standardization of Soil-Related Ecosystem Service Assessments. *Eur. J. Soil Sci.* **2021**, *72*, 1543–1558. [CrossRef]
71. Thiele-Bruhn, S.; Bloem, J.; de Vries, F.T.; Kalbitz, K.; Wagg, C. Linking Soil Biodiversity and Agricultural Soil Management. *Curr. Opin. Environ. Sustain.* **2012**, *4*, 523–528. [CrossRef]
72. Schulten, H.; Hempfling, R. Influence of Agricultural Soil-Management on Humus Composition and Dynamics—Classical and Modern Analytical Techniques. *Plant Soil* **1992**, *142*, 259–271. [CrossRef]
73. Laghari, M.; Naidu, R.; Xiao, B.; Hu, Z.; Mirjat, M.S.; Hu, M.; Kandhro, M.N.; Chen, Z.; Guo, D.; Jogi, Q.; et al. Recent Developments in Biochar as an Effective Tool for Agricultural Soil Management: A Review. *J. Sci. Food Agric.* **2016**, *96*, 4840–4849. [CrossRef] [PubMed]
74. Liu, L.; Huang, X.; Zhao, J.; Zhang, J.; Cai, Z. Characterizing the Key Agents in a Disease-Suppressed Soil Managed by Reductive Soil Disinfestation. *Appl. Environ. Microbiol.* **2019**, *85*, e02992-18. [CrossRef]
75. Wiesmeier, M.; Mayer, S.; Burmeister, J.; Huebner, R.; Koegel-Knabner, I. Feasibility of the 4 per 1000 Initiative in Bavaria: A Reality Check of Agricultural Soil Management and Carbon Sequestration Scenarios. *Geoderma* **2020**, *369*, 114333. [CrossRef]
76. Gustavo Belduma Belduma, R.; Barrezueta-Unda, S.; Vargas Gonzales, O.; Sanchez Romero, O. Management and Use of Agricultural Land in the Rural Area of the Canton Chilla from a Socioeconomic Perspective. *Rev. Univ. Soc.* **2020**, *12*, 299–306.
77. Prager, K.; Schuler, J.; Helming, K.; Zander, P.; Ratinger, T.; Hagedorn, K. Soil Degradation, Farming Practices, Institutions and Policy Responses: An Analytical Framework. *Land Degrad. Dev.* **2011**, *22*, 32–46. [CrossRef]
78. Van der Ploeg, R.R.; Gieska, M.; Schweigert, P. Impact of postwar agricultural soil management on surface hydrology and river peak discharge. *Ber. Landwirtsch.* **2001**, *79*, 447–465.
79. Paul, C.; Techen, A.-K.; Robinson, J.S.; Helming, K. Rebound Effects in Agricultural Land and Soil Management: Review and Analytical Framework. *J. Clean Prod.* **2019**, *227*, 1054–1067. [CrossRef]
80. FAOSTAT Several Statistics. Available online: https://www.fao.org/faostat/en/#home (accessed on 8 January 2022).
81. World Bank Several Statistics and Information. Available online: https://www.worldbank.org/en/home (accessed on 16 January 2022).
82. Anselin, L.; Syabri, I.; Kho, Y. GeoDa: An Introduction to Spatial Data Analysis. *Geogr. Anal.* **2006**, *38*, 5–22. [CrossRef]
83. GeoDa. GeoDa Software. Available online: https://geodacenter.github.io/ (accessed on 8 January 2022).
84. Moran, P.A.P. Notes on Continuous Stochastic Phenomena. *Biometrika* **1950**, *37*, 17–23. [CrossRef] [PubMed]
85. Eurostat Countries Shapefiles. Available online: https://ec.europa.eu/eurostat/web/gisco/geodata/reference-data/administrative-units-statistical-units/countries (accessed on 8 January 2022).
86. QGIS.org. QGIS Geographic Information System. QGIS Association. 2022. Available online: https://qgis.org/en/site/ (accessed on 8 January 2022).
87. Vincent, J.L. *Factor Analysis in International Relations: Interpretation, Problem, Areas, and an Application*; Univ of Florida Press: Gainesville, FL, USA, 1971; ISBN 978-0-8130-0315-3.
88. Torres-Reyna, O. Getting Started in Factor Analysis (Using Stata) (Ver. 1.0 Beta/Draft) n.d. Available online: https://www.princeton.edu/~{}otorres/Factor.pdf (accessed on 8 January 2022).
89. StataCorp. *Stata 15 Base Reference Manual*; Stata Press: College Station, TX, USA, 2017.
90. StataCorp. *Stata Statistical Software: Release 15*; StataCorp LLC: College Station, TX, USA, 2017.
91. Stata Stata: Software for Statistics and Data Science. Available online: https://www.stata.com/ (accessed on 8 January 2022).
92. Baltagi, B.H. *Econometric Analysis of Panel Data*, 4th ed.; Wiley: Chichester, UK; Hoboken, NJ, USA, 2008; ISBN 978-0-470-51886-1.
93. Torres-Reyna, O. *Panel Data Analysis Fixed and Random Effects Using Stata (v. 4.2)*; Priceton University: Princeton, NJ, USA, 2007.
94. Spearman, C. The Proof and Measurement of Association between Two Things. *Am. J. Psychol.* **1904**, *15*, 72–101. [CrossRef]
95. Granger, C.W.J. Investigating Causal Relations by Econometric Models and Cross-Spectral Methods. *Econometrica* **1969**, *37*, 424–438. [CrossRef]

soil systems

MDPI

Article

Impacts of Landfill Leachate on the Surrounding Environment: A Case Study on Amin Bazar Landfill, Dhaka (Bangladesh)

Rysul Hasan Hredoy [1], Md. Abu Bakar Siddique [2], Md. Ahedul Akbor [2,*], Md. Aftab Ali Shaikh [2,3] and Md. Mostafizur Rahman [1,4,*]

1 Department of Environmental Science, Bangladesh University of Professionals, Mirpur Cantonment, Dhaka 1216, Bangladesh
2 Institute of National Analytical Research and Service (INARS), Bangladesh Council of Scientific and Industrial Research (BCSIR), Dhanmondi, Dhaka 1205, Bangladesh
3 Department of Chemistry, University of Dhaka, Dhaka 1000, Bangladesh
4 Laboratory of Environmental Health and Ecotoxicology, Department of Environmental Sciences, Jahangirnagar University, Dhaka 1342, Bangladesh
* Correspondence: akborbcsir@gmail.com (M.A.A.); rahmanmm@juniv.edu (M.M.R.)

Abstract: Currently, a total of about 15,000 tons/day of waste is generated in the entire Dhaka city with an average per capita waste generation of 0.641 kg/day. Only 37% of this waste is collected and dumped into the two sanitary landfill sites, which is the only waste management system in Dhaka. To investigate the impact of landfill leachate of Amin Bazar landfill on the environmental compartments, a total of 14 composite samples (two leachates, three surface water, three groundwater, three soil, and three plants) were collected and analyzed for physicochemical parameters and heavy metal(loid)s concentration. Based on the result of physicochemical parameters, all results were found higher in the leachate samples than the permissible limit. The heavy metal(loid)s in leachate samples have a value of high levels of contamination. Surface water, groundwater, soil, and vegetation are all polluted as a result of high levels of metal contamination. Although the Water Quality Index values of the samples based on heavy metal(loid)s concentrations were within the acceptable range, heavy metal concentrations in the soil and plants were quite high. The concentrations of lead (Pb—8 mg/kg), cadmium (Cd—0.4 mg/kg), chromium (Cr—2.26 mg/kg), and cobalt (Co—1.72 mg/kg) in all plant samples were found to be higher than the allowable limit. The individual concentration of arsenic (As—0.021 mg/L) in the leachate was higher than the maximum allowed limit. Inverse Distance Weighted analysis through ArcGIS showed that landfill leachate has the maximum probability of contaminating the surrounding environment with heavy metal(loid)s. Results showed that samples collected near the landfill have higher concentrations of heavy metal(loid)s than others, which establishes the contribution of landfill leachate in contaminating the environment with heavy metal(loid)s. The improper leachate management of landfill has a high impact on the environment.

Keywords: landfill leachate; heavy metal pollution; surface water; groundwater; soil; plants

1. Introduction

Landfill leachate is the liquid residuals of a landfill resulting from a combination of the physical, chemical, and biological processes that transfer pollutants from the waste materials [1]. Landfill leachate pollution creates alarming stress for developing countries due to rapid and improper urbanization and industrialization.

The capital of Bangladesh, Dhaka, is one of the most populous cities in the world with a total of about 15,000 tons per day of waste generation, and it is increasing rapidly [2,3]. The average per capita waste generation of Dhaka North City Corporation (DNCC) is 0.641 Kg/day [4]. Only 37% of this total waste is collected and dumped into the Amin Bazar and Matuail sanitary landfill sites, which is the only solid waste management system currently running in Dhaka [5]. The waste management system of Dhaka city had been working on its improvement with the technical aid of the Japan International Cooperation Agency since 2000 [6]. Leachate collection and gas venting systems, improved surface drainage, daily covering of waste disposal facilities, slope reformation, working roadways, weighbridge operation, and car washing facilities are all included in the Matuail and Amin Bazar sanitary landfill project [7]. Waste dumped in this process undergoes slow anaerobic decomposition for 30 to 50 years, which produces a considerable amount of leachate along with heavy metals and the hazardous chemical compounds, which can seep from the landfill and contaminate the nearby water body along with the groundwater through percolation, soil, and plants through bioaccumulation [8].

Landfill leachate is an unavoidable substance, and the management of leachate is one of the major difficulties [9]. Leachate has a negative impact on groundwater, surface water, soil, and plants [10,11]. Landfill leachate alters the physicochemical parameters and heavy metal concentration in surface water, groundwater, soil, and plants [8,12]. Therefore, faulty management of landfill leachate could have an adverse effect on the environment as well as human health. Water samples with lower depth (30 ft) and distance (1 km) from the landfill had greater concentrations of chemical oxygen demand (COD), chloride (Cl-), sodium (Na), and potassium (K) which were (128 mg/L), (115 mg/L), (98 mg/L), and (42.2 mg/L), respectively, in the Chandigarh, Mohali, and Panchkula landfill sites in India [10]. Leachate from the Matuail landfill site has a high concentration of total dissolved solids (734 mg/L), COD (1631 mg/L), ammonium (1253 mg/L), hydrogen carbonate (27,962 mg/L), and some heavy metals such as Ni (1.05 mg/L) and Cr (0.74 mg/L), and it has a significant potential for polluting groundwater and surface water [8]. Haque et al. [5] conducted a study on the Aminbazar landfill area on seasonal effects on heavy metal concentration in leachate and converted soil, which is our study area. Kamal et al. [2] conducted a study in the same area on the bioaccumulation of trace metals in plants. Both of these studies found high concentrations of heavy metals in soil and plants which are considered to be polluted.

Altering the natural quality of soil, surface water, and groundwater has a major impact on the environment as well as on human health [9,13,14]. Therefore, leachate pollution is an alarming concern for the environment and has become one of the major concerns for the current age. Leachate is an unavoidable substance of a landfill, but its impact on the environment can be avoided. The assessment of possible risks from the landfill leachate is crucial for sustainable environmental management [15]. The study on all four components of the environment (surface water, groundwater, soil, and plants) is essential to understand the impact of landfill leachate on the environment. Thus, this study aims to investigate the environmental impact occurring from the landfill leachate by analyzing some physical and chemical parameters such as electrical conductivity (EC), total dissolved solids (TDS), dissolved oxygen (DO), pH, turbidity, salinity, temperature, total hardness, lead (Pb), nickel (Ni), chromium (Cr), cadmium (Cd), cobalt (Co), and arsenic (As) in leachate, surface water (SW), groundwater (GW), surrounding soil, and plants. To determine the contribution of landfill leachate in heavy metal(loid)s concentration in soil, water, and plants, Inverse Distance Weighted (IDW) analysis in GIS was performed.

2. Materials and Methods

2.1. Study Area

Our study area, the sanitary landfill at Amin Bazar, is located at Savar Upazilla in Dhaka, Bangladesh, near the low-lying floodplain of the Karnatali River. The landfill lies between the latitude 23°48′0.86″ and 23°47′44.33″ N, and longitude 90°17′51.03″ and 90°18′12.03″ E. Since 2007, the location has been used as a dumpsite, with the first phase of its operation being an open dumpsite (Figure 1). Currently, this facility is a semi-aerobic sanitary landfill site with a total size of roughly 20.23 hectares that operates with the fast breakdown of wastes. Between June and October of the year, the region is submerged during every rainstorm [16].

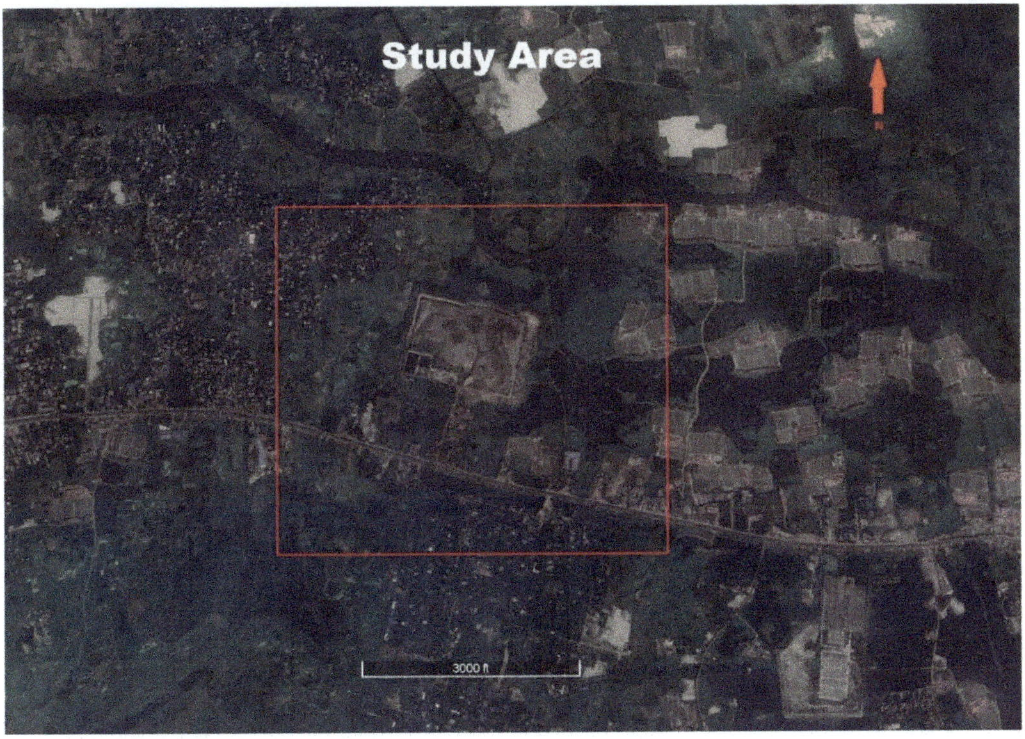

Figure 1. Satellite map of the study area.

2.2. Sample Collection

To collect the samples, the study area was selected within a 1 km of radius from the center of the landfill. A total of 42 samples were collected from the landfill site of Amin Bazar and the surrounding area within a radius of 1 Km following Ahsan et al. [17] and Siddique et al. [18]. A total of 5 types of samples were collected from the study area which includes 2 leachate samples, 3 surface water samples, 3 groundwater samples, 3 soil samples, and 3 plant samples. The details of the collected plant samples are given in Table 1.

For each type of sample, triplicate samples (n = 3) were collected to obtain a total of 42 samples (14 × 3 = 42). However, the collected triplicate samples from each type of sample of the study area were mixed together and homogenized well to obtain a total of 14 composite samples (42/3 = 14) from the study area. The samples were collected randomly from different locations within the study region, which can be addressed as cluster area sampling (Figure 2).

2.3. Sample Preparation

For the preparation of leachate, surface water, and groundwater samples, 50 mL of each liquid sample was taken into a 250 mL beaker, and 2–3 mL of concentrated nitric acid (HNO_3, 65%) was added into it [17,18]. The mixture was then heated at 90 °C in a hot plate until the volume of the solution reach around 3 to 5 mL after evaporation. After that, the solution was cooled and filtered into a volumetric flask with Whatman 42 filter paper rinsing the sample beaker with deionized water to make the final solution volume of 50 mL. The collected soil and plant samples were dried at 60 °C in an oven for 24 h. After that, samples were grounded to powder using a hand grinder and stored in a zip-lock plastic bag. The digestion of the soil sample was performed following Siddique et al. [18]

and Hasan et al. [19]. About 5 gm of powdered sample was taken into a 250 mL beaker and 10 mL of conc. HNO_3 and 5 mL of conc. $HClO_4$ was added to it. The mixture was then heated in a hot plate at 90 °C for 3–4 h. The solution becomes almost transparent, and then, it was cooled and filtered into a 50 mL volumetric flask as mentioned above. The plant samples were prepared following Nasrin et al. [20]. In summary, a muffle furnace was used to gradually heat 5 g of powdered plant sample to 600 °C, and that temperature was maintained for 6 h. Then, the ash sample was treated with concentrated HNO_3 and $HClO_4$ (ratio: 2:1) and boiled on a hot plate at approximately 150 °C to produce a colorless clear solution after cooling. In a volumetric flask, the solution was prepared to a final volume of 50 mL using deionized water. Until analysis in the lab, the entire prepared sample solution was stored in plastic bottles at 4 °C.

Table 1. Details of collected samples.

Sample	Sample Type	Location		Details of Sample
		Latitude	Longitude	
01	Untreated leachate	23°47′51.842″	90°17′52.571″	Collected from the leachate pond.
02	Treated leachate	23°47′51.842″	90°17′52.576″	Collected from the leachate pond.
03-01	Surface water	23°47′39.361″	90°17′47.176″	Collected from the open lake exposed to the landfill at a depth of 50 cm.
03-02	Surface water	23°47′34.842″	90°18′01.633″	Collected from a pond at a depth of 50 cm.
03-03	Surface water	23°47′33.541″	90°18′13.691″	Collected from a pond at a depth of 50 cm.
04-01	Groundwater	23°47′39.947″	90°17′44.797″	At a depth of 70 ft (21.3 m).
04-02	Groundwater	23°47′33.844″	90°18′07.443″	At a depth of 280 ft (85.3 m).
04-03	Groundwater	23°47′42.151″	90°18′15.884″	At a depth of 400 ft (121.9 m).
05-01	Soil	23°47′39.947″	90°17′44.797″	Collected from 10 cm beneath the surface.
05-02	Soil	23°47′33.547″	90°18′13.691″	Collected from 10 cm beneath the surface.
05-03	Soil	23°47′42.153″	90°18′15.882″	Collected from 10 cm beneath the surface.
06-01	Plant	23°47′39.947″	90°17′44.797″	1. *Artocarpus heterophyllus* (Jackfruit), collected raw leaves from the plant. 2. *Carica papaya* (Papaya plant), collected raw leaves from the plant. 3. *Musa acuminate* (Banana plant), collected raw leaves from the plant.
06-02	Plant	23°47′34.845″	90°18′01.631″	1. *Ipomoea aquatica* (Water spinach) collected raw leaves from the plants. 2. *Ocimum tenuiflorum* (Tulsi) collected raw leaves from the plants.
06-03	Plant	23°47′33.545″	90°18′13.694″	1. *Ocimum tenuiflorum* (Tulsi) collected raw leaves from the plants.

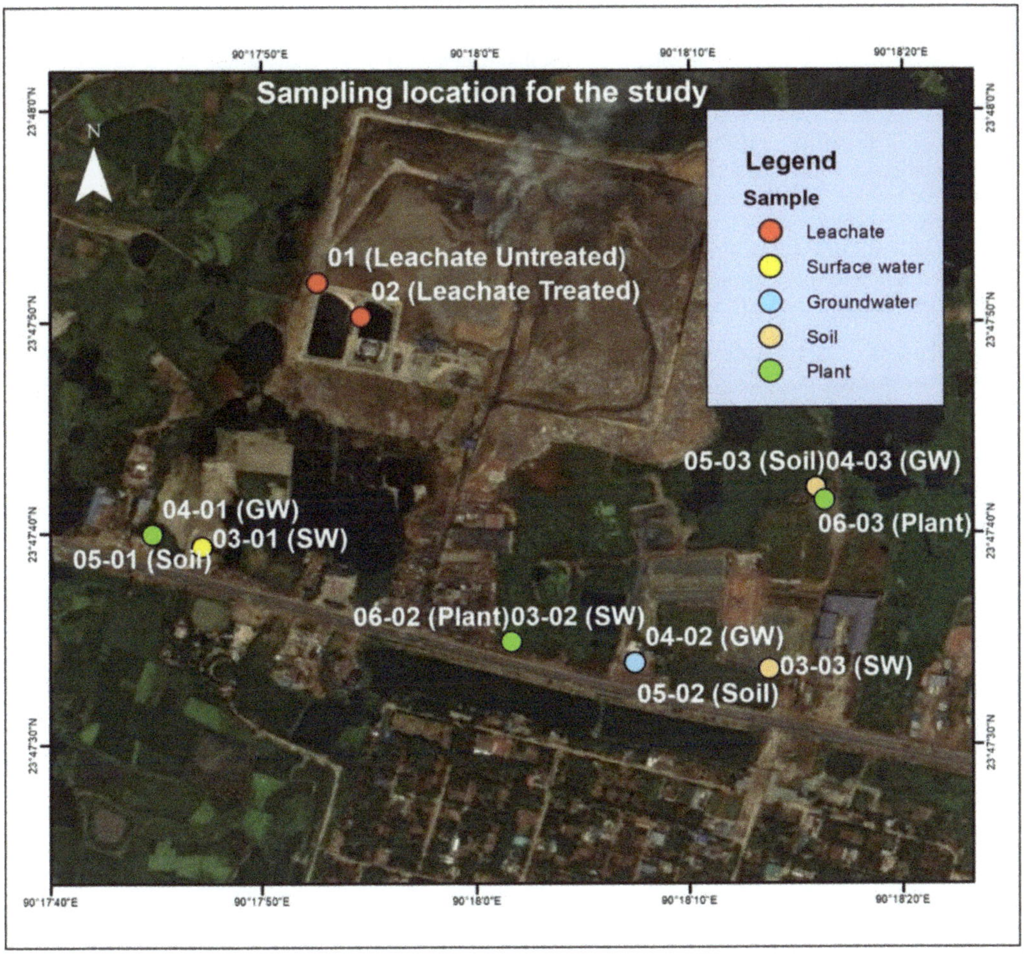

Figure 2. Sampling locations of the study area.

2.4. Sample Analysis

All samples were analyzed following the methods described by American Public Health Association [21] along with the in-house laboratory methods. Physiochemical parameters of water samples such as temperature, electrical conductivity (EC), total dissolved solids (TDS), pH, salinity, and dissolved oxygen (DO) were measured in the field using a multi-parameter meter (Hanna HI-9829). Total hardness (TH) was measured by the conventional titration method. The concentrations of several trace metal(loid)s such as Co, Cd, Ni, Pb, As, and Cr was measured using an Atomic Absorption Spectrophotometer (AAS) (Model: AA240FS, Varian, Agilent, Victoria, Australia; Software: SpectrAA version 5.1). The details of the analytical procedure and quality control to produce reliable data in AAS are reported earlier [17–19]. In brief, the concentration of trace metal(loid)s was measured against a prepared calibration curve using certified reference materials (CRM, Fluka Analytical, Sigma-Aldrich, Darmstadt, Germany) for individual elements. The accuracy and precision of the analysis were checked through triplicate analysis of the CRM and samples. To ensure further analytical quality, the CRM, method blank, and sample blank were analyzed sequentially. The detection limit for the analyzed trace metal(loid)s viz., Co, Cd, Ni, Pb, As, and Cr were 2.13, 0.29, 2.17, 1.20, 0.16, and 0.41 ppb, respectively. The spike recovery in the analysis was within 94–104%. All samples, standards, and blanks were measured three times, and mean results were taken into consideration.

2.5. Statistical Analysis and Spatial Distribution of Metal(loid)s

The results of metal(loid)s concentrations were analyzed statistically for principal component and Pearson's correlation analysis using SPSS software (version 25) to identify the source of pollutants through their associations. Inverse Distance Weighted (IDW) through ArcGIS has been performed to identify the spatial distribution of the elemental concentration [22]. IDW is an interpolation tool of ArcGIS which is used for a better understanding of the surface grid and predicting the values of cells at locations that lack sampled points. ArcMap 10.3 has been used to analyze the spatial distribution.

2.6. Indices for Water Quality, Pollution, and Feasibility Assessment

There are several indices that are widely used in different water research (surface water, groundwater, drinking water) to assess the water quality, water suitability, and pollution degree to various extents [23–27]. In this work, the water quality index (WQI) [28], degree of contamination (CD) [29], heavy metals evaluation index (HEI) [30], and heavy metal pollution index (HPI) [31,32] were used to evaluate the water quality and the level of water pollution in the study area. The indices were calculated using the following equations.

$$\mathrm{WQI} = \sum\left[\left(\frac{\mathrm{W}}{\sum \mathrm{Wi}}\right) \times \left(\frac{\mathrm{Ci}}{\mathrm{Si}} \times 100\right)\right] \quad (1)$$

$$\mathrm{CD} = \sum\nolimits_{i=1}^{n} \mathrm{Cn} \quad (2)$$

$$\mathrm{HEI} = \sum\nolimits_{i=1}^{n} \frac{\mathrm{Mi}}{\mathrm{Si}} \quad (3)$$

$$\mathrm{HPI} = \frac{\sum \mathrm{WiQi}}{\sum \mathrm{Wi}} \quad (4)$$

Here, assigned weights were according to their relative significance (W), relative weight (Wi), the concentration of each variable (Ci), standard values (Si), contamination factor (Cn), measured value (Mi), and Sub-index (Qi). The required parameters for the calculation methods of various indices are given in Table 2.

Table 2. Several standard values used to calculate different indices of water.

	MAC (μg/L)	CNi (μg/L)	Ii (μg/L)	Si (μg/L)	Wi (μg/L)	References
Pb	10	1.5	10	50	0.02	[33,34]
Cd	3	3	3	5	0.20	[34,35]
Cr	50	50	50	100	0.01	[34,36]
Co	1000	1000	50	100	0.01	[34,36]
Ni	70	20	20	70	0.01	[33,37,38]
As	10	10	10	50	0.02	[33,35]

Note. Maximum admissible concentration (MAC), Upper permissible concentration (CNi), Ideal concentration value for the ith parameter (Ii), Standard concentration value for the ith parameter (Si), and Unit weight (Wi).

3. Results and Discussion

3.1. Physicochemical Characterization of Leachate and Water Samples

The results of the physicochemical parameters of leachate, surface water, and groundwater samples of the study area are summarized in Table 3. The pH values of all samples are found within the standard limit. The mean pH value found in the leachate sample was 7.85, which refers to a mature landfill leachate based on the study of Tchobanoglous et al. (1993) [39]. The study showed that the pH value for new landfills normally varies from 4.5 to 7.5, and for mature landfills, it varies from 6.6 to 7.5. It is important to note that landfill leachate may raise the pH of drinking water and may help in producing trihalomethane (THM), which is a chemical that is hazardous to humans [40]. The DO level in leachate samples is found to be very low, suggesting a significant link with EC, which also shown to be at high levels in landfill leachates, reflecting a high presence of inorganic components [13]. However, in the water samples, the level of DO is found within the standard limit. The mean turbidity of groundwater samples was found far lower than the leachate and surface water, which is plausible. The salinity of the leachate sample was extremely higher than the surface water and groundwater. This highly saline leachate can contribute to increasing the salinity of surface water and groundwater. This is corroborated by the discharge of domestic waste in landfill. The mean TDS and EC values of the leachate samples are found to be much higher than the water samples for which the values are within the standard limit. Extremely high conductivity values are caused by an abundance of cations and anions. The intensity and overall pollutant load of the leachate are further reflected in the total mineral content. The leachate contains salt because it contains potassium, sodium, chloride, nitrate, sulfate, ammonia, and other chemicals [13]. The mean temperature of leachate, surface water, and groundwater was 33.12, 28.87, and 28.8 °C, which was higher than the recommended value of EU. The mean values of the total hardness of leachate are also found to be much higher than the surface and groundwater. The total hardness of all the samples was found to be more than the permissible limit recommended by the ECR and WHO [35,38].

Table 3. Physicochemical parameters of leachate and water samples.

Sample ID	Sample Type	pH	DO (mg/L)	Turbidity (FNU)	Salinity (mg/L)	TDS (mg/L)	EC (μS/cm)	Temperature (°C)	TH (mg/L)
01	Leachate	7.85	1.47	6.58	10.4	5847	11,694	33.87	2805.6
02	Leachate	7.85	1.26	3.19	19.9	2980	5960	32.38	1132.3
	Mean	7.85	1.36	4.88	15.2	4413.5	8827	33.12	1968.9
	SD	0.00	0.15	2.40	6.72	2027.3	4054.6	1.05	836.7
03-01	SW	7.85	15.5	0.37	19.7	385	770	31.38	501
03-02	SW	7.05	3.50	7.30	0.22	231	458	27.63	498
03-03	SW	7.11	4.20	3.40	0.17	182	370	27.62	492
	Mean	7.33	7.73	3.69	6.70	266	532.7	28.87	497
	SD	0.44	5.49	3.47	11.3	105.9	210.2	2.16	3.74
04-01	GW	7.38	15.3	0.14	1.01	149	297	30.85	330
04-02	GW	7.30	4.50	0.40	0.16	171	342	27.74	288
04-03	GW	7.08	4.60	2.10	0.31	326	653	27.82	316
	Mean	7.25	8.13	0.88	0.49	215.3	430.7	28.80	311.5
	SD	0.15	5.06	1.06	0.45	96.47	193.9	1.77	17.6
ECR, 1997 [28]		6.5–8.5	6	10	-	1000	350	20–30	-
WHO, 2017 [30]		6.5–8.0	4–6	5	-	500	250	-	-

Note. SD = Standard Deviation.

3.2. Concentration of Metal(loid)s in Leachate, Water, Soil, and Plants Samples

The results of heavy metal(loid)s concentration of leachate, surface water, and groundwater samples of the study area are summarized in Supplementary Table S1. The mean concentration of Pb in leachate, surface water, groundwater, soil, and plants were 0.05 mg/L, 0.01 mg/L, 0.003 mg/L, 16 mg/kg, and 8 mg/kg, respectively. The concentration of Pb found in previous studies on various landfills' leachate was ranging from 0.01 to 0.45 mg/L [41]. Alam et al. (2020) detected a high concentration of Pb in surface water, ground water and soil around a landfill site in Sylhet, Bangladesh [42]. Although the usefulness of lead in human physiology is unknown [43], prolonged exposure at high levels could harm vital organs and systems such as the nervous, digestive, hematopoietic, cardiovascular, reproductive, and immune systems, as well as the skeleton and kidneys [44]. In contrast, the mean concentration of Cd was 0.003 mg/L, 0.002 mg/L, 0.002 mg/L, 0.11 mg/kg, and 0.4 mg/kg, respectively. The concentration of Pb and Cd in plants was higher than the maximum permissible limit set by the WHO but in the water and soil, the concentration was within the standard limits. The mean concentration of Cr in leachate, surface water, groundwater, soil, and plants were 0.0179 mg/L, 0.044 mg/L, 0.0052 mg/L, 47.73 mg/kg, and 2.26 mg/kg, respectively. These trace elements are considered as potentially harmful pollutants. Because they may make strong metallic bonds with several functional macromolecules at once, leading to clump development, they can interfere with a cell's basic functioning in a biological system. Pb is harmful even at low doses and can induce anemia, brain damage, anorexia, mental deficit, vomiting, and even death in people [13]. The Co concentrations in all these samples were 0.057 mg/L, 0.022 mg/L, 0.0096 mg/L, 9.808 mg/kg, and 1.725 mg/kg, respectively. The concentration of Cr and Co were higher than the maximum permissible limit in plant samples set by the WHO, but in the water and soil samples, the concentration was within the limits. The Cr concentration detected in previous studies on various landfills ranged from 0.005 to 2 mg/L [41]. Since Cr does not significantly affect plant metabolism, growth, or productivity, it is found that Cr accumulation in plants is highly toxic [45]. Long-term Cr accumulation in the soil lowers agricultural production and crop quality [46]. The mean concentration of Ni in leachate, surface water, groundwater, soil, and plants was 0.16 mg/L, 0.053 mg/L, 0.0463 mg/L, 28.98 mg/kg, and 4.76 mg/kg, respectively. In contrast, the concentration of As in all these samples was 0.0129 mg/L, 0.0047 mg/L, 0.0025 mg/L, 1.6 mg/kg, and 0.382 mg/kg, respectively. The leaching of As, Cr, and Cu from wood wastes such as building and demolition projects, utility poles, furniture, landscape structures, and wood products industries, which is often treated with chromated copper arsenate (CCA) preservatives, may result in higher metal levels in wood [47]. The Ni concentration in leachate was above the maximum permissible limit, but in the rest of the samples, the concentration was within the permissible limit. On the other hand, the concentrations of As were slightly above the permissible limit in the leachate. The concentration of As in treated leachate was found higher than in untreated leachate. This may be occurring because of the aerobic treatment of leachate, which may result in oxidative dissolution [48]. The concentration of all the analyzed heavy metal(loid)s has been graphically represented in Figure 3.

3.3. Principal Component Analysis

The principal components that have the maximum probability of polluting the environment by being present were analyzed by principal component analysis (PCA). The PCA extracted two controlling factors from the analytical data set of heavy metal(loid)s concentration with eigenvalues >1 (Figure 4). The extracted two factors contain about 95% of the total variance, which is explained by whole factors. Component 1 (PC1) comprises about 78% of the total variance with strong positive loadings of the factors due to lead (Pb) only. PC2 accounted for 17% of the total variance, which represented the strong positive loading of Cadmium (Cd). Therefore, analyzing these two components will be enough to acquire 95% of all these metals' analyses.

3.4. Pearson's Correlation Analysis

According to the correlation matrix of physicochemical parameters of leachate, surface water, and groundwater in Table 4, the pH has the maximum positive proportional relation with all the other parameters, except for turbidity. In contrast, DO has an inversely proportional relation among turbidity, TDS, EC, and TH.

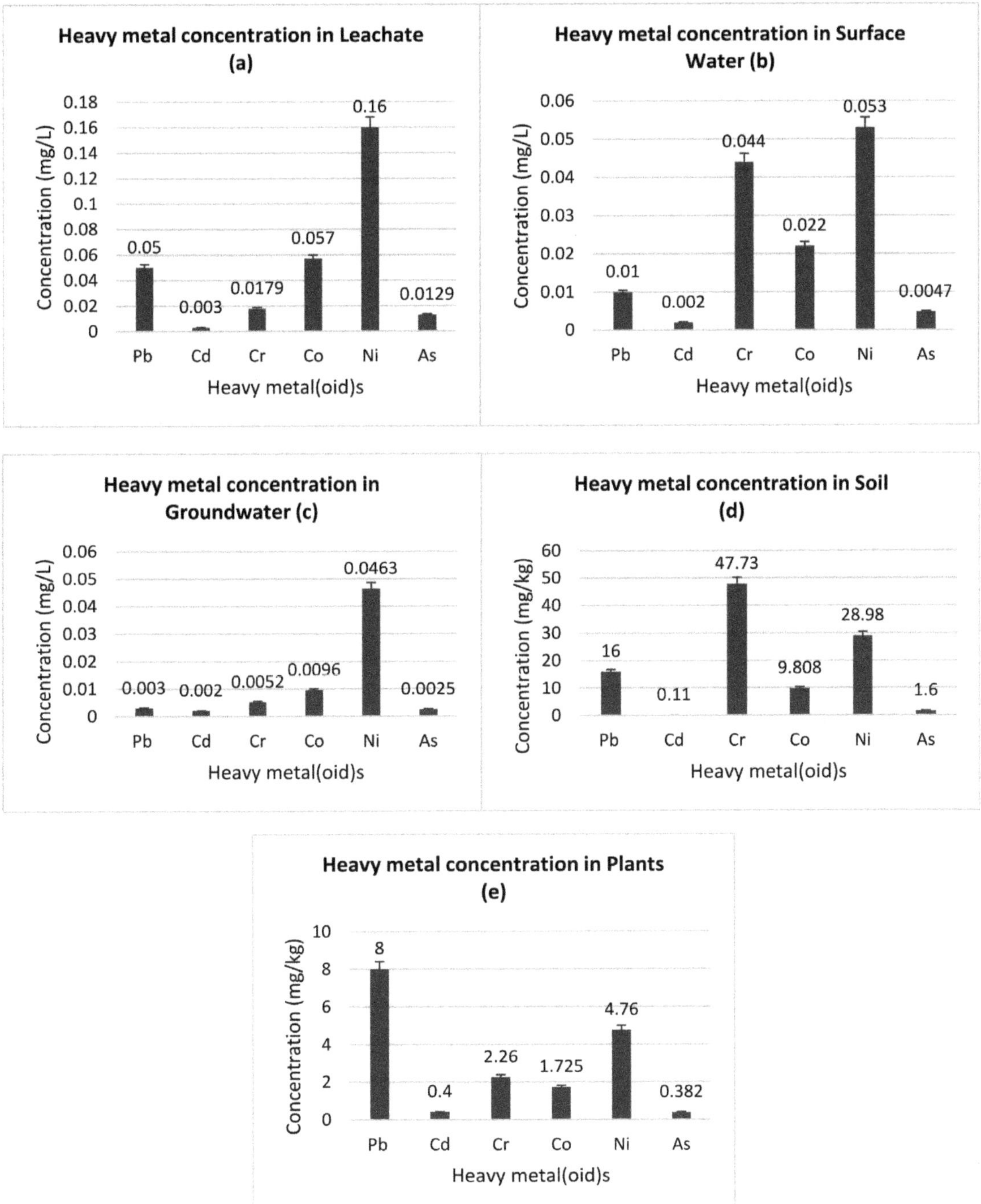

Figure 3. Heavy metal(loid)s concentration in (**a**) leachate, (**b**) surface water, (**c**) groundwater, (**d**) soil, and (**e**) plant sample.

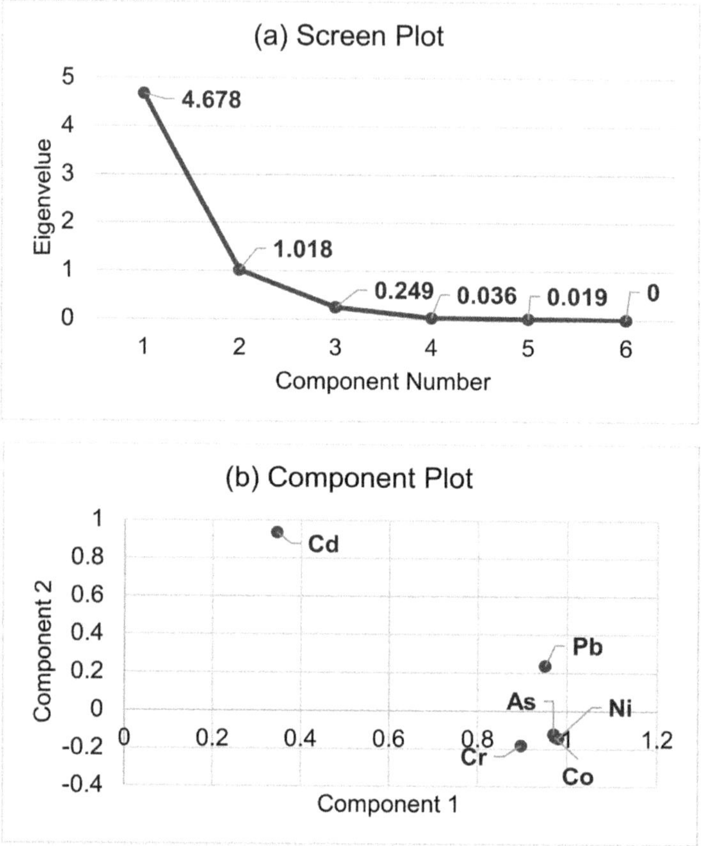

Figure 4. Principal component analysis of the analyzed heavy metal concentration by (**a**) scree plot and (**b**) component plot.

Table 4. Pearson's correlation matrix among the physicochemical parameters of leachate, surface water, and groundwater.

Parameters	pH	DO	Turbidity	Salinity	TDS	EC	Temperature	TH
pH	1							
DO	0.125	1						
Turbidity	−0.069	−0.667	1					
Salinity	**0.906**	0.107	−0.058	1				
TDS	**0.672**	−0.479	**0.520**	0.473	1			
EC	**0.672**	−0.479	**0.520**	0.473	**1**	1		
Temperature	**0.922**	0.089	0.100	**0.753**	**0.797**	**0.797**	1	
TH	**0.613**	−0.437	**0.583**	0.393	**0.980**	**0.980**	**0.756**	1

Note. Bold values refer to significant correlation.

According to Pearson's correlation matrix of heavy metal(loid)s concentration (Table 5), Pb has a positive proportional relation with all the other parameters. On the other hand, Cd does not comply with any positive proportional relation with other metals except for Pb. This also complies with the PCA result where Pb is the first component comprising 78% of total variance, as it has the maximum positive correlation with all the other metals except Cd, which as a result becomes the second component of PCA.

Table 5. Pearson's correlation among the heavy metal(loid)s.

Parameters	Pb	Cd	Cr	Co	Ni	As
Pb	1					
Cd	**0.533**	1				
Cr	**0.858**	0.146	1			
Co	**0.831**	0.188	**0.823**	1		
Ni	**0.861**	0.180	**0.854**	**0.995**	1	
As	**0.830**	0.178	**0.822**	**0.978**	**0.983**	1

Note. Bold values refer to significant correlation.

3.5. Water Quality and Pollution Assessment

According to the determined values listed in Table 6, the WQI value of leachate is higher than that of both surface and groundwater. Although the WQI of leachate shows that they are of good quality, the CD value of leachate shows that the leachate is heavily contaminated. The obtained value of WQI of the leachate sample was 50.01, which is not in the excellent range [49,50], and the CD value of treated leachate was 4.663, which denoted that the leachate was heavily contaminated according to the pollution level [29].

Table 6. Water quality index (WQI) and pollution indices including degree of contamination (CD), heavy metals evaluation index (HEI), and heavy metal pollution index (HPI) for leachate, surface water (SW), and groundwater (GW).

Sample ID	Sample Types	WQI	CD	HEI	HPI
01	Leachate untreated	50.01	3.317	9.317	16.445
02	Leachate treated	28.41	4.663	10.663	−5.159
	Mean	39.21	3.990	9.990	5.643
03-01	SW	43.63	−2.061	3.938	10.059
03-02	SW	25.33	0.162	6.162	−8.232
03-03	SW	17.82	−4.075	1.924	−15.750
	Mean	23.56	−2.204	3.795	−10.002
04-01	GW	29.15	−2.966	3.033	−4.421
04-02	GW	31.41	−3.878	2.121	−2.158
04-03	GW	33.25	−3.980	2.019	−0.312
	Mean	21.89	−4.008	1.991	−11.677

3.6. Spatial Comparison of Heavy Metal(loid)s Concentration and Distance from Landfill Site

As the results revealed, landfill leachate has the maximum concentration of the analyzed heavy metal(loid)s than other water samples. Water, soil, and plant samples also have a higher concentration of contaminants than the permissible limit. It was essential to analyze whether the surrounding environment had been contaminated because of the landfill leachate or if there were other reasons. According to the IDW analysis, sample 03-01 of surface water was the closest to the leachate pond, which was followed by sample 03-02 and sample 03-03. The cadmium concentration in sample 03-01 was the highest (0.004 mg/L), which was followed by sample 03-02 (0.002 mg/L) and sample 03-03 (0.0016 mg/L). Therefore, the Cd concentration in the surface water is inversely proportional to the distance from the landfill site. The smaller the distance, the higher the concentration. The same goes for Ni in surface water, as shown in Figure 5. In Figure 5, the color from green to red denotes the concentration from lower to higher. The concentration of Ni and As in groundwater is highest in samples 04-01, which was at the nearest distance from the landfill site. In Figure 5, the color from blue to red denotes the lower to higher concentration of Ni and As in groundwater.

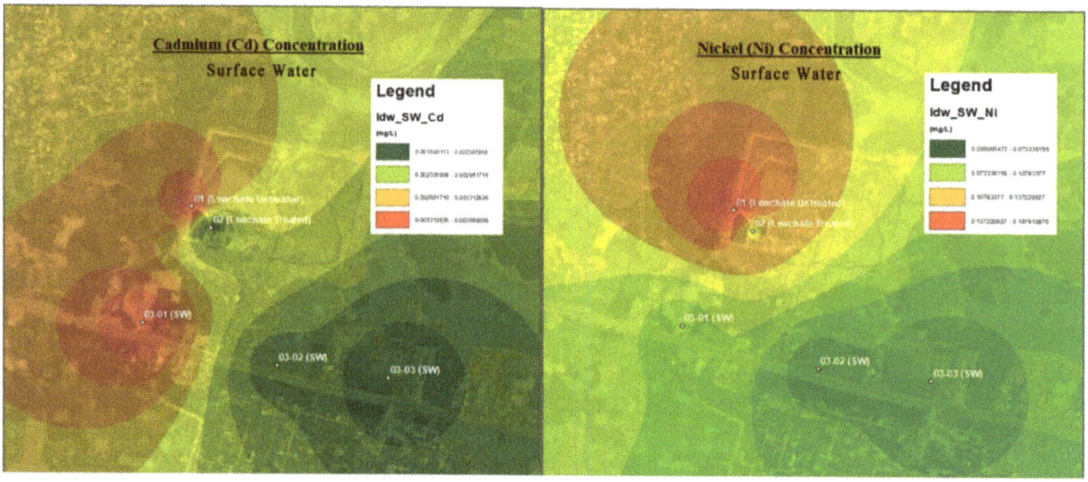

(a)

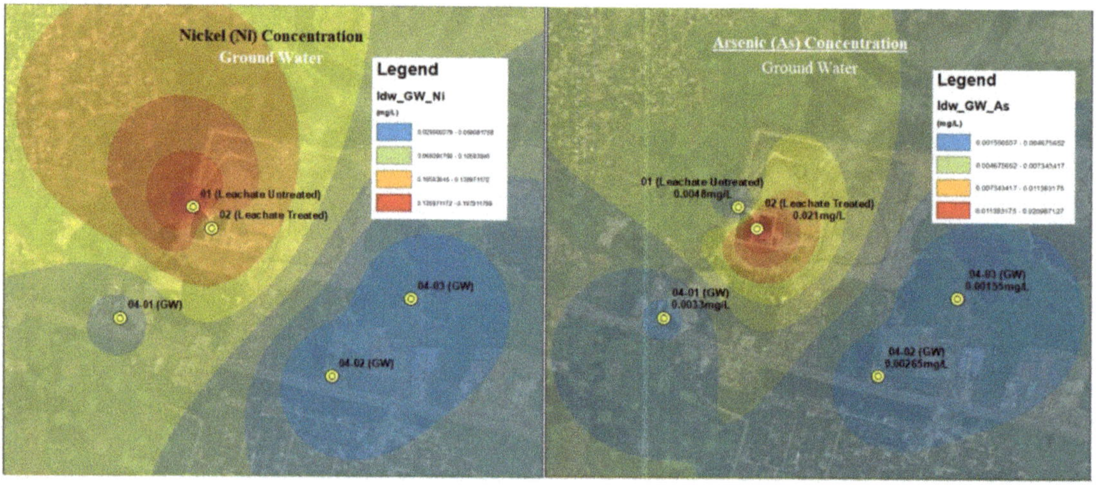

(b)

Figure 5. IDW of (a) Cd and Ni concentration in surface water and (b) Ni and As concentration in groundwater.

The same goes for the Pb and Cd in the soil and plant samples (Figure 6). The concentration of Pb, Cd, and Ni shows the most acceptable result based on IDW. Sample 06-01 and sample 06-03 were closer to the landfill site and sample 06-02 was farther. Therefore, the concentration of all these heavy metal(loid)s was higher in samples 06-01 and 06-03. More GIS analysis of soil and plant has been added in Supplementary Figure S1.

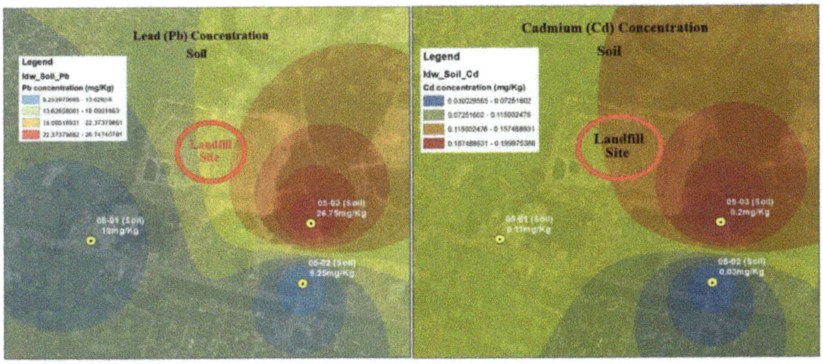

(a)

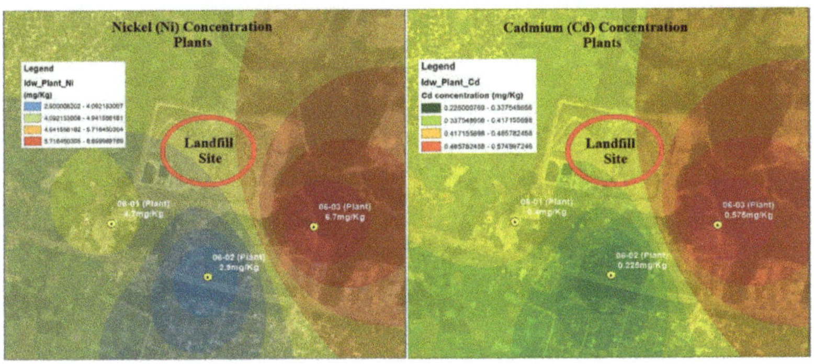

(b)

Figure 6. IWD of (**a**) Pb and Cd concentration in soil samples and (**b**) Ni and Cd concentration in plant samples.

4. Conclusions

This study has been conducted focusing on the impacts of Amin Bazar landfill, Dhaka, Bangladesh on the surrounding environment. To assess the impact of landfill leachate on the quality of the environmental compartments, physical and chemical characterization of the leachate and its surrounding environmental samples including the surface water, groundwater, soil, and plants were carried out. From the investigations, it can be concluded that the leachate (both treated and untreated) from the landfill site has a higher degree of contamination with respect to the analyzed parameters, which contribute to the surrounding environmental components including the surface water, groundwater, soil, and plants by polluting them adversely. Although the WQI of the samples based on the heavy metal(loid)s concentration was found within the standard limit, the concentration of heavy metal(loid)s in the soil and plants was found to be very high. This indicates a considerable deposition and accumulation of heavy metal(loid)s in soil and plants, respectively. Thus, all the plant samples have been found to accumulate with a higher concentration of Pb, Cd, Cr, and Co than the permissible limit. Again, the individual concentration of As was higher than the maximum permissible limit in the leachate. According to the Inverse Distance Weighted (IDW) analysis, landfill leachate has the maximum probability to contaminate the surrounding environment with heavy metal(loid)s. The results of this study support the need for continuous monitoring of the environmental components around the Amin Bazar landfill along with the landfill management system. The investigation also showed that despite being treated with conventional aeration, the quality of the leachate sample in this area does not meet Bangladesh's inland surface water quality criteria. As a result, mitigation measures are critical for preventing soil and water contamination. Solid waste should be converted to reusable items by solid waste adjustment and cementing for cost-effective management. The findings of this flow study can be used to supplement the next stage of research, which will investigate the feasibility of getting involved in the development of a waste management system as well as ensuring ideal environmental standards for municipal solid waste and balancing environmental quality.

Supplementary Materials: The following supporting information can be downloaded at: https://www.mdpi.com/article/10.3390/soilsystems6040090/s1, Table S1: Heavy metal(loid)s concentration in all samples; Figure S1: IWD of (a) Cr concentration in soil samples and (b) Pb concentration in plant samples.

Author Contributions: R.H.H.: Conceptualization, methodology, Investigation, Software, Writing—original draft preparation. M.A.B.S.: Methodology, Validation, Formal analysis, Investigation, Data curation, Writing—original draft preparation, Writing—reviewing and editing, Visualization. M.A.A.: Methodology, Resources, Writing—reviewing and editing, Supervision. M.A.A.S.: Resources, Writing—reviewing and editing. M.M.R.: Conceptualization, Investigation, Resources, Writing—reviewing and editing, Supervision. All authors have read and agreed to the published version of the manuscript.

Funding: This research received no external funding.

Institutional Review Board Statement: Not applicable.

Informed Consent Statement: Not applicable.

Data Availability Statement: All the data are included in this manuscript.

Conflicts of Interest: The authors declare no conflict of interest.

References

1. Christensen, T.H.; Kjeldsen, P. Basic biochemical processes in landfills. In *Sanitary Landfilling: Process, Technology and Environmental Impact*; Christensen, T.C., Ed.; Academic Press: London, UK, 1989; p. 29.
2. Kamal, A.K.I.; Islam, R.; Hassan, M.; Ahmed, F.; Rahman, M.A.T.M.T.; Moniruzzaman, M. Bioaccumulation of Trace Metals in Selected Plants within Amin Bazar Landfill Site, Dhaka, Bangladesh. *Environ. Process.* **2016**, *3*, 179–194. [CrossRef]
3. Mukti, S.A. Solid Waste Management in Dhaka City: Problems and Prospects. *Int. J. Innov. Res. Dev.* **2015**, *2*, 33–37.
4. DNCC. *Waste Report*; Dhaka North City Corporation: Dhaka, Bangladesh, 2020.
5. Hoque, M.A.; Haque, M.A.; Mondal, M.S.A. Seasonal Effects on Heavy Metal Concentration in Decomposed Solid Waste of DNCC and DSCC Landfill Sites. *Civ. Eng. Arch.* **2014**, *2*, 52–56. [CrossRef]
6. Kabir, M.R. Municipal solid waste management system: A study on Dhaka north and South City corporations. *Bangladesh Inst. Plan* **2015**, *2075*, 9363.
7. Yousuf, T.B.; Rahman, M. Monitoring quantity and characteristics of municipal solid waste in Dhaka City. *Environ. Monit. Assess* **2007**, *135*, 3–11. [CrossRef] [PubMed]
8. Azim, M.D.; Rahman, M.M.; Khan, R.H.; Kamal, A. Characteristics of Leachate Generated at Landfill Sites and Probable Risks of Surface and Groundwater Pollution in The Surrounding Areas: A Case Study of Matuail Landfill Site, Dhaka. *J. Bangladesh Acad. Sci.* **1970**, *35*, 153–160. [CrossRef]
9. Gavrilescu, M.; Schiopu, A.-M.; Robu, B.M.; Apostol, I. Impact of landfill leachate on soil quality in Iasi county. *Environ. Eng. Manag. J.* **2009**, *8*, 1155–1164. [CrossRef]
10. Negi, P.; Mor, S.; Ravindra, K. Impact of landfill leachate on the groundwater quality in three cities of North India and health risk assessment. *Environ. Dev. Sustain.* **2018**, *22*, 1455–1474. [CrossRef]
11. Jahan, E.; Nessa, A.; Hossain, F.; Parveen, Z. Characteristics of municipal landfill leachate and its impact on surrounding agricultural land. *Bangladesh J. Sci. Res.* **2016**, *29*, 31–39. [CrossRef]
12. El-Salam, M.M.A.; Abu-Zuid, G.I. Impact of landfill leachate on the groundwater quality: A case study in Egypt. *J. Adv. Res.* **2014**, *6*, 579–586. [CrossRef]
13. Naveen, B.; Mahapatra, D.M.; Sitharam, T.; Sivapullaiah, P.; Ramachandra, T. Physico-chemical and biological characterization of urban municipal landfill leachate. *Environ. Pollut.* **2017**, *220*, 1–12. [CrossRef]
14. Adhikari, B.; Khanal, S.N.; Manandhar, D.R. Study of leachate and waste composition at different landfill sites of Nepal. *Kathmandu Univ. J. Sci. Eng. Technol.* **2013**, *9*, 15–21.
15. Chang, A.C.; Page, A.L. Trace elements slowly accumulating, depleting in soils. *Calif. Agric.* **2000**, *54*, 49–55. [CrossRef]
16. Alam, H. Second Modern Sanitary Landfill at Amin Bazar Awaits Green Signal. The Daily Star. 2008. Available online: http://archive.thedailystar.net/newDesign/news-details.php?nid=52541 (accessed on 20 December 2021).

17. Ahsan, A.; Satter, F.; Siddique, A.B.; Akbor, A.; Ahmed, S.; Shajahan, M.; Khan, R. Chemical and physicochemical characterization of effluents from the tanning and textile industries in Bangladesh with multivariate statistical approach. *Environ. Monit. Assess.* **2019**, *191*, 575. [CrossRef]
18. Siddique, A.B.; Alam, K.; Islam, S.; Diganta, M.T.M.; Akbor, A.; Bithi, U.H.; Chowdhury, A.I.; Ullah, A.K.M.A. Apportionment of some chemical elements in soils around the coal mining area in northern Bangladesh and associated health risk assessment. *Environ. Nanotechnol. Monit. Manag.* **2020**, *14*, 100366. [CrossRef]
19. Hasan, A.B.; Reza, A.H.M.S.; Kabir, S.; Siddique, A.B.; Ahsan, A.; Akbor, A. Accumulation and distribution of heavy metals in soil and food crops around the ship breaking area in southern Bangladesh and associated health risk assessment. *SN Appl. Sci.* **2020**, *2*, 155. [CrossRef]
20. Nasrin, S.; Islam, M.N.; Abu Tayab, M.; Nasrin, M.S.; Siddique, A.B.; Bin Emran, T.; Reza, A.A. Chemical profiles and pharmacological insights of *Anisomeles indica* Kuntze: An experimental chemico-biological interaction. *Biomed. Pharmacother.* **2022**, *149*, 112842. [CrossRef]
21. APHA (American Public Health Association). *Standard Methods for the Examination of Water and Wastewater*, 23rd ed.; Baird, R.B., Eaton, A.D., Rice, E.W., Eds.; American Public Health Association, American Water Works Association, Water Environment Federation: Washington, DC, USA, 2017; Available online: https://secure.apha.org/imis/ItemDetail?iProductCode=978-087553-2875&CATEGORY=BK (accessed on 20 December 2021).
22. Childs, C. Interpolating surfaces in ArcGIS spatial analyst. *ArcUser* **2004**, *3235*, 32–35.
23. Gao, B.; Gao, L.; Gao, J.; Xu, D.; Wang, Q.; Sun, K. Simultaneous evaluations of occurrence and probabilistic human health risk associated with trace elements in typical drinking water sources from major river basins in China. *Sci. Total Environ.* **2019**, *666*, 139–146. [CrossRef]
24. Rahman, M.A.T.M.T.; Paul, M.; Bhoumik, N.; Hassan, M.; Alam, K.; Aktar, Z. Heavy metal pollution assessment in the groundwater of the Meghna Ghat industrial area, Bangladesh, by using water pollution indices approach. *Appl. Water Sci.* **2020**, *10*, 186. [CrossRef]
25. Abrahim, G.M.S.; Parker, R.J. Assessment of heavy metal enrichment factors and the degree of contamination in marine sediments from Tamaki Estuary, Auckland, New Zealand. *Environ. Monit. Assess.* **2007**, *136*, 227–238. [CrossRef] [PubMed]
26. Sharmin, S.; Mia, J.; Miah, M.S.; Zakir, H. Hydrogeochemistry and heavy metal contamination in groundwaters of Dhaka metropolitan city, Bangladesh: Assessment of human health impact. *J. Hydro-Environ. Res.* **2020**, *3*, 106–117. [CrossRef]
27. Doza, B.; Islam, S.D.-U.; Rume, T.; Quraishi, S.B.; Rahman, M.S.; Bhuiyan, M.A.H. Groundwater quality and human health risk assessment for safe and sustainable water supply of Dhaka City dwellers in Bangladesh. *Groundw. Sustain. Dev.* **2020**, *10*, 100374. [CrossRef]
28. Islam, A.R.M.T.; Islam, H.T.; Mia, U.; Khan, R.; Habib, A.; Doza, B.; Siddique, A.B.; Chu, R. Co-distribution, possible origins, status and potential health risk of trace elements in surface water sources from six major river basins, Bangladesh. *Chemosphere* **2020**, *249*, 126180. [CrossRef] [PubMed]
29. Backman, B.; Bodiš, D.; Lahermo, P.; Rapant, S.; Tarvainen, T. Application of a groundwater contamination index in Finland and Slovakia. *Environ. Geol.* **1998**, *36*, 55–64. [CrossRef]
30. Edet, A.E.; Offiong, O.E. Evaluation of water quality pollution indices for heavy metal contamination monitoring. A study case from Akpabuyo-Odukpani area, Lower Cross River Basin (Southeastern Nigeria). *GeoJournal* **2002**, *57*, 295–304. [CrossRef]
31. Horton, R.K. An index system for rating water quality. *J. Water Pollut. Control Ed.* **1965**, *37*, 300–306.
32. Mohan, S.V.; Nithila, P.; Reddy, S.J. Estimation of heavy metals in drinking water and development of heavy metal pollution index. *J. Environ. Sci. Health Part A Environ. Sci. Eng. Toxicol.* **1996**, *31*, 283–289. [CrossRef]
33. Xiao, J.; Wang, L.; Deng, L.; Jin, Z. Characteristics, sources, water quality and health risk assessment of trace elements in river water and well water in the Chinese Loess Plateau. *Sci. Total. Environ.* **2018**, *650*, 2004–2012. [CrossRef]
34. WHO (World Health Organization). *Guidelines for Drinking Water Quality, Library Cataloguing-in-Publication Data*, 4th ed.; NLM classification: WA 675; World Health Organization: Geneva, Switzerland, 2011.
35. *ECR (The Environment Conservation Rules)*; Government of the People's Republic of Bangladesh, Ministry of Environment and Forest: Dhaka, Bangladesh, 1997.
36. USEPA. *US Environmental Protection Agency. Risk Assessment Guidance for Superfund Volume I: Human Health Evaluation Manual (part E, Supplemental Guidance for Dermal Risk Assessment) Final*; EPA/540/R/99/005 OSWER 9285.702EP PB99-963312 July 2004; Office of Super fund Remediation and Technology Innovation: Washington, DC, USA, 2004.
37. WHO (World Health Organization). *Guidelines for Drinking Water Quality, Library Cataloguing in Publication Data*, 3rd ed.; NLM classification: WA 675; World Health Organization: Geneva, Switzerland, 2004.
38. World Health Organization. *Guidelines for Drinking-Water Quality: Fourth Edition Incorporating the First Addendum*; World Health Organization: Geneva, Switzerland, 2017; ISBN 9789241549950.
39. Tchobanoglous, G.; Theisen, H.; Vigil, S.A. *Integrated Solid Waste Management: Engineering Principles and Management Issues*; McGraw-Hill: New York, NY, USA, 1993. [CrossRef]
40. Kumar, D.; Alappat, B.J. Analysis of leachate pollution index and formulation of sub-leachate pollution indices. *Waste Manag. Res.* **2005**, *23*, 230–239. [CrossRef]
41. Parvin, F.; Tareq, S.M. Impact of landfill leachate contamination on surface and groundwater of Bangladesh: A systematic review and possible public health risks assessment. *Appl. Water Sci.* **2021**, *11*, 100. [CrossRef]

42. Alam, R.; Ahmed, Z.; Howladar, M.F. Evaluation of heavy metal contamination in water, soil and plant around the open landfill site Mogla Bazar in Sylhet, Bangladesh. *Groundw. Sustain. Dev.* **2019**, *10*, 100311. [CrossRef]
43. Raviraja, A.; Babu, G.; Bijoor, A.; Menezes, G.; Venkatesh, T. Lead Toxicity in a Family as a Result of Occupational Exposure. *Arch. Ind. Hyg. Toxicol.* **2008**, *59*, 127–133. [CrossRef]
44. Riess, M.L.; Halm, J.K. Lead poisoning in an adult: Lead mobilization by pregnancy? *J. Gen. Intern. Med.* **2007**, *22*, 1212–1215. [CrossRef]
45. Prasad, S.; Yadav, K.K.; Kumar, S.; Gupta, N.; Cabral-Pinto, M.M.; Rezania, S.; Radwan, N.; Alam, J. Chromium contamination and effect on environmental health and its remediation: A sustainable approaches. *J. Environ. Manag.* **2021**, *285*, 112174. [CrossRef]
46. Wakeel, A.; Xu, M. Chromium Morpho-Phytotoxicity. *Plants* **2020**, *9*, 564. [CrossRef]
47. Hussein, M.; Yoneda, K.; Mohd-Zaki, Z.; Amir, A.; Othman, N. Heavy Metals in Leachate, Impacted Soils and Natural Soils of Different Landfills in Malaysia: An Alarming Threat. *Chemosphere* **2020**, 128874. [CrossRef] [PubMed]
48. Battistel, M.; Stolze, L.; Muniruzzaman, M.; Rolle, M. Arsenic release and transport during oxidative dissolution of spatially-distributed sulfide minerals. *J. Hazard. Mater.* **2020**, *409*, 124651. [CrossRef]
49. Tandel, B.N.; Macwan, J.E.M.; Soni, C.K. Assessment of water quality index of small lake in south Gujarath region, India. In Proceedings of the ISEM-2011, Bangkok, Thailand, 23–24 December 2011.
50. Siddique, A.B.; Islam, A.R.M.T.; Hossain, S.; Khan, R.; Akbor, A.; Hasanuzzaman; Sajid, W.M.; Mia, Y.; Mallick, J.; Rahman, M.S.; et al. Multivariate statistics and entropy theory for irrigation water quality and entropy-weighted index development in a subtropical urban river, Bangladesh. *Environ. Sci. Pollut. Res.* **2021**, *29*, 8577–8596. [CrossRef]

soil systems

Article

Large Chestnut Trees Did Not Respond to Annual Fertiliser Applications, Requiring a Long-Term Approach to Establishing Effective Fertilisation Plans

Margarida Arrobas [1,2], João Silva [3], Mariah Ruaro Busato [4], Ana Carolina Ferreira [4], Soraia Raimundo [1,2], Abel Pereira [5], Taciane Finatto [4], Nilvania Aparecida de Mello [4], Carlos Manuel Correia [6] and Manuel Ângelo Rodrigues [1,2,*]

1 Centro de Investigação de Montanha (CIMO), Instituto Politécnico de Bragança, Campus de Santa Apolónia, 5300-253 Bragança, Portugal
2 Laboratório para a Sustentabilidade e Tecnologia em Regiões de Montanha, Instituto Politécnico de Bragança, Campus de Santa Apolónia, 5300-253 Bragança, Portugal
3 Cadubal-Companhia de Adubos, S.A. Zona Industrial Fase 1, S/N, 4935-231 São Romão Do Neiva, Viana Do Castelo, Portugal
4 Campus Pato Branco, Universidade Tecnológica Federal do Paraná, Pato Branco 85503-390, Brazil
5 ARBOREA—Associação Agro-Florestal e Ambiental da Terra Fria Transmontana, Rua Álvaro Leite, No. 3, 5320-332 Vinhais, Portugal
6 Centre for the Research and Technology of Agro-Environmental and Biological Sciences (CITAB), University of Trás-os-Montes e Alto Douro, 5000-801 Vila Real, Portugal
* Correspondence: angelor@ipb.pt; Tel.: +351-273303260

Abstract: Due to the high value of the fruit, the European chestnut (*Castanea sativa* Mill.), usually grown in agroforestry systems, has been planted as a single species in orchards managed with increasingly intensive cropping practices, such as the regular use of fertilisers. This justifies research into establishing fertilisation programmes oriented towards ecological intensification. In this study, the results of fruit production, plant nutritional status and soil properties are reported from a field trial in which three NPK fertilisers (20:7:10, 13:11:21 and 7:14:14) and a control treatment were used. Chestnut yields did not vary significantly between treatments, although the mean values of the control showed a clear downward trend. N supplied by the fertilisers seems to have been the most important factor in the difference between the fertilised and control treatments, since leaf N concentrations were lower in the control and often below the lower limit of the sufficiency range. Soil inorganic N levels in the autumn, and tissue N concentrations of the herbaceous vegetation developing beneath the trees, indicated risks of N loss to the environment and highlighted the importance of this vegetation remaining during the winter. The chestnuts' poor response to fertiliser applications was attributed to the buffering effect of the large perennial structure of the trees on the distribution of nutrients to the growing plant parts. In large trees, it seems appropriate to base the annual fertilisation plan on leaf nutrient concentration. Thus, farmers probably should avoid spending money on fertilizer applications as long as leaf nutrient concentrations do not approach the lower limits of sufficiency ranges.

Keywords: chestnut tree; *Castanea sativa*; chestnut yield; plant nutritional status; soil inorganic nitrogen

1. Introduction

Chestnuts are the main source of income for farmers in the upland areas of the north of Portugal. However, farmers are facing a quite complex situation due to a set of pests and diseases that weaken the trees, thereby reducing their productivity and, in some cases, causing their death. Currently, ink disease (*Phytophthora* sp.pl.), chestnut blight (*Cryphonectria parasitica* (Murrill) Barr.) and the Asian gall wasp (*Dryocosmus kuriphilus* Yasumatsu) are the main health problems affecting chestnut trees [1–3]. Notwithstanding this, chestnuts have maintained very good market prices [4], which has led farmers to devote great attention and care to their crops, either replacing dead trees or establishing new orchards [5].

Chestnut is grown all over the world as part of agroforestry systems with little phytotechnical intensification [6]. In the mountainous areas of the north of Portugal, the lack of other crop options has raised chestnut to the status of the main crop, having been grown in monoculture and integrated into increasingly intensive farming systems, in a similar way to orchards of other important fruit trees [7–9]. One of the practices that has received greater attention from producers is fertilisation, with trees currently being fertilised regularly [5,10,11].

Crop fertilisation, being essential for obtaining high productivity in any species [12–14], can also be associated with high risks of environmental contamination, especially the use of N fertilisers that can lead to the eutrophication of ground water [15,16] and the emission of greenhouse gases into the atmosphere, in particular, N oxides [17,18]. Thus, crop fertilisation must be managed judiciously, in order to apply the appropriate nutrient rates, thereby reducing the risk of environmental damage [19–22].

The prospects of a growing global population and the need to feed it, associated with the risks of environmental contamination, have led to the need to develop farming practices based on the concept of ecological intensification [23,24], which, in practice, means maintaining high productivity, but by using production factors in a more rational way. Thus, as with the main world crops, but also with chestnut, it is necessary to manage resources properly, using them in the smallest amounts necessary to maintain productivity.

Previous work carried out in NE Portugal has shown that in chestnut groves, nutrients are often below the lower limit of the sufficiency range [25] and that trees generally tend to respond to fertiliser applications [11,26], although in some studies, they did not [5]. However, there are still only a few studies on chestnut fertilisation, and the use of fertilisers is far from being optimized, with more data being required to establish adequate fertilisation programmes. It is therefore necessary to establish better guidelines for the fertilisation of these trees, to try to keep them healthy and productive, so that these magnificent ecosystems may persist, allowing man to continue to occupy these mountain territories which are showing concerning signs of depopulation [27]. This study reports the results from a field experiment of chestnut fertiliser application using NPK fertilisers with different combinations of macronutrients, trying as best as possible to replicate the diversity of fertilisers found on the market, which farmers have access to. The objectives of the study are to understand better how these huge trees respond to fertiliser application so as to help farmers make better decisions when they need to acquire them.

2. Materials and Methods

2.1. Experimental Conditions

The field experiment took place in Vinhais (41°50′15.8″ N; 7°03′40.4″ W, 800 m above sea level), northeastern Portugal, in a 50-year-old chestnut orchard of the cultivar Judia with trees spaced at 10 m × 10 m. The region benefits from a warm-summer Mediterranean climate (Csb), according to the Köppen–Geiger classification [28]. The annual mean temperature and the accumulated annual precipitation are 11.9 °C and 880 mm, respectively. Average monthly temperatures and precipitation of the climatological normal (1981–2010), together with those recorded during the experimental period (2018–2021), are presented in Figure 1.

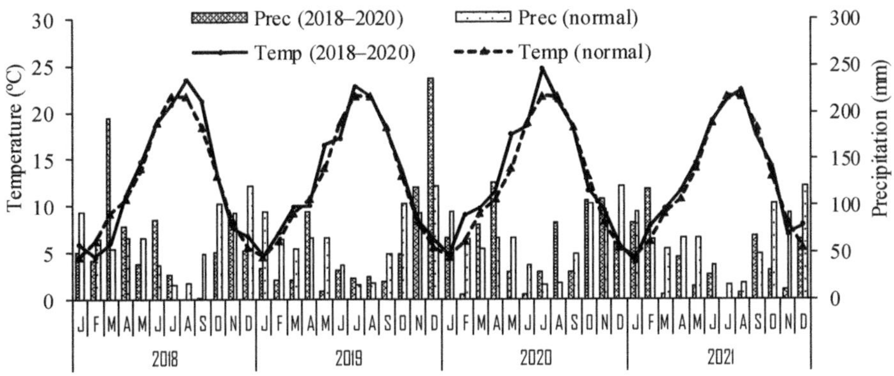

Figure 1. Average monthly temperature and accumulated monthly precipitation during the experimental period, and climatological normal values for the region.

The soil where the chestnut orchard is planted is a Leptosol, sandy-loam textured. It is a very shallow soil (~0.20 m deep), which separates, determined from composite soil samples (n = 3) taken from the 0.0–0.2 m soil layer at the beginning of the study, were 11.8% clay, 17.3% silt and 70.9% sand. Soil organic C was low, pH acidic and extractable P and K were medium and very high, respectively. Some other soil properties determined at the beginning of the field trial are presented in Table 1.

Table 1. Selected soil properties (average ± standard deviation, n = 3) from composite samples (10 cores per composite sample) taken at 0–0.20 m depth at the beginning of the study.

Soil Properties		Soil Properties (Cont.)	
[1] Organic carbon (g kg^{-1})	13.4 ± 0.50	[4] Exchang. sodium ($cmol_c$ kg^{-1})	0.1 ± 0.02
[2] pH (H_2O)	5.3 ± 0.19	[5] Exchang. acidity ($cmol_c$ kg^{-1})	0.7 ± 0.08
[2] pH (KCl)	4.2 ± 0.15	[6] CEC ($cmol_c$ kg^{-1})	6.0 ± 0.26
[3] Extract. phosphorus (mg kg^{-1}, P_2O_5)	93.1 ± 15.75	[7] Extract. boron (mg kg^{-1})	0.4 ± 0.06
[3] Extract. potassium (mg kg^{-1}, K_2O)	344.7 ± 20.63	[8] Extract. iron (mg kg^{-1})	62.2 ± 4.78
[4] Exchang. calcium ($cmol_c$ kg^{-1})	3.1 ± 0.20	[8] Extract. zinc (mg kg^{-1})	2.5 ± 0.31
[4] Exchang. magnesium ($cmol_c$ kg^{-1})	1.1 ± 0.11	[8] Extract. copper (mg kg^{-1})	1.2 ± 0.25
[4] Exchang. potassium ($cmol_c$ kg^{-1})	1.0 ± 0.13	[8] Extract. manganese (mg kg^{-1})	132.6 ± 18.59

[1] Wet digestion (Walkley–Black); [2] Potentiometry; [3] Ammonium lactate; [4] Ammonium acetate; [5] Potassium chloride; [6] Cation Exchange Capacity; [7] Hot water, azomethine-H; [8] Ammonium acetate and EDTA (ethylenediaminetetraacetic acid).

2.2. Experimental Design and Management of the Field Trial

Sixteen large-sized trees with similar, spherical canopies (~270 m^3) were selected for the study. They were randomly distributed into four groups, corresponding to four fertilisation treatments, with four trees (replicates) in each treatment, in a completely randomized design. The treatments consisted of three compound NPK fertilisers with different levels of N, P and K, and an unfertilised control.

One of the treatments, named 7:14:14, consisted of the application of a 7:14:14 NPK compound fertiliser that doses 7% N (5% ammoniacal-N and 2% ureic-N), 14% P_2O_5 (11% water soluble) and 14% K_2O. This fertiliser also contains 4% CaO, 2% MgO, 15% SO_3, and 0.02% B. Another treatment named YA20:7:10, corresponds to the application of the commercial fertiliser Yara MilaTM Actyva 20:7:10, with 20% N (9.4% nitric-N, 10.6% ammoniacal-N), 7% P_2O_5 (25 to 30% as polyphosphates) and 10% K_2O. The fertiliser also contains other important nutrients, namely S (10% SO_3) and Mg (3% MgO). The third treatment, named YS13:11:21, consisted of the application of the NPK compound fertiliser

Yara MilaTM Solán 13:11:21, which doses 13% N (5.5% nitric-N, 7.7% ammoniacal-N), 11% P_2O_5 (20 to 30% as polyphosphates) and 21% K_2O. This fertiliser also contains relevant amounts of Mg (2% MgO) and B (0.2%).

All fertilisers were applied at a rate of 4 kg per tree (~400 kg ha^{-1}). Thus, the YA20:7:10 treatment corresponded to an application of 80, 28 and 40 kg ha^{-1} of N, P_2O_5 and K_2O, respectively; the 7:14:14 treatment to an application of 28, 56 and 56 kg ha^{-1} of N, P_2O_5 and K_2O, respectively, and the YS13:11:21 to an application of 52, 44, 84 kg ha^{-1} of N, P_2O_5 and K_2O, respectively. Similar fertilisations are commonly used by local farmers, and these particular fertilisers were chosen for the trial because they present a good variation in the levels of macronutrients.

The fertilisers were evenly applied beneath the canopy of the trees in the first week of April over the four years of the study and incorporated with a cultivator. The orchard was tilled a second time every year at the end of May to control the weeds. No further cropping practices were carried out in the orchard during the four years of the study.

2.3. Measurements in the Field

The effect of the treatments was assessed in the field by measuring the greenness of the leaves and by chlorophyll a fluorescence analysis.

The SPAD (Soil and Plant Analysis Development)-502 Plus chlorophyll meter (Spectrum Technologies, Inc., Aurora, IL, USA) was used to measure leaf greenness. SPAD-502 provides *dimensionless* readings, proportional to the chlorophyll content of the leaves, by measuring the transmittance of light through the leaves at 650 nm (red light, absorbed by chlorophyll) and 940 nm (infrared light, non-absorbed by chlorophyll). Each mean value was obtained after 30 individual readings taken around the crown on fully expanded young leaves.

Chlorophyll a fluorescence was assessed using the dark adaptation protocols with the OS-30p+ fluorometer (Opti-sciences, Inc., Hudson, NH, USA). F_M, F_0 and F_V are, respectively, maximum, minimum and variable fluorescence from dark-adapted leaves. F_V/F_M is estimated as $(F_M - F_0)/F_M$.

To harvest the chestnuts, it is necessary to wait for them to fall to the ground and then pick them up manually or mechanically. In this experiment, the fruits were harvested manually, in three passes during the autumn, to allow individual weighing per tree. In 2021, the COVID-19 pandemic restrictions did not allow the completion of harvest records, and therefore, only the values for 2018–2020 are available.

2.4. Soil and Plant Tissue Sampling and Analytical Determinations

Three composite samples were taken at the beginning of the experiment to characterize the experimental plot. The soil was sampled again in October 2021 to evaluate the effect of the treatments on soil properties. All soil samples taken to the laboratory were composite samples, taken at six different sampling points. Sampling was carried out in the 0.0–0.20 m soil layer, beneath the canopy of the trees, where the fertilisers had been applied.

Soil samples were oven-dried at 40 °C and sieved (2 mm mesh). Thereafter, the samples were analysed for pH (H_2O and KCl) (soil: solution, 1:2.5), cation-exchange capacity (ammonium acetate, pH 7.0), organic C (wet digestion, Walkley-Black method) and extractable P and K (Egner–Riehm method, ammonium lactate extract). Soil B was extracted by hot water and determined by the method of azomethine-H. For more details on these analytical procedures, the reader is referred to van Reeuwijk [29]. The availability of other micronutrients (Cu, Fe, Zn, and Mn) in the soil was determined by atomic absorption spectrometry after extraction with ammonium acetate and EDTA, according to the method described by Lakanen and Erviö [30]. Soil inorganic-N was determined in soil extracts prepared from 20 g of soil and 40 mL 2 M KCl. The suspension was shaken for 1 h and filtered through Watmann No. 42 filter paper. Nitrate and ammonium concentrations in the extracts were analysed in an UV–Vis spectrophotometer [31].

By the end of July, in each of the four years, samples of young, fully developed leaves were taken for elemental analysis. Following each harvest, samples of 50 nuts per tree were randomly taken to evaluate their size and also for elemental analysis. After counting and weighing, the kernel was separated from shell and pellicle, and the two parts analysed separately. In April 2022, the spontaneous vegetation which had developed beneath the canopy of the trees was mowed to serve as a biological index of soil-available nutrients. The samples were collected by randomly placing a grid of 0.5 m × 0.5 m on the vegetation.

The samples of leaves, fruit kernels, shells and pellicles and spontaneous vegetation, were oven-dried at 70 °C until they reached a constant weight and ground (1 mm mesh). Elemental analyses of tissue samples were performed by Kjeldahl (N), colorimetry (B and P) and atomic absorption spectrophotometry (K, Ca, Mg, Fe, Mn, Cu, Zn) methods [32] after tissue samples had been previously digested with nitric acid in a microwave.

2.5. Data Analysis

The data was analysed for normality and homogeneity of variance using the Shapiro–Wilk and Bartlett's test, respectively. The analysis of variance was performed as a one-way ANOVA, using the Statistical Package for the Social Sciences (SPSS) version 25 (IBM Corporation, New York, NY, USA). When significant differences were found, the means were separated by the Tukey HSD post hoc test ($\alpha = 0.05$).

3. Results

3.1. Chestnut Yield

Chestnut yield did not vary significantly in the three years in which it was possible to collect the fruits (Figure 2). However, the accumulated yield of the three years showed a clear tendency towards reduction in the control in comparison to the fertilised treatments. In addition, the probability values decreased over time, nearing significant differences between treatments in the last year (2020) for which it was possible to obtain records. In the YS13:11:21 treatment, an accumulated average nut yield of 94.9 kg tree^{-1} was recorded, while in the control treatment, the value was 80 kg tree^{-1}.

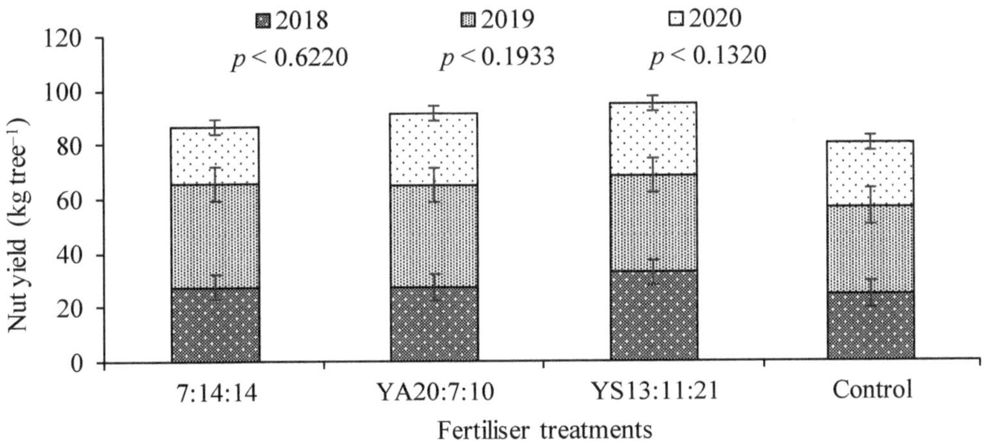

Figure 2. Average annual nut yield as a response to fertilisation treatments. Error bars are the standard errors.

3.2. Nutrients in Plant Tissues and Chlorophyll a Fluorescence

Leaf nitrogen concentrations varied significantly between treatments only in the 2021 sampling (Figure 3). In the last year, the control treatment displayed an average concentration of N in the leaves (18.9 g kg^{-1}) much lower than the values recorded in the fertilised treatments (22.4 to 22.9 g kg^{-1}) and below the lower limit of the sufficiency range. Fertiliser YA20:7:10, being more concentrated in N, showed average concentrations of leaf N tending to be higher than the other treatments. Leaf P concentrations did not vary significantly between treatments and remained above the lower limit of the sufficiency range. No sign of any coherence was observed between the application of P and the concentration of the nutrient in the leaves. The concentrations of K in leaves fluctuated inconsistently with treatments over the years, although there were found to be significant differences between treatments in the last sampling. As observed for P, there seem to have been more important variables other than the application of these nutrients determining their concentration in the leaves. The values of K tended to remain above the lower limit of the sufficiency range. Leaf concentrations of Ca and Mg did not vary significantly between treatments. In the case of Ca, the values were close to the lower limit of the sufficiency range and those of Mg were clearly within the interval of adequate concentrations.

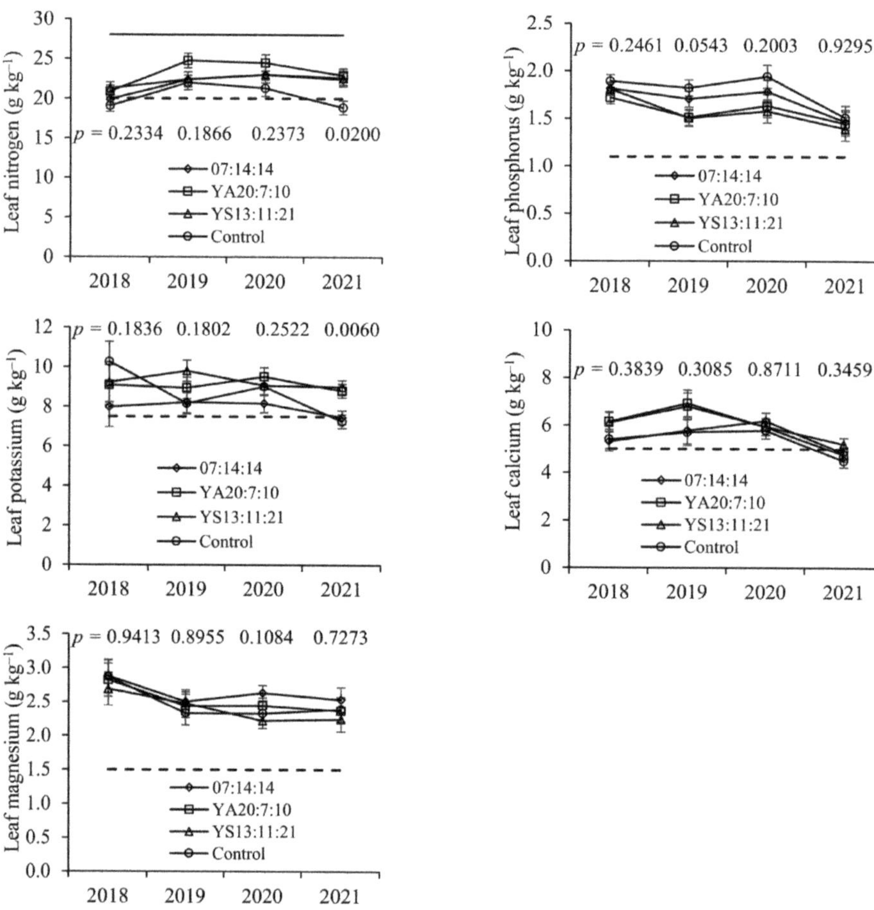

Figure 3. Leaf concentrations of nitrogen, phosphorus, potassium, calcium and magnesium as a response to fertilisation treatments. Dashed and solid lines are, respectively, the lower and upper limits of the sufficiency ranges. Error bars are the standard errors.

Leaf B concentrations remained low in three treatments (07:14:14, YA20:7:10 and control) for all sampling dates, but without going down to the deficiency zone (Figure 4). The fertiliser YS13:11:21 which, in addition to the macronutrients N, P and K, also contains B (0.2%), maintained a nutrient concentration in the leaves higher than the other treatments, especially in the last two samplings. Fe concentrations in the leaves fluctuated within the sufficiency range, but without a clear coherence between treatments. The average levels of Mn in the leaves appeared in the upper part of the sufficiency range, although they never reached the toxicity zone. Control treatment values remained consistently lower than those for the fertilised treatments. The concentrations of Zn and Cu in the leaves showed no relationship with the fertilisation treatments and were, in general, within their sufficiency ranges.

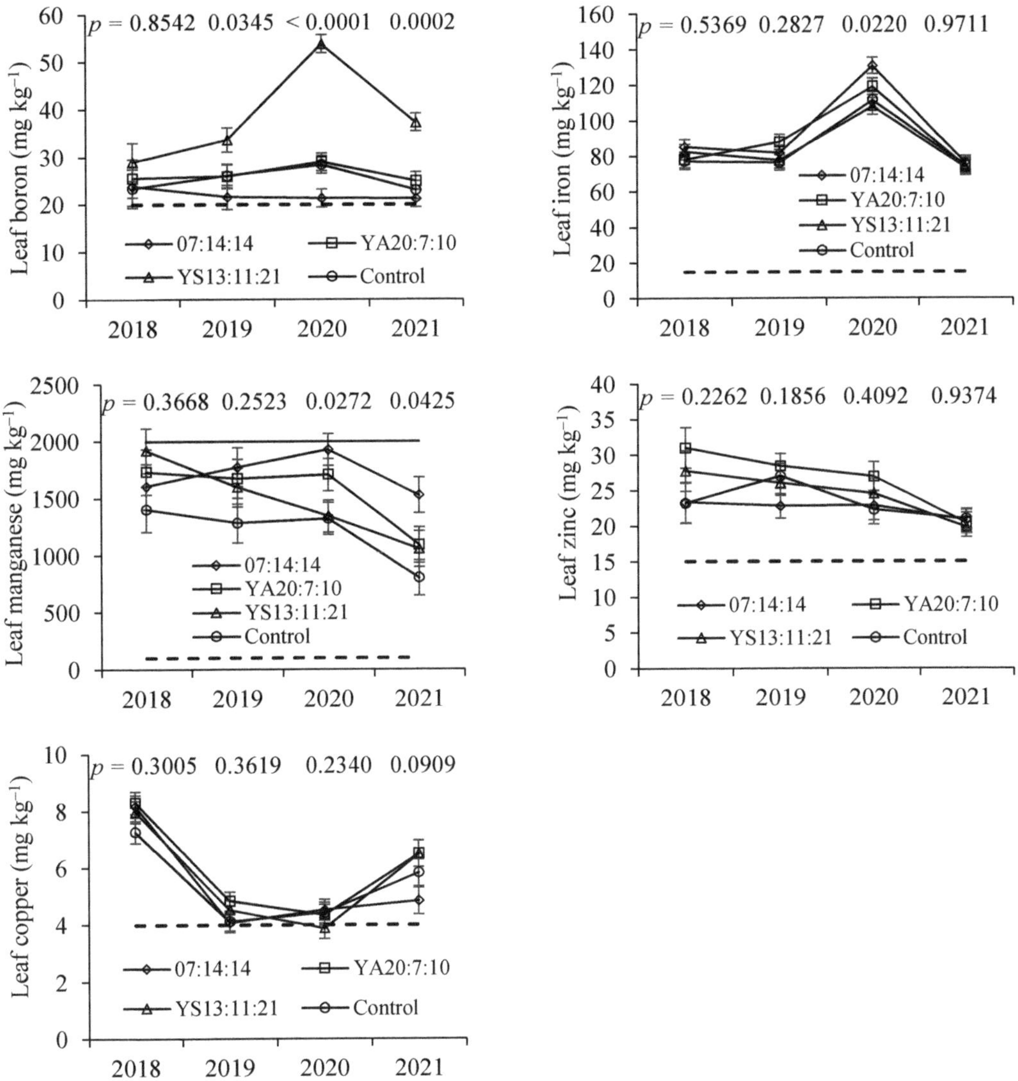

Figure 4. Leaf concentration of boron, iron, manganese, zinc and copper as a response to fertilisation treatments. Dashed and solid lines are, respectively, the lower and upper limits of the sufficiency ranges. Error bars are the standard errors.

The concentration of the majority of nutrients in the kernel and shell did not vary significantly with fertilisation treatments (data not shown). Only the concentration of B in these tissues showed a pattern that is worth reporting. In the kernel, significant differences between treatments were found in the last sampling (2020), with the highest values recorded in treatment YS13:11:21 (Figure 5). In the shell, the same pattern of the kernel was maintained, but with more accentuated average differences between the YS13:11:21 and other treatments. In the shell, the average B concentrations were also higher than in the kernel for the same treatment and sampling date.

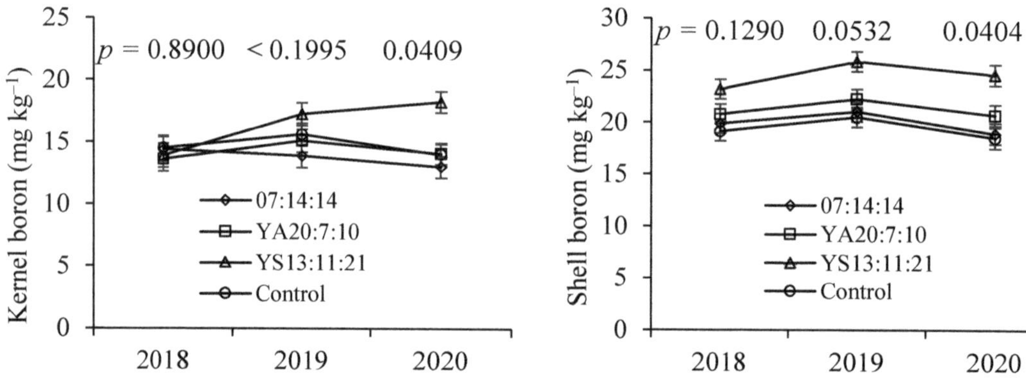

Figure 5. Boron concentrations in kernel and shell as a response to fertilisation treatments. Error bars are the standard errors.

Mean SPAD values showed a tendency to be lower in the control treatment compared to the fertilised treatments (Table 2). In the 2021 reading, significant differences were found between the values of the YA20:710 fertiliser (47.5), the most concentrated in N, and the control (44.2). A somewhat similar trend showed F_V/F_M, although for this variable, the differences were only significant in the 2019 readings, with the mean value of the YA20:7:10 treatment (0.834) being higher than that of the control (0.807).

Table 2. SPAD-readings and F_V/F_M (ratio of variable fluorescence/maximum fluorescence) as a function of fertilisation treatments.

Fertilisation		SPAD			F_V/F_M	
Treatment	2019	2020	2021	2019	2020	2021
7:14:14	44.4 a *	43.6 a	44.8 ab	0.814 ab	0.833 a	0.821 a
YA20:7:10	45.4 a	45.6 a	47.5 a	0.834 a	0.845 a	0.828 a
YS13:11:21	46.3 a	45.2 a	45.9 ab	0.824 ab	0.829 a	0.829 a
Control	43.9 a	43.4 a	44.2 b	0.807 b	0.827 a	0.817 a
Prob > F	0.0762	0.1837	0.0432	0.0331	0.1797	0.3712
St. error	0.63	0.82	0.77	0.006	0.005	0.005

* In columns, means followed by the same letter are not significantly different by Tukey HSD test ($\alpha = 0.05$).

3.3. Chemical Soil Properties

Fertilisation treatments carried out during the four consecutive years did not influence soil organic C content (Table 3). However, fertilisation acidified the soil ($pH_{H}2O$ and pHKCl) compared to the control treatment. Extractable P also varied significantly with the fertilisation treatment, with the highest mean values appearing in the treatments corresponding to the fertilisers more concentrated in P. Soil K levels also varied significantly with treatments and, as for P, there was also a good consistency between the application of the nutrient and its resulting level in the soil.

Table 3. Soil organic carbon (C), pH and extractable phosphorus (P) and potassium (K) (Egner-Riehm) as a function of fertilisation treatments.

Fertilisation	Organic C			Extractable P	Extractable K
Treatment	g kg^{-1}	pH(H_2O)	pH(KCl)	mg kg^{-1}, P_2O_5	mg kg^{-1}, K_2O
7:14:14	13.8 a *	5.08 bc	4.11 b	142.6 a	414.0 ab
YA20:7:10	13.4 a	4.89 c	4.02 b	95.0 b	327.8 b
YS13:11:21	13.5 a	5.11 bc	4.13 b	121.1 ab	519.0 a
Control	12.5 a	5.41 a	4.33 a	81.3 b	303.8 b
Prob > F	0.6947	<0.0001	0.0017	0.0064	0.0006
St. error	0.79	0.048	0.043	10.5	28.1

* In columns, means followed by the same letter are not significantly different by Tukey HSD test ($\alpha = 0.05$).

Soil exchangeable Ca^{2+} and Mg^{2+} did not vary significantly with fertilisation treatments (Table 4). The exchangeable K^+ (extracted by ammonium acetate) varied significantly with the treatments and the mean values were related to the amount of nutrient provided by the fertilisers, as had been verified with the K extracted by the Egner-Riehm method (ammonium lactate). Soil Na+ levels also varied with treatments, with mean values being significantly higher in the 07:14:14 and YS13:11:21 fertiliser plots compared to the values in the YA20:7:10 and control plots. Exchangeable acidity also varied significantly between treatments, with the lowest mean value being recorded in the control treatment. The cation-exchange capacity did not vary significantly with the treatments, maybe because no significant differences were found between two important bases, Ca^{2+} and Mg^{2+}.

Table 4. Soil exchangeable bases, exchangeable acidity (EA) and cation-exchange capacity (CEC) as a function of fertilisation treatments.

			Exchangeable Complex			
Fertilisation	Ca^{2+}	Mg^{2+}	K^+	Na^+	EA	CEC
Treatment			cmol$_c$ kg^{-1}			
7:14:14	3.43 a *	1.00 a	1.08 ab	0.44 a	0.85 ab	6.73 a
YA20:7:10	2.88 a	1.02 a	0.81 b	0.07 b	1.15 a	5.95 a
YS13:11:21	2.36 a	0.78 a	1.37 a	0.44 a	1.17 a	6.00 a
Control	3.25 a	1.16 a	0.80 b	0.16 b	0.70 b	6.34 a
Prob > F	0.3233	0.2371	0.0017	0.0002	0.0073	0.7907
St. error	0.42	0.12	0.09	0.05	0.09	0.61

* In columns, means followed by the same letter are not significantly different by Tukey HSD test ($\alpha = 0.05$).

Soil B levels were significantly higher in the YS13:11:21 treatment than in the other fertilisation treatments and in the control (Table 5). In contrast, soil Fe, Zn and Cu levels did not vary significantly with treatments. In the case of Mn, significant differences between treatments were observed, with the lowest mean values being recorded in the control treatment.

Table 5. Soil boron, iron, zinc, copper and manganese as a function of fertilisation treatments.

Fertilisation	Boron	Iron	Zinc	Copper	Manganese
Treatment			mg kg^{-1}		
7:14:14	0.35 b *	65.2 a	2.7 a	2.0 a	152.5 a
YA20:7:10	0.43 b	72.6 a	2.3 a	1.9 a	136.2 ab
YS13:11:21	1.28 a	59.2 a	2.8 a	1.8 a	149.1 ab
Control	0.32 b	60.5 a	2.2 a	2.6 a	127.8 b
Prob > F	<0.0001	0.2244	0.0852	0.0654	0.0240
St. error	0.102	4.67	0.17	0.17	5.40

* In columns, means followed by the same letter are not significantly different by Tukey HSD test ($\alpha = 0.05$).

Soil ammonium levels extracted by hot or cold KCl did not vary significantly with the fertilisation treatments, although the average values were higher in treatment YA20:7:10, the fertiliser being more concentrated in N (Table 6). The hydrolysable NH_4^+ showed significant differences between treatments, with the mean value of YA20:7:10 being higher than that of the other treatments. Soil nitrate levels also varied significantly between treatments, with YA20:7:10 and the control recording the highest and lowest mean values, respectively.

Table 6. Soil ammonium (NH_4^+) extracted by hot and cold potassium chloride, NH_4^+ hydrolysable (Hyd) (NH_4^+ hot − NH_4^+ cold) and nitrate extracted by cold KCl.

Fertilisation	NH_4^+ Hot	NH_4^+ Cold	NH_4^+ Hyd	NO_3^- Cold
Treatment		mg kg^{-1}		
7:14:14	82.1 a *	69.0 a	13.1 b	59.0 b
YA20:7:10	108.1 a	92.20 a	15.9 a	100.5 a
YS13:11:21	72.3 a	58.1 a	14.2 b	78.0 ab
Control	65.4 a	51.7 a	13.6 b	44.9 b
Prob > F	0.4803	0.5649	0.9213	0.0038
St. error	19.99	21.12	2.99	8.65

* In columns, means followed by the same letter are not significantly different by Tukey HSD test ($\alpha = 0.05$).

The development of spontaneous vegetation beneath the canopy of trees, where fertilisers were applied, showed significant differences between treatments (Table 7). Dry matter yield appeared in three response groups, in which the values were higher in the YA20:7:10 treatment, followed by the YS13:11:21 and 7:14:14 treatments and finally, the control treatment. N concentrations in the dry matter followed exactly the same trend, while the concentrations of the macronutrients P and K did not vary significantly between treatments. Tissue B concentration was significantly higher in YS13:11:21 than in the other treatments. Tissue Mn levels did not differ significantly between treatments, although the control treatment had the lowest mean value, following the trend observed in chestnut leaves and the soil. For the other nutrients analysed (Ca, Mg, Fe, Cu and Zn), there were no significant differences or any trend that deserve to be reported.

Table 7. Dry matter yield and nutrient concentrations in the herbaceous vegetation developing under the canopy of chestnut trees where the fertilisation treatments were applied.

		Tissue Nutrient Concentration				
Fertilisation	DM Yield	N	P	K	B	Mn
Treatment	Mg ha^{-1}		g kg^{-1}		mg kg^{-1}	
7:14:14	2.3 b *	24.5 b	2.6 a	32.2 a	20.5 b	636.3 a
YA20:7:10	3.5 a	29.6 a	2.4 a	33.8 a	20.4 b	722.9 a
YS13:11:21	2.6 b	26.2 b	2.5 a	35.3 a	43.0 a	681.9 a
Control	1.8 c	19.5 c	2.4 a	27.0 a	18.4 b	534.8 a
Prob > F	0.0001	< 0.0001	0.0898	0.3458	0.0003	0.1152
St. error	0.16	0.45	0.07	3.19	2.51	49.14

* In columns, means followed by the same letter are not significantly different by Tukey HSD test ($\alpha = 0.05$).

4. Discussion

The annual and accumulated (2018–2020) chestnut yields did not vary significantly with fertilisation treatments. However, the average accumulated yield showed a clear tendency of reduction in the control in relation to the fertilised treatments. Chestnut trees are particularly large, with a huge perennial structure and canopy. In previous studies, it has already been observed that chestnut tends to respond poorly to fertilisers applied to the soil, probably due to the buffering effect that the perennial structure exerts in regulating the supply of nutrients for the growth of the aerial plant parts [5].

Leaf N concentrations tended to be higher in the fertilised treatments, which were more concentrated in N compared to the control, although significant differences only occurred on the last sampling date. In the control treatment, the values were close to the lower limit of the sufficiency range, having even fallen into the deficiency range on some sampling dates. The relevant structural role of N in plant tissues is undeniable [33], and it is still recognized as the main nutrient that limits plant productivity in both natural ecosystems and cultivated fields [19]. This result points to N as the most likely cause for the apparent drop in productivity in the control treatment. The SPAD values, which have been used mainly as an index of the N nutritional status of crops [5,29,34], agreed with tissue N concentrations, and in 2021, significant differences were found between the YA20:7:10 treatment and the control. The F_V/F_M ratio, a widely used indicator of photoinhibition or other injuries at the PSII complexes [35], followed the same trend as the N nutritional status indices, with values in the control being lower than in the treatment YA20:7:10 in the 2019 reading. However, the values never dropped below 0.78, which is the threshold limit below which most plants are considered to be under clear environmental stress [26,36–38]. Thus, the values of the maximum quantum efficiency of PSII also highlight the poor response of the photochemical reactions of photosynthesis of these huge trees to nutrient supply.

Leaf P concentrations did not vary significantly with treatments and always remained within the sufficiency range established for this species (1.1 to 3.0%) [25,39]. Initial soil P levels were at a level classified as medium (Table 1), and in the region, it has been difficult to obtain a response of different crop species to P applications [5,40,41]. In chestnut, the lack of response may be due to those reasons but also to the buffering effect of the perennial parts, already mentioned for N, and to a possible role of mycorrhizal fungi. Chestnut is recognized as a plant that establishes symbiotic relationships with several mycorrhizal fungi [42]. One of the main benefits for mycorrhizal plants is the access to sparingly soluble P sources that non-mycorrhizal plants do not have [43–47]. Thus, whatever the reason, the results seem to indicate a reduced importance of P in chestnut tree fertilisation programmes.

Although soil K levels increased with the application of fertilisers, no significant differences were observed in the K concentrations of the chestnut leaves. Leaf K concentrations varied greatly over the years and between treatments, although they generally remained within the sufficiency range. This pattern of K is common in shrub and tree species [13,21,38,48] and may be due to source/sink relationships and/or environmental constraints. Growing fruits are a primary sink for available K, the nutrient being remobilized from leaves [33]. In chestnut, fruit growth coincides with the end of summer, a period in which there is often little soil moisture, which limits the movement of nutrients in the soil by mass flow and diffusion, making nutrient uptake difficult [19]. In addition, the original levels of K in the soil were relatively high, which would have reduced the impact of applying K as a fertiliser. Finally, the buffering effect of the perennial tree structure may have moderated the effects of the fertiliser applications, as mentioned for N and P.

Tissue B levels differed between treatments on three dates, with YS13:11:21 fertiliser (B-rich) values being significantly higher than in the other treatments. Additionally in the fruits, B concentrations were the highest in the YA13:11:21 treatment, in particular, in the shell. Boron is very important in dicots, where it plays an important role in cell wall and membrane integrity, with these plants requiring greater amounts of B than monocots [48–50]. In the region, dicots often respond to the application of B [12,41,51]. In this field trial, however, tissue B levels were never below the sufficiency range even though they were close to the limit. Perhaps for this reason, B was not determinant in crop productivity, in contrast to what has been shown in other studies with chestnut [11,52–54].

In the control treatment, mean leaf Mn levels were lower than in the fertilised treatments, and the differences were statistically significant on the last sampling date. A tendency for lower Mn levels in soils was also observed in the control treatment. Soil pH in the control was higher than in the fertilised treatments, particularly in those that had a greater amount of applied N. Nitrification can decrease soil pH [19], and the increase of available Mn levels in the fertilised treatments may have been a reflection of pH reduction [19,49]. Even so, leaf Mn levels never exceeded the upper limit of the sufficiency range, which is set at 2000 mg kg^{-1} [25,39], so its effect on crop productivity must not have been relevant.

In the autumn, the availability of inorganic N in the soil as measured by hydrolysable NH_4^+ and cold-extracted NO_3^- was higher in the treatments with more N-concentrated fertilisers. This may indicate a greater risk of N loss through leaching and/or denitrification, since the rainy season follows, a precondition for the occurrence of these phenomena [19]. However, in April, dry matter and tissue N concentrations in the spontaneous vegetation were also higher in the treatments that received more N as a fertiliser. These plants, which appear after the first autumn rains and develop during the winter, can play important roles by controlling soil erosion [55,56], increasing soil organic matter [57,58] and developing ecosystem biodiversity [59,60]. They also act as an N catch crop [21,61], justifying the promotion of their presence in orchard soils [62]. This result also shows that the effect of N applications is easier to obtain in herbaceous vegetation than in a tree, probably due to the latter's large perennial structure. It seems clear that in large trees, it is more difficult to get a response to fertilisation and therefore, more difficult to optimize a fertilisation programme. In trees, a dynamic optimization method should always be used [63], by which, based on a given annual fertilisation plan, nutrient concentrations in leaves are monitored and fertiliser rates adjusted according to increasing (reduce fertilisation) or decreasing (increase fertilisation) concentrations of a particular nutrient being observed in leaves. This is a programme optimized for long-term monitoring and not just based on annual observations, which is the procedure currently used in fruit crops.

In Mediterranean climates, with rainfall concentrated in the winter, and the summer being particularly dry, in rainfed orchards, where there are no fertigation practices, there is only one window of opportunity to apply fertilizers, which is in early spring, just before the regrowth of vegetation. If applying earlier, there is a risk of loss of mobile nutrients, such as N, by leaching and denitrification, whereas if applying later, there is a risk of loss of effectiveness due to reduced soil moisture [21]. In addition, slow and controlled release fertilizers tend to be less effective, as they delay nutrient availability for the summer, when the opportunity for root uptake is low [64].

5. Conclusions

The results of this study indicate that these large trees had a poor response to the annual application of fertilisers. Even so, the nutrient that had the greatest effect on the plant was N, since on some dates, significant differences were observed between treatments in leaf N concentrations, and the nutrient in the control treatment was close to, or even below the lower limit of the sufficiency range. The poor response of trees to fertiliser applications was probably due to the buffering effect that the huge perennial structure has on the redistribution of nutrients by the growing plant parts. Thus, in large trees, the fertilisation plan must be based on monitoring leaf nutrient concentration over time and on evaluating the trend of the nutrient concentrations in the leaves. As long as there is no decrease of leaf nutrient concentration that approaches the lower limit of the sufficiency ranges, the farmer should probably avoid spending money on fertiliser applications. In contrast, the herbaceous vegetation developing beneath the canopy responded to the application of fertilisers, in particular to N, with increased dry matter yield and tissue nutrient concentrations. This vegetation, which begins to develop with the first autumn rains, seems to play an important role in protecting the soil and acts as a catch crop, reducing the risk of N loss during the winter.

Author Contributions: Conceptualization, M.Â.R. and C.M.C.; methodology, M.Â.R., M.A., T.F. and N.A.d.M.; investigation, M.R.B., A.C.F. and S.R.; data curation, M.Â.R.; writing—original draft preparation, M.A.; writing—review and editing, M.Â.R. and C.M.C.; project administration, J.S. and A.P.; funding acquisition, J.S. and M.Â.R. All authors have read and agreed to the published version of the manuscript.

Funding: This work was funded by the Operational Group "EGIS–Estratégias de Gestão Integrada do Solo e da Água em Espécies Produtoras de Frutos Secos", funded by PT2020 and EAFRD (European Agricultural Fund for Rural Development). The authors are also grateful to the Foundation for Science and Technology (FCT, Portugal) for financial support through national funds FCT/MCTES (PIDDAC) to CIMO (UIDB/00690/2020 and UIDP/00690/2020), SusTEC (LA/P/0007/2021) and CITAB (UIDB/04033/2020).

Institutional Review Board Statement: Not applicable.

Informed Consent Statement: Not applicable.

Data Availability Statement: The data that support the findings of this study are available from the corresponding author upon reasonable request.

Conflicts of Interest: The authors declare no conflict of interest.

References

1. Gouveia, E. Doenças [Crop diseases]. In *Manual de boas Práticas do Castanheiro [Handbook of Good Management Practices on Chestnut Orchards]*; Bento, A., Ribeiro, A.C., Eds.; Terras de Trás-os-Montes: Bragança, Portugal, 2020; pp. 191–203. (In Portuguese)
2. Santos, A.; Marrão, R.; Bento, A. Pragas [Pests]. In *Manual de Boas Práticas do Cas-Tanheiro [Handbook of Good Management Practices on Chestnut Orchards]*; Bento, A., Ribeiro, A.C., Eds.; Terras de Trás-os-Montes: Bragança, Portugal, 2020; pp. 205–227. (In Portuguese)
3. Rosário, J.N.; Coelho, V.; Rodrigues, M.; Raimundo, S.; Afonso, S.; Arrobas, M.; Gouveia, M.E. Metalaxyl-M, phosphorous acid and potassium silicate applied as soil drenches show different chestnut seedling performance and protection against Phytophthora root rot. *Eur. J. Plant Pathol.* **2021**, *161*, 147–159. [CrossRef]
4. INE (Instituto Nacional de Estatística); [National Institute of Statistics]. *Estatísticas Agrícolas–2020 [Agricultural Statistics–2020]*; INE: Lisboa, Portugal, 2021. (In Portuguese)
5. Rodrigues, M.; Raimundo, S.; Pereira, A.; Arrobas, M. Large Chestnut Trees (Castanea sativa) Respond Poorly to Liming and Fertilizer Application. *J. Soil Sci. Plant Nutr.* **2020**, *20*, 1261–1270. [CrossRef]
6. Aguiar, C.F. Sistemática, morfologia, fenologia e biologia da reprodução [Systematics, morphology, phenology and biology of reproduction]. In *Manual de boas Práticas do Castanheiro [Handbook of Good Management Practices on Chestnut Orchards]*; Bento, A., Ribeiro, A.C., Eds.; Terras de Trás-os-Montes: Bragança, Portugal, 2020; pp. 31–72. (In Portuguese)
7. Almeida, A.C.F. Instalação da cultura [Crop planting]. In *Manual de boas Práticas do Castanheiro [Handbook of Good Management Practices on Chestnut Orchards]*; Bento, A., Ribeiro, A.C., Eds.; Terras de Trás-os-Montes: Bragança, Portugal, 2020; pp. 85–92. (In Portuguese)
8. Rodrigues, M.A.; Arrobas, M. Gestão do solo [Soil management]. In *Manual de boas Práticas do Castanheiro [Handbook of Good Management Practices on Chestnut Orchards]*; Bento, A., Ribeiro, A.C., Eds.; Terras de Trás-os-Montes: Bragança, Portugal, 2020; pp. 119–129. (In Portuguese)
9. Patrício, M.A. Sistemas de condução e poda [Training and pruning systems]. In *Manual de boas Práticas do Castanheiro [Handbook of Good Management Practices on Chestnut Orchards]*; Bento, A., Ribeiro, A.C., Eds.; Terras de Trás-os-Montes: Bragança, Portugal, 2020; pp. 149–170. (In Portuguese)
10. Arrobas, M.; Rodrigues, M.A. Fertilização [Crop fertilization]. In *Manual de boas Práticas do Castanheiro [Handbook of Good Management Practices on Chestnut Orchards]*; Bento, A., Ribeiro, A.C., Eds.; Terras de Trás-os-Montes: Bragança, Portugal, 2020; pp. 131–148. (In Portuguese)
11. Rodrigues, M.; Grade, V.; Barroso, V.; Pereira, A.; Cassol, L.C.; Arrobas, M. Chestnut Response to Organo-mineral and Controlled-Release Fertilizers in Rainfed Growing Conditions. *J. Soil Sci. Plant Nutr.* **2019**, *20*, 380–391. [CrossRef]
12. Arrobas, M.; Ribeiro, A.; Barreales, D.; Pereira, E.L.; Rodrigues, M. Soil and foliar nitrogen and boron fertilization of almond trees grown under rainfed conditions. *Eur. J. Agron.* **2019**, *106*, 39–48. [CrossRef]
13. Lopes, J.I.; Gonçalves, A.; Brito, C.; Martins, S.; Pinto, L.; Moutinho-Pereira, J.; Raimundo, S.; Arrobas, M.; Rodrigues, M.; Correia, C.M. Inorganic Fertilization at High N Rate Increased Olive Yield of a Rainfed Orchard but Reduced Soil Organic Matter in Comparison to Three Organic Amendments. *Agronomy* **2021**, *11*, 2172. [CrossRef]
14. Ferreira, I.Q.; Arrobas, M.; Moutinho-Pereira, J.M.; Correia, C.M.; Rodrigues, M. The effect of nitrogen applications on the growth of young olive trees and nitrogen use efficiency. *Turk. J. Agric. For.* **2020**, *44*, 278–289. [CrossRef]
15. Yang, X.; Zhang, P.; Li, W.; Hu, C.; Zhang, X.; He, P. Evaluation of four seagrass species as early warning indicators for nitrogen overloading: Implications for eutrophic evaluation and ecosystem management. *Sci. Total Environ.* **2018**, *635*, 1132–1143. [CrossRef]
16. Poikane, S.; Phillips, G.; Birk, S.; Free, G.; Kelly, M.G.; Willby, N.J. Deriving nutrient criteria to support 'good' ecological status in European lakes: An empirically based approach to linking ecology and management. *Sci. Total Environ.* **2019**, *650*, 2074–2084. [CrossRef]
17. Coyne, M.S. Biological denitrification. In *Nitrogen in Agricultural Systems, Agrono-My Monograph n.o 49*; Schepers, J.S., Raun, W.R., Eds.; ASA, CSSA, SSSA: Madison, WI, USA, 2008; pp. 201–253.

18. Pelster, D.E.; Larouche, F.; Rochette, P.; Chantigny, M.H.; Allaire, S.; Angers, D.A. Nitrogen fertilization but not soil tillage affects nitrous oxide emissions from a clay loam soil under a maize–soybean rotation. *Soil Tillage Res.* **2011**, *115–116*, 16–26. [CrossRef]
19. Weil, R.R.; Brady, N.C. *The Nature and Properties of Soils*, 15th ed.; Global Edition: London, UK, 2017.
20. Bryson, G.; Mills, H.A.; Sasseville, D.N.; Jones, J.B., Jr.; Barker, A.V. *Plant Analysis Handobook III. A Guide to Sampling, Preparation, Analysis and Interpretation for Agronomic and Horticultural Crops*; Micro-Macro Publishing, Inc.: Athens, GA, USA, 2014.
21. Silva, E.; Arrobas, M.; Gonçalves, A.; Martins, S.; Raimundo, S.; Pinto, L.; Brito, C.; Moutinho-Pereira, J.; Correia, C.M.; Rodrigues, M. A controlled-release fertilizer improved soil fertility but not olive tree performance. *Nutr. Cycl. Agroecosyst.* **2021**, *120*, 1–15. [CrossRef]
22. Rodrigues, M.; Torres, L.D.N.D.; Damo, L.; Raimundo, S.; Sartor, L.; Cassol, L.C.; Arrobas, M. Nitrogen Use Efficiency and Crop Yield in Four Successive Crops Following Application of Biochar and Zeolites. *J. Soil Sci. Plant Nutr.* **2021**, *21*, 1053–1065. [CrossRef]
23. Tilman, D.; Balzer, C.; Hill, J.; Befort, B.L. Global food demand and the sustainable intensification of agriculture. *Proc. Natl. Acad. Sci. USA* **2011**, *108*, 20260–20264. [CrossRef] [PubMed]
24. Tittonell, P. Ecological intensification of agriculture—Sustainable by nature. *Curr. Opin. Environ. Sustain.* **2014**, *8*, 53–61. [CrossRef]
25. Arrobas, M.; Afonso, S.; Rodrigues, M. Diagnosing the nutritional condition of chestnut groves by soil and leaf analyses. *Sci. Hortic.* **2018**, *228*, 113–121. [CrossRef]
26. Arrobas, M.; Afonso, S.; Ferreira, I.Q.; Moutinho-Pereira, J.; Correia, C.M.; Rodrigues, M. Liming and application of nitrogen, phosphorus, potassium, and boron on a young plantation of chestnut. *Turk. J. Agric. For.* **2017**, *41*, 441–451. [CrossRef]
27. INE (Instituto Nacional de Estatística); [National Institute of Statistics]. *Recenseamento Agrícola–2019 [Agricultur-al Census–2019]*; INE: Lisboa, Portugal, 2019.
28. IPMA (Instituto Português do Mar e da Atmosfera); [Portuguese Institute of the Sea and the Atmosphere]. Normais Climatológicas [Climate Normals]. 2022. Available online: https://www.ipma.pt/pt/oclima/normais.clima/ (accessed on 15 April 2022). (In Portuguese).
29. Van Reeuwijk, L. *Procedures for Soil Analysis*; Technical Paper 9; Int Soil Ref Inform Center: Wageningen, MI, USA, 2002.
30. Lakanen, E.; Erviö, R. A Comparison of eight extractants for the determination of plant available micronutrients in soils. *Acta Agral. Fenn.* **1971**, *123*, 223–232.
31. Baird, R.B.; Eaton, A.D.; Rice, E.W. Nitrate by ultraviolet spectrophotometric method. In *Standard Methods for the Examination of Water and Wastewater*; American Public Health Association, American Water Works Association, Water Environment Federation: Washington, DC, USA, 2017.
32. Temminghoff, E.E.J.M.; Houba, V.G. *Plant Analysis Procedures*, 2nd ed.; Kluwer Acad Publ.: Dordrecht, The Netherlands, 2004. [CrossRef]
33. Hawkesford, M.; Horst, W.; Kichey, T.; Lambers, H.; Schjoerring, J.; Skrumsager Møller, I.; White, P. Functions of macronutrients. In *Marschner's Mineral Nutrition of Higher Plants*, 3rd ed.; Marschner, P., Ed.; Elsevier Ltd.: Amsterdam, The Netherlands, 2012; pp. 135–189. [CrossRef]
34. Afonso, S.; Arrobas, M.; Ferreira, I.Q.; Rodrigues, M. Assessing the potential use of two portable chlorophyll meters in diagnosing the nutritional status of plants. *J. Plant Nutr.* **2017**, *41*, 261–271. [CrossRef]
35. Maxwell, K.; Johnson, G.N. Chlorophyll fluorescence—A practical guide. *J. Exp. Bot.* **2000**, *51*, 659–668. [CrossRef]
36. Opti-Sciences. *Desktop Plant Stress Guide*; Edition 3.0; Opti-Sciences: Hudson, WI, USA, 2014; Available online: www.optisci.com (accessed on 15 April 2022).
37. Rodrigues, M.; Afonso, S.; Ferreira, I.Q.; Arrobas, M. Response of stevia to nitrogen fertilization and harvesting regime in northeastern Portugal. *Arch. Agron. Soil Sci.* **2017**, *63*, 626–637. [CrossRef]
38. Ferreira, I.Q.; Arrobas, M.; Moutinho-Pereira, J.; Correia, C.; Rodrigues, M. Olive response to potassium applications under different water regimes and cultivars. *Nutr. Cycl. Agroecosyst.* **2018**, *112*, 387–401. [CrossRef]
39. Portela, E.; Martins, A.; Pires, A.L.; Raimundo, F.; Marques, G. Práticas culturais no souto: O manejo do solo [Cropping practices in chestnut orchards: Soil management]. In *Castanheiros [Chestnut Groves]*; Gomes-Laranjo, J., Ferreira-Cardoso, J., Portela, E., Abreu, C.G., Eds.; Universidade de Trás-os-Montes e Alto Douro: Vila Real, Portugal, 2007; pp. 207–264. (In Portuguese)
40. Ferreira, I.Q.; Rodrigues, M.; Moutinho-Pereira, J.M.; Correia, C.M.; Arrobas, M. Olive tree response to applied phosphorus in field and pot experiments. *Sci. Hortic.* **2018**, *234*, 236–244. [CrossRef]
41. Rodrigues, M.; Ferreira, I.Q.; Afonso, S.; Arrobas, M. Sufficiency ranges for lemon balm and nutrient removals in aboveground phytomass. *J. Plant Nutr.* **2018**, *41*, 996–1008. [CrossRef]
42. Pereira, E.; Coelho, V.; Tavares, R.M.; Lino-Neto, T.; Baptista, P. Effect of competitive interactions between ectomycorrhizal and saprotrophic fungi on Castanea sativa performance. *Mycorrhiza* **2012**, *22*, 41–49. [CrossRef] [PubMed]
43. Mechri, B.; Attia, F.; Tekaya, M.; Cheheb, H.; Hammami, M. Colonization of olive trees (*Olea europaea* L.) with the arbuscular mycorrhizal fungus *Glomus* sp. modified the glycolipids biosynthesis and resulted in accumulation of unsaturated fatty acids. *J. Plant Physiol.* **2014**, *171*, 1217–1220. [CrossRef] [PubMed]
44. Lanfranco, L.; Bonfante, P.; Genre, A. The mutualistic interaction between plants and arbuscular mycorrhizal fungi. *Microbiol. Spectr.* **2016**, *4*, 1–20. [CrossRef] [PubMed]

45. Tekaya, M.; Mechri, B.; Mbarki, N.; Cheheb, H.; Hammami, M.; Attia, F. Arbuscular mycorrhizal fungus Rhizophagus irregularis influences key physiological parameters of olive trees (*Olea europaea* L.) and mineral nutrient profile. *Photosynthetica* **2017**, *55*, 308–316. [CrossRef]
46. Ortaş, I.; Bykova, A. The Effect of Mycorrhiza Inoculation and Phosphorus Application on Phosphorus Efficiency of Wheat Plants. *Commun. Soil Sci. Plant Anal.* **2018**, *49*, 1199–1207. [CrossRef]
47. Lopes, J.; Arrobas, M.; Brito, C.; Gonçalves, A.; Silva, E.; Martins, S.; Raimundo, S.; Rodrigues, M.; Correia, C. Mycorrhizal Fungi were More Effective than Zeolites in Increasing the Growth of Non-Irrigated Young Olive Trees. *Sustainability* **2020**, *12*, 10630. [CrossRef]
48. Afonso, S.; Arrobas, M.; Morais, J.S.; Rodrigues, M. Hop dry matter yield and cone quality responses to amino acid and potassium-rich foliar spray applications. *J. Plant Nutr.* **2021**, *44*, 2042–2056. [CrossRef]
49. Broadley, M.; Brown, P.; Cakmak, I.; Rengel, Z.; Zhao, F. Function of nutrients: Micronutrients. In *Marschner's Mineral Nutrition of Higher Plants*, 3rd ed.; Marschner, P., Ed.; Elsevier Ltd.: Amsterdam, The Netherlands, 2012; pp. 191–248. [CrossRef]
50. Wimmer, M.A.; Eichert, T. Review: Mechanisms for boron deficiency-mediated changes in plant water relations. *Plant Sci.* **2013**, *203–204*, 25–32. [CrossRef]
51. Ferreira, I.Q.; Rodrigues, M.; Arrobas, M. Soil and foliar applied boron in olive: Tree crop growth and yield, and boron remobilization within plant tissues. *Span. J. Agric. Res.* **2019**, *17*, e0901. [CrossRef]
52. Portela, E.A.C.; Ferreira-Cardoso, J.V.; Louzada, J.L. Boron application on a chestnut orchard: Effect on yield and quality of nuts. *J. Plant Nutr.* **2011**, *34*, 1245–1253. [CrossRef]
53. Portela, E.; Ferreira-Cardoso, J.; Louzada, J.; Gomes-Laranjo, J. Assessment of Boron Application in Chestnuts: Nut Yield and Quality. *J. Plant Nutr.* **2015**, *38*, 973–987. [CrossRef]
54. Portela, E.M.A.C.; Louzada, J.L.P. Early diagnosis of boron deficiency in chestnut. *J. Plant Nutr.* **2012**, *35*, 304–310. [CrossRef]
55. Keesstra, S.; Pereira, P.; Novara, A.; Brevik, E.C.; Azorin-Molina, C.; Parras-Alcántara, L.; Jordán, A.; Cerdà, A. Effects of soil management techniques on soil water erosion in apricot orchards. *Sci. Total Environ.* **2016**, *551–552*, 357–366. [CrossRef] [PubMed]
56. Repullo-Ruibérriz De Torres, M.A.; Ordóñez-Fernández, R.; Giráldez, J.V.; Márquez-García, J.; Laguna, A.; Carbonell-Bojollo, R. Efficiency of four different seeded plants and native vegetation as cover crops in the control of soil and carbon losses by water erosion in olive orchards. *Land Degrad. Dev.* **2018**, *29*, 2278–2290. [CrossRef]
57. Márquez-García, F.; Sánchez, E.J.G.; Castro-Garcia, S.; Ordóñez-Fernández, R. Improvement of soil carbon sink by cover crops in olive orchards under semiarid conditions. Influence of the type of soil and weed. *Span. J. Agric. Res.* **2013**, *11*, 335. [CrossRef]
58. Rodrigues, M.; Dimande, P.; Pereira, E.L.; Ferreira, I.Q.; Freitas, S.; Correia, C.M.; Moutinho-Pereira, J.; Arrobas, M. Early-maturing annual legumes: An option for cover cropping in rainfed olive orchards. *Nutr. Cycl. Agroecosyst.* **2015**, *103*, 153–166. [CrossRef]
59. Nicholls, C.I.; Parrella, M.P.; Altieri, M.A. Reducing the abundance of leafhoppers and thrips in a northern California organic vineyard through maintenance of full season floral diversity with summer cover crops. *Agric. For. Entomol.* **2000**, *2*, 107–113. [CrossRef]
60. Irvin, N.A.; Bistline-East, A.; Hoddle, M.S. The effect of an irrigated buckwheat cover crop on grape vine productivity, and beneficial insect and grape pest abundance in southern California. *Biol. Control* **2016**, *93*, 72–83. [CrossRef]
61. De Notaris, C.; Rasmussen, J.; Sørensen, P.; Olesen, J.E. Nitrogen leaching: A crop rotation perspective on the effect of N surplus, field management and use of catch crops. *Agric. Ecosyst. Environ.* **2018**, *255*, 1–11. [CrossRef]
62. Rodrigues, M.A.; Arrobas, M. Cover cropping for increasing fruit production and farming sustainability. In *Fruit Crops: Diagnosis and Management of Nutrient Constraints*; Srivastava, A.K., Hu, C., Eds.; Elsevier: Amsterdam, The Netherlands, 2020; pp. 279–295.
63. Arrobas, M.; Rodrigues, M.A. *Fertilização do Pomar Baseada no Método de Otimização Dinâmica. Revista Voz do Campo*; Voz do Campo Editora, Lda.: Castelo Branco, Portugal, 2021; pp. 61–63.
64. Arrobas, M.; Belotto, L.B.; Marchetti, J.A.; Barroso, V.; Raimundo, S.; Cassol, L.C.; Correia, C.M.; Rodrigues, M.Â. Excessive delay in nutrient release by controlled-release fertilizers can reduce chestnut yield. *Horticulturae* **2022**, *8*, 1067. [CrossRef]

soil systems

MDPI

Article

Effect of Natural Phytohormones on Growth, Nutritional Status, and Yield of Mung Bean (*Vigna radiata* L.) and N Availability in Sandy-Loam Soil of Sub-Tropics

Aasma Parveen [1,†], Muhammad Mahran Aslam [2,3,†], Rashid Iqbal [4,*], Muhammad Ali [3], Muhammad Kamran [5,*], Mona S. Alwahibi [6], Muhammad Akram [7] and Mohamed S. Elshikh [6]

1 Department of Soil Science, Faculty of Agriculture & Environment, The Islamia University of Bahawalpur, Bahawalpur 63100, Pakistan

2 State Key Laboratory of Agrobiotechnology/Beijing Key Laboratory of Crop Genetic Improvement, College of Agronomy and Biotechnology, China Agricultural University, Beijing 100193, China

3 Nuclear Institute of Agriculture (NIA), Tandojam 70060, Pakistan

4 Department of Agronomy, Faculty of Agriculture & Environment, The Islamia University of Bahawalpur, Bahawalpur 63100, Pakistan

5 School of Agriculture, Food and Wine, The University of Adelaide, Adelaide, SA 5005, Australia

6 Department of Botany and Microbiology, College of Science, King Saud University, Riyadh 11451, Saudi Arabia

7 Department of Botany, The Islamia University of Bahawalpur, Bahawalpur 63100, Pakistan

* Correspondence: rashid.iqbal@iub.edu.pk (R.I.); muhammad.kamran@adelaide.edu.au (M.K.)

† These authors contributed equally to this work.

Abstract: Climate changes and poor soil nutrient profiles in sub-tropics are determinant factors to estimate crop productivity. This study aims to evaluate the impact of phytohormones, e.g., indole acetic acid (IAA) and gibberellic acid (GA_3), on mung bean yield, seed nutritional profile, and soil N availability in the sub-tropical region of Pakistan. The mung bean plants were treated with three levels (0, 30, and 60 mg L^{-1}) of IAA and GA_3 individually and/or in combination using a hydraulic sprayer. The amendments were applied in the flowering stage (approximately 25 days after germination) in a randomized complete block design. The results revealed that the 60 mg L^{-1} concentration of IAA and GA_3 led to significant changes in the growth and yield traits compared to non-treated plants. For example, GA_3 positively influenced the biological yield (35.0%), total carbohydrate (7.0%), protein (16.0%), and nitrogen (14.0%) contents in mung bean seeds, compared to the control (CK). Additionally, the combined foliar treatment of IAA and GA_3 (IAA_2 + GA_2) displayed a much stronger influence on yield attributes, such as the number of pods by 66.0%, pods' weights by 142.0%, and seed yield by 106.5%, compared with the CK. Mung bean plants showed a significant improvement in leaf photosynthetic pigments under a higher level (60 mg L^{-1}) of sole and combined treatments of IAA and GA_3. Moreover, except abscisic acid, the endogenous concentration of IAA, GA_3, and zeatin was enhanced by 193.0%, 67.0%, and 175.0% after the combined application of IAA and GA_3 (IAA_2 + GA_2) compared to the CK treatment. In addition, soil N availability was increased by 72.8% under the IAA_2 treatment and 61.5% under IAA_2 + GA_2, respectively, compared with the control plot. It was concluded that the combined treatment of IAA and GA_3 (IAA_2 + GA_2) followed by the sole application of GA_3 and IAA at a 60 mg L^{-1} concentration were most effective treatments to improve the morpho-physiology and nutrient profile of mung beans; however, the underlying molecular mechanisms need to be explored further.

Keywords: indole acetic acid; gibberellic acid; *Vigna radiate* L.; photosynthetic pigments; protein; economic yield

1. Introduction

The current cultivation pattern depends on 30 crops which are responsible for providing 95% of the daily caloric requirements of world population [1]. Among them, four crops are a major part of the diet, namely wheat, rice, maize, and potatoes. However, minor crops are still very significant at the national, regional, and local levels. Pulses are rich in protein and contain more than three times higher quality protein than cereals [2]. In addition, pulse crops preserve and improve soil fertility, especially soil N availability, through biological nitrogen fixation, and hence have a significant role in sustainable agriculture. Pulses are popularly known as the "poor man's meat" in most developing countries due to being cheaper and more widely available than animal protein [3]. Mung bean (*Vigna radiata* L.), a short-duration crop (70–90 days), is an important pulse all over the world with admirable economic importance. The grains of mung bean contain 25.67% protein, 1–3% fat, 5.4% carbohydrates, 3.5–4.5% fibers, and 4.5–5.5% ash with very minimal flatulent effects [4], and they are rich in folate and iron [5]. They contain a high amount of protein with a diversity of essential amino acids and are especially rich in lysine [6]. They also contain important forms of fatty acids (linoleic acid and linolenic acid), which are essential in an organism's growth. It is a vital crop to Asian farmers with small land holdings [7].

Mung bean (*Vigna radiata* L.) fixes atmospheric nitrogen and contains high nutritious value for forage and seed purposes. The crop fits well in multi-cropping systems because of its rapid growth and early maturity. Subsequently, the crop is widely grown in marginal and abiotically stressed agro-ecosystems [8], but it experiences considerable yield losses. Worldwide food insecurity affects more than 800 million people. Almost 60% of them live in South Asia and sub-Saharan Africa [9], whereby half of them are livestock keepers and small land holders [10]. Small land holder farmers in dryland areas or areas with erratic rainfall usually lack technologies to diversify their production, making them particularly vulnerable. In Pakistan, the production of pulses has stagnated and has not kept the pace needed to meet consumption demand. The gap has been fulfilled by massive imports [7]. The size of areas suitable for the growth of pulses has been steady since the 1960s, with chickpeas being the dominant pulse grown in Pakistan [11]. However, the area in which mung beans are cultivated increased substantially during 2019–20 by 35%, in which production increased by 65% in Punjab province, 6% in Khyber Pakhtunkhwa province, and 17% in Baluchistan province. About 88% of the area in which mung beans are cultivated is in the Punjab province, which produces 85% of the total production in the country [12]. Still, a lot needs to be done in Pakistan to enhance mung bean production in terms of quality and yield. More recently, the Australian Centre for International Agricultural Research, in collaboration with the Pakistani government, supported research for improving the productivity and marketing of pulses in Pakistan [11]. Therefore, considering this scenario, the total pulse production must urgently be increased to meet the consumption and protein demands [7].

Many artificial and natural soil amendments, e.g., phytohormones [13,14], C-rich organic amendments [15–17], and synthetic minerals nutrients [18–21] have been recognized to govern soil fertility; facilitate progressive methods, from the germination to the harvesting of crop plants; and facilitate improvements in plant production [22]. The natural form of auxin found in plants is indole-3-acetic acid (IAA), which is primarily present in the young leaves and stem apex of a plant. Indole-3-acetic acid (IAA) is one of the main important enzymes, which is non-toxic for plants even at high concentrations [23]. IAA has been recognized in root growth, cell division, cell elongation, adventitious root formation, tissue swelling, embryogenesis induction, callus initiation, as well as the loosening of cell walls even at lower levels of this hormone [24,25]. Gibberellic acid (GA_3) is an important natural plant growth regulator (PGR) which enhances the growth, development, and yield of different crops. GA_3 is a phyto-hormone, and terpenoid compounds containing 19–20 carbon atoms, naturally produced in new leaves and germinated seed embryos, and more than 136 species have been identified [26]. A very small amount of GA_3 enhances stem elongation [27] by increasing cell divisions and cell size [28]; improves plant growth and development [29,30] by inducing metabolic activities [31,32] of many key enzymes such as carbonic anhydrase (CA) and nitrate reductase (NR) [33] and regulating nitrogen utilization [34]; and consequently, increases the dry weight and yield [35,36]. It also induces growth and development by enhancing water uptake in plant tissues [37,38], facilitates the production of photosynthetic pigments and photosynthesis [39], and enables flower formation and fruit set in legumes [36,40].

Keeping in view the importance of mung beans in sub-tropics, this study aimed to improve their morpho-physiology, yield, and nutritional profile when grown in a sandy-loam soil with low-fertility status. In the literature, many researchers have evaluated the individual beneficial effects of indole acetic acid (IAA) and gibberellic acid (GA_3) on many crops under stressful and non-stressful environments [14,35,41]; however, their combinations with various dosage levels still have to be evaluated. Therefore, the individual and/or combined effectiveness of various levels of IAA and GA_3 was assessed in the present study to evaluate mung bean (Var. AEM-96) performance in the sub-tropical region of Pakistan for two consecutive years (2018–19/2019–20). This experiment has huge importance as a reference for the impact of two important phytohormones on mung bean growth, yield, quality, and nutritional contents in the sub-tropical region of Pakistan.

2. Materials and Methods

2.1. Experiment Site and Climate Conditions

The two-year field experiment was carried out in the agronomic research area of The Islamia University of Bahawalpur, Punjab, Pakistan, during the 2019–20 growing seasons. The meteorological data were obtained from Bahawalpur meteorological station and are described in Table 1. The experiment was carried out to determine the properties of the soil at the experimental site. The soil sample test results showed that the soil was sandy loam and had an acidic pH (7.2). The detailed statuses of the soil's physio-chemical characteristics are depicted in Table 2.

Table 1. Meteorological data recorded at the experimental site during the study period of 2019–2020.

Month	T_{max} °C		T_{min} °C		Total Rain Fall (mm)		R.H (%)	
	2019	2020	2019	2020	2019	2020	2019	2020
October	36.1	37.4	21.6	20.5	0	0	58	54
November	31.6	32.20	13.9	11.6	0	0	53	52
December	27.5	26.6	12.2	9.0	0	0	57	53
January	26.4	22.1	11.3	8.3	0	2.5	61	62
February	22.4	24.2	10.2	11.0	0	0	45	50
March	35.1	34.4	18.1	15.5	0	0	48	48
April	38.0	40.2	21.9	18.9	0	0	44	43
May	39.5	46.0	29.0	28.0	0	0	58	56
June	41.0	42.0	28.0	27.0	0	0	47	45
July	45.0	47.0	27.0	25.0	0.7	0	56	53
August	42.0	43.0	28.0	23.0	0	2.6	47	49
September	38.0	41.0	26.0	24.5	0	0	50	52

T_{max} = maximum temperature, T_{min} = minimum temperature, and R.H = relative humidity.

Table 2. The detailed fertility status of the soil at the agronomic research field station used in the current experiment.

Soil Properties	Units	Values
Texture	-	Sandy loam
pH	-	7.2
SOM	%	1.43
Total N	%	0.07
Na	meq/60 g soil	0.07
K	meq/60 g soil	0.12

Table 2. *Cont.*

Soil Properties	Units	Values
Mg	meq/60 g soil	2.1
P	ug/g soil	12.3
S	ug/g soil	16.09
Ca	ug/g soil	3.02
Zn	ug/g soil	1.43
B	ug/g soil	0.38

2.2. Experiment Design and Treatments

The field experiment was conducted under a randomized complete block design (RCBD) with three replications per block. The area was divided into different plots, and each plot size was 10 m^2 containing 6 rows. To fulfill the nutrient profile of the growth media of the experimental site, it was fortified with potassium (K), phosphorus (P), and nitrogen (N) in the form of calcium superphosphate (12%), potassium sulfate (50%), and urea (46%), respectively. The above ratio of fertilizer was maintained, with one fertilizer bag containing 50 kg of DAP (diammonium phosphate) and 10 kg of urea, and it was incorporated in the soil before planting the mung bean genotypes.

The mung bean seeds (Var. AEM-96) were collected and washed three times with distilled water. After being washed, these seeds were spread to dry overnight at room temperature. Later on, the mung bean seeds were sown and thinned after three weeks of germination for the maintenance of the plant-to-plant distance. The application solutions of IAA and GA_3 were prepared according to the treatment specification (0, 30, and 60 mg L^{-1} concentrations), and they were applied on the plant foliage at flowering time after 25 days of germination (DAG) thrice, with an 11-day interval each time. The foliar application with a hand hydraulic sprayer was carried out at dawn and dusk to avoid evaporation losses. Optimal cultural and agronomic practices that affect the yield of the crop were applied efficiently during the whole plant growth cycle. The details of the foliar treatment of IAA and GA_3 are described in Table 3.

Table 3. Treatment layout of the experiment conducted in 2019–2020 in a sandy-loam soil in sub-tropics.

Treatment Number	Treatment Labels	IAA Conc. (mg L^{-1})	GA_3 Conc. (mg L^{-1})
T1	CK	00.0	00.0
T2	IAA_1	30.0	00.0
T3	IAA_2	60.0	00.0
T4	GA_1	00.0	30.0
T5	GA_2	00.0	60.0
T6	$IAA_1 + GA_1$	30.0	30.0
T7	$IAA_1 + GA_2$	30.0	60.0
T8	$IAA_2 + GA_1$	60.0	30.0
T9	$IAA_2 + GA_2$	60.0	60.0

IAA: indole-3-acetic acid; GA_3: gibberellic acid.

2.3. Determination of Growth Traits

The mung bean plant samples which were to be assessed to determine morphological parameters were collected 60 days after sowing (DAS) by randomly selecting plants from each plot for the estimation of growth characteristics. Plant height was recorded with the help of a measuring scale from the ground to the top of the leaf in centimeters (cm). Other growth traits, including leaf fresh and dry weight (wt.), shoot length, number of leaves/plant, and number of branches/plant, were also calculated accordingly.

2.4. Determination of Agronomic Traits

Post-harvesting data and seed-related traits were measured 75 days after sowing (DAS) from the agronomic research area, The Islamia University of Bahawalpur, Pakistan. During harvesting, the mung bean plant samples were collected for the estimation of agronomic and yield-related parameters. The pod weight/plant, the number of pods/plant, and the number of seeds/pod were counted from ten randomly selected plants at harvest. Four rows from each treatment condition were harvested and weighed to compute the biological yield (kg ha^{-1}). However, the seed yield was recorded at harvest, where the mass of seeds collected per square meter (g/m^2) was recorded and converted to kg/ha.

2.5. Estimation of Photosynthetic Traits

The mung bean fresh leaf samples were collected 60 DAS. The determination of chlorophyll pigments was carried out by randomly selecting ten plants/plots. Fresh leaf samples were collected from each treatment and subjected to grinding with 80% acetone. The semi-liquid extract was filtered and centrifuged at 10,000 rpm for 5 min. The supernatant was then subjected to a spectrophotometer (Model Analytikjena Spekol 1500 Germany). For acetone (80% v/v) extraction, the following equations given by Lichtenthaler [42] were used to estimate chlorophyll contents (mg/g FW) in the leaves.

$$\text{Chl. a} = \{(12.25A_{663.2} - 2.79A_{646.8}) \times V\} \div W$$

$$\text{Chl. b} = \{(21.50A_{646.8} - 5.10A_{663.2}) \times V\} \div W$$

$$\text{Carotenoids} = \{(1000A_{470} - 1.82_{\text{Chl.a}} - 85.02_{\text{Chl. b}}) \times V\} \div (198 \times W)$$

where V represents the final volume of chlorophyll extract in 80% acetone, and W is 0.1 g.

2.6. Determination of Total Carbohydrates, Nitrogen, and Protein Contents

The leaves of 3 randomly selected plants 60 DAS were subjected to the estimation of total carbohydrates, nitrogen, and protein contents. The phenol–sulfuric acid method was applied to measure the total carbohydrates from the IAA and GA_3 treatment individually or in combination. Micro Kjeldahl's apparatus was used to estimate the nitrogen contents [43]. The crude protein was determined using a previously reported method [44]. In this method, crude protein was measured by multiplying protein contents with a 5.75 factor.

2.7. Estimation of Endogenous Growth Regulators

The mung bean plant samples were collected randomly from each plot and treatment 75 DAS for the determination of endogenous critical plant regulators. For this particular trait, the plants were screened based on data obtained from morphological and yield attributes. The most promising treatments were then selected for the estimation of endogenous regulators, including IAA, GA_3, ABA, and zeatin. The hormones were extracted using the previously reported method described by Shindy et al. [45]. Briefly, 2 g of fresh leaves was ground in cold 80% (v/v) aqueous methanol with a mortar and pestle for extraction. After extraction, the gas–liquid chromatography (GLC) technique was used to determine the IAA, GA, and ABA contents [45]. Furthermore, the cytokinin contents were estimated using the method previously reported by Müller [46] via HPLC (high-performance liquid chromatography).

2.8. Statistical Analysis

The data were pooled for both the growing years. All the experimental data were recorded and subjected to a one-way analysis of variance (ANOVA) with three replicates to record statistically significant/non-significant differences among different traits of mung bean through a computer program, Statistix Version 8.1 (Analytical Software,

2005). Moreover, Bonferroni's statistical analysis test was applied to verify the level of significance (5%) among different treatment means.

3. Results

3.1. Influence of Natural Phytohormones on Mung Bean Growth

The statistical data of morphological/vegetative parameters depicted that the application of two phytohormones, i.e., IAA (T_2–T_3) and GA_3 (T_4–T_5), individually or in combination (T_6–T_9), significantly influenced the growth of mung bean plants (Figure 1A–F). A significant increase was observed under IAA_2 + GA_2 treatments in shoot length by 33.8%, leaf fresh weight by 56.0%, leaf dry weight by 57.7%, no. of leaves by 34.8%, and no. of branches per plant by 22.9%, as compared to the control (CK) treatment. The maximum growth performance was observed in the IAA_2 + GA_2 (T_9) treatment compared to the other treatments (T_1–T_8) and the control (T_0).

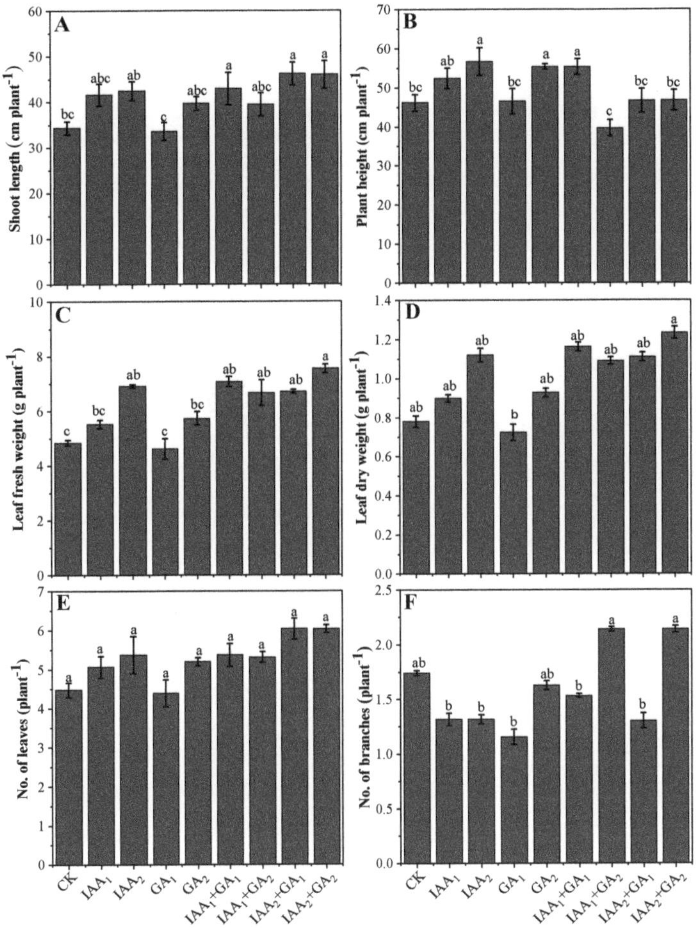

Figure 1. Influence of individual and/or combined foliar application of natural phytohormones on growth and biomass characteristics, e.g., shoot length (**A**), total plant height (**B**), leaf fresh weight (**C**), leaf dry weight (**D**), no. of leaves (**E**), and no. of branches (**F**) of mung bean plants grown in sandy-loam soil in sub-tropics. The bar values are an average of three replicates (n = 3), and the bars not sharing the same lowercase letters indicate significant differences from each other according to Bonferroni's statistical analysis at $p < 0.05$.

3.2. Influence of Natural Phytohormones on Yield and Yield Related Parameters

The foliar application of the singular or combined treatment of IAA and GA_3 hormones positively influenced the mung bean yield and yield-related parameters, as presented in Figure 2A–F. The data analysis showed that the highest yield of each parameter was obtained with the IAA_2 alone and IAA_2 + GA_2 treatment. However, the most effective treatment was found to be IAA_2 + GA_2 (T_9), which led to the highest mung bean yield in terms of seed yield (>230.3%), straw yield (>29.5%), and biological yield (>43.4%), as compared with the CK treatment in which no phytohormone was supplied. In most of the yield attributes, however, there was no statistical difference among the effect of the IAA_2 (T_3) and IAA_2 + GA_2 (T_9) treatment levels. On the other hand, it was noticed that a lower concentration of GA_3 hormone (GA_1 treatment) was less effective in improving mung bean's yield-related traits when compared with the higher dosages of the studied hormones.

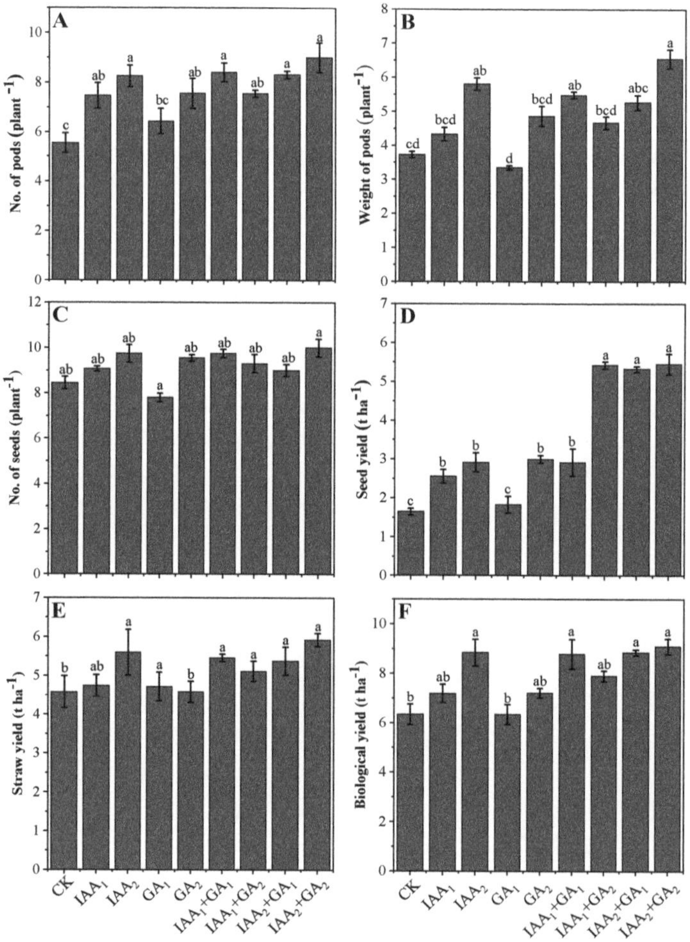

Figure 2. Influence of individual and/or combined foliar application of natural phytohormones on yield-related characteristics, e.g., no. of pods (**A**), weight of pods (**B**), no. of seeds per plant (**C**), seed yield (**D**), straw yield (**E**), and biological yield (**F**) of mung bean plants grown in a sandy-loam soil in sub-tropics. The bar values are an average of three replicates (n = 3), and the bars not sharing the same lowercase letters indicate significant differences from each other according to Bonferroni's statistical analysis at $p < 0.05$ level.

3.3. Influence of Natural Phytohormones on Photosynthetic Pigments

The influence of different concentrations of the spraying of IAA and GA_3, combined or alone, on total photosynthetic pigments, carotenoids, chlorophyll a (Chl a), and chlorophyll b (Chl b) are presented in (Figure 3A–D). Overall, the photosynthetic pigments and carotenoid contents were influenced positively by the foliar application of IAA and GA_3. The analysis of the results showed that IAA and GA_3 treatments caused a significant increase in photosynthetic pigments. Briefly, maximum enhancements of 16.5%, 19.7%, and 17.9% was reported in chlorophyll a, chlorophyll b, and carotenoid contents, respectively, under the combined application of phytohormones (IAA_2 + GA_2), as compared with the plants grown in the control plot. Moreover, quite similar results were observed for the total concentration of pigments in the leaves of mung bean under the application of phytohormones.

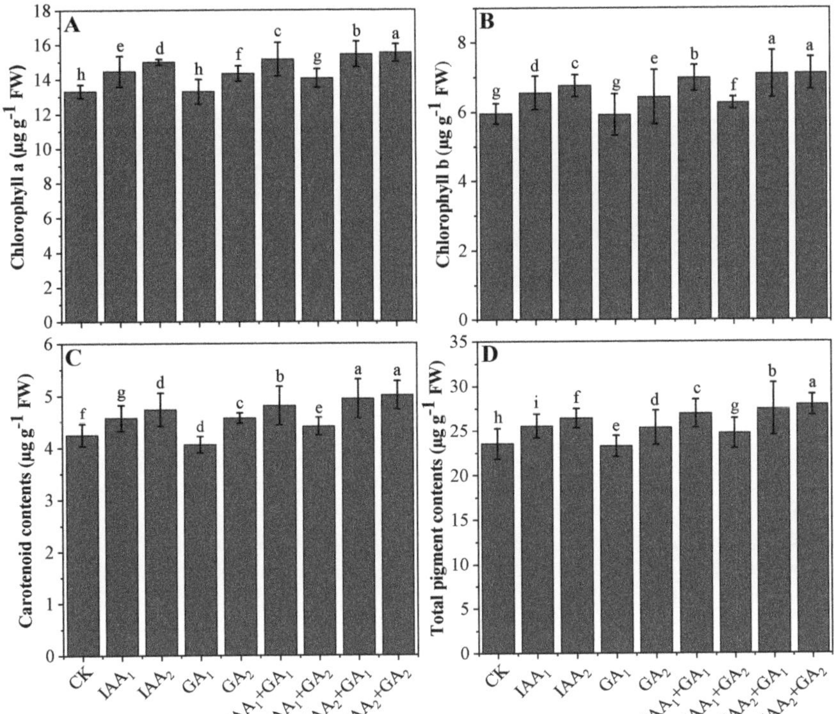

Figure 3. Influence of individual and/or combined foliar application of natural phytohormones on the physiological parameters, e.g., chlorophyll a (**A**), chlorophyll b (**B**), carotenoid contents (**C**), and total pigment contents (**D**) of mung bean plants grown in a sandy-loam soil of sub-tropics. The bar values are an average of three replicates (n = 3), and the bars not sharing the same lowercase letters indicate significant differences from each other according to Bonferroni's statistical analysis at $p < 0.05$ level.

3.4. Influence of Natural Phytohormones on Nutritional Status of Mung Bean Seeds

The findings related to the nutritional status of the seeds of mung bean plants, e.g., total carbohydrates, protein contents, and nitrogen contents, from the current trial are given in Figure 4A–C. The results revealed that the singular or combined foliar application of IAA and GA_3 significantly improved the total carbohydrate, protein, and nitrogen contents in the yielded mung bean seeds as compared with the control (CK) plants. The highest increases in the contents of total carbohydrates (6.6% and 7.5%), protein (17.0% and 17.3%), and nitrogen (13.2% and 14.5%) were observed under the IAA_2 and $IAA_2 + GA_1$ treatments, compared to the other treatments and control plants. On the contrary, it was noticed that the lower concentration of GA_3 hormone (GA_1 treatment) was not effective in improving nutritional traits of mung bean seeds as compared to the other treatments.

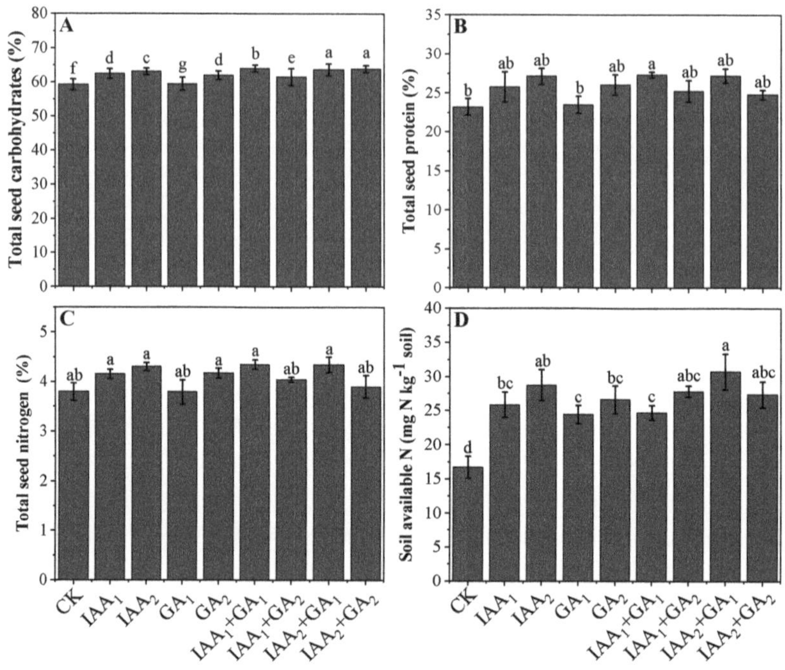

Figure 4. Influence of individual and/or combined foliar application of natural phytohormones on the nutritional status of seeds of mung bean plants, e.g., total carbohydrates (**A**), protein contents (**B**), nitrogen contents (**C**), and soil N availability (**D**) of the studied sandy-loam soil in sub-tropics. The bar values are an average of three replicates (n = 3), and the bars not sharing the same lowercase letters indicate significant differences from each other according to Bonferroni's statistical analysis at $p < 0.05$ level.

3.5. *Influence of Natural Phytohormones on Soil N Availability*

The influence of different concentrations of the spraying of IAA and GA_3, combined or alone, on soil N availability was determined after harvesting in the experiment (75 DAS), and the data are presented in Figure 4D. Overall, the availability of N in the soil was positively influenced with the foliar application of IAA and GA_3. It was observed that the IAA and GA_3 treatments caused a significant increase in N availability in sandy-loam soil. The maximum increase was reported under the IAA_2 (71.8%), $IAA_1 + GA_2$ (66.5%), and $IAA_2 + GA_1$ (83.8%) treatments.

3.6. *Influence of Natural Phytohormones on Endogenous Phytohormone Production*

The current study determined the changes in the endogenous IAA GA_3, zeatin, and ABA contents after the foliar application of varying concentrations of natural phytohormones on seedlings 60 days after sowing (DAS). The corresponding results are presented in Figure 5A–D. The endogenous production of IAA GA_3, zeatin, and ABA was only evaluated under the higher application doses of IAA and GA_3 based on the results obtained for growth- and yield-related characteristics. It was noticed that the exogenous applications of GA_3 and IAA were correlated with the changes in their endogenous contents of studied phytohormones in mung bean. The combined foliar application of IAA and

GA_3 ($IAA_2 + GA_2$) markedly increased the IAA, GA_3, and zeatin concentrations, while it declined the ABA contents in the leaves of mung bean seedlings.

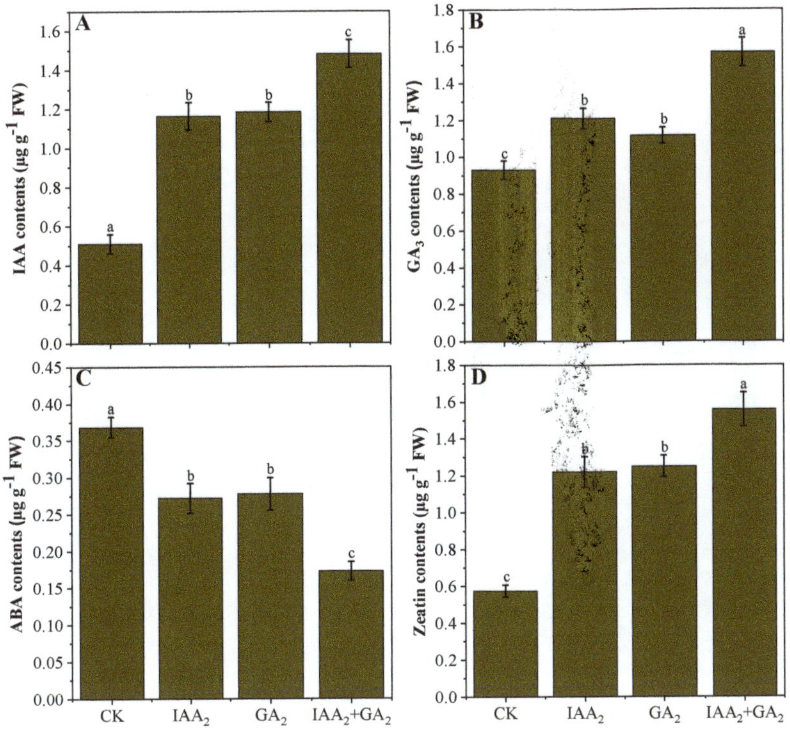

Figure 5. The influence of individual and/or combined foliar application of natural phytohormones on the endogenous concentration of phytohormones, e.g., IAA contents (**A**), GA_3 contents (**B**), ABA contents (**C**), and zeatin contents (**D**) in the leaves of mung bean plants grown in a sandy-loam soil in sub-tropics. The bar values are an average of three replicates ($n = 3$), and the bars not sharing the same lowercase letters indicate significant differences from each other according to Bonferroni's statistical analysis at $p < 0.05$ level.

3.7. Hierarchical Agglomerative Cluster Analysis

A two-dimensional relationship analysis was carried out to elaborate the distinction between the studied treatments and various physio-biochemical attributes of mung bean plants (Figure 6). The double hierarchical heatmap presented the correlation between the various studied traits of mung bean (row) and different singular and/or combined application levels of natural phytohormones (columns). The column hierarchical dendrogram clearly showed that the effect of IAA_2 and GA_2 and $IAA_2 + GA_2$ on the studied variables of mung bean seedlings under a subtropical environment was much more superior compared to all the other treatments. On the other hand, the analysis also showed that the effect of these treatments was very useful to boost the growth, yield, and biochemical attributes of mung bean seedlings.

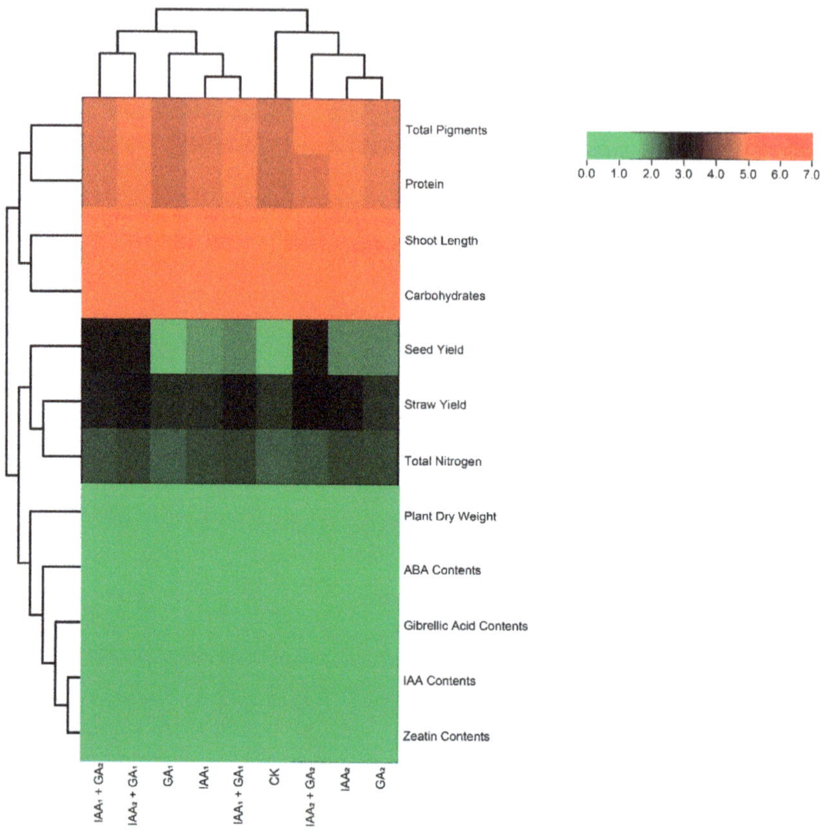

Figure 6. Heatmap analysis based on the correlation matrix of the growth and physio-biochemical variables measured in mung beans (*Vigna radiata* L.) grown in a sandy-loam soil sprayed with two natural phytohormones, viz., IAA and GA_3. The double hierarchical dendrogram reveals the relationship among treatments (column) and among various plant characteristics of mung bean seedlings (row).

4. Discussion

In a natural system, the ratio of various phytohormones is maintained to a required level by finely regulating their synthesis, transport, metabolism, and/or destruction to ensure the coordinated growth of various tissues/organs along a defined pattern of growth and development in the life span of the plant. A limited desired deviation in this set pattern of growth and development may, however, be possible by enhancing the level of any of these regulators by their exogenous application to intact plants or their parts [47]. Therefore, it has been very well established that exogenous hormonal treatment alters plant growth and development by modifying their growth, physiology, and endogenous contents [48,49]. However, it is still ambiguous as to whether the effects of exogenous hormones on growth are direct or whether they are related with changes induced in endogenous hormones [50]. With reference to the comparison of two hormones, IAA foliar application was more efficient in improving the growth attributes of mung bean than GA_3. The present increase in growth is in parallel to the earlier reported studies on various plants, including faba-bean and mung bean [51,52] with the exogenous treatment of plant growth regulators (GA_3 and IAA). Another study also supported the current findings that GA_3, IAA, and the interaction of both phytohormones significantly contributed towards plant height, the number of pods/plant, biological yield, straw yield, seed yield, total carbohydrates, protein and N contents in the seeds [35]. IAA has been recognized to increase growth and photosynthetic pigments' concentration in the leaves of plants, as shown in the current study [53], and stimulate cell division and enhance biochemical traits, i.e., total carbohydrates content and polysaccharides [54]. Furthermore, previously, an important role of GA_3 on the growth traits of the *Simmondsia chinensis* plant was observed under a sub-tropical environment [55]. In the current study, the statistical results revealed that all the treatments, including the singular and/or combined application of phytohormones, had a promotive effect on the growth and yield of mung bean and faba-bean, as reported by [53] and [54], respectively. For example, it was suggested that a combined dose of auxin (1.0 mg L^{-1}) and gibberellin (200 mg L^{-1}) is recommended for the enhancement of seed yield, whereas a 0.5 mg L^{-1} dose of auxin is recommended for the enhancement of vegetative growth. This enhancement of mung bean growth (Figure 1), photosynthetic assimilates (Figure 3), and endogenous phytohormones (Figure 5) after the spraying of plant growth regulators, alone or combined, might lead to the accretion of photo-assimilates in the seed and enhance the transfer rate of these assimilates to boost the yield of mung bean (Figure 2). Additionally, the exogenous foliar application of IAA at a 100 ppm concentration previously resulted in maximum plant height, chlorophyll content, spike length, 1000-grain weight, and grain yield in bread wheat [56]. Moreover, an increase in grain yield could be positively correlated to enhanced grain numbers per spike of two growth hormones (IAA $\times$ 6-BAP) in the booting stage of wheat [57].

Our results were in accordance with the "acid growth theory" of auxin, e.g., the indole-3-acetic acid (IAA) causes acidification (excretion of protons into the apoplast) of cell walls which ultimately increases stem growth by loosening 0f cell walls via cleavage of the bonds [47,58] and also has an impact on the functioning of ionic channels, thereby affecting the direction of the movement of ions and solutes and the turgor of cells [47]. Moreover, auxins also involve genes in inducing certain expressions by altering the type, activity, and level of the proteins [47]. This is possibly the reason for the enhanced rate of photosynthesis in the leaves of mung bean in the current study, and therefore, it may have gave a boost to the growth of the root and the shoot, as expressed in the form of increased fresh and dry mass (Figure 1). On the other hand, GA_3 has been reported to increase the pigment content of Vicia faba [59], and the water use efficiency of wheat [60], and it increased photosynthesis by increasing the carboxylase activity of Rubisco in broad bean and soybean [61] and regulated the transport of ions in plants. Additionally, it might increase water uptake in plant tissues, causing cell expansion and the dilution of sugars in tissues under harsh conditions [13]. Likewise, GA_3 tends to boost the protein content by increasing the nitrate reductase activity in cowpea [62], wheat [60], black cumin [63], and mung bean [41]. The growth-promoting effect of GA_3 may be attributed to the stimulation of the mobility of soil nutrients towards the buds, thereby increasing cell division and/or increasing the differentiation of the vascular tissues. Additionally, the increases in the yield of mung bean plants via the application of different growth hormones might result by breaking the apical dominance of mung bean plants, leading to the increase in flowering, branches, and consequently, the number of fruits. The increase in seed weights might be because of the promotive effect of IAA and/or GA_3 in increasing the assimilates and their translocations from leaves to fruits, where the seed weight increases [64].

Hormonal coordination is an important aspect which regulates leaf growth processes [22]. The current findings related to pigment formation are in line with those obtained for maize [65], wheat [66] and pulses [53,67]. Moreover, the plant growth regulator IAA presumably acts as a coenzyme in the metabolism of higher plants, which directly affects the synthesis or formation of photosynthetic pigments [68]. These increments in photosynthetic pigments may play their role in improving photosynthesis and the retardation of its degradation. It was confirmed that GA_3 is involved in the photosynthates machinery of the *Simmondsia chinensis* plant, and irrespective of the dosage of GA_3 used in the current experiment, it was shown that 300, 200 and 100 ppm of GA_3 differentially modulated the studied traits of mung bean (photosynthetic pigments: carotenoids, chlorophyll a, and chlorophyll b): yield-contributing traits such as pod length, the number of grains pod^{-1}, plant height, the number of pods plant^{-1}, and 100-grain weight; and quality parameters including seed protein and nitrogen [55]. It is important to increase protein rich crops such as mung bean in developing countries, because they can serve both as food and feed. Our findings from the current experiment are also in parallel with several scientific results that state that GA_3 improves the photosynthetic attributes in *V. radiate* [35,41], wheat [69], cotton [70], broccoli leaves [71], *Simmondsia chinensis* [55], and *Visiafaba* [13]. It has also been scientifically proven that the application of GA_3 alters the specific components of plastids that affect the retention of chlorophyll apparatus in maize [72]. Among all photosynthetic species, carotenoids act as a non-enzymatic antioxidant, protecting the "antenna complex" from photo-oxidative damage, enabling wavelengths of light to be available for photosynthesis [73]. Our findings are supported by the previous results from *V. radiata* that GA_3 treatment significantly increased leaf carotenoid contents [35]. It was then recommended that the improvement in carotenoid capacity by GA_3 may be attributed to the ultrastructural morphogenesis of plastids [74].

It has been well established that to prevent oxidative plant injury, plants have evolved adaptive mechanisms, including the upregulation of the antioxidant defense system, which includes ROS-scavenger enzymes, e.g., ascorbate peroxidase (APX), catalase (CAT), and superoxide dismutase (SOD), and non-enzymatic antioxidants such as glutathione, α-tocopherol, ascorbic acid, and phenolic compounds [75–79]. On the other hand, many researchers have explored whether the application of plant growth regulators degrade the activities of various enzymes which are directly related to improving endogenous phytohormone production [13,80]. Therefore, the increased endogenous phytohormone production in our study might have been due to the alleviation of oxidative stress and the improvement in antioxidant capacity, as suggested by various researchers [81,82]. These results are also comparable to those obtained for cowpea [70], wheat [45], and faba-bean [46]. The previous findings suggest that isopentenyl pyrophosphate is a common precursor for the biosynthesis of cytokinin and/or gibberellins and ABA [50]; apparently, the exogenous application of IAA and GA_3 may have caused a shift into cytokinin or gibberellin biosynthesis instead of ABA, which resulted in a decline in ABA content in the current research (Figure 5).

The exogenous application of natural phytohormones such as IAA and GA_3 has been shown to be an emerging trend which can positively regulate growth parameters and total carbohydrate, polysaccharide, proline, free amino acid, and total phenolic contents in faba-beans [54]. Moreover, many studies have shown that foliar and seed-priming-based applications of phytohormones have increased the availability of N in soil by enhancing nitrogenase activity in mung bean crops [83]. Higher nitrogenase activity signifies an increase in the rate of the reduction of nitrogen to ammonia. Therefore, more and more organic forms of nitrogen are made available in plants to be incorporated with keto acids to generate additional quantities of the required amino acids/amides [84]. Simultaneously, the synthesis of additional proteins (Figure 4) could have also sped up the availability of the enzymes (glutamin synthetase and glutamate synthase) involved in the glutamin synthase cycle determining the incorporation of ammonia [85]. It seems quite natural from these observations that the hormones might have elevated the useable form of nitrogen (ammonia) to produce a larger pool of amino acids/amides.

5. Conclusions

The application of natural phytohormones, e.g., IAA and GA_3, affected plant growth, and it enabled the mung bean plants to survive in a sandy-loam soil with nutrient deficiency. Moreover, these phytohormones also enhanced the yield of mung bean in two different growing seasons in 2019–2020. The most effective treatment was 60 mg L^{-1} IAA + 60 mg L^{-1} GA_3 ($IAA_2 + GA_2$), which improved the growth characteristics, photosynthetic pigments formation, yield attributes, endogenous phytohormone production, and biochemical composition of mung bean seeds. In the current investigation, it was therefore suggested that the mung bean performed better and gave the maximum capability of yield after the singular or combined application of high doses of IAA (60 mg L^{-1}) and GA_3 (60 mg L^{-1}) in a subtropical region of Pakistan. However, the recommended dosages, in future experiments, should be tested in various field environments in different parts of the world to confirm the potential of these two hormones.

Author Contributions: Conceptualization, M.K. and M.M.A.; methodology, A.P. and M.M.A.; software, R.I. and M.A. (Muhammad Ali); validation, M.K., M.S.E. and M.S.A.; formal analysis, A.P., M.K. and M.A. (Muhammad Ali); investigation, A.P.; resources, M.K. and M.S.E.; data curation, R.I.; writing—original draft preparation, A.P., M.M.A. and M.K.; writing—review and editing, M.S.E., M.S.A. and M.A. (Muhammad Akram); visualization, M.A. (Muhammad Akram) and M.S.A.; supervision, M.K. and M.S.E.; project administration, R.I.; funding acquisition, M.S.E. and M.K. All authors have read and agreed to the published version of the manuscript.

Funding: The authors extend their appreciation to the Researchers supporting project number (RSP2023R173), King Saud University, Riyadh, Saudi Arabia.

Institutional Review Board Statement: Not applicable.

Informed Consent Statement: Not applicable.

Data Availability Statement: All the data are available upon request from the corresponding author(s).

Conflicts of Interest: The authors declare no conflict of interest.

References

1. Konuma, H. Status and Outlook of Global Food Security and the Role of Underutilized Food Resources: Sago Palm. In *Sago Palm: Multiple Contributions to Food Security and Sustainable Livelihoods*; Ehara, H., Toyoda, Y., Johnson, D., Eds.; Springer: Singapore, 2018; pp. 3–16.
2. Hilger, T.; Lewandowski, I.; Winkler, B.; Ramsperger, B.; Kageyama, P.; Colombo, C. Seeds of Change—Plant Genetic Resources and People's Livelihoods. In *Agroecology*; IntechOpen: London, UK, 2015.
3. Sehrawat, N.; Yadav, M.; Sharma, A.K.; Kumar, V.; Bhat, K.V. Salt Stress and Mungbean [*Vigna radiata* (L.) Wilczek]: Effects, Physiological Perspective and Management Practices for Alleviating Salinity. *Arch. Agron. Soil Sci.* **2019**, *65*, 1287–1301. [CrossRef]
4. Abbas, G.; Ahmed, A.; Amer, M.; Abbas, Z.; Rehman, M.; Hussain, A.; Khan, G.A. Impact of Pre-Emergence Herbicides for the Control of Weeds in Chick Pea (*Cicer arietinum* L.) under Hot Arid Climate. *J. Bioresour. Manag.* **2016**, *3*, 7.
5. Noble, T.J.; Tao, Y.; Mace, E.S.; Williams, B.; Jordan, D.R.; Douglas, C.A.; Mundree, S.G. Characterization of Linkage Disequilibrium and Population Structure in a Mungbean Diversity Panel. *Front. Plant Sci.* **2018**, *8*, 2102. [CrossRef] [PubMed]
6. Mubarak, A.E. Nutritional Composition and Antinutritional Factors of Mung Bean Seeds (*Phaseolus aureus*) as Affected by Some Home Traditional Processes. *Food Chem.* **2005**, *89*, 489–495. [CrossRef]
7. Rani, S.; Schreinemachers, P.; Kuziyev, B. Mungbean as a Catch Crop for Dryland Systems in Pakistan and Uzbekistan: A Situational Analysis. *Cogent Food Agric.* **2018**, *4*, 1499241. [CrossRef]
8. Raina, S.K.; Govindasamy, V.; Kumar, M.; Singh, A.K.; Rane, J.; Minhas, P.S. Genetic Variation in Physiological Responses of Mungbeans (*Vigna radiata* (L.) Wilczek) to Drought. *Acta Physiol. Plant.* **2016**, *38*, 263. [CrossRef]
9. FAO; WFP; IFAD; UNICEF; WHO. *The State of Food Security and Nutrition in the World 2017. Building Resilience for Peace and Food Security*; FAO: Rome, Italy, 2017.
10. Ogunkunle, A. *Management of Nigerian Soil Resources: An Imperative for Sustainable Development*; Ibadan University Press: Ibadun, Nigeria, 2016.
11. Vanzetti, D.; Petersen, E.H.; Rani, S. Economic Review of the Pulses Sector and Pulses-Related Policies in Pakistan. In Proceedings of the Mid-Project Workshop of ACIAR Project ADP/2016/140 "How Can Policy Reform Remove Constraints and Increase Productivity in Pakistan?", Islamabad, Pakistan, 3 April 2017; Volume 3.
12. Ullah, A.; Shah, T.M.; Farooq, M. Pulses Production in Pakistan: Status, Constraints and Opportunities. *Int. J. Plant Prod.* **2020**, *14*, 549–569. [CrossRef]
13. Rady, M.M.; Boriek, S.H.K.; El-Mageed, T.A.A.; El-Yazal, M.A.S.; Ali, E.F.; Hassan, F.A.S.; Abdelkhalik, A. Exogenous Gibberellic Acid or Dilute Bee Honey Boosts Drought Stress Tolerance in Vicia Faba by Rebalancing Osmoprotectants, Antioxidants, Nutrients, and Phytohormones. *Plants* **2021**, *10*, 748. [CrossRef]
14. Sadiq, R.; Maqbool, N.; Hussain, M.; Tehseen, S.; Naseer, M.; Rafique, T.; Zikrea, A.; Naqve, M.; Mahmood, A.; Javaid, A.; et al. Boosting Antioxidant Defense Mechanism of Mungbean with Foliar Application of Gibberellic Acid to Alleviate Cadmium Toxicity. *Plant Physiol. Rep.* **2021**, *26*, 741–748. [CrossRef]
15. Malik, Z.; Jamil, M.; Abassi, G.H.; Nafees, M.; Rafey, M.; Kamran, M. Biochar and Fly Ash Role in Improving Mechanical and Physical Properties of Vertisol. *Sarhad J. Agric.* **2017**, *33*, 151–161. [CrossRef]

16. Kamran, M.; Malik, Z.; Parveen, A.; Huang, L.; Riaz, M.; Bashir, S.; Mustafa, A.; Abbasi, G.H.; Xue, B.; Ali, U. Ameliorative Effects of Biochar on Rapeseed (*Brassica napus* L.) Growth and Heavy Metal Immobilization in Soil Irrigated with Untreated Wastewater. *J. Plant Growth Regul.* **2020**, *39*, 266–281. [CrossRef]
17. Kamran, M.; Huang, L.; Nie, J.; Geng, M.; Lu, Y.; Liao, Y.; Zhou, F.; Xu, Y. Effect of Reduced Mineral Fertilization (NPK) Combined with Green Manure on Aggregate Stability and Soil Organic Carbon Fractions in a Fluvo-Aquic Paddy Soil. *Soil Tillage Res.* **2021**, *211*, 105005. [CrossRef]
18. Chen, G.; Shi, L. A Multi-Element Mineral Conditioner: A Supplement of Essential Cations for Red Soil and Crops Growth. *Clean* **2016**, *44*, 1690–1699. [CrossRef]
19. Chen, G.; Shi, L. Removal of Cd(II) and Pb(II) Ions from Natural Water Using a Low-Cost Synthetic Mineral: Behavior and Mechanisms. *RSC Adv.* **2017**, *7*, 43445–43454. [CrossRef]
20. Chen, G.; Shah, K.J.; Shi, L.; Chiang, P.C.; You, Z. Red Soil Amelioration and Heavy Metal Immobilization by a Multi-Element Mineral Amendment: Performance and Mechanisms. *Environ. Pollut.* **2019**, *254*, 112964. [CrossRef] [PubMed]
21. Huang, G.; Ding, C.; Guo, N.; Ding, M.; Zhang, H.; Kamran, M.; Zhou, Z.; Zhang, T.; Wang, X. Polymer-Coated Manganese Fertilizer and Its Combination with Lime Reduces Cadmium Accumulation in Brown Rice (*Oryza sativa* L.). *J. Hazard. Mater.* **2021**, *415*, 125597. [CrossRef]
22. Iqbal, N.; Khan, N.A.; Ferrante, A.; Trivellini, A.; Francini, A.; Khan, M.I.R. Ethylene Role in Plant Growth, Development and Senescence: Interaction with Other Phytohormones. *Front. Plant Sci.* **2017**, *8*, 475. [CrossRef] [PubMed]
23. Duca, D.R.; Glick, B.R. Indole-3-Acetic Acid Biosynthesis and Its Regulation in Plant-Associated Bacteria. *Appl. Microbiol. Biotechnol.* **2020**, *104*, 8607–8619. [CrossRef]
24. Pérez-Alonso, M.M.; Ortiz-García, P.; Moya-Cuevas, J.; Lehmann, T.; Sánchez-Parra, B.; Björk, R.G.; Karim, S.; Amirjani, M.R.; Aronsson, H.; Wilkinson, M.D.; et al. Endogenous Indole-3-Acetamide Levels Contribute to the Crosstalk between Auxin and Abscisic Acid, and Trigger Plant Stress Responses in Arabidopsis. *J. Exp. Bot.* **2021**, *72*, 459–475. [CrossRef]
25. Taiz, L.; Zeiger, E. Auxin: The Growth Hormone. *Plant Physiol.* **2006**, *4*, 468–507.
26. Hedden, P. The Current Status of Research on Gibberellin Biosynthesis. *Plant Cell Physiol.* **2020**, *61*, 1832–1849. [CrossRef] [PubMed]
27. Keykha, M.; Ganjali, H.R.; Mobasser, H.R. Effect of Salicylic Acid and Gibberellic Acid on Some Characteristics in Mungbean (*Vigna radiata* L.). *Int. J. Biosci. (IJB)* **2014**, *5*, 70–75. [CrossRef]
28. Tan, H.; Man, C.; Xie, Y.; Yan, J.; Chu, J.; Huang, J. A Crucial Role of GA-Regulated Flavonol Biosynthesis in Root Growth of Arabidopsis. *Mol. Plant* **2019**, *12*, 521–537. [CrossRef]
29. Camara, M.C.; Vandenberghe, L.P.S.; Rodrigues, C.; de Oliveira, J.; Faulds, C.; Bertrand, E.; Soccol, C.R. Current Advances in Gibberellic Acid (GA_3) Production, Patented Technologies and Potential Applications. *Planta* **2018**, *248*, 1049–1062. [CrossRef]
30. Rahman, M.; Khan, A.; Hasan, M.; Banu, L.; Howlader, M. Effect of Foliar Application of Gibberellic Acid on Different Growth Contributing Characters of Mungbean. *Progress. Agric.* **2018**, *29*, 233–238. [CrossRef]
31. Saleem, M.H.; Fahad, S.; Adnan, M.; Ali, M.; Rana, M.S.; Kamran, M.; Ali, Q.; Hashem, I.A.; Bhantana, P.; Ali, M.; et al. Foliar Application of Gibberellic Acid Endorsed Phytoextraction of Copper and Alleviates Oxidative Stress in Jute (*Corchorus capsularis* L.) Plant Grown in Highly Copper-Contaminated Soil of China. *Environ. Sci. Pollut. Res.* **2020**, *27*, 37121–37133. [CrossRef]
32. Wang, Y.H.; Zhang, G.; Chen, Y.; Gao, J.; Sun, Y.R.; Sun, M.F.; Chen, J.P. Exogenous Application of Gibberellic Acid and Ascorbic Acid Improved Tolerance of Okra Seedlings to NaCl Stress. *Acta Physiol. Plant* **2019**, *41*, 93. [CrossRef]
33. Ahmad Dar, T.; Uddin, M.; Khan, M.M.A.; Ali, A.; Hashmi, N.; Idrees, M. Cumulative Effect of Gibberellic Acid and Phosphorus on Crop Productivity, Biochemical Activities and Trigonelline Production in *Trigonella foenum-graecum* L. *Cogent Food Agric.* **2015**, *1*, 995950. [CrossRef]
34. Miceli, A.; Moncada, A.; Sabatino, L.; Vetrano, F. Effect of Gibberellic Acid on Growth, Yield, and Quality of Leaf Lettuce and Rocket Grown in a Floating System. *Agronomy* **2019**, *9*, 382. [CrossRef]
35. el Karamany, M.F.; Sadak, M.S.; Bakry, B.A. Improving Quality and Quantity of Mungbean Plant via Foliar Application of Plant Growth Regulators in Sandy Soil Conditions. *Bull. Natl. Res. Cent.* **2019**, *43*, 61. [CrossRef]
36. Sanjida, T.; Alam, M.J.; Rahman, M.M.; Islam, M.S.; Sikdar, M.S.I. Response of Mungbean Growth and Yield to GA_3 Rate and Time of Application. *Asian J. Crop Soil Sci. Plant Nutr.* **2020**, *1*, 28–36. [CrossRef]
37. Ghani, M.A.; Mushtaq, A.; Ziaf, K.; Ali, B.; Jahangir, M.M.; Khan, R.W.; Khan, I.; Azam, M.; Noor, A. Exogenously Applied GA_3 Promotes Plant Growth in Onion by Reducing Oxidative Stress under Saline Conditions. *Tarim Bilim. Derg.* **2021**, *27*, 122–128. [CrossRef]
38. Ahmad, F.; Kamal, A.; Singh, A.; Ashfaque, F.; Alamri, S.; Siddiqui, M.H.; Khan, M.I.R. Seed Priming with Gibberellic Acid Induces High Salinity Tolerance in Pisum Sativum through Antioxidants, Secondary Metabolites and up-Regulation of Antiporter Genes. *Plant Biol.* **2021**, *23*, 113–121. [CrossRef] [PubMed]
39. Karamany, M.F.E.; Sadak, M.S.; Bakry, B.A. Synergistic Effect of Indole Acetic Acid and Gibberellic Acid on Mung Bean Grown under Sandy Soil Conditions. *J. Appl. Sci.* **2019**, *19*, 718–724. [CrossRef]
40. Nandan, R.; Yadav, R.; Singh, S.P.; Singh, A.K.; Singh, A. Effect of Seed Priming with Plant Growth Regulators on Growth, Biochemical Changes and Yield of Mung Bean (*Vigna radiata* L.). *Int. J. Chem. Stud.* **2021**, *9*, 2922–2927. [CrossRef]

41. Islam, M.S.; Hasan, M.K.; Islam, B.; Renu, N.A.; Hakim, M.A.; Islam, M.R.; Chowdhury, M.K.; Ueda, A.; Saneoka, H.; Ali Raza, M.; et al. Responses of Water and Pigments Status, Dry Matter Partitioning, Seed Production, and Traits of Yield and Quality to Foliar Application of GA_3 in Mungbean (*Vigna radiata* L.). *Front. Agron.* **2021**, *2*, 596850. [CrossRef]
42. Lichtenthaler, H.K. Chlorophyll and Carotenoid Determination: Pigments of Photosynthetic Biomembranes. *Method. Ensymol.* **1987**, *8*, 350–382.
43. Miller, L.; Houghton, J.A. The MicroKjeldahl Determination of the Nitrogen Content of Amino Acids and Proteins. *J. Biol. Chem.* **1945**, *169*, 373–383. [CrossRef]
44. Bradford, M.M. A Rapid and Sensitive Method for the Quantitation of Microgram Quantities of Protein Utilizing the Principle of Protein-Dye Binding. *Anal. Biochem.* **1976**, *72*, 248–254. [CrossRef]
45. Shindy, W.W.; Smith, O.E. Identification of Plant Hormones from Cotton Ovules. *Plant Physiol.* **1975**, *55*, 550–554. [CrossRef]
46. Müller, P.; Hilgenberg, W. Isomers of Zeatin and Zeatin Riboside in Clubroot Tissue: Evidence for Trans-zeatin Biosynthesis by Plasmodiophora Brassicae. *Physiol. Plant* **1986**, *66*, 245–250. [CrossRef]
47. Chapman, E.J.; Estelle, M. Mechanism of Auxin-Regulated Gene Expression in Plants. *Annu. Rev. Genet.* **2009**, *43*, 265–285. [CrossRef] [PubMed]
48. Ahmad, S.; Su, W.; Kamran, M.; Ahmad, I.; Meng, X.; Wu, X.; Javed, T.; Han, Q. Foliar Application of Melatonin Delay Leaf Senescence in Maize by Improving the Antioxidant Defense System and Enhancing Photosynthetic Capacity under Semi-Arid Regions. *Protoplasma* **2020**, *257*, 1079–1092. [CrossRef] [PubMed]
49. Parveen, A.; Ahmar, S.; Kamran, M.; Malik, Z.; Ali, A.; Riaz, M.; Abbasi, G.H.; Khan, M.; Sohail, A.B.; Rizwan, M. Abscisic Acid Signalling Reduced Transpiration Flow, Regulated Na^+ Ion Homeostasis and Antioxidant Enzyme Activities to Induce Salinity Tolerance in Wheat (*Triticum aestivum* L.) Seedlings. *Environ. Technol. Innov.* **2021**, *24*, 101808. [CrossRef]
50. Szalai, G.; Horgosi, S.; Soós, V.; Majláth, I.; Balázs, E.; Janda, T. Salicylic Acid Treatment of Pea Seeds Induces Its de Novo Synthesis. *J. Plant Physiol.* **2011**, *168*, 213–219. [CrossRef] [PubMed]
51. Choudhury, S.; Islam, N.; Ali, M. Growth and Yield of Summer Tomato as Influenced by Plant Growth Regulators. *Int. J. Sustain. Agric.* **2013**, *5*, 25–28. [CrossRef]
52. Hossain, M.E.; Amin, R.; Sani, M.N.H.; Ahamed, K.U.; Hosain, M.T.; Nizam, R. Impact of Exogenous Application of Plant Growth Regulators on Growth and Yield Contributing Attributes of Summer Tomato. *Int. J. Plant Soil Sci.* **2018**, *24*, 1–14. [CrossRef]
53. Naeem, M.; Bhatti, I.; Ahmad, R.H.; Ashraf, M.Y. Effect of Some Growth Hormones (GA_3, IAA and Kinetin) on the Morphology and Early or Delayed Initiation of Bud of Lentil (*Lens culinaris Medik*). *Pak. J. Bot.* **2004**, *36*, 801–809.
54. Sadak, M.S.; Dawood, M.G.; Bakry, B.A.; El-Karamany, M.F. Synergistic Effect of Indole Acetic Acid and Kinetin on Performance, Some Biochemical Constituents and Yield of Faba Bean Plant Grown under Newly Reclaimed Sandy Soil. *World J. Agric. Sci.* **2013**, *9*, 335–344.
55. Atteya, A.K.G.; Genaidy, E.A.E.; Zahran, H.A. Chemical Constituents and Yield of Simmondsia Chinensis Plants as Affected by Foliar Application of Gibberellic Acid and Zinc Sulphate. *Biosci. Res.* **2018**, *15*, 1528–1541.
56. Hanaa, H.; Safaa, A. Foliar Application Foliar Application of IAA at Different Growth Stages and Their Influenced on Growth and Productivity of Bread Wheat (*Triticum aestivum* L.). *J. Phys. Conf. Ser.* **2019**, *1294*, 092029. [CrossRef]
57. Jalali-Honarmand, S.; Rasaei, A.; Saeidi, M.; Ghobadi, M.E.; Khanizadeh, S. The Effects of Foliar Application of Plant Hormones at Booting Stage Non Wheat Yield Components. *Thai J. Agric. Sci.* **2015**, *48*, 35–38.
58. Cleland, R.E. Auxin-Induced Hydrogen Ion Excretion: Correlation with Growth, and Control by External PH and Water Stress. *Planta* **1975**, *127*, 233–242. [CrossRef]
59. Aldesuquy, H.S.; Gaber, A.M. Effect of Growth Regulators on Vicia Faba Plants Irrigated by Sea Water Leaf Area, Pigment Content and Photosynthetic Activity. *Biol. Plant* **1993**, *35*, 519–527. [CrossRef]
60. Aldesuquy, H.S.; Ibrahim, A.H. Interactive Effect of Seawater and Growth Bioregulators on Water Relations, Abscisic Acid Concentration and Yield of Wheat Plants. *J. Agron. Crop Sci.* **2001**, *187*, 185–193. [CrossRef]
61. Yuan, L.; Xu, D.Q. Stimulation Effect of Gibberellic Acid Short-Term Treatment on Leaf Photosynthesis Related to the Increase in Rubisco Content in Broad Bean and Soybean. *Photosynth. Res.* **2001**, *68*, 39–47. [CrossRef] [PubMed]
62. Miri, M.R.; Ghooshchi, F.; Tohidi-Moghadam, H.R.; Larijani, H.R.; Kasraie, P. Ameliorative Effects of Foliar Spray of Glycine Betaine and Gibberellic Acid on Cowpea (*Vigna unguiculata* L. Walp.) Yield Affected by Drought Stress. *Arab. J. Geosci.* **2021**, *14*, 830. [CrossRef]
63. Azadi, F.; Hatami, A.; Salek Mearaji, H. The Effect of Cytokinin Foliar on Morpho-Physiological Traits, Yield and Yield Components of Black Cumin (*Nigella sativa* L.) under Salinity Stress Conditions. *Environ. Stress. Crop Sci.* **2022**, *15*, 975–990.
64. Mostafa, H.A.M.; El-Bassiouny, H.M.S.; Khattab, H.K.I.; Sadak, M.S. Improving the Characteristics of Roselle Seeds as a New Source of Protein and Lipid by Gibberellin and Benzyladenine Application. *J. Appl. Sci. Res.* **2005**, *1*, 161–167.
65. Parida, A.K.; Dagaonkar, V.S.; Phalak, M.S.; Umalkar, G.V.; Aurangabadkar, L.P. Alterations in Photosynthetic Pigments, Protein and Osmotic Components in Cotton Genotypes Subjected to Short-Term Drought Stress Followed by Recovery. *Plant Biotechnol. Rep.* **2007**, *1*, 37–48. [CrossRef]
66. Farhat, F.; Arfan, M.; Wang, X.; Tariq, A.; Kamran, M.; Tabassum, H.N.; Tariq, I.; Mora-Poblete, F.; Iqbal, R.; El-Sabrout, A.M.; et al. The Impact of Bio-Stimulants on Cd-Stressed Wheat (*Triticum aestivum* L.): Insights Into Growth, Chlorophyll Fluorescence, Cd Accumulation, and Osmolyte Regulation. *Front. Plant Sci.* **2022**, *13*, 850567. [CrossRef]

67. Rashid, N.; Khan, S.; Wahid, A.; Ibrar, D.; Hasnain, Z.; Irshad, S.; Bashir, S.; Al-Hashimi, A.; Elshikh, M.S.; Kamran, M. Exogenous Application of Biostimulants and Synthetic Growth Promoters Improved the Productivity and Grain Quality of Quinoa Linked with Enhanced Photosynthetic Pigments and Metabolomics. *Agronomy* **2021**, *11*, 2302. [CrossRef]
68. Mir, A.R.; Siddiqui, H.; Alam, P.; Hayat, S. Foliar Spray of Auxin/IAA Modulates Photosynthesis, Elemental Composition, ROS Localization and Antioxidant Machinery to Promote Growth of Brassica Juncea. *Physiol. Mol. Biol. Plants* **2020**, *26*, 2503–2520. [CrossRef]
69. Pazuki, A.; Sedghi, M.; Aflaki, F. Interaction of Salinity and Phytohormones on Wheat Photosynthetic Traits and Membrane Stability. *Agriculture/Pol'nohospodárstvo* **2013**, *59*, 33–41. [CrossRef]
70. da Costa, W.A.; Ribeiro, V.T.; da Silva, D.C.; Neto, A.; Wanderley Neto, A.d.O.; de Castro Dantas, T.N.; Ferrari, M.; dos Santos, E.S. Low-Energy Nanoemulsified Systems Containing Antioxidant Eutectic Extract from Rhodotorula Mucilaginosa Yeast Cells. *Colloids Surf. A Physicochem. Eng. Asp.* **2022**, *651*, 129715. [CrossRef]
71. Majeed, A.; Kaleem Abbasi, M.; Hameed, S.; Imran, A.; Rahim, N. Isolation and Characterization of Plant Growth-Promoting Rhizobacteria from Wheat Rhizosphere and Their Effect on Plant Growth Promotion. *Front. Microbiol.* **2015**, *6*, 198. [CrossRef] [PubMed]
72. Ashour, N.I.; Neumann, D.; Nieden, U. zur Gibberellic Acid Induced Changes in the Ultrastructure of Chloroplasts and the Content of Chlorophyll in Leaves of Dwarf Maize (*Zea mays* L.). *Biochem. Physiol. Der Pflanz.* **1973**, *164*, 402–413. [CrossRef]
73. Vichnevetskaia, K.D.; Roy, D.N. Oxidative Stress and Antioxidative Defense with an Emphasis on Plants Antioxidants. *Environ. Rev.* **1999**, *7*, 31–51. [CrossRef]
74. Arteca, R.N. *Plant Growth Substances: Principles and Applications*; Springer Science & Business Media: Berlin/Heidelberg, Germany, 2013; ISBN 1475724519.
75. Sabir, F.; Noreen, S.; Malik, Z.; Kamran, M.; Riaz, M.; Dawood, M.; Parveen, A.; Afzal, S.; Ahmad, I.; Ali, M. Silicon Improves Salinity Tolerance in Crop Plants: Insights into Photosynthesis, Defense System, and Production of Phytohormones. In *Silicon and Nano-Silicon in Environmental Stress Management and Crop Quality Improvement*; Elsevier: Amsterdam, The Netherlands, 2022; pp. 91–103.
76. Malik, Z.; Afzal, S.; Dawood, M.; Abbasi, G.H.; Khan, M.I.; Kamran, M.; Zhran, M.; Hayat, M.T.; Aslam, M.N.; Rafay, M. Exogenous Melatonin Mitigates Chromium Toxicity in Maize Seedlings by Modulating Antioxidant System and Suppresses Chromium Uptake and Oxidative Stress. *Environ. Geochem. Health* **2021**, *44*, 1451–1469. [CrossRef]
77. Parveen, A.; Hamzah Saleem, M.; Kamran, M.; Zulqurnain Haider, M.; Chen, J.T.; Malik, Z.; Shoaib Rana, M.; Hassan, A.; Hur, G.; Tariq Javed, M.; et al. Effect of Citric Acid on Growth, Ecophysiology, Chloroplast Ultrastructure, and Phytoremediation Potential of Jute (*Corchorus capsularis* L.) Seedlings Exposed to Copper Stress. *Biomolecules* **2020**, *10*, 592. [CrossRef]
78. Kamran, M.; Danish, M.; Saleem, M.H.; Malik, Z.; Parveen, A.; Abbasi, G.H.; Jamil, M.; Ali, S.; Afzal, S.; Riaz, M.; et al. Application of Abscisic Acid and 6-Benzylaminopurine Modulated Morpho-Physiological and Antioxidative Defense Responses of Tomato (*Solanum lycopersicum* L.) by Minimizing Cobalt Uptake. *Chemosphere* **2021**, *263*, 128169. [CrossRef] [PubMed]
79. Ali, M.; Kamran, M.; Abbasi, G.H.; Saleem, M.H.; Ahmad, S.; Parveen, A.; Malik, Z.; Afzal, S.; Ahmar, S.; Dawar, K.M. Melatonin-Induced Salinity Tolerance by Ameliorating Osmotic and Oxidative Stress in the Seedlings of Two Tomato (*Solanum lycopersicum* L.) Cultivars. *J. Plant Growth Regul.* **2020**, *40*, 2236–2248. [CrossRef]
80. Hashem, M.; Abo-Elyousr, K.A. Management of the Root-Knot Nematode Meloidogyne Incognita on Tomato with Combinations of Different Biocontrol Organisms. *Crop Prot.* **2011**, *30*, 285–292. [CrossRef]
81. Moravcová, Š.; Tůma, J.; Dučaiová, Z.K.; Waligórski, P.; Kula, M.; Saja, D.; Słomka, A.; Bąba, W.; Libik-Konieczny, M. Influence of Salicylic Acid Pretreatment on Seeds Germination and Some Defence Mechanisms of *Zea mays* L. Plants under Copper Stress. *Plant Physiol. Biochem.* **2018**, *122*, 19–30. [CrossRef] [PubMed]
82. Kim, S.K.; Son, T.K.; Park, S.Y.; Lee, I.J.; Lee, B.H.; Kim, H.Y.; Lee, S.C. Influences of Gibberellin and Auxin on Endogenous Plant Hormone and Starch Mobilization during Rice Seed Germination under Salt Stress. *J. Environ. Biol.* **2006**, *27*, 181.
83. Ali, B.; Hayat, S.; Aiman Hasan, S.; Ahmad, A. A Comparative Effect of IAA and 4-Cl-IAA on Growth, Nodulation and Nitrogen Fixation in *Vigna radiata* (L.) Wilczek. *Acta Physiol. Plant* **2008**, *30*, 35–41. [CrossRef]
84. Beevers, L. *Nitrogen Metabolism in Plants*; Springer: Berlin/Heidelberg, Germany, 1981; ISBN 3642679218.
85. Hopkins G, W.; A, H.N.P. *Introduction to Plant Physiology*, 4th ed.; John Wiley & Sons: Hoboken, NJ, USA, 2009.

soil systems

Article

Arbuscular Mycorrhizal Fungi Enhance Growth and Increase Concentrations of Anthocyanin, Phenolic Compounds, and Antioxidant Activity of Black Rice (*Oryza sativa* L.)

Sabaiporn Nacoon [1], Wasan Seemakram [1], Jindarat Ekprasert [1], Piyada Theerakulpisut [2,3], Jirawat Sanitchon [4], Thomas W. Kuyper [5] and Sophon Boonlue [1,3,*]

1 Department of Microbiology, Faculty of Science, Khon Kaen University, Khon Kaen 40002, Thailand
2 Department of Biology, Faculty of Science, Khon Kaen University, Khon Kaen 40002, Thailand
3 Salt-Tolerant Rice Research Group, Khon Kaen University, Khon Kaen 40002, Thailand
4 Department of Agronomy, Faculty of Agriculture, Khon Kaen University, Khon Kaen 40002, Thailand
5 Soil Biology Group, Wageningen University & Research, P.O. Box 47, 6700 AA Wageningen, The Netherlands
* Correspondence: bsopho@kku.ac.th; Tel.: +66-89-186-2646

Abstract: Black rice (*Oryza sativa* L.) contains high concentrations of bioactive compounds that are associated with human-health benefits. Arbuscular mycorrhizal fungi (AMF) can increase plant performance and concentrations of these bioactive compounds. In a pot experiment, the effects of four different species of AMF (*Claroideoglomus etunicatum; Rhizophagus variabilis; Rhizophagus nov. spec.; Acaulospora longula*) were assessed on growth performance, grain yield, concentrations of phenolic compounds and anthocyanin, and antioxidant activity of two black-rice cultivars. The experiment was a completely randomized factorial design with two factors, viz. cultivar (Niew Dam Hmong and Maled Phai) and treatment (four different species of AMF and two non-inoculated treatments, without and with mineral fertilizer). Results showed that cultivar, treatment, and their interaction were almost always significant sources of variation for both plant performance parameters and concentrations of bioactive compounds. Maled Phai showed higher performance and higher concentrations of phenolics and anthocyanins but lower antioxidant activity than Niew Dam Hmong. The non-inoculated treatment without mineral fertilizer showed the lowest performance. The non-inoculated treatment with mineral fertilizer resulted in larger root and shoot biomass than the mycorrhizal treatments, but grain yield was higher in the mycorrhizal treatments. Inoculation with *R. variabilis* resulted in the highest concentration of phenolics and anthocyanins. We conclude that *R. variabilis* was the best inoculum for increasing grain yield and bioactive compounds, especially in Maled Phai.

Keywords: AMF; *Rhizophagus variabilis*; Maled Phai; Niew Dam Hmong; phytochemical; rice productivity

1. Introduction

Rice (*Oryza sativa* L.) is a major staple cereal that feeds more than 50% of the global population. Almost all rice (more than 90%) is being produced in Asia. The most commonly consumed rice cultivars have a white kernel. However, specialty rice cultivars exist with red, purple, and black colored kernels, and these cultivars are known as black rice [1]. Their grains contain high amounts of phenolic compounds, especially anthocyanins and antioxidant substances. These substances are considered health-promoting and are especially beneficial for memory enhancement and for strengthening the human immune system [2–4]. Black-rice cultivars contain the highest concentrations of anti-oxidants [5]. Although the demand for black rice in Thailand is rapidly increasing because of its high nutritional value, the market share of this specialty rice is still low. In Thailand, the production of black rice is clustered around the northern and northeastern parts of the country. However, rice production in these locations is characterized by low productivity because soils are mostly sandy and have unfavorable physicochemical conditions, including low organic-matter content, amounts of available nitrogen and phosphorus, and cation exchange capacity [6]. Because of low inherent fertility, farmers have constantly and extensively used mineral fertilizer to increase rice production. Injudicious use of mineral fertilizer has caused deterioration in the physical and chemical properties of these soils as well as environmental pollution.

Because of declining soil fertility, other forms of field management are clearly needed, and these include a larger focus on the soil biota that contribute to improved performance of black rice. Such improved performance could go hand in hand with a reduced environmental footprint through reduced water demand and lower greenhouse gas emissions. Among such beneficial soil biota, there is a major role for arbuscular mycorrhizal fungi (AMF), a group of mutualistic root and soil-inhabiting fungi. The large majority (over 80%) of crops are able to form mycorrhizal symbiosis, and rice, especially when grown under aerobic conditions, also benefits from the mycorrhizal symbiosis [7,8]. AMF enhance the acquisition of growth-limiting nutrients and improve the acquisition of water and drought tolerance through various mechanisms. Moreover, AMF often increase resistance against belowground and aboveground pathogens and herbivores, although a recent study indicated increased susceptibility of mycorrhizal rice against various pests [9]. In many soils, there is sufficiency of AMF inoculum and the potential of the mycorrhizal symbiosis to contribute to enhanced plant performance then depends on the forms of management applied by the farmer. However, past injudicious management could have reduced AMF inoculum to too low levels, and in such conditions, the use of commercial inoculum could be considered. Such commercial inocula consist of one or more species of AMF, whose provenance may either be local or may have been imported from elsewhere, with the potential risk of these organisms becoming invasive in their new habitat. Experimental research towards the mycorrhizal symbiosis, on the other hand, often involves laboratory or greenhouse experiments, in which control plants are cultivated in sterilized soil and in which various AMF species, alone or in combination, are tested. Such studies allow ascertaining the relative benefits of individual species of AMF under specific conditions, and such knowledge can subsequently be used for upscaling under field conditions.

The main objective of this study was to test the effects of different species of AMF on the growth performance and concentrations of bioactive substances of black rice in a pot experiment under laboratory conditions. The effects of these AMF species were compared with two control treatments, viz. a treatment at inherent soil fertility and a treatment where mineral fertilizer has been added. By executing the experiment with two different controls, the study does allow to the assess both AMF under the inherent soil fertility and the extent to which management of AMF can be an economic substitute for the use of mineral fertilizer. As mycorrhizal management is often cheaper than the acquisition of mineral fertilizer, the study has the potential to demonstrate options to reduce farmer expenses by refraining from mineral-fertilizer use and thereby increase their income. Farmers could then also benefit from other ecosystem services from their rice fields due to improved soil quality.

2. Materials and Methods

2.1. AMF Identification

2.1.1. Isolation of Species of AMF

A total of 15 soil samples were collected from the rhizosphere of black-rice plants grown as upland (aerobic) rice in different regions in the Northeast of Thailand, including Khon Kaen Province (Nam Phong district and Muang district), Roi-Et Province (Nong Phok district) and Kalasin Province (Kuchinarai district). Fifty grams of each rhizosphere sample were thoroughly suspended in 500 mL water in a beaker and allowed to settle afterwards. AMF spores were separated from these soils by wet-sieving and decanting techniques [10] using a series of sieves that were arranged from top to bottom in the following order: 250 μm, 125 μm, and 90 μm. The trapped spores were filtered througWhatman No. 1 filter paper by repeated washing with water. Spores were gently picked using forceps and placed on a glass slide for visualization under a stereomicroscope (Nikon SMZ745 model LC-LEDS, Zhejiang, China.

2.1.2. Multiplication of Spores of AMF

Only the spores of the most abundant AMF species obtained from each soil sample were subjected to spore multiplication. Multiplication of AMF spores was carried out using a pot culture technique as described by Boonlue et al. [11]. Briefly, soil was twice sterilized by autoclave at 121 °C for 2 h and then added to 20-cm-diameter plastic pots. Maize (*Zea mays* L.) seeds were surface-sterilized by soaking in 10% sodium hypochlorite solution for 30 min before adding them to the pots. Then, individual spores of each morphologically distinct AMF species were added to the pots containing these maize seeds. Maize was subsequently grown in a greenhouse at 30–35 °C and irrigated with tap water every day. After 90 days, irrigation was stopped, and the plants were allowed to dry out, which causes sporulation by the AMF. The plants were cut off at a position just above the soil surface. After that, the soil was air-dried and then ground into fine particles (<0.2 mm). The purity of spores and the total spore number in the soil were determined using the sucrose centrifugation method [12]. Dried soils containing AMF spores, mycelia, and colonized root fragments were then used as the inoculum in the pot experiment.

2.1.3. Identification of AMF Species

DNA of AMF was extracted from single spores obtained from the cultures described above. The spore surface was cleaned by sonication for 10 s and sterilized thoroughly using Chloramine–T solution (2% *w/v*) for 5 min. The spore was then rinsed with sterilized deionized water 3 times. A sterilized spore was transferred into a PCR tube containing 7 µL of TE (10 mM Tris-HCl, pH 8, 1 mM EDTA) [13] and then broken using a sterilized microtip under a stereomicroscope. DNA extracts were submitted to a nested PCR protocol [14]. The Thermo Scientific Phire Plant Direct PCR Master Mix kit was used for a nested PCR amplification in a total volume of 20 µL. The primer pairs SSUmAf1 and LSUmAr3 were used in the first amplification step. The second amplification step was performed using the mixed primer SSUmCf3 and SSUmCf1, LSUmBr5 and LSUmBr1, which targets only glomeromycotan fungi. PCR products were purified using the QIAGEN PCR Purification Kit (Thermo Scientific, Vilnius, Lithuania). The rDNA sequences were submitted for sequencing at the U2Bio Thailand. After DNA sequencing, the DNA sequences were compared with similar sequences on GenBank. Phylogenetic analyses was executed through the program MEGA 7, using maximum likelihood [15]. The obtained sequences were deposited to the National Center for Biotechnology Information (NCBI) (for accession numbers, see Section 3.1).

2.2. Greenhouse Experiment

2.2.1. Soil Preparation for Rice Cultivation

The soil for growing black rice in this study was a sandy loam soil with a pH of 7.26, electrical conductivity (EC) of 0.043 dS m^{-1}, organic matter (OM) content of 6.4 g kg^{-1}, total nitrogen (N) content of 240 mg kg^{-1} (C:N $\approx$ 13), total phosphorus content of 146 mg kg^{-1}, total potassium (K) content of 428 mg kg^{-1}, available P (method Bray 2) 61 mg kg^{-1}, exchangeable K 50 mg kg^{-1}, calcium (Ca) 655 mg kg^{-1}, and sodium (Na) of 50 mg kg^{-1}. Stones, wood chips, and plant debris in a soil sample were removed. The soil samples were sterilized by autoclaving at 121 °C at a pressure of 15 psi for 120 min. The soils were left at room temperature overnight and then sterilized again at the same condition before being packed into 20-cm-diameter plastic pots (5 kg pot^{-1}).

2.2.2. Preparation of Rice Seedlings

Two different cultivars of black upland rice (*Oryza sativa* subsp. *indica*) named as Niew Dam Hmong and Maled Phai were provided by the group of Rice project, Faculty of Agriculture, Khon Kaen University, Thailand. Niew Dam Hmong is a glutinous, pigmented, upland rice with a black seed coat, light-sensitive behavior, and short life span. Generally, it is cultivated around August. Maled Phai is a non-glutinous, pigmented, upland rice that originated from southern Thailand. It has brown-purple seed, and its cultivation period is from the beginning of June–November. Its seeds are somewhat smaller, and the cultivar is somewhat less productive. However, it has stronger bioactivities than Niew Dam Hmong [16]. Note that the authors of that publication refer to the cultivars as Ma-Led-Fy and Dam-Mong, respectively. Both rice cultivars have been developed at the Agronomy Station, Khon Kaen University and are widely distributed to the community. Rice seeds were surface-sterilized by soaking in 6% sodium hypochlorite solution for 10 min. Twice-sterilized soil was used as the plant substrate contained in 20-cm-diameter plastic pots. The rice was grown in a greenhouse at 30−35 °C and irrigated with tap water every day. Seven-day-old rice seedlings with relatively similar sizes and having true leaves were selected for the experiment.

2.2.3. Experimental Design and Pot Preparation

The experiment was conducted in enclosed greenhouses at Khon Kaen University, Khon Kaen, Thailand. The experiment was arranged as 2-factorial experiment (factors: rice cultivar–two cultivars; mycorrhiza–six treatments) in a Completely Randomized Design (CRD) with 7 replicates per treatment, resulting in a total of 84 pots. The experiment was carried out for 4 months after which the plants were harvested. The mycorrhizal factor included six treatments, viz. (T1) control, sterilized soil without AMF; (T2) control with mineral fertilizer, sterilized soil without AMF, with addition of mineral nutrients (per pot in total 150 mg N, 90 mg P, and 25 mg K, applied in three doses, at 20–25 days after planting (DAP), 45–50 DAP, and 70–75 DAP); (T3) inoculation with AMF isolate ROI-ET2-02; (T4) inoculation with isolate ROI-ET2-01; (T5) inoculation with isolate ROI-ET1-01; (T6) inoculation with isolate KS-02. The AMF inoculum was applied adjacent to plant roots at a rate of approximately 200 spores pot^{-1}.

2.2.4. Determination of Plant and AMF Performance

The following plant performance parameters were measured at 120 days after transplantation (at harvesting stage). SPAD chlorophyll meter reading (SCMR) was recorded using a chlorophyll meter SPAD-502 plus (Konica Minolta, Japan). Plant biomass (shoots, panicles with grain, roots) was determined after the samples were dried at 80 °C for 3 days. Shoot samples were also analyzed for concentrations of N, P, and K.

The following functional compounds were analyzed: anthocyanin concentration (TAC), total phenolic compound (TPC), and 1,1-diphenyl-2-picrylhydrazyl radical (DPPH). TAC, TPC, and DDPH were extracted according to Kapcum et al. [17] with some modifications. The samples of dried seed were finely ground. An amount of 1.0 g of samples was extracted with 10 mL methanol, then shaken for 2 h, and centrifuged at 3000 rpm for 10 min. The mixture was filtered (Whatman No.1 filter paper), and the residues were re-extracted twice with 5 mL methanol using the same procedure. The three aliquots were combined and stored at −40 °C in the dark until analyzed.

TAC was determined using two aliquots of 50 µL of extracts to which 3 mL 0.025 M of KCl buffer at pH 1.0 and 0.4 M sodium acetate buffer at pH 4.5 were added. The mixture was then allowed to stand for 20 min before measuring absorbance at 520 and 700 nm. Total anthocyanin concentration was calculated using the following equation and expressed as cyanidin-3-glucoside equivalent per 100 g sample [6]:

$$\text{Total anthocyanins (mg/100 g)} = \frac{\Delta A \times MW \times D \times (V/G) \times 100}{\epsilon \times L} \quad (1)$$

where ΔA is absorbance = (A520 nm–A700 nm) pH 1.0–(A520 nm–A700 nm) pH 4.5, ϵ is molar extinction coefficient of Cy-3-G = 29,600 M^{-1} cm^{-1}, L is cell path length of cuvette = 1 cm, MW is molecular weight of anthocyanins = 449.2 g mol^{-1}, D is a dilution factor, V is a final volume (mL), and G is weight of sample (g).

The TPC of the extracts was determined using 125 μL of extracts and 250 μL Folin-Ciocalteu's reagent, followed by the addition of 3 mL distilled water. The solution was mixed well and then allowed to stand for 6 min, after which 2.5 mL 7% sodium carbonate solution was added. The reaction mixture was allowed to stand for 90 min at room temperature before measuring absorbance at 760 nm (Hitachi High-Tech Science Corporation, Tokyo, Japan). Gallic acid was used as a calibration standard, and results were expressed as mg gallic acid equivalent per 100 g sample.

DPPH free radical scavenging activity was determined according to the method described by Leong & Shui [18], with some modifications. Freshly prepared solution of 0.1 mM solution of DPPH in methanol was prepared with absorbance 517 nm. An aliquot of 100 μL of each sample (with appropriate dilution) was mixed with 4.0 mL of DPPH solution, then allowed to stand at room temperature for 30 min before measurement. The percentage of radical-scavenging ability was calculated by using the formula:

$$\text{Scavenging ability } (\%) = \frac{(\text{Absorbance 515 nm of control}) \times (\text{Absorbance 515 nm of sample})}{(\text{Absorbance 515 nm of control})} \times 100 \quad (2)$$

To assess the intensity of AMF root colonization, fresh root samples were stained with 0.05% lactoglycerol trypan blue solution according to the method described by Koske & Gemma [19]. The stained root segments were observed under a microscope. Mycorrhizal colonization intensity was measured according to Trouvelot et al. [20].

2.2.5. Statistical Analysis

Data were analyzed by two-way analysis of variance (ANOVA). Data were tested for normality and homogeneity of variances. The least significant difference (LSD) test was applied to test for significant differences among the means of different treatments at p-value < 0.05. The correlation between parameters was calculated by Pearson's correlation coefficient and evaluated at p-value < 0.05. All statistical analyses were performed using the Statistix 10.0 version software.

3. Results

3.1. Molecular Identification of AMF Species and Performance on Pot Culture

Molecular identification, based on a phylogenetic analysis of the newly generated sequences, showed that isolate ROI-ET2-01 belonged to *Claroideoglomus etunicatum* (Accession No. OQ466528), KS-02 in the clade of *Rhizophagus*, with best matches to *R. variabilis* (Accession No. OQ456401) and especially with a sequence (FR873160) from the French Antilles [21], ROI-ET1-01 belonged to the genus *Rhizophagus* (Accession No. OQ755166), likely constituting a new species, conspecific with a sequence (JX683735) of Glomus Agro-03-S [22], and isolate ROI-ET2-02 to *Acaulospora longula* (Accession No. OQ455726). In the remainder of this paper, these fungal species will be indicated by their scientific names.

These four isolates, which are stored at the Mycorrhiza and Microtechnology Lab, Department of Microbiology, Khon Kaen University, were subjected to spore multiplication. *Acaulospora longula* produced the highest amount of spores, viz. 32 spores g^{-1} soil, and C. etunicatum the lowest amount, viz. 2 spores g^{-1} soil. All four species colonized roots of maize, with fractional colonization ranging between 18% (*Rhizophagus nov. spec.*) and 35% (*R. variabile*) (Table 1). Based on adequate colonization and sufficient spore production, these four species were used to study rice growth in the greenhouse experiment.

Table 1. Total number of spores (g^{-1} soil) and fractional root colonization of maize of the isolated AMF species.

AMF Isolates	Total Spore (Spore/g Soil)	Root Colonization (%)
A. longula	32	25
C. etunicatum	2	19
R. nov. spec.	6	18
R. variabilis	13	35

3.2. Fractional Root Colonization and Spore Number of AMF on Maled Phai and Niew Dam Hmong

Fraction colonization by the four AMF species in black-rice roots and the number of AMF spores are shown in Table 2. Analysis of variance showed that mycorrhizal species, rice cultivar, and the interaction were significant sources of variation (Table 2). The control plants of Maled Phai remained free of mycorrhizal colonization, whereas the control plants of Niew Dam Hmong showed low colonization, lower than the inoculation treatments. No AMF spores were found in the controls of both rice cultivars. All four AMF successfully colonized the roots of both rice cultivars. Fractional colonization and spore number were significantly higher in Maled Phai than in Niew Dam Hmong. Among the AMF species investigated, *R. variabile* exhibited the highest fractional colonization and spore number.

Table 2. The number of AMF spores in soil and the percentage of AMF colonization in roots of Maled Phai and Niew Dam Hmong rice cultivars at the harvest stage.

Treatments	Root Colonization (%)	Total Spore (Spore g^{-1} Soil)
	Maled Phai	
Control	0 i	0 f
NPK fertilizer	0 i	0 f
A. longula	19 c	2 d
C. etunicatum	25 b	3 c
R. nov. spec.	14 d	3 c
R. variabile	28 a	6 a
	Niew Dam Hmong	
Control	2 h	0 f
NPK fertilizer	1 h	0 f
A. longula	7 f	2 d
C. etunicatum	9 e	1 e
R. nov. spec.	5 g	2 d
R. variabile	14 d	4 b
% CV	10	18
Treatment	**	**
Rice cultivar	**	**
Treatment × Rice cultivar	**	**

Numbers followed by the same letter in each column are not significantly different according to the LSD test. **, Significant difference at $p \leq 0.01$.

3.3. Effects of AMF Species on the Promotion of Growth and Yield of Maled Phai Niew Dam Hmong

The results of plant growth and yield parameters of both rice cultivars are provided in Table 3. For all nine growth parameters, mycorrhiza (treatment) was a significant source of variation, rice cultivar was a significant source of variation for six parameters (excluding tiller number, panicle number, and seed weight), while the interaction term was significant for five parameters (excluding harvest index [HI], the ratio of grain mass over total aboveground mass, tiller number, panicle number, and SPAD). The control treatment without fertilizer resulted in the smallest plants, but with addition of mineral fertilizer, these plants achieved largest plant height and total biomass. One exception was noted. The total biomass of Maled Phai, inoculated with *R. variabilis*, was larger than that of plants that had received mineral fertilizer. Those mycorrhizal plants were characterized by a particularly large root dry weight. The effect of mineral fertilizer, compared to the mycorrhizal treatment, was especially visible in increases in shoot and root dry weight but not in grain yield. Non-inoculated (control) plants, both without and with mineral fertilizer, had lower grain weight than AMF-inoculated plants. As a consequence, the harvest index was significantly higher for AMF-inoculated than non-inoculated plants. Nutrient concentration of shoots was lowest in the non-inoculated control without fertilizer and highest in the non-inoculated control with mineral fertilizer. The N:P ratios of all plants were below 5, indicating several N limitations. In addition, AMF, rice cultivar, and the interaction were also significant sources of variation for antioxidant activity, and concentrations of phenolic compounds and anthocyanin (Table 4). Concentrations of phenolic compounds and anthocyanin were higher in Maled Phai than in Niew Dam Hmong, consistent with previous reports [16]. Antioxidant activity showed a more variable pattern, without consistent differences between cultivars. Among the AMF species, *R. variabilis* showed the largest positive effect on phenolic and anthocyanin concentrations in both cultivars (Table 4).

Table 3. Effects of AMF on plant growth parameters of Maled Phai and Niew Dam Hmong cultivars at harvesting stage.

Treatments	Root Dry Weight (g)	Shoot Dry Weight (g)	Grain Weight (g)	Aboveground Weight (G)	Harvest Index (HI)	Number of Panicles	Number of Tillers	Height (cm)	SPAD
				Maled Phai cultivar					
Control	16 i	44 e	4 g	48 g	0.08 g	3 e	8 d	92 de	35 d
NPK fertilizer	75 b	121 a	14 f	135 a	0.10 g	6 cd	18 a	102 b	45 a
A. longula	39 g	58 d	17 cde	65 de	0.23 de	6 cd	11 cd	97 bcd	45 a
C. etunicatum	46 e	60 d	18 def	78 d	0.22 e	6 cd	14 bc	90 e	42 abc
R. nov. spec.	35 h	71 c	20 bcd	91 c	0.22 e	6 cd	13 c	96 b–e	44 a
R. variabilis	104 a	94 b	25 a	119 b	0.21 e	8 a	13 bc	98 bcd	45 a
				Niew Dam Hmong cultivar					
Control	7 c	23 f	5 g	28 h	0.16 f	3 e	9 d	95 cde	37 cd
NPK fertilizer	16 i	76 c	15 ef	91 c	0.17 f	6 bc	17 ab	114 a	43 ab
A. longula	50 d	43 e	22 b	65 ef	0.34 a	6 bc	14 bc	95 cde	38 bcd
C. etunicatum	42 fg	45 e	19 bcd	64 ef	0.30 bc	5 d	12 cd	99 bc	38 bcd
R. nov. spec.	33 h	43 e	20 bc	63 f	0.32 ab	6 cd	14 abc	96 b–e	45 a
R. variabilis	45 ef	51 de	18 cde	69 de	0.26 cd	7 ab	13 bc	97 b–e	42 abc
% CV	8	15	19	13	16	16	27	6	14
Treatment	**	**	**	**	**	**	**	**	**
Rice cultivar	**	**	ns	**	**	ns	ns	*	*
Treatment × Rice cultivar	**	**	**	**	ns	ns	ns	*	ns

Numbers followed by the same letter in each column are not significantly different according to LSD test. ns, not significant *, Significant difference at $p \leq 0.05$; **, Significant difference at $p \leq 0.01$.

Table 4. Effects of AMF on phenolic compounds, total antioxidant and anthocyanin of rice seeds and nutrient concentrations in shoots.

Treatments	Phenolic Compound (mg 100 g^{-1} DW)	Antioxidant (% DPPH Radical Scavenging)	Anthocyanin (mg 100 g^{-1} DW)	N (mg g^{-1})	P (mg g^{-1})	K (mg g^{-1})
			Maled Phai cultivar			
Control	138 d	25 f	40 d	1.6 g	0.5 e	4.6 de
NPK fertilizer	196 b	39 de	98 a	6.0 a	1.2 b	11.2 a
A. longula	184 c	27 f	70 c	2.3 ef	0.6 de	5.6 de
C. etunicatum	140 d	58 c	62 c	2.1 f	0.5 e	5.5 de
R. nov. spec.	188 bc	43 d	66 c	2.7 de	0.7 cde	7.3 c
R. variabilis	209 a	68 b	82 b	3.5 c	0.9 bc	9.7 b
			Niew Dam Hmong cultivar			
Control	53 h	32 ef	36 d	1.6 g	0.5 e	2.7 f
NPK fertilizer	69 g	80 a	21 e	4.7 b	1.7 a	9.1 b
A. longula	87 f	55 c	34 d	2.9 d	0.8 cd	4.4 e
C. etunicatum	71 g	41 de	23 e	2.8 d	0.8 cd	5.1 de
R. nov. spec.	70 g	58 c	36 d	2.8 d	0.7 cde	5.4 de
R. variabilis	116 e	43 d	94 a	3.6 c	0.8 cd	5.8 d
%CV	7	18	18	9	20	13
AMF	**	**	**	**	**	**
Rice cultivars	**	**	**	ns	**	**
AMF × Rice cultivar	**	**	**	**	**	*

Numbers followed by the same letter in each column were not significantly different according to LSD test. ns, not significant; *, Significant difference at $p \leq 0.05$; **, Significant difference at $p \leq 0.01$.

Correlations between AMF colonization, plant performance parameters, and concentrations of secondary compounds are shown in Table 5. Mycorrhizal colonization was significantly positively correlated with seed weight, root weight, and total biomass, whereas the correlation between mycorrhizal colonization and shoot biomass was not significant. Mycorrhizal colonization was also significantly positively correlated with concentrations of phenolics and anthocyanin. Seed weight was significantly positive correlated with root weight, and less so with shoot weight. Both shoot and root weight were positively correlated with concentrations of phenolics and anthocyanin, whereas the correlation between seed weight and concentrations of phenolics and anthocyanin were barely significant.

Table 5. Correlation between AMF and % colonization with plant growth parameters of rice at harvest stage.

Correlation	Root Colonization	No. of Spore	Root Dry Weight	Shoot Dry Weight	Grain Weight	Biomass	HI	No. of Pani-cles	No. of Tillers	Height	SPAD	Anti-Oxidant	Phenolic Compound	Antho-Cyanin	N	P
No. of spore	0.86 **															
Root dry weight	0.39 **	0.45 **														
Shoot dry weight	0.16 ns	0.14 ns	0.75 **													
Grain weight	0.53 **	0.66 **	0.55 **	0.28 *												
Biomass	0.34 **	0.38 **	0.94 **	0.92 **	0.54 **											
HI	0.20 ns	0.36 **	−0.04 ns	−0.41 **	0.69 **	−0.14 ns										
No. of panicles	0.47 **	0.60 **	0.64 **	0.40 **	0.86 **	0.63 **	0.46 **									
No. of tillers	−0.03 ns	0.03 ns	0.45 **	0.47 **	0.32 **	0.50 **	0.07 ns	0.39 **								
Height	−0.22 *	−0.18 ns	0.33 **	0.30 *	0.10 ns	0.33 **	−0.06 ns	0.20 ns	0.18 ns							
SPAD	0.23 *	0.21 *	0.32 **	0.36 **	0.26 *	0.37 **	0.00 ns	0.24 *	0.34 **	0.00 ns						
Antioxidant	0.19 ns	0.25 *	0.55 **	0.29 *	0.46 **	0.47 **	0.19 ns	0.43 **	0.39 **	0.26 *	0.26 *					
Phenolic	0.52 **	0.42 **	0.45 **	0.66 **	0.21 ns	0.59 **	−0.35 **	0.27 *	0.09 ns	−0.11	0.31 **	−0.13 ns				
Anthocyanin	0.41 **	0.45 **	0.41 **	0.57 **	0.23 *	0.52 **	−0.20 ns	0.38 **	0.22 *	−0.10	0.30 *	−0.13 ns	0.73 **			
N	−0.20 ns	−0.06 ns	0.69 **	0.75 **	0.28 *	0.76 **	−0.14 ns	0.44 **	0.57 **	0.51 **	0.30 *	0.37 **	0.22 *	0.37 **		
P	−0.26 *	−0.16 ns	0.58 **	0.49 **	0.24 *	0.56 **	−0.08 ns	0.38 **	0.47 **	0.65 **	0.18 ns	0.57 **	−0.09 ns	−0.05 ns	0.8 **	
K	0.11 ns	0.14 ns	0.79 **	0.89 **	0.32 **	0.88 **	−0.31 **	0.46 **	0.51 **	0.40 **	0.39 **	0.39 **	0.56 **	0.46 **	0.8 **	0.7 **

**, Significant difference at $p \leq 0.01$; *, Significant difference at $p \leq 0.05$; ns, not significant.

4. Discussion

At present, there are only a few studies that provide information on the effects of AMF species on black rice. Surendirakumar et al. [23] reported seven species of AMF associated with black rice in paddy soils in India, including one species of *Acaulospora* and two species of *Rhizophagus*. However, the very different environmental conditions in that study (paddy rice) and this study (aerobic rice) would make a direct comparison difficult. Wangiyana et al. [24] reported increased grain yield of black rice after inoculation with a commercial inoculum, whose composition was not specified but likely consists of a mixture of different species of AMF and ectomycorrhizal fungi. A similar positive mycorrhizal effect was reported by Anugrah et al. [25], again without reporting the identity of the AMF species that might have caused this yield increase. Tisarum et al. [26] reported that inoculation with AMF, a mixture of three different species, increased drought tolerance and enhanced anthocyanin concentrations of black rice. Fractional colonization in that study was approximately 30%, comparable with our results (Table 2) for Maled Phai (14–28%) but considerably higher than for Niew Dam Hmong (5–14%). Other studies reported even higher fractional mycorrhizal colonization of rice than reported here, for instance, a colonization rate of 26–40% [27], <5–40% [28], and 12–27% under flooded, and 22–43% under non-flooded conditions [29].

This study showed that colonization by AMF had a beneficial effect on the growth performance of two black-rice cultivars. Biomass production of plants, inoculated with one of four species of AMF, outperformed non-inoculated plants in the absence of mineral fertilizer. The application of mineral fertilizer also boosted plant performance, but the effect was especially noteworthy in shoot biomass. In fact, seed weight and harvest index were less for non-inoculated fertilized plants than for inoculated, unfertilized plants. These non-inoculated plants showed very low fractional colonization. Considering the differences in seed yield and harvest index, we argue that the low levels of mycorrhizal colonization had only a minor impact on plant performance. High levels of fertilizer application have been reported before to result in plant investment in vegetative growth rather than in seed production. Mycorrhizal effects on cereal yields have been reported before by Zhang et al. [30] in a meta-analysis that showed a yield increase in grain yield of rice of 17% due to the mycorrhizal symbiosis, indicating an effect that was noted both under field and greenhouse conditions.

The increase in yield in the mycorrhizal treatments, as compared with the non-inoculated fertilized treatment, coincided with lower concentrations of nitrogen, phosphorus, and potassium in shoots (Table 4). However, compared with the unfertilized control, mycorrhizal plants had higher shoot concentrations of these three macronutrients. These mycorrhizal plants also showed increases in concentrations of phenolics and anthocyanins compared with unfertilized, non-inoculated plants. Fertilizer application also increased the concentration of these bioactive compounds. In the case of inoculated Maled Phai, unfertilized mycorrhizal plants exhibited similar concentrations of phenolics and anthocyanins than non-inoculated, fertilized plants. In the case of Niew Dam Hmong, inoculated, unfertilized plants even outperformed these non-inoculated, fertilized plants, demonstrating an effect that was most conspicuous in plants that were inoculated with *R. variabilis*. This recently described fungal species also achieved highest fractional root colonization in both cultivars and also significantly increased root biomass in Maled Phai. This effect is particularly interesting considering the significant positive correlations between root biomass, root colonization, and concentrations of bioactive compounds. Plant species-specific differences between AMF species in effects on plant performance have been regularly published, whereas similar effects on different cultivars of the same species have been less frequently described. Wang et al. [31] demonstrated how two cultivars of maize, a landrace and a hybrid, responded differently to two different AMF species, with the landraces being more responsive to *Funneliformis mosseae*, and the modern hybrid showing a slight preference for *Claroideoglomus etunicatum*. In this study, *Rhizophagus variabilis* had a more positive effect than the other three AMF species on the performance of Maled Phai, whereas Niew Dam Hmong performed equally with these four AMF species. Further studies on *R. variabilis*, which possibly has a global distribution [32], are planned.

The concentrations of phenolics and anthocyanins of Maled Phai were higher than Niew Dam Hmong, which is in agreement with the earlier study by Sripanidkulchai et al. [16], but the higher productivity of Maled Phai than of Niew Dam Hmong is contrary to study. However, comparing productivity data from field studies with those of individual plants in pots is inherently difficult. Beneficial effects of the AMF symbiosis on secondary compounds have been reported frequently [33–35]. Fiorilli et al. [36] reported upregulation of a gene involved in anthocyanin biosynthesis in mycorrhizal rice, inoculated with *R. irregularis*, compared with the non-mycorrhizal condition, but did not provide further details. Upregulation of anthocyanin biosynthesis in black rice as a consequence of the AMF symbiosis has also been reported by Tisarum et al. [26] and Wangiyana et al. [24]. Soltaniband et al. [37] reported increase in anthocyanin levels after inoculation with the AMF *R. irregularis* in strawberry (*Fragaria x ananassa*), and a similar effect was noted for anthocyanin concentrations in the berries of tempranillo grapevine (*Vitis vinifera*) [38]. Lee & Scagel [39] observed an increase approximately 35% in the concentrations of anthocyanins in leaves of *Ocimum basilicum* associated with *R. intraradices* in comparison with non-mycorrhizal controls in greenhouse cultivation. Baslam & Goicoechea [40] found that levels of anthocyanins in leaves were very sensitive to the presence of AMF colonizing roots of both cultivars of lettuce (*Lactuca sativa*), Batavia Rubia Munguía and Maravilla de Verano. AMF-species-dependent effects on anthocyanins were found in two weed species, *Solanum nigrum* and *Digitaria sanguinaria*, where inoculation with *F. mosseae* increased anthocyanin concentrations, whereas inoculation with *R. intraradices* or *R. fasciculatus* decreased it.

In this study, as in many other studies [33–35], there is a double mycorrhizal effect, both a direct effect through increased plant or seed yield and an additional indirect effect through higher concentrations of bioactive compounds such as phenolics and anthocyanins than in non-mycorrhizal plants. The underlying mechanism of enhanced synthesis of these bioactive compounds is currently unknown. In a study on *Cannabis sativa* [41], the increase in bioactive compounds was correlated with increases in phosphorus acquisition. However, in this study, mineral fertilizer plants showed the highest shoot concentration of N, P, and K, but this did not result in the highest concentrations of bioactive compounds. Both nutritional and non-nutritional factors have therefore been proposed to explain the increased production of secondary metabolites in AMF-colonized plants [35]. Nutritional mechanisms refer to the improvement of the nutritional condition of the host [41–43]. Zhao et al. [35] summarized current knowledge on potential non-nutritional mechanisms by stating that AMF colonization results in the activation of plant defense mechanisms with the production of phenolics and flavonoids. AMF also induce the production of signaling molecules, such as nitric oxide, salicylic acid, and hydrogen peroxide, which influence the activation of key enzymes such as l-phenylalanine ammonia lyase and chalcone synthase, for the biosynthesis of phenolic compounds [44]. Non-nutritional mechanisms may further involve the activation of metabolic routes [45], production of signaling molecules, alterations in the activity of key-enzymes for the production of these compounds [46–48], and hormonal alterations [49]. The results obtained in this study partly supported a nutritional mechanism, as evidenced by the correlations between nutrient concentrations and concentrations of bioactive compounds, both phenolics (for N, P, and K) and anthocyanins (for N and K) (Table 5). However, the comparison between non-inoculated mineral-fertilized plants and mycorrhizal, unfertilized plants indicates the importance of non-nutritional mechanisms as well.

5. Conclusions

Inoculation of black rice with four different species of AMF promoted plant growth compared with non-inoculated plants under unfertilized conditions. While mineral fertilization under non-inoculated conditions enhanced plant performance, this was mainly expressed through larger shoot biomass, whereas seed weight and harvest index were smaller under fertilization than under mycorrhizal fungal inoculation. The importance of AMF was also evident by the significant correlations between mycorrhizal colonization, seed weight and concentrations of bioactive compounds. Higher correlations between seed mass and root mass than between seed mass and shoot biomass also support this important role of AMF. Plant inoculated with AMF also possessed larger concentrations of phenolics and anthocyanins compared with uninoculated controls. The largest beneficial effect was noted for the AMF *Rhizophagus variabilis*, and this species deserves further scrutiny.

Author Contributions: Conceptualization, S.N., W.S. and S.B.; Methodology and investigation, S.N., W.S. and S.B.; Validation, J.E., J.S., P.T., T.W.K. and S.B.; Formal analysis J.E., T.W.K. and S.B.; Writing—original draft preparation, S.N.; Writing—review and editing, J.E., T.W.K. and S.B.; Supervision, S.B.; Funding acquisition, S.B. All authors have read and agreed to the published version of the manuscript.

Funding: This research was funded by the Fundamental Fund of Khon Kaen University through the research on Effects of arbuscular mycorrhizal fungi combination with endophytic fungi for enhancing rice growth and accumulation of anthocyanin and phenolic compound of black upland rice seeds by Khon Kaen University, Department of Microbiology, Faculty of Science has received funding support from the National Science, Research and Innovation Fund (NSRF) (Kh: FRB660052/0161).

Institutional Review Board Statement: Not applicable.

Informed Consent Statement: Not applicable.

Data Availability Statement: Not applicable.

Conflicts of Interest: The authors declare no conflict of interest.

References

1. Sompong, R.; Siebenhandl-Ehn, S.; Linsberger-Martin, G.; Berghofer, E. Physicochemical and antioxidative properties of red and black rice varieties from Thailand, China and Sri Lanka. *Food Chem.* **2011**, *124*, 132–140. [CrossRef]
2. Hu, C.; Zawistowski, J.; Ling, W.; Kitts, D.D. Black Rice (*Oryza sativa* L. *indica*) Pigmented Fraction Suppresses both Reactive Oxygen Species and Nitric Oxide in Chemical and Biological Model Systems. *J. Agric. Food Chem.* **2003**, *51*, 5271–5277. [CrossRef]
3. Konczak, I.; Zhang, W. Anthocyanins—More than nature's colours. *J. Biomed. Biotechnol.* **2004**, *2004*, 307613. [CrossRef]
4. Tanaka, J.; Nakanishi, T.; Ogawa, K.; Tsuruma, K.; Shimazawa, M.; Shimoda, H.; Hara, H. Purple rice extract and anthocyanidins of the constituents protect against light-induced retinal damage in vitro and in vivo. *J. Agric. Food Chem.* **2011**, *59*, 528–536. [CrossRef]
5. Goufo, P.; Trindade, H. Rice antioxidants: Phenolic acids, flavonoids, anthocyanins, proanthocyanidins, tocopherols, tocotrienols, c-oryzanol, and phytic acid. *Food Sci. Nutr.* **2014**, *2*, 75–104. [CrossRef]
6. Yodmanee, S.; Karrila, T.T.; Pakdeechanuan, P. Physical, chemical and antioxidant properties of pigmented rice grown in Southern Thailand. *Int. Food Res. J.* **2011**, *18*, 901–906.
7. Smith, S.E.; Facelli, E.; Pope, S.; Smith, F.A. Plant performance in stressful environments: Interpreting new and established knowledge of the roles of arbuscular mycorrhizas. *Plant Soil* **2010**, *326*, 3–20. [CrossRef]
8. Mbodj, D.; Effa-Effa, B.; Kane, A.; Manneh, B.; Gantet, P.; Laplaze, L.; Diedhiou, A.G.; Grondin, A. Arbuscular mycorrhizal symbiosis in rice: Establishment, environmental control and impact on plant growth and resistance to abiotic stresses. *Rhizosphere* **2018**, *8*, 12–26. [CrossRef]
9. Bernaola, L.; Cosme, M.; Schneider, R.W.; Stout, M. Belowground inoculation with arbuscular mycorrhizal fungi increases local and systemic susceptibility of rice plants to different pest organisms. *Front. Plant Sci.* **2018**, *9*, 747. [CrossRef]
10. Gerdemann, J.W.; Nicolson, T.H. Spores of mycorrhizal *Endogone* species extracted from soil by wet sieving and decanting. *Trans. Br. Mycol. Soc.* **1963**, *46*, 235–244. [CrossRef]
11. Boonlue, S.; Surapat, W.; Pukahuta, C.; Suwanarit, P.; Suwanarit, A.; Morinaga, T. Diversity and efficiency of arbuscular mycorrhizal fungi in soils from organic chili (*Capsicum frutescens*) farms. *Mycoscience* **2012**, *53*, 10–16. [CrossRef]
12. Daniels, B.A.; Skipper, H.D. Methods for the Recovery and Quantitative Estimation of Propagules from Soil [Vesicular-Arbuscular Mycorrhizal Fungi]. In *Methods and Principles of Mycorrhizal Research*; Schenck, N.C., Ed.; American Phytopathological Society: St. Paul, MN, USA, 1982; pp. 29–35.
13. Mosbah, M.; Philippe, D.L.; Mohamed, M. Molecular identification of arbuscular mycorrhizal fungal spores associated to the rhizosphere of *Retama raetam* in Tunisia. *Soil Sci. Plant Nutr.* **2018**, *64*, 335–341. [CrossRef]
14. Krüger, M.; Stockinger, H.; Krüger, C.; Schüßler, A. DNA-based species level detection of Glomeromycota: One PCR primer set for all Arbuscular Mycorrhizal Fungi. *New Phytol.* **2009**, *183*, 212–223. [CrossRef]
15. Kumar, S.; Stecher, G.; Tamura, K. MEGA7: Molecular evolutionary genetics analysis version 7.0 for bigger datasets. *Mol. Biol. Evol.* **2016**, *33*, 1870–1874. [CrossRef]

16. Sripanidkulchai, B.; Junlatat, J.; Tuntiyasawasdikul, S.; Fangkrathok, N.; Sanitchon, J.; Chankaew, S. Phytochemical and bioactivity investigation of Thai pigmented-upland rice: Dam-Mong and Ma-Led-Fy varieties. *Agric. Nat. Resour.* **2021**, *55*, 209–212. [CrossRef]
17. Kapcum, N.; Uriyapongson, J.; Alli, I.; Phimphilai, S. Anthocyanins, phenolic compounds and antioxidant activities in colored corn cob and colored rice bran. *Int. Food Res. J.* **2016**, *23*, 2347–2356.
18. Leong, L.P.; Shui, G. An Investigation of antioxidant capacity of fruits in Singapore markets. *Food Chem.* **2002**, *76*, 69–75. [CrossRef]
19. Koske, R.E.; Gemma, J.N. A modified procedure for staining roots to detect VA Mycorrhizas. *Mycol. Res.* **1989**, *92*, 486–488. [CrossRef]
20. Trouvelot, A.; Kough, J.L.; Gianinazzi-Pearson, V. Evaluation of VA infection levels. Research for estimation methods having a functional significance. In *Physiological and Genetical Aspect of Mycorrhiza*; Gianinazzi-Pearson, V., Gianinazzi, S., Eds.; INRA Edition: Paris, France, 1986; pp. 217–221.
21. Jalonen, R.; Timonen, S.; Sierra, J.; Nygren, P. Arbuscular mycorrhizal symbioses in a cut-and-carry forage production system of legume tree *Gliricidia sepium* and fodder grass *Dichanthium aristatum*. *Agrofor. Syst.* **2013**, *87*, 319–330. [CrossRef]
22. Gai, J.; Gao, W.; Liu, L.; Chen, Q.; Feng, G.; Zhang, J.; Christie, P.; Li, X. Infectivity and community composition of arbuscular mycorrhizal fungi from different soil depths in intensively managed agricultural ecosystems. *J. Soils Sediments* **2015**, *15*, 1200–1211. [CrossRef]
23. Surendirakumar, K.; Pandey, R.R.; Muthukumar, T. Arbuscular Mycorrhizal Fungi in roots and rhizosphere of black rice in terrace fields of North-East India. *Proc. Natl. Acad. Sci. India Sect. B–Biol. Sci.* **2021**, *91*, 277–287. [CrossRef]
24. Wangiyana, W.; Farida, N.; Aryana, I.G.P.M. Yield performance of several promising lines of black rice as affected by application of mycorrhiza biofertilizer and additive intercropping with soybean under aerobic irrigation system on raised beds. *IOP Conf. Ser. Earth Environ. Sci.* **2021**, *913*, 012005. [CrossRef]
25. Anugrah, A.M.; Wangiyana, W.; Aryana, I.G.P.M. Mycorrhizal colonization, growth and yield of several promising lines of Black Rice between sterilized and non-sterilized soil. *Int. J. Environ. Agric. Biotechnol.* **2019**, *4*, 462–468. [CrossRef]
26. Tisarum, R.; Theerawitaya, C.; Samphumphuang, T.; Phisalaphong, M.; Singh, H.P.; Cha-Um, S. Promoting water deficit tolerance and anthocyanin fortification in pigmented rice cultivar (*Oryza sativa* L. subsp. indica) using arbuscular mycorrhizal fungi inoculation. *Physiol. Mol. Biol. Plants* **2019**, *25*, 821–835. [CrossRef] [PubMed]
27. Dhillion, S.S.; Ampornpan, L. The influence of inorganic nutrient fertilization on the growth, nutrient composition and vesicular-arbuscular mycorrhizal colonization of pretransplant rice (*Oryza sativa* L.) Plants. *Biol. Fertil. Soils* **1992**, *13*, 85–91. [CrossRef]
28. Maiti, D.; Toppo, N.N.; Variar, M. Integration of crop rotation and Arbuscular Mycorrhiza (AM) inoculum application for enhancing AM activity to improve phosphorus nutrition and yield of upland rice (*Oryza sativa* L.). *Mycorrhiza* **2011**, *21*, 659–667. [CrossRef] [PubMed]
29. Hajiboland, R.; Aliasgharzad, N.; Barzeghar, R. Influence of Arbuscular Mycorrhizal Fungi on uptake of Zn and P by two contrasting rice genotypes. *Plant Soil Environ.* **2009**, *55*, 93–100. [CrossRef]
30. Zhang, S.; Lehmann, A.; Zheng, W.; You, Z.; Rillig, M.C. Arbuscular Mycorrhizal Fungi increase grain yields: A Meta-Analysis. *New Phytol.* **2019**, *222*, 543–555. [CrossRef] [PubMed]
31. Wang, X.X.; Hoffland, E.; Feng, G.; Kuyper, T.W. Phosphate uptake from phytate due to hyphae-mediated phytase activity by arbuscular mycorrhizal maize. *Front. Plant Sci.* **2017**, *8*, 684. [CrossRef]
32. Song, J.; Chen, L.; Chen, F.; Ye, J. Edaphic and host plant factors are linked to the composition of arbuscular mycorrhizal fungal communities in the root zone of endangered *Ulmus chenmoui* Cheng in China. *Ecol. Evol.* **2019**, *9*, 8900–8910. [CrossRef]
33. Avio, L.; Turrini, A.; Giovannetti, M.; Sbrana, C. Designing the ideotype mycorrhizal symbionts for the production of healthy food. *Front. Plant Sci.* **2018**, *9*, 1089. [CrossRef] [PubMed]
34. Noceto, P.A.; Bettenfeld, P.; Boussageon, R.; Hériché, M.; Sportes, A.; van Tuinen, D.; Courty, P.E.; Wipf, D. Arbuscular mycorrhizal fungi, a key symbiosis in the development of quality traits in crop production, alone or combined with plant growth-promoting bacteria. *Mycorrhiza* **2021**, *31*, 655–669. [CrossRef] [PubMed]
35. Zhao, Y.Y.; Cartabia, A.; Lalaymia, I.; Declerck, S. Arbuscular mycorrhizal fungi and production of secondary metabolites in medicinal plants. *Mycorrhiza* **2022**, *32*, 221–256. [CrossRef] [PubMed]
36. Fiorilli, V.; Vallino, M.; Biselli, C.; Faccio, A.; Bagnaresi, P.; Bonfante, P. Host and non-host roots in rice: Cellular and molecular approaches reveal differential responses to arbuscular mycorrhizal fungi. *Front. Plant Sci.* **2015**, *6*, 636. [CrossRef]
37. Soltaniband, V.; Brégard, A.; Gaudreau, L.; Dorais, M. Biostimulants promote plant development, crop productivity, and fruit quality of protected strawberries. *Agronomy* **2022**, *12*, 1684. [CrossRef]
38. Torres, N.; Antolín, M.C.; Garmendia, I.; Goicoechea, N. Nutritional properties of tempranillo grapevine leaves are affected by clonal diversity, mycorrhizal symbiosis and air temperature regime. *Plant Physiol. Biochem.* **2018**, *130*, 542–554. [CrossRef]
39. Lee, J.; Scagel, C.F. Chicoric acid found in Basil (*Ocimum basilicum* L.) Leaves. *Food Chem.* **2009**, *115*, 650–656. [CrossRef]
40. Baslam, M.; Goicoechea, N. Water deficit improved the capacity of arbuscular mycorrhizal fungi (AMF) for inducing the accumulation of antioxidant compounds in lettuce leaves. *Mycorrhiza* **2012**, *22*, 347–359. [CrossRef]
41. Seemakram, W.; Paluka, J.; Suebrasri, T.; Lapjit, C.; Kanokmedhakul, S.; Kuyper, T.W.; Ekprasert, J.; Boonlue, S. Enhancement of growth and cannabinoids content of hemp (*Cannabis sativa*) using arbuscular mycorrhizal fungi. *Front. Plant Sci.* **2022**, *13*, 845794. [CrossRef]

42. Kumar, S.; Arora, N.; Upadhyay, H. *Arbuscular Mycorrhizal Fungi: Source of Secondary Metabolite Production in Medicinal Plants*; Elsevier: Amsterdam, The Netherlands, 2021. [CrossRef]
43. Dos Santos, E.L.; Falcão, E.L.; Barbosa da Silva, F.S. Mycorrhizal technology as a bioinsumption to produce phenolic compounds of importance to the herbal medicine industry. *Res. Soc. Dev.* **2021**, *10*, e54810212856. [CrossRef]
44. Zhang, R.-Q.; Zhu, H.-H.; Zhao, H.-Q.; Yao, Q. Arbuscular Mycorrhizal fungal inoculation increases phenolic synthesis in clover roots via hydrogen peroxide, salicylic acid and nitric oxide signaling pathways. *J. Plant Physiol.* **2013**, *170*, 74–79. [CrossRef] [PubMed]
45. Netto, A.F.R.; Freitas, M.S.M.; Martins; de Carvalho, A.J.C.; Filho, J.A.V. Efeito de fungos micorrízicos arbusculares na bioprodução de fenóis totais e no crescimento de *Passiflora alata* Curtis. *Rev. Bras. Plantas Med.* **2014**, *16*, 1–9. [CrossRef]
46. Lohse, S.; Schliemann, W.; Ammer, C.; Kopka, J.; Strack, D.; Fester, T. Organization and metabolism of plastids and mitochondria in arbuscular mycorrhizal roots of *Medicago truncatula*. *Plant Physiol.* **2005**, *139*, 329–340. [CrossRef] [PubMed]
47. Mandal, S.; Upadhyay, S.; Wajid, S.; Ram, M.; Jain, D.C.; Singh, V.P.; Abdin, M.Z.; Kapoor, R. Arbuscular mycorrhiza increase artemisinin accumulation in *Artemisia annua* by higher expression of key biosynthesis genes via enhanced jasmonic acid levels. *Mycorrhiza* **2015**, *25*, 345–357. [CrossRef] [PubMed]
48. Mandal, S.; Upadhyay, S.; Singh, V.P.; Kapoor, R. Enhanced production of steviol glycosides in mycorrhizal plants: A concerted effect of arbuscular mycorrhizal symbiosis on transcription of biosynthetic genes. *Plant Physiol. Biochem.* **2015**, *89*, 100–106. [CrossRef]
49. Ran, Z.; Ding, W.; Cao, S.; Fang, L.; Zhou, J.; Zhang, Y. Arbuscular mycorrhizal fungi: Effects on secondary metabolite accumulation of traditional Chinese medicines. *Plant Biol.* **2022**, *24*, 932–938. [CrossRef]

soil systems

MDPI

Article

Rotational Tillage Practices to Deal with Soil Compaction in Carbon Farming

Chris Cavalaris *, Theofanis Gemtos and Christos Karamoutis

Department of Agriculture, Crop Production and Rural Environment, University of Thessaly, Fytokou Str., 38500 Volos, Greece; gemtos@uth.gr (T.G.); chkaram@uth.gr (C.K.)
* Correspondence: chkaval@uth.gr; Tel.: +30-2421-093229

Abstract: Conservation tillage practices, such as reduced tillage and no-tillage, have recently garnered significant attention as core elements of the regenerative agriculture and carbon farming concepts. By minimizing mechanical soil disturbance, these practices preserve soil carbon and facilitate CO_2 fixation in the soil. Despite the widely acknowledged benefits, many farmers still approach no-tillage with skepticism. Their primary concerns are weed management and soil compaction. While weeds can be effectively controlled with the deployment of integrated weed management strategies, urgent soil compaction problems can be rapidly resolved only with mechanical interventions. That is why many no-till farmers resort to occasional heavy tillage, in a scheme characterized as rotational tillage, inadvertently sacrificing their regenerative assets in soil carbon. This is also a pivotal issue within carbon farming: the fate of soil carbon at the end of a compliant scheme focused on carbon fixation. The present study explores data of soil organic matter (SOM), soil penetration resistance (PR), and dry bulk density (DBD) from the initial, six-year period of a long-term tillage experiment in Greece. During that period, modifications to the experimental design allowed diverse combinations of five tillage methods (conventional tillage, 3 reduced tillage methods, and no-tillage). The findings indeed underscore the farmers' concerns about soil compaction. High levels of PR and DBD were observed even at the topsoil layer of the no-tillage (NT). Conventional, moldboard plowing (MP) or reduced, chisel plowing (CP) applied after four years of uninterrupted no-tillage ameliorated most of the soil compaction; however, at the same time, this induced unfavorable consequences for SOM. In contrast, NT applied permanently for six years resulted in a substantial enhancement in SOM that reached 2.24%, for a sampling depth 0–0.30 m compared to 1.54% for permanent MP. When no-tillage was rotated with plowing in the fifth year, almost 50% of the sequestered carbon was lost and the SOM dropped to 1.87%. Nevertheless, the amount of SOM observed at the deeper 0.15–0.30 m layer was greater compared to permanent NT. This suggests that while plowing induced some loss of SOM, it also facilitated the uniform distribution into the soil profile, in contrast with the accumulation in the topsoil at prolonged NT. The permanent CP method and the NT/CP rotation provided comparative outcomes in terms of both soil compaction and soil carbon sequestration with the rotational NT/MP scheme, while all the other tillage combinations were inferior.

Keywords: carbon farming; soil tillage; no-tillage; soil compaction; soil penetration resistance; soil dry bulk density

1. Introduction

Conservation tillage is a management approach that aims to minimize the frequency or intensity of tillage operations in order to leave at least 30% of plant residues on the soil surface for erosion control and moisture conservation [1]. It encompasses various forms such as reduced tillage, vertical tillage, strip tillage, ridge tillage, mulch tillage, no-tillage, and others. Many of these practices have been implemented worldwide for many decades. But, despite the wide acknowledgement of their profits, the rate of adoption, especially in areas with traditional agricultural cultures, such as Europe, Asia, and Africa, is still relatively low [2,3]. Farmers generally appear cautious and skeptical about problems regarding efficient crop establishment, soil compaction, weed management, yield losses, and other [4–7]. Nevertheless, conservation tillage and, mainly, no-tillage have been recently brought to the foreground again as core components of the contemporary regenerative agriculture and carbon farming concepts [8–12].

Reduced tillage is a conservative practice that involves minimum soil disturbance utilizing various tillage implements that avoid soil inversion, operate at shallow depths, and avoid intensive soil crumbling. No-tillage, on the other hand, implicates minimal soil disturbance by performing direct drilling into mulches and stable or natural vegetation [7]. Beyond-surface disturbance is "banned" to avoid distortion of the soil biota and the destruction of its natural habitat [13,14]. That way, the soil is gradually enriched with soil organic matter and the soil structure becomes improved, containing more stable aggregates and a soil porosity with higher continuity and diffusivity [15–19]. Nevertheless, quite often, farmers complain about increased soil compaction that poses a negative impact on crop yields [6,20]. Soil compaction is a combined result of physical and artificially induced processes. Physical processes include natural soil consolidation under the impact of its own mass and suppression from heavy precipitation (snow, rain) [21,22]. Artificially induced compaction is the most severe and comes from the movement of heavy agricultural vehicles in the field (tractors, sprayers, harvesters, etc.), especially under wet soil conditions [23–25]. The above effects, in combination with the absence of annual mechanical soil loosening due to the elimination or abortion of tillage, result in accumulated soil compaction that, after several years, may act adversely to plant root development [26]. Consequently, crop yields are lower [27,28], with dicots being more sensitive compared to monocots [29]. Nonetheless, soils have the ability to employ a self-repairing action through physical shrinking-swelling cycles induced by natural wet/drying and freezing/thawing processes [30–32], but this is not sufficiently effective for the semi-arid Mediterranean climate of southern Europe [33,34]. Moreover, poorly structured soils with low amounts of organic matter are most susceptible to compaction and may also suffer from poor infiltration and aeration [35,36].

To deal with the above problems, farmers employing no-tillage have two options: either to implement an integrated strategy using special cover crops with strong tap roots that perform a "biological" kind of tillage in the soil [37,38], which is, however, a long-term process requiring persistence and patience, or, to perform periodically some kind of shallow or deep tillage operation which has rapid outcomes [39,40]. Certainly, the second solution is favored by the majority of farmers, with plowing, chiseling, or subsoiling being among the most frequently employed methods to tackle soil compaction [41–46]. This strategy is also referred to as rotational tillage [47–50] or occasional tillage [51]. However, introducing mechanical disturbance to a soil that has gradually developed its natural structure over years of uninterrupted no-tillage comes with a host of adverse outcomes. Among these, the most significant is the oxidation of soil organic matter (SOM), resulting in loss of soil carbon due to the aeration of the soil [52], especially when a soil inversive kind of tillage is performed, such as moldboard or disc plowing [53,54]. Destruction of the continuous pores and losses in soil biodiversity are also noteworthy [55,56]. Losses of soil carbon implies that CO_2 is released to the atmosphere, and it is added to the GHGs that are responsible for climate change [57–60]. On the other hand, carbon farming aims at mitigating climate change through capturing atmospheric carbon and fixing it in organic pools into the soil [10,61,62]. As realized, introducing a periodic soil tillage in carbon farming conflicts with the aims and scopes of the strategy. Nonetheless, hybrid systems that intermittently combine intensive and less intensive tillage practices could provide a practical solution for soil compaction, easing the farmers' apprehensions and facilitating a smoother transition toward regenerative soil management systems [47,51,63,64]. The present study capitalizes on the above hypothesis and explores data from the first period of a long-term tillage experiment established in Greece, during which some changes were imposed to the experimental design, leading to diverse combinations of conventional and conservation tillage practices. The study examines the impacts on soil compaction, along with the changes in soil carbon, and aims to identify the optimum tillage schemes that compromise benefits and drawbacks.

2. Materials and Methods

2.1. Experiment Description

The data shown in the present work are obtained from the long-term tillage experiment named THESUSTILL that was established in 1997 at the University of Thessaly Farm in Velestino, Central Greece. The data concern the first six years of the experiment and give a special focus on the fourth, fifth, and sixth years, during which a rotation in tillage treatments was introduced. The initial experimental design established in 1997 was a randomized complete block (RCB) with five methods of tillage:

- **Conventional, moldboard-based tillage (MP).** This method included moldboard plowing at a depth of 0.25–0.30 m while seedbed preparation was accomplished with two or three passes of a disk harrow or a field cultivator at a depth of 0.07–0.09 m. A moldboard plow with four 13-inch plowshares was used. The working speed ranged from 4 to 6 km/h, according to the soil conditions. A tandem disk harrow was used for secondary tillage, with disks of 0.5 m diameter operating at a speed of 7–8 km/h. The field cultivator had spring-type tines, 0.3 m long in 0.07 m spaces, and was operating also at 7–8 km/h. Plowing was usually performed in autumn and secondary tillage was accomplished a few days prior to planting the winter or summer crops. This is the most common method for soil preparation in Greece.
- **Reduced, chisel plowplow-based tillage (CP).** The primary tillage was performed with a chisel plow (also referred as a "heavy cultivator") at a depth of 0.20–0.25 m, and seedbed preparation was accomplished with one or two passes of a disk harrow or a field cultivator. The chisel plow had rigidly mounted tines, 0.80 m long, placed at 0.23 m space intervals. It operated at a speed of 5–6 km/h, according to the soil conditions. This method is common for the establishment of winter crops in Greece.
- **Reduced, power harrow-based tillage (PH).** A single tillage was performed with one pass of a power harrow (also referred to as a "rotary cultivator") at a depth of 0.12–0.15 m, close to planting. In spring-sown crops, one pass of a disk harrow was performed during the previous autumn to control natural vegetation. The implement had tandem vertical tines, 0.30 m long, placed on rotating plates, with a frequency of 180 rpm. The working speed was 4 km/h.
- **Reduced, disc harrow-based tillage (DH).** Shallow tillage was performed with the same disk harrow used for secondary tillage in conventional tillage. The implement operated at a depth of 0.06–0.08 m with a speed of 8 km/h. In autumn-sown crops, two to three passes were made a few days before planting. In spring-sown crops, one pass was made in autumn for residue management and weed destruction and two passes were made in the spring, prior to planting.
- **No-tillage (NT).** Direct sowing was applied using a row crop seeder for the summer crops and a drill seeder for winter wheat. All crop and natural vegetation residues were left on the soil surface. Weeds were destroyed with glyphosate (5–6 kg·ha^{-1}) within one week prior to or after planting the crops.

The original experimental plots were 6 m wide by 60 m long (Figure 1) (the corridors shown on the image were not yet formed). This RCB design was kept constant for a four-year period. In the fifth year (2001), a modification of the original design was introduced, and the tillage strips were performed perpendicular to the original ones (Figure 1). That way, a new strip plot design was formed, with the initial tillage strips comprising the horizontal factor A and the newly introduced perpendicular tillage comprising the vertical factor B. The combinations of factors A and B provided sub-plots, where a new tillage operation was introduced over the previous four-year ones, along with sub-plots still continuing the initial treatments for a fifth year. The dimensions of the newly formed sub-plots were 6 × 10 m. This formation allowed the study of the effects of the lastly introduced tillage operations over the previous ones, as well as the study of the residual effects from the previous operations on the last ones. A total of 5 × 5 = 25 tillage combinations were formed and compared (Table 1).

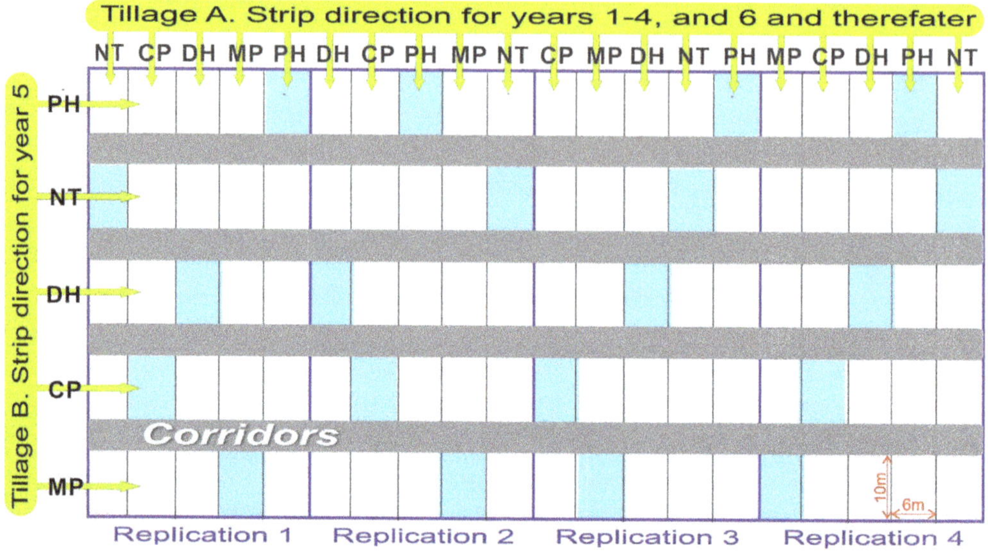

Figure 1. Experimental design. (MP = conventional moldboard plow tillage, CP, chisel plow tillage, PH = power harrow tillage, DH = disk harrow tillage, NT = no-tillage). Blue shaded polygons indicate plots with an all-year constant tillage regime.

Table 1. Tillage combinations formed on the fifth year of the experiment.

Tillage A (Years 1–4 & 6): Tillage B (5th Year)	MP	CP	PH	DH	NT
MP	5 * MP	4CP + 1MP	4PH + 1MP	4DH + 1MP	4NT + 1MP
CP	4MP + 1CP	5CP	4PH + 1CP	4DH + 1CP	4NT + 1CP
PH	4MP + 1PH	4CP + 1PH	5PH	4DH + 1PH	4NT + 1PH
DH	4MP + 1DH	4CP + 1DH	4PH + 1DH	5DH	4NT + 1DH
NT	4MP + 1NT	4CP + 1NT	4PH + 1NT	4DH + 1NT	5NT

* The numbers indicate the years of applied tillage treatment.

In the sixth year (2002), the direction of tillage was reversed to its original design. In that way, 25 tillage combinations were formed again with plots of constant tillage for six years, as well as plots where tillage was interrupted by another method for one year (such as 4MP + 1CP + 1MP, 4NT + 1MP + 1NT, and so on). The experimental design was again a strip plot.

The present work examines results from the first six years of the trial, focusing particularly on the transition period 2000–2002.

A regional common crop rotation scheme was followed, including annual summer and winter crops (Figure 2). Multiple rotations of sugar beet, maize, and cotton were followed during the 1997–1999 summer periods, winter wheat was introduced in 1999–2000 and 2000–2001, and the summer cycle started again with sugar beet in 2002, continued with maize in the following year, and other crops thereafter (the experiment is still running). Both fields were fallow for a long period before starting the trials.

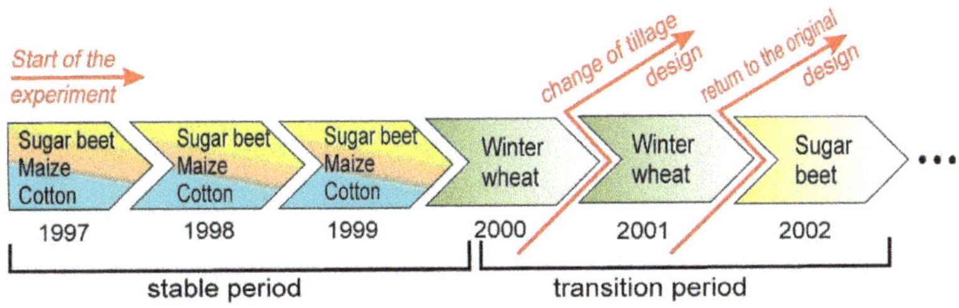

Figure 2. The crop rotation that was followed during the first six years of the experiment.

The experiment had four replicates and was repeated at two fields with different soil types. Field 1 (39°23′43.87″ N, 22°45′25.23″ E) was a silty-clayey Vertisol (sand 9.7%, silt 41%, clay 49.2%) and Field 2 (39°24′1.37″ N, 22°45′34.11″ E) was a clayey Vertisol (sand 20.1, silt 32.7, clay 47.1). Both fields present a calcaric Fluvisol origin.

2.2. Field Measurements

Soil properties were measured once per year at selected periods. Multiple sampling was performed on each plot. During 1997–2000, each experimental field consisted of $5 \times 4 = 20$ main plots. In 2001 and 2002, after changing the tillage direction, each field consisted of $25 \times 4 = 100$ sub-plots. The properties examined in the present study are soil penetration resistance, soil dry bulk density, and soil organic matter.

2.2.1. Soil Penetration Resistance

Soil penetration resistance (PR) was measured with a handheld soil penetrometer (Findley-Irvine Ltd., Penicuik, Midlothian, Scotland UK) with a cone base diameter of 12.83 mm and a tip angle of 30°, following the standards of ASABE [65]. The instrument recorded the soil penetration resistance at intervals of 0.01 m, down to a depth of 0.50 m. Five insertions were made randomly at each plot and average values were estimated at intervals of 0.05 m down to 0.40 m. The 0.40–0.50 m interval was aborted because it was not possible to always achieve the full (0.50 m) depth, due to the extreme resistance. Measurements were made once per year, two to three months after the last tillage operation, to allow adequate time for the soil to consolidate. The mean soil water content during the measurements were obtained from four soil cores obtained at random places from each replication, at three depth intervals: 0–0.15 m, 0.15–0.30 m, and 0.30–0.45 m (Supplementary Materials Table S5). Since the soil resistance on the penetration of a metal tip depends highly on soil moisture, and since water regime always differs in time, the penetration resistance values were normalized to allow a year-by-year comparison. The normalization was performed over the annual *MP* values that were used as a common basis according to the formula:

$$N_{PRi} = 100 + 100\frac{(PR_{MP} - PR_i)}{(PR_{MP} + PR_i)} \quad (1)$$

where:

N_{PRi} = the normalized value of penetration resistance for a specific treatment i (i refereeing to *MP*, CP, PH, DH, and NT) for a particular depth interval.
PR_{MP} = the penetration resistance value for the *MP* treatment at the corresponding depth.
PR_i = the penetration resistance value for a treatment (i) at the corresponding depth.

From the above expression (1) it is conceived that the PR_i value for the *MP* treatment is always estimated to be 100, being the common base to construct a comparable time series.

2.2.2. Soil Dry Bulk Density

Dry bulk density (DBD) of the soil was assessed from undisturbed core samples taken at a depth of 0.30 m. A sampling device consisting of an outer metal cylinder containing retractable and exchangeable inner plastic tubes for holding the soil cores was used (Figure 3). The device was designed and constructed by the Laboratory of Agricultural Engineering, University of Thessaly. The metal cylinder with the inner plastic tube was pressed into the soil with the help of a farm tractor hydraulic system. After each sampling, the plastic tube containing the undisturbed soil sample was removed and replaced with a new, empty one. The plastic tubes have an inner base diameter of 0.048 m and a height of 0.30 m. These tubes were carried to the lab where the soil was passed gently to another plastic tube of the same diameter that had predefined perpendicular slots at distances 0.025 m, 0.10 m, 0.20 m, and 0.275 m from the top (Figure 4). These slots were used as a guide to separate with a knife the 0–0.30 m original soil core at three fractions, A: 0.025–0.100 m, B: 0.100–0.200 m, and C: 0.200–0.275 m. For the border, 0.025 m-long core pieces were omitted from the top and the bottom of the tube to secure any soil losses during extraction and transportation. As a result, from each original sample, three sub-samples of constant volume were obtained. The volumes of "A" and "C" samples were 1.3565×10^{-4} m^3, and "B" was 1.8086×10^{-4} m^3. The "A" sample represented the 0–0.10 m soil depth, B the 0.10–0.20 m, and C the 0.20–0.30 m depth. The sub-samples were crumbled manually, oven dried for 48 h at 104 °C, and weighted. Dry bulk density of the soil was expressed as the ratio of the mass to the soil volume (Mg·m^{-3}). From each plot, three 0–0.30 m samples were taken randomly. From these, average dry bulk densities were estimated for the three depth intervals. The sampling was performed close to the penetration resistance measurements.

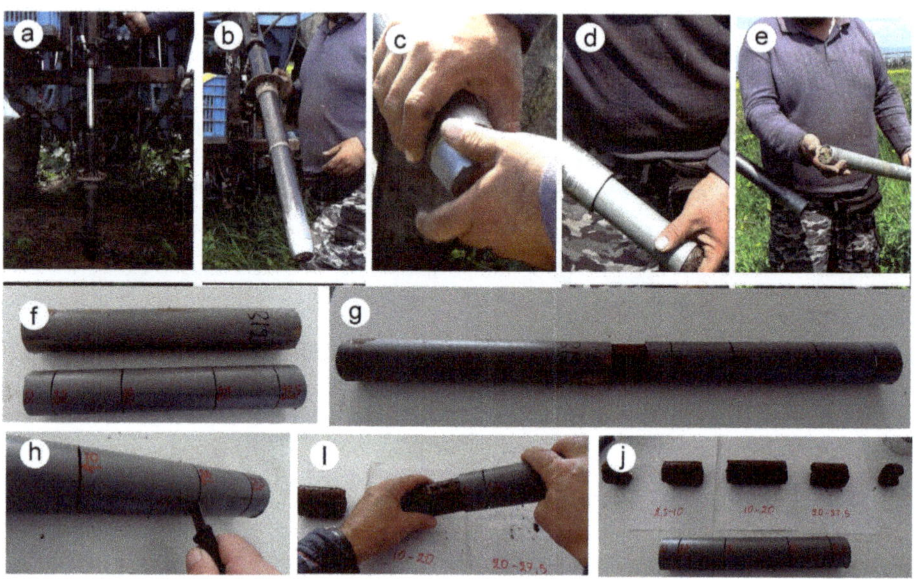

Figure 3. Soil sampling: (**a**) Tractor-powered metal cylinder inserting the soil; (**b**) the cylinder retracted from the soil; (**c**) removing the cylinder tip; (**d**) extracting the inner plastic tube; (**e**) the extracted plastic tube containing the soil core sample and core segmentation in the lab; (**f**) the plastic tube holding the soil core (up) and the empty tube with the predefined slots (down); (**g**) transferring the soil from the holding tube (left) to the slotted tube (right); (**h**) cutting the soil core with a knife at the predefined dimensions; (**i**) extracting the core segments from the tube; (**j**) the three main core segments (0.025–0.10 m, 0.10–0.20 m, and 0.20–0.275 m) and the two omitted pieces (left and right).

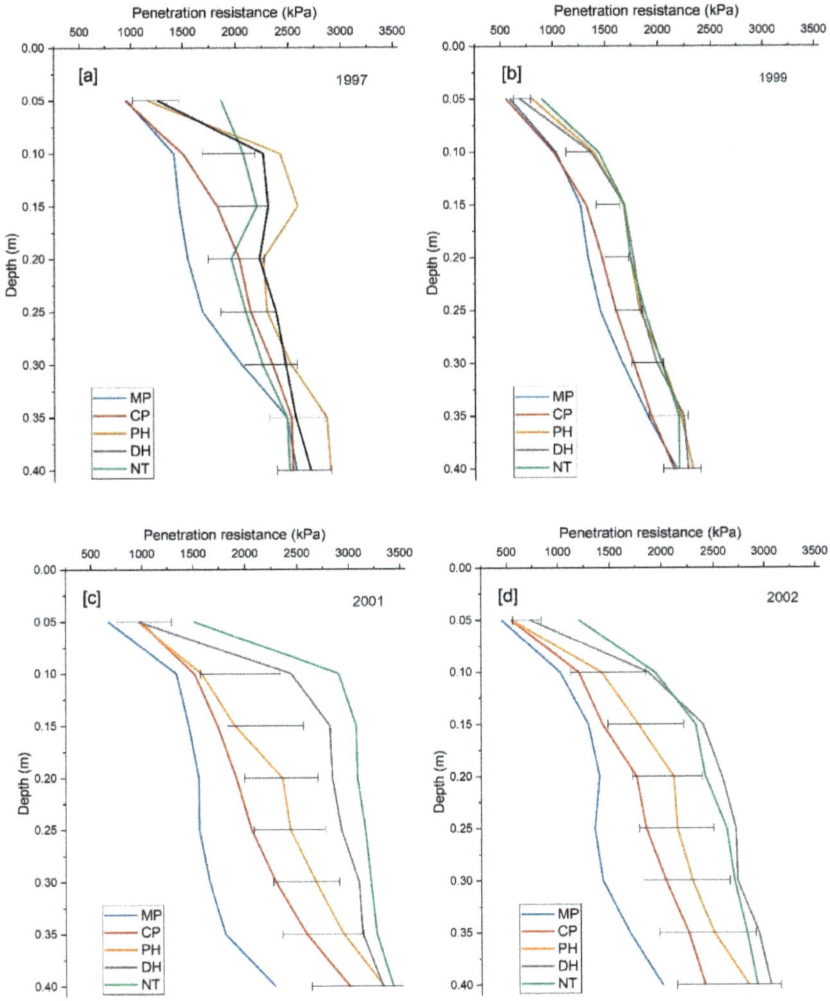

Figure 4. Soil penetration resistance at the permanent tillage treatments (error bars indicate 95% confidence intervals range, numerical results from the statistical analysis are presented in Supplementary Materials Tables S1–S4). (**a**–**d**) indicate results from different years.

2.2.3. Soil Organic Matter

Soil Organic Matter (SOM) was measured at two depths, 0–0.15 m and 0.15–0.30 m. Preliminary measurements were taken in 1997, before the introduction of any tillage treatments, to address the start conditions. The measurements were repeated in 2001 after the introduction of the tillage combinations and in 2002 when the treatments were revered to the initial design. The sampling was performed with a screw-type open auger in 1997, while, for 2001 and 2002, the hydraulic device used for the dry bulk density was utilized. Three samples were taken from every plot. The same depths were thoroughly mixed and homogenized to give one composite sample for each plot and depth. The soil organic matter was estimated with the Walkley and Black method [66].

2.3. Statistical Analysis

Two statistical models were used for the analysis of the data. For the first four years, a one-way analysis of variance (ANOVA) was performed for the tillage treatment A. For years five and six, where a second vertical factor (Tillage B) was introduced, a custom univariate linear model was built with the following sources of variation: Location, Replication, Tillage A, error 1, Tillage B, error 2, Tillage AxB, error 3, total. The statistical analysis was performed in SPSS (IBM SPSS Statistics v29.0. Armonk, NY, USA: IBM Corp). Post hoc tests for mean comparisons were conducted using Tukey correction with a p level 0.05. The graphs were built with OriginPro, 2022 (OriginLab Corporation, Northampton, MA, USA).

3. Results

The statistical analysis showed no significant interaction between the site (Field 1 and Field 2) and the tillage treatments. Therefore, the average values from the two fields are henceforth presented.

3.1. Soil Penetration Resistance

The soil penetration resistance measurements were carried out at least two months after the last soil tillage on already established crops. Figure 4 presents measurements obtained from the plots with a constant tillage regime (blue shaded plots in Figure 2) for years one, three, five, and six. There are significant differences among the tillage treatments, as revealed by the error bars and the corresponding p-values in Tables S1–S4 (Supplementary Materials). The graphs reveal also the efficient decompaction effect of the moldboard plow in the MP treatment. Deep soil plowing provided essential soil loosening, surprisingly expanding even beyond the 0.30-m tillage depth. The differences are statistically significant from DH and NT, and occasionally from PH, but not from CP (see also Supplementary Materials Tables S1–S4). The deep layer differences cannot be attributed to a mechanical action (conversely, moldboard plows are known for causing a hardpan underneath), but rather to differences in soil water content due to altered infiltration and water-holding capacity. This evidence of course requires further investigation. The CP treatment also provided considerable soil loosening that was significantly lower than DH and NT below 0.15 m. The use of a chisel plow in that case, with shanks working also on a deep depth (about 0.25 m), introduced cracks and soil crumbling without inverting the soil, as did the moldboard plow. The soil loosening effect shows a gradual decrease with depth. The PH and DH methods utilized machinery that operated at shallower depths (around 0.10 m). As a result, soil penetration resistance increased rapidly beyond the disturbed top layer. In 1997, these methods introduced a soil compaction right below the tillage depth due to the rather wet conditions during the soil preparation (Figure 4a). The NT method presented an increased soil penetration resistance, even from the topsoil 0–0.05 m layer. The differences were not significant from CP, PH, and DH, but were statistically significant compared to MP (Table S1, Supplementary Materials). This is somehow anticipated, due to the absence of soil loosening at any depth. Aside from the year 2001, the soil compaction in NT depicted through penetration resistance presents a rather uniform vertical distribution. The smaller differences among the treatments in 1999 are attributed to the higher soil water content during the measurements (Supplementary Materials Table S5).

Soil penetration resistance is a useful and simple measurement to address soil compaction, but it is highly affected and too sensitive to differences in soil moisture. The results may be useful for a 'snapshot' comparison during a singular time, but cannot be used to compare data from different dates or periods with alternative soil water regimes. To overcome this difficulty, the data were normalized using the MP results as a basis, to construct a timeseries of soil penetration resistance and access through that the evolution of soil compaction in time (Formula (1)). For each date, MP were given an arbitrary stable value of $N_{PR} = 100$ and the rest of the treatment levels were expressed as a change % (Figure 5). Average PR values are estimated for three depths: 0–0.10 m, 0.10–0.20 m, and 0.20–0.30 m. The NT and DH treatments presented a gradual increase in soil penetration resistance in time. After the fifth year of constant application, the PR reached a maximum plateau at the 0.10–0.20 m and 0.20–0.30 m layers, but was still increasing at the top 0–0.10 m layer. In particular, the NT method had a 25–30% higher PR in the top layer, even from the first year that was gradually increasing in time, and exceeded 40% in the sixth year. At the 0.10–0.20 m layer, the PR was 15–20% higher in the beginning and became 30–35% higher in the fifth year, after which it remained constant. At the deeper, 0.20–0.30 m, layer, PR was only about 8% higher at the beginning and became 30–35% higher after the fifth year. The pattern was similar for DH, besides some lower PR levels at the top layer. The PH method presented a 10–20% increased PR from the first year at the 0–0.10 m depth, and a 15–25% higher PR at the 0.10–0.20 m depth, which remained relatively constantly higher in time. At the lower 0.20–0.30 m layer, however, the PR showed again a gradual increase from 10% to 20–25% during a five-year period. Finally, the CP treatment presented an almost similar PR with the control MP during the first year at the 0–0.10 m layer, but it had an around 10% higher PR at the 0.10–0.20 m and 0.20–0.30 m depths. Compared with MP, the differences in PR remained constant at the 0.10–0.20 m layer but presented an increase in time of about 10% for the upper 0–0.10 m and the deeper 0.20–0.30 m layers.

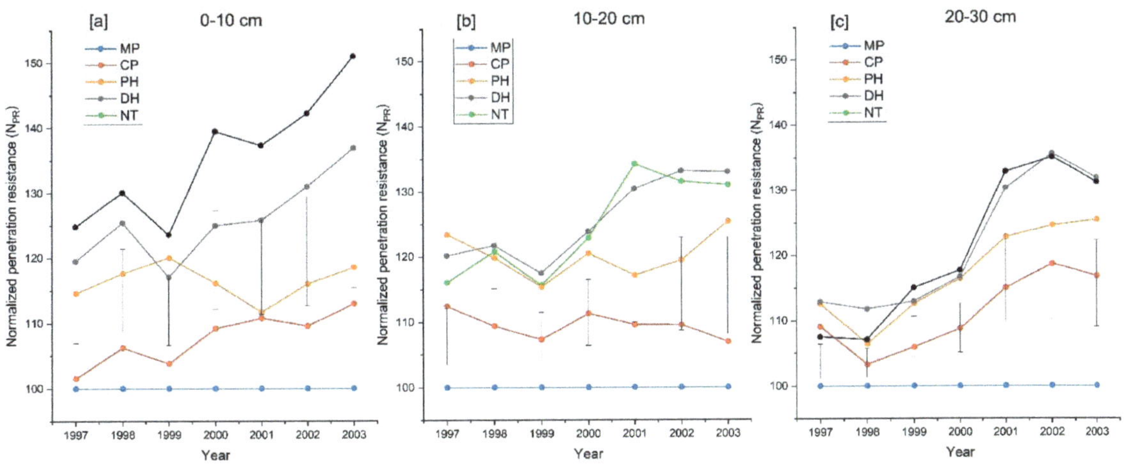

Figure 5. Timeseries of soil compaction accessed through normalized penetration resistance measurements over a period of six years for the permanent tillage schemes (error bars indicate 95% confidence intervals range). (**a–c**) indicate normalized penetration resistance for different depths.

The previous findings confirm the farmers' concerns for increased soil compaction after some years of reducing or ceasing soil tillage. As revealed in Figures 6a and 7a, the increase in soil compaction during the first year of non-plowing is relatively low because the residual effects of the plowshares are still present. Therefore, the farmers do not encounter serious compaction problems at the beginning. Notably, some residual effects of the moldboard plow and the chisel plow were detectable in NT, even after four years (Figure 6c). However, when the intensive plow tillage was abandoned, soil compaction gradually built in the soil (Figure 5). On the other hand, introducing some kind of deep soil tillage after a constant no-tillage period was capable of ameliorating most of the built-in compaction (Figures 6b and 7c). It is also remarkable in Figure 7b that using a disk harrow after no-tillage was the worst combination for exacerbating the compaction problems.

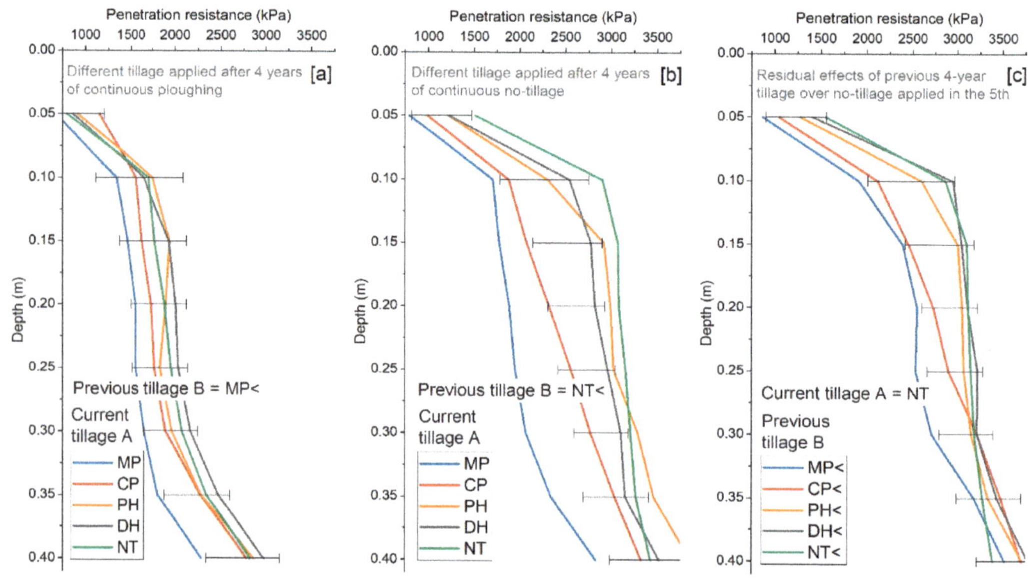

Figure 6. Tillage rotation effects in soil penetration resistance in 2001 (error bars indicate 95% confidence intervals range, numerical results from the statistical analysis are presented in Supplementary Materials Table S3).

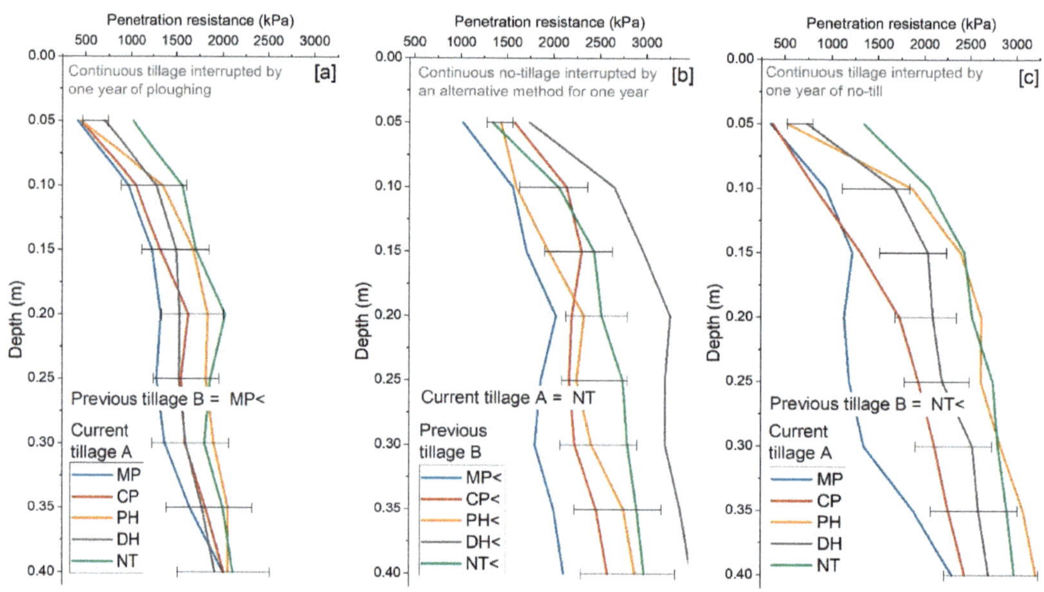

Figure 7. Tillage rotation effects in soil penetration resistance in 2002 (error bars indicate 95% confidence intervals range, numerical results from the statistical analysis are presented in Supplementary Materials Table S4).

3.2. Dry Bulk Density

Dry bulk density was monitored three times; first in 2000, prior to the tillage change, in 2001 after the tillage alternation, and in 2002 when the tillage treatments were reversed to the original pattern. The measurements were made in parallel to the penetration resistance monitoring. The average results for the two fields for 2000 are presented in Figure 8. After four years of continuous reduced and no-tillage application, compaction is built into the soil. At the 0–0.10 m depth, no-tillage presents significantly higher dry bulk density (1.56 Mg·m^{-3}), and so do the DH and PH methods (1.42 and 1.30 Mg·m^{-3}, respectively, Supplementary Materials Table S6). Even though the 0–0.10 m depth is within the active range of the power harrow, the implement caused much less soil loosening compared to the moldboard plow (1.13 Mg·m^{-3}) or the chisel plow (1.18 Mg·m^{-3}). The disc harrow was operating at a shallower depth of 0.08–0.10 m; therefore, it presented a higher bulk density. At a greater depth of 0.10–0.20 m, the DH had almost the same dry bulk density with no-tillage (1.57 and 1.58 Mg·m^{-3}, respectively). At that depth, the bulk density was generally high for all the treatment levels, although MP and CP still presented lower values (1.29 Mg·m^{-3}) compared to PH, and especially DH and NT. The same applies to the third sampling depth of 0.20–0.30 m and, although the differences among the treatment levels are smaller, they are still statistically significant (p-value = 0.000). The results of the dry bulk density agree with the penetration resistance measurements shown in Figure 4.

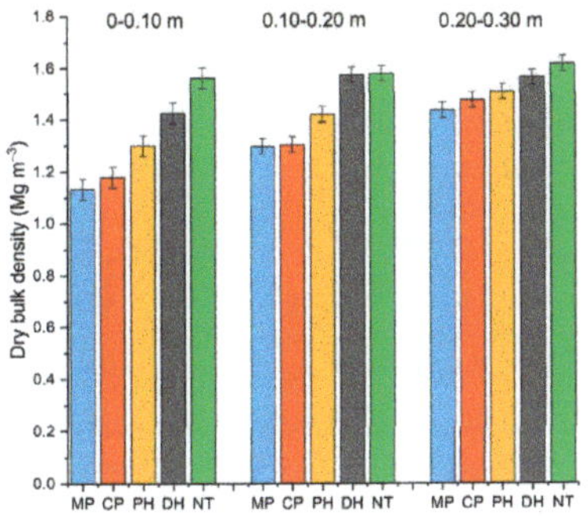

Figure 8. Dry bulk density of the soil for the five permanent methods of tillage in 2000 error bars indicate 95% confidence intervals range, numerical results from the statistical analysis are presented in Supplementary Materials Table S6).

In the following year, rotations to the tillage treatments were introduced (Figure 9). The plots receiving the same tillage treatment for a fifth year continue to present the same behavior. MP had a DBD of 1.11 Mg·m^{-3} at the 0–0.10 m sampling depth, which increased to 1.33 Mg·m^{-3} at the 0.10–0.20 m depth and reached 1.42 Mg·m^{-3} at a depth of 0.20–0.30 m. NT, on the other hand, presented an increased DBD of 1.60 Mg·m^{-3}, even from the topsoil layer that reached a peak of 1.64 Mg·m^{-3} at the 0.20–0.30 m depth. Intermediate values were found for the rest of the tillage treatments. It is remarkable that the previous year's tillage presented a considerable residual effect on soil compaction. Whenever a deep kind of tillage was performed during the previous year (including moldboard plowing or the use of a chisel plow), the soil compaction was lower. The effect was more profound at the less intensive methods of DH and NT. For instance, the DBD at the DH method for the 0–0.10 m layer was 1.41 Mg·m^{-3} when the method was applied constantly, but dropped to 1.12 Mg·m^{-3} if moldboard plowing had been performed during the previous year. At a greater depth of 0.10–0.20 m, DBD was 1.64 Mg·m^{-3} and 1.33 Mg·m^{-3}, respectively. The same applied when no-tillage was performed after moldboard plowing. These differences were statistically significant (p-values < 0.001, Supplementary Materials Table S7).

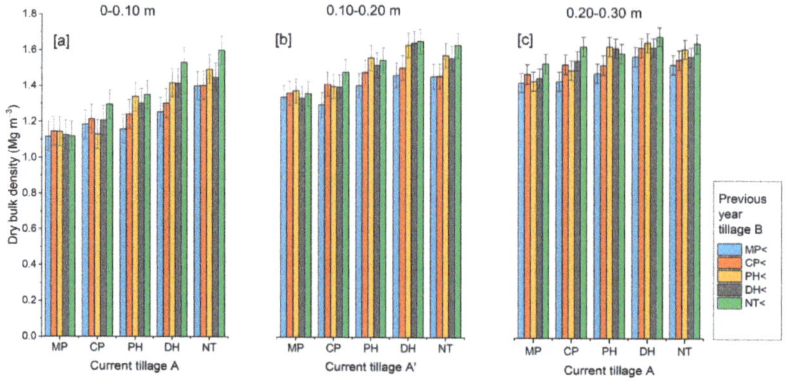

Figure 9. Dry bulk density in permanent and rotational schemes of tillage for 2001 error bars indicate 95% confidence intervals range, numerical results from the statistical analysis are presented in Supplementary Materials Table S7.

In 2002, tillage direction was reverted to the initial direction, so combinations of the five currently applied methods and the methods employed in the previous year were formed again (Figure 10). The results follow a similar pattern with Figure 9, indicating that the most significant impact comes primarily from the recent tillage and secondly from the tillage performed in the previous year (see F-statistic in Table S7, Supplementary Materials). The permanent application of no-tillage increased DBD at around 1.57 Mg·m^{-3} at the topsoil layer and at 1.64–1.66 Mg·m^{-3} at greater depths. The increased soil compaction was relieved whenever a deep tillage operation was intervened. Plowing or chisel plowing, for instance, during the previous year kept the DBD of the soil at a 0–0.10 m depth below 1.40 Mg·m^{-3}. The benefits, however, are limited to greater depths. On the other hand, soil crumbling was easier whenever a more intensive tillage was applied in the previous year. For instance, the power harrow resulted in a DBD of 1.15 Mg·m^{-3} when it was used after plowing, but raised to 1.38 Mg·m^{-3} when it was used after no-tillage. The results again comply with the corresponding PR measurements.

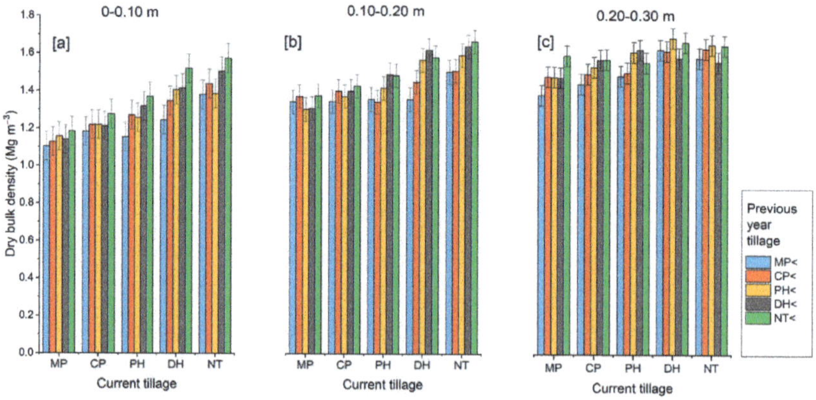

Figure 10. Dry bulk density in permanent and rotational schemes of tillage for 2002 error bars indicate 95% confidence intervals range, numerical results from the statistical analysis are presented in Supplementary Materials Table S7).

3.3. Soil Organic Matter

The preliminary measurements from 1997 revealed an average SOM content of 2.27% at the topsoil (0–0.15 m) layer and 2.07% at the 0.15–0.30 m depth (Figure 11a,b). No statistical differences among the plots were detected during that time because the measurements were conducted prior to the introduction of the tillage treatments. The trend was similar for the two fields, although Field 1 had a slightly higher SOM at both depths (Supplementary Materials Table S8).

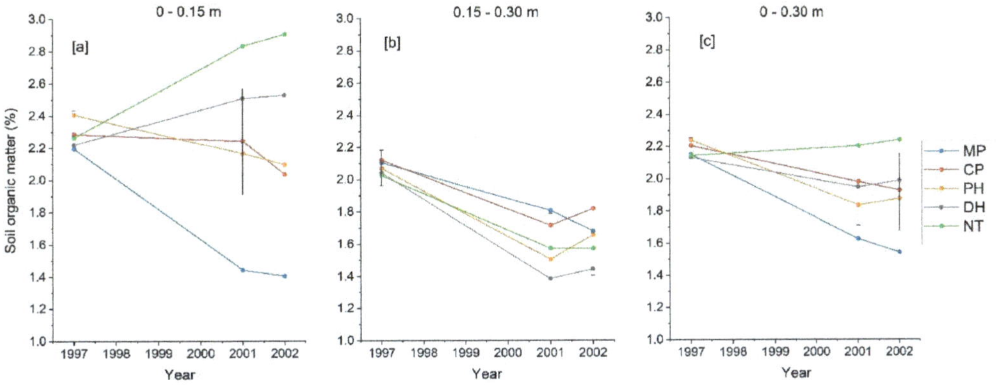

Figure 11. Change in soil organic matter during a five-year period with permanent tillage treatments (error bars indicate 95% confidence intervals range). (**a**–**c**) indicate results from different depths.

Five years later, in 2001, no-tillage significantly improved the SOM from 2.26% to 2.83% at the topsoil layer (Figure 11a and Supplementary Materials Table S9). DH also raised SOM from 2.22% to 2.50%. On the contrary, continuous moldboard plowing caused a significant reduction in SOM from 2.20% to 1.44%, while PH had an also negative, but weaker effect, reducing SOM from 2.31% to 2.16%. Finally, CP didn't show any important impact at the topsoil SOM until 2001, but caused a 0.2% reduction in the next year. At a greater depth of 0.15–0.30 m (Figure 11b), all the methods that imposed a reduction in SOM from 1997 to 2001 remained stable or slightly increased in the following year, except for the conventional MP, where SOM continued to drop. The findings, however, were rather opposite compared to the topsoil. Permanent MP presented greater SOM than permanent NT. While NT retains all the plant residue on the soil surface, the moldboard plow causes soil inversion and mixing of the plant residue into deeper layers, where they decompose and enrich the organic deposits. The overall impacts at a 0–0.30 m arable layer were positive for NT, slightly negative for the reduced tillage methods of CP, DH, and PH, and strongly negative for the intensive MP method (Figure 11c). NT resulted in an increase in SOM from 1997 to 2002 of 0.09%, while MP caused a decrease of 0.61%.

Figures 12 and 13 examine the combined effects of previous and current tillage treatments in 2001 and 2002, respectively. The results prove that soil disturbance enhances SOM decay. Most of the beneficial effects in SOM were inverted whenever a more intensive tillage system was introduced over a less intensive one. Plowing in 2001 after four years of no-tillage, for instance, significantly decreased SOM from 2.83% to 1.54% at the topsoil layer. Less intensive methods such as CP and PH were less destructive, decreasing SOM after NT by 0.43–0.45%, while DH had almost a null effect (Figure 12a). At a deeper depth, the impacts were less profound (Figure 12b) and less statically important (lower F-statistic, Table S9, Supplementary Materials). It is remarkable, nonetheless, that the greatest SOM content is found when plowing was performed after a four-year no-tillage period (MP/NT < combination). In that case, the enrichment in SOM was greater because all the enhanced organic matter built into the surface of NT was buried into deeper layers. Regarding the overall impacts at the 0–0.30 m layer (Figure 12c and

Table S9, Supplementary Materials) it is found that continuous long-term no-tillage was the most beneficial in terms of gains in SOM. The difference was statistically significant with all the other treatments that involved another tillage method in the previous year. On the other hand, moldboard plowing had the most negative residual effect. The worst combination was the application of a power harrow after a four-year moldboard plowing period (PH/MP<).

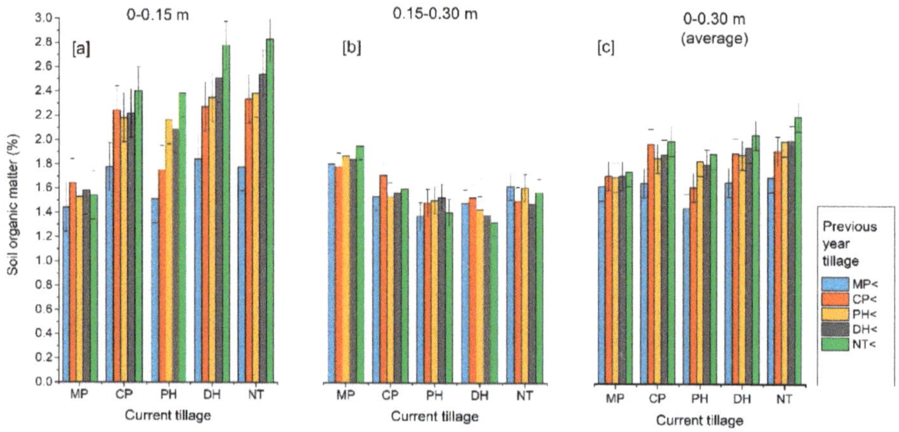

Figure 12. Soil organic matter in permanent and rotational schemes of tillage in 2001 (error bars indicate 95% confidence intervals range, numerical results from the statistical analysis are presented in Supplementary Materials Table S9).

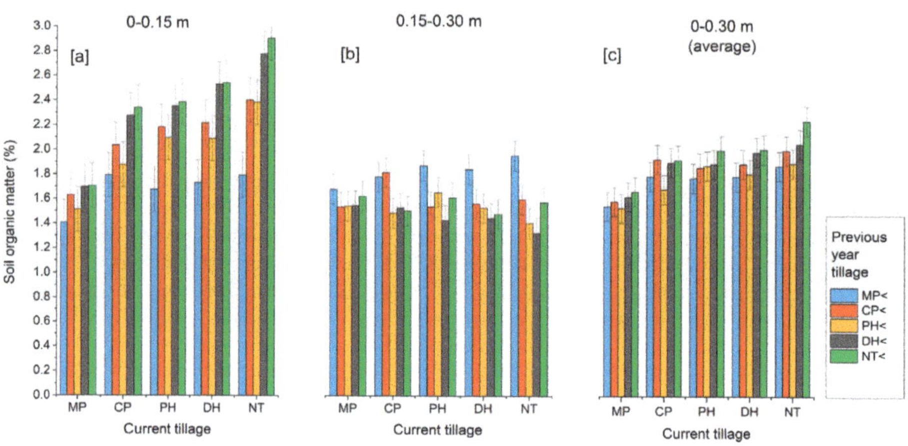

Figure 13. Soil organic matter in permanent and rotational schemes of tillage in 2002 error bars indicate 95% confidence intervals range, numerical results from the statistical analysis are presented in Supplementary Materials Table S9).

During the next year (2002), tillage was revered to the initial direction. That way, tillage rotations of the type 4MP-1CP-1MP, 4NT-1MP-1NT etc. were formed. As shown in Figure 13a, there is a much greater amount of SOM at the topsoil layer for the reduced and the no-tillage methods, either applied recently or during the previous year. On the other hand, one year of moldboard plowing was able to compensate for almost all the benefits obtained in the top layer from the continuous application of a conservation tillage system. For example, while six years of continuous NT raised the SOM at 2.90%, the intervention of moldboard plowing in the fifth year caused a significant reduction to 1.79%. Nevertheless, not all the fixed SOM was lost, but a part of it was reallocated at a greater depth, as shown in Figure 13b. At the 0.15–0.30 m depth, the NT/MP tillage rotation presented a SOM of 1.95% compared to 1.57% found on the permanent NT. The benefits apply not only for NT, but also for all the other reduced tillage methods that facilitate the accumulation of the SOM on the topsoil layer. Overall, at the 0–0.30 m layer, the greatest advantage was for the permanent NT that significantly increased the SOM at 2.24% compared to 1.87% for the NT/MP tillage rotation, or 1.54% for permanent MP (Figure 13c and Table S9, Supplementary Materials). It is also remarkable that the permanent CP facilitated both the increase in SOM and its uniform distribution on the soil profile. The permanent CP method presented an increased SOM content, both at the top and the deeper sampling depths (2.03% and 1.81%, respectively) (Figure 13a,b) that led to an increased total amount at the whole profile (1.92%) (Figure 13c and Table S9, Supplementary Materials).

4. Discussion

The results of the present study confirm the farmers' concerns about soil compaction issues in no-till systems, particularly when applied to soils low in organic carbon content, as seen in southern Europe. Both penetration resistance (PR) and dry bulk density (DBD) measurements reveal that fact. The values increase as the tillage system becomes less intensive and is consistently practiced over a greater period. Similar findings were reported by other authors [67–70]. In our study, the PR exceeded the 2500 kPa after five years of constant NT and DH application, and dry bulk density reached 1.60 $Mg \cdot m^{-3}$, even at the top 0–0.10 m soil layer. PR values greater than 2500 kPa are considered to hinder the elongation of the roots [70]. Increased PR for the conservation tillage treatments, however, may also imply enhanced soil aggregate stability [71]. On the other hand, avoiding soil disturbance led to an important improvement in the SOM content. While the introduction of soil inversion through moldboard plowing in the two fields that were previously at fallow caused a considerable loss in SOM of 0.79% in the topsoil (0–0.15 m) layer, the NT led to a further improvement of 0.64%. As a result, the two methods differed 1.43% in the topsoil after five years. Organic matter in the topsoil has a critical role in soil protection. It promotes the binding of inorganic soil particles, facilitating the formation of more stable soil aggregates that exhibit enhanced resistance to both soil erosion and compaction [70–72]. Furthermore, this topsoil organic matter serves as a vital resource for the soil biota that is predominantly concentrated in the uppermost soil layer [73]. Soil microorganisms transform the soil into a thriving ecosystem that fosters nutrient cycling and improves fertility [14,74–76]. Nonetheless, it is important to note that soil inversion induced by the plow did not result in a total loss of the fixed carbon. Instead, a portion of it underwent a redistribution into deeper soil layers, as evidenced by the measurements in the years 2001 and 2002. Plowing after four years of no-tillage, for instance, resulted in the highest amount of SOM for the lower, 0.15–0.30 m, layer. Subsequently, when returning to a no-tillage system in the consequent year, the benefits were even higher (SOM = 1.96% compared to 1.57% for permanent NT). Moreover, the plowing intervention had an additional advantage by relieving soil compaction in NT. The effects of this soil loosening persisted even after transitioning back to no-tillage in the subsequent year, and they could be detected for another two years.

Examining the overall impacts in the 0–0.30 m soil layer, the most beneficial method for fixing carbon into the soil was the permanent application of no-tillage. The SOM content after six years of continuous NT was 2.24%, the highest of any other tillage combination. In contrast, continuous MP resulted in a SOM of 1.54%. However, it could be raised to 1.87% if plowing was applied only once every four years to deal with soil compaction in a rotational NT/MP scheme. Besides, as shown in weed measurements during the 2001 period, deep plowing also provided benefits in weed suppression [77].

The CP method also showed interesting results. The chisel plow demonstrated a soil loosening capacity compatible with MP. It also facilitated both the increase of SOM and its uniform distribution on the soil profile. The occasional use of a chisel plow in a NT system and the permanent CP system provided similar results. Studies in Spain and China demonstrated also that the occasional use of tools with shanks, such as chisel plows or subsoilers, could serve as an alternative to the moldboard plow for relieving soil compaction in no-tillage without destroying its biochemical and biological benefits [46,47,63,69]. Occasional subsoiling in no-till systems have been also proved to improve water storage and facilitate higher crop yields [45]. Nevertheless, subsoiling is a high-cost operation requiring a lot of energy, while its effects decrease over time, making new subsoiling necessary after few years [44]. On the other hand, chisel plows require almost half the energy compared to conventional moldboard plowing and subsoilers, as depicted from tractor draft force measurements on the same trials [78]. The present study indicates that a chisel plow rotational system may provide substantial benefits in relieving soil compaction, at a considerably lower cost and lower soil carbon losses compared to moldboard plowing. The worst combination for fixing soil carbon was the application of a power harrow after moldboard plowing, while the worst combination for exacerbating soil compaction was the use of a disc harrow after a period of no-tillage.

Yield data from the same experiments concerning the 1997–2002 period are published in several articles [79–83] and demonstrate that crop yields in the permanent no-tillage and reduced tillage treatments were 4–37% lower compared to conventional plowing. The most negative effects were observed in permanent NT. Nevertheless, when a plowing operation was intervened in the previous year, the yield losses were considerably diminished, or even equaled MP [83].

The above findings suggest that adoption of rotational soil tillage schemes, which combine no-tillage and occasional plowing or no-tillage and chiseling, are effective alternatives that strike a balance between soil compaction and carbon sequestration. Constant chisel plowing is also another option. Despite providing approximately half the carbon credits compared to permanent no-tillage, these systems offer practical solutions for mitigating farmers' concerns about soil compaction, which is a prominent obstacle to the realization of more sustainable farming practices in the context of southern Europe. Notably, a meta-analysis study published in 2020 [51] tracked down only two relevant studies in Europe (Spain) out of 68 worldwide.

Eventually, the appropriate implementation of regenerative agriculture practices encapsulating the principles of minimum soil disturbance, permanent soil coverage, and crop diversification is expected to optimize the system's performance, but farmers certainly need well-informed advisory services and particular guidance to overcome all the buriers by increasing their resource-use efficiency and soil conserving ability [10,84].

5. Conclusions

Permanent no-tillage was the most effective method for sequestering soil carbon. Over a span of six continuous years of no-tillage application, the SOM content was increased by 0.70% compared to continuous plowing. Nevertheless, this accomplishment was obscured by emerging soil compaction identified from increased bulk density and soil penetration resistance measurements. When constant no-tillage was interrupted by a moldboard plowing or a chisel plowing operation in a rotational tillage scheme, the gains in SOM were 0.33–0.46%. Notably, plowing introduced a more uniform distribution of SOM into the soil profile compared to the accumulated carbon at the topsoil layer in no-tillage. Residual soil loosening effects were still noticeable for up to four years after plowing, with the most prominent effects observed within the initial two years. A permanent, chisel plow-based system had similar impacts with the rotational, no-tillage/moldboard plowing approach. The results indicate that carbon credits in carbon farming may be halved if periodic deep tillage operations should be introduced to counteract the consequences of extreme soil compaction.

Supplementary Materials: The following supporting information can be downloaded at: https://www.mdpi.com/article/10.3390/soilsystems7040090/s1, Table S1: Results from the statistical analysis for the 1997 penetration resistance measurements, Table S2: Results from the statistical analysis for the 1999 penetration resistance measurements, Table S3: Results from the statistical analysis for the 2001 penetration resistance measurements, Table S4: Results from the statistical analysis for the 2002 penetration resistance measurements, Table S5: Volumetric, average soil water content during the penetration resistance measurements, Table S6: Results from the statistical analysis for the 2000 soil dry bulk density measurements, Table S7: Results from the statistical analysis for 2001 and 2002 soil dry bulk density measurements, Table S8: Results from the statistical analysis for the preliminary (1997) soil organic matter measurements, Table S9: Results from the statistical analysis for 2001 and 2002 soil organic matter measurements.

Author Contributions: Conceptualization, C.C. and T.G.; methodology, C.C. and T.G.; formal analysis, C.C.; investigation, C.C. and C.K.; data curation, C.C. and C.K.; writing—original draft preparation, C.C.; writing—review and editing, C.C. and T.G.; visualization, C.C.; supervision, C.C. All authors have read and agreed to the published version of the manuscript.

Funding: This research received no external funding.

Institutional Review Board Statement: Not applicable.

Informed Consent Statement: Not applicable.

Data Availability Statement: Data are provided from the authors on demand.

Conflicts of Interest: The authors declare no conflict of interest.

References

1. Claassen, R.; Bowman, M.; McFadden, J.; Smith, D.; Wallander, S. *Tillage Intensity and Conservation Cropping in the United States*; United States Department Agriculture USDA: Washington DC, USA, 2018; pp. 1–18.
2. Kassam, A.; Friedrich, T.; Derpsch, R. Global spread of Conservation Agriculture. *Int. J. Environ. Stud.* **2019**, *76*, 29–51. [CrossRef]
3. Kertész, Á.; Madarász, B. Conservation Agriculture in Europe. *Int. Soil Water Conserv. Res.* **2014**, *2*, 91–96. [CrossRef]
4. Farooq, M.; Flower, K.C.; Jabran, K.; Wahid, A.; Siddique, K.H.M. Crop yield and weed management in rainfed conservation agriculture. *Soil Tillage Res.* **2011**, *117*, 172–183. [CrossRef]
5. Derrouch, D.; Dessaint, F.; Fried, G.; Chauvel, B. Weed community diversity in conservation agriculture: Post-adoption changes. *Agric. Ecosyst. Environ.* **2021**, *312*, 107351. [CrossRef]
6. Van den Putte, A.; Govers, G.; Diels, J.; Gillijns, K.; Demuzere, M. Assessing the effect of soil tillage on crop growth: A meta-regression analysis on European crop yields under conservation agriculture. *Eur. J. Agron.* **2010**, *33*, 231–241. [CrossRef]
7. Soane, B.D.; Ball, B.C.; Arvidsson, J.; Basch, G.; Moreno, F.; Roger-Estrade, J. No-till in northern, western and south-western Europe: A review of problems and opportunities for crop production and the environment. *Soil Tillage Res.* **2012**, *118*, 66–87. [CrossRef]
8. Landers, J.N.; de Freitas, P.L.; de Oliveira, M.C.; Neto, S.P.d.S.; Ralisch, R.; Kueneman, E.A. Next steps for conservation agriculture. *Agronomy* **2021**, *11*, 2496. [CrossRef]
9. Khangura, R.; Ferris, D.; Wagg, C.; Bowyer, J. Regenerative Agriculture—A Literature Review on the Practices and Mechanisms Used to Improve Soil Health. *Sustainability* **2023**, *15*, 2338. [CrossRef]
10. Sharma, M.; Kaushal, R.; Kaushik, P.; Ramakrishna, S. Carbon farming: Prospects and challenges. *Sustainability* **2021**, *13*, 11122. [CrossRef]
11. Lal, R. Carbon farming by re-carbonization of agroecosystems. *Pedosphere* **2021**, *33*, 676–679. [CrossRef]
12. Page, C.; Witt, B. A Leap of Faith: Regenerative Agriculture as a Contested Worldview Rather Than as a Practice Change Issue. *Sustainability* **2022**, *14*, 14803. [CrossRef]
13. Li, Y.; Chang, S.X.; Tian, L.; Zhang, Q. Conservation agriculture practices increase soil microbial biomass carbon and nitrogen in agricultural soils: A global meta-analysis. *Soil Biol. Biochem.* **2018**, *121*, 50–58. [CrossRef]
14. Roger-Estrade, J.; Anger, C.; Bertrand, M.; Richard, G. Tillage and soil ecology: Partners for sustainable agriculture. *Soil Tillage Res.* **2010**, *111*, 33–40. [CrossRef]
15. Lou, Y.; Xu, M.; Chen, X.; He, X.; Zhao, K. Stratification of soil organic C, N and C:N ratio as affected by conservation tillage in two maize fields of China. *Catena* **2012**, *95*, 124–130. [CrossRef]
16. López, M.V.; Blanco-Moure, N.; Limón, M.Á.; Gracia, R. No tillage in rainfed Aragon (NE Spain): Effect on organic carbon in the soil surface horizon. *Soil Tillage Res.* **2012**, *118*, 61–65. [CrossRef]
17. Xue, J.F.; Qi, Z.W.; Chen, J.L.; Cui, W.H.; Lin, W.; Gao, Z.Q. Dynamic of Soil Porosity and Water Content under Tillage during Summer Fallow in the Dryland Wheat Fields of the Loess Plateau in China. *Land* **2023**, *12*, 230. [CrossRef]

18. Eden, M.; Bachmann, J.; Cavalaris, C.; Kostopoulou, S.; Kozaiti, M.; Böttcher, J. Soil structure of a clay loam as affected by long-term tillage and residue management. *Soil Tillage Res.* **2020**, *204*, 104734. [CrossRef]
19. Weidhuner, A.; Hanauer, A.; Krausz, R.; Crittenden, S.J.; Gage, K.; Sadeghpour, A. Tillage impacts on soil aggregation and aggregate-associated carbon and nitrogen after 49 years. *Soil Tillage Res.* **2021**, *208*, 104878. [CrossRef]
20. Munkholm, L.J.; Heck, R.J.; Deen, B. Long-term rotation and tillage effects on soil structure and crop yield. *Soil Tillage Res.* **2013**, *127*, 85–91. [CrossRef]
21. Bauer, T.; Strauss, P.; Grims, M.; Kamptner, E.; Mansberger, R.; Spiegel, H. Long-term agricultural management effects on surface roughness and consolidation of soils. *Soil Tillage Res.* **2015**, *151*, 28–38. [CrossRef]
22. Kuhwald, M.; Dörnhöfer, K.; Oppelt, N.; Duttmann, R. Spatially explicit soil compaction risk assessment of arable soils at regional scale: The SaSCiA-Model. *Sustainability* **2018**, *10*, 1618. [CrossRef]
23. Jamali, H.; Nachimuthu, G.; Palmer, B.; Hodgson, D.; Hundt, A.; Nunn, C.; Braunack, M. Soil compaction in a new light: Know the cost of doing nothing—A cotton case study. *Soil Tillage Res.* **2021**, *213*, 105158. [CrossRef]
24. Lamandé, M.; Greve, M.H.; Schjønning, P. Risk assessment of soil compaction in Europe–Rubber tracks or wheels on machinery. *Catena* **2018**, *167*, 353–362. [CrossRef]
25. Shahgholi, G.; Moinfar, A.; Khoramifar, A.; Maciej, S.; Szymanek, M. Investigating the Effect of Tractor's Tire Parameters on Soil Compaction Using Statistical and Adaptive Neuro-Fuzzy Inference System (ANFIS) Methods. *Agriculture* **2023**, *13*, 259. [CrossRef]
26. Kadžienž, G.; Munkholm, L.J.; Mutegi, J.K. Root growth conditions in the topsoil as affected by tillage intensity. *Geoderma* **2011**, *166*, 66–73. [CrossRef]
27. Radford, B.J.; Yule, D.F.; McGarry, D.; Playford, C. Crop responses to applied soil compaction and to compaction repair treatments. *Soil Tillage Res.* **2001**, *61*, 157–166. [CrossRef]
28. Anazodo, U.G.N.; Raghavan, G.S.V.; McKyes, E.; Norris, E.R. Physico-mechanical properties and yield of silage corn as affected by soil compaction and tillage methods. *Soil Tillage Res.* **1983**, *3*, 331–345. [CrossRef]
29. Arvidsson, J.; Håkansson, I. Response of different crops to soil compaction-Short-term effects in Swedish field experiments. *Soil Tillage Res.* **2014**, *138*, 56–63. [CrossRef]
30. Chinn, C.; Pillai, U.P.P. Self-repair of compacted Vertisols from Central Queensland, Australia. *Geoderma* **2008**, *144*, 491–501. [CrossRef]
31. Wang, X.; Wang, C.; Wang, X.; Huo, Z. Response of soil compaction to the seasonal freezing-thawing process and the key controlling factors. *Catena* **2020**, *184*, 104247. [CrossRef]
32. Sarmah, A.K.; Pillai-McGarry, U.; McGarry, D. Repair of the structure of a compacted Vertisol via wet/dry cycles. *Soil Tillage Res.* **1996**, *38*, 17–33. [CrossRef]
33. Radford, B.J.; Yule, D.F.; McGarry, D.; Playford, C. Amelioration of soil compaction can take 5 years on a Vertisol under no till in the semi-arid subtropics. *Soil Tillage Res.* **2007**, *97*, 249–255. [CrossRef]
34. Lopez-Bellido, R.J.; Muñoz-Romero, V.; Lopez-Bellido, F.J.; Guzman, C.; Lopez-Bellido, L. Crack formation in a mediterranean rainfed Vertisol: Effects of tillage and crop rotation. *Geoderma* **2016**, *281*, 127–132. [CrossRef]
35. Barzegar, A.R.; Rengasamy, P.; Oades, J.M. Effects of clay type and rate of wetting on the mellowing of compacted soils. *Geoderma* **1995**, *68*, 39–49. [CrossRef]
36. Parvin, N.; Sandin, M.; Larsbo, M. Seedbed consolidation and surface sealing for soils of different texture and soil organic carbon contents. *Soil Tillage Res.* **2021**, *206*, 104849. [CrossRef]
37. Torres, J.L.R.; Leal Júnior, A.L.B.; Barreto, A.C.; Carvalho, F.J.; de Assis, R.L.; Loss, A.; Lemes, E.M.; da Silva Vieira, D.M. Mechanical and Biological Soil Decompaction for No-Tillage Maize Production. *Agronomy* **2022**, *12*, 2310. [CrossRef]
38. Jabro, J.D.; Allen, B.L.; Rand, T.; Dangi, S.R.; Campbell, J.W. Effect of Previous Crop Roots on Soil Compaction in 2 Yr Rotations under a No-Tillage System. *Land* **2021**, *10*, 202. [CrossRef]
39. Ferreira, C.J.B.; Tormena, C.A.; Severiano, E.d.C.; Nunes, M.R.; de Menezes, C.C.E.; Antille, D.L.; Preto, V.R.d.O. Effectiveness of narrow tyne and double-discs openers to overcome shallow compaction and improve soybean yield in long-term no-tillage soil. *Soil Tillage Res.* **2023**, *227*, 105622. [CrossRef]
40. Peralta, G.; Alvarez, C.R.; Taboada, M.Á. Soil compaction alleviation by deep non-inversion tillage and crop yield responses in no tilled soils of the Pampas region of Argentina. A meta-analysis. *Soil Tillage Res.* **2021**, *211*, 105022. [CrossRef]
41. Qiang, X.; Sun, J.; Ning, H. Impact of Subsoiling on Cultivated Horizon Construction and Grain Yield of Winter Wheat in the North China Plain. *Agriculture* **2022**, *12*, 236. [CrossRef]
42. Tim Chamen, W.C.; Moxey, A.P.; Towers, W.; Balana, B.; Hallett, P.D. Mitigating arable soil compaction: A review and analysis of available cost and benefit data. *Soil Tillage Res.* **2015**, *146*, 10–25. [CrossRef]
43. Botta, G.F.; Jorajuria, D.; Balbuena, R.; Ressia, M.; Ferrero, C.; Rosatto, H.; Tourn, M. Deep tillage and traffic effects on subsoil compaction and sunflower (*Helianthus annus* L.) yields. *Soil Tillage Res.* **2006**, *91*, 164–172. [CrossRef]
44. Martínez, I.G.; Prat, C.; Ovalle, C.; del Pozo, A.; Stolpe, N.; Zagal, E. Subsoiling improves conservation tillage in cereal production of severely degraded Alfisols under Mediterranean climate. *Geoderma* **2012**, *189–190*, 10–17. [CrossRef]
45. Qin, H.-L.; Gao, W.-S.; Ma, Y.-C.; Ma, L.; Yin, C.-M.; Chen, Z.; Chen, C. Effects of Subsoiling on Soil Moisture Under No-Tillage for Two Years. *Agric. Sci. China* **2008**, *7*, 88–95. [CrossRef]

46. Wang, Q.; Lu, C.; Li, H.; He, J.; Sarker, K.K.; Rasaily, R.G.; Liang, Z.; Qiao, X.; Li, H.; Mchugh, A.D.J. The effects of no-tillage with subsoiling on soil properties and maize yield: 12-Year experiment on alkaline soils of Northeast China. *Soil Tillage Res.* **2014**, *137*, 43–49. [CrossRef]
47. Wang, R.; Ma, L.; Lv, W.; Li, J. Rotational Tillage: A Sustainable Management Technique for Wheat Production in the Semiarid Loess Plateau. *Agriculture* **2022**, *12*, 1582. [CrossRef]
48. Hou, X.; Li, R.; Jia, Z.; Han, Q.; Wang, W.; Yang, B. Effects of rotational tillage practices on soil properties, winter wheat yields and water-use efficiency in semi-arid areas of north-west China. *Field Crops Res.* **2012**, *129*, 7–13. [CrossRef]
49. Hou, X.Q.; Li, R.; Jia, Z.K.; Han, Q.F. Effect of Rotational Tillage on Soil Aggregates, Organic Carbon and Nitrogen in the Loess Plateau Area of China. *Pedosphere* **2013**, *23*, 542–548. [CrossRef]
50. Zhang, Y.; Tan, C.; Wang, R.; Li, J.; Wang, X. Conservation tillage rotation enhanced soil structure and soil nutrients in long-term dryland agriculture. *Eur. J. Agron.* **2021**, *131*, 126379. [CrossRef]
51. Peixoto, D.S.; da Silva, L.D.C.M.; De Melo, L.B.B.; Azevedo, R.P.; Araújo, B.C.L.; De Carvalho, T.S.; Moreira, S.G.; Curi, N.; Silva, B.M. Occasional tillage in no-tillage systems: A global meta-analysis. *Sci. Total Environ.* **2020**, *745*, 140887. [CrossRef]
52. Murindangabo, Y.T.; Kopecký, M.; Konvalina, P.; Ghorbani, M.; Perná, K.; Nguyen, T.G.; Bernas, J.; Baloch, S.B.; Hoang, T.N.; Eze, F.O.; et al. Quantitative Approaches in Assessing Soil Organic Matter Dynamics for Sustainable Management. *Agronomy* **2023**, *13*, 1776. [CrossRef]
53. Gajda, A.M.; Czyż, E.A.; Klimkowicz-Pawlas, A. Effects of different tillage intensities on physicochemical and microbial properties of a eutric fluvisol soil. *Agronomy* **2021**, *11*, 1497. [CrossRef]
54. Jakab, G.; Madarász, B.; Masoudi, M.; Karlik, M.; Király, C.; Zacháry, D.; Filep, T.; Dekemati, I.; Centeri, C.; Al-Graiti, T.; et al. Soil organic matter gain by reduced tillage intensity: Storage, pools, and chemical composition. *Soil Tillage Res.* **2023**, *226*, 105584. [CrossRef]
55. Schwen, A.; Bodner, G.; Scholl, P.; Buchan, G.D.; Loiskandl, W. Temporal dynamics of soil hydraulic properties and the water-conducting porosity under different tillage. *Soil Tillage Res.* **2011**, *113*, 89–98. [CrossRef]
56. Zuber, S.M.; Villamil, M.B. Meta-analysis approach to assess effect of tillage on microbial biomass and enzyme activities. *Soil Biol. Biochem.* **2016**, *97*, 176–187. [CrossRef]
57. Franco-Luesma, S.; Cavero, J.; Plaza-Bonilla, D.; Cantero-Martínez, C.; Arrúe, J.L.; Álvaro-Fuentes, J. Tillage and irrigation system effects on soil carbon dioxide (CO_2) and methane (CH^4) emissions in a maize monoculture under Mediterranean conditions. *Soil Tillage Res.* **2020**, *196*, 104488. [CrossRef]
58. Dachraoui, M.; Sombrero, A. Effect of tillage systems and different rates of nitrogen fertilisation on the carbon footprint of irrigated maize in a semiarid area of Castile and Leon, Spain. *Soil Tillage Res.* **2020**, *196*, 104472. [CrossRef]
59. González-Sánchez, E.J.; Ordóñez-Fernández, R.; Carbonell-Bojollo, R.; Veroz-González, O.; Gil-Ribes, J.A. Meta-analysis on atmospheric carbon capture in Spain through the use of conservation agriculture. *Soil Tillage Res.* **2012**, *122*, 52–60. [CrossRef]
60. Schwengbeck, L.; Hölting, L.; Witing, F. Modeling Climate Regulation of Arable Soils in Northern Saxony under the Influence of Climate Change and Management Practices. *Sustainability* **2023**, *15*, 11128. [CrossRef]
61. Van Hoof, S. Climate Change Mitigation in Agriculture: Barriers to the Adoption of Carbon Farming Policies in the EU. *Sustainability* **2023**, *15*, 10452. [CrossRef]
62. Dumbrell, N.P.; Kragt, M.E.; Gibson, F.L. What carbon farming activities are farmers likely to adopt? A best-worst scaling survey. *Land Use Policy* **2016**, *54*, 29–37. [CrossRef]
63. Melero, S.; Panettieri, M.; Madejón, E.; Macpherson, H.G.; Moreno, F.; Murillo, J.M. Implementation of chiselling and mouldboard ploughing in soil after 8 years of no-till management in SW, Spain: Effect on soil quality. *Soil Tillage Res.* **2011**, *112*, 107–113. [CrossRef]
64. López-Garrido, R.; Madejón, E.; Murillo, J.M.; Moreno, F. Soil quality alteration by mouldboard ploughing in a commercial farm devoted to no-tillage under Mediterranean conditions. *Agric. Ecosyst. Environ.* **2011**, *140*, 182–190. [CrossRef]
65. *ASABE Stand. S313.3. FEB1999*; Soil Cone Penetrometer. ASABE, American Society of Agricultural Engineers: Saint Joseph, MI, USA, 1999.
66. Walkley, A.; Black, I.A. An examination of the Degtjareff method for determining soil organic matter, and a proposed modification of the chromic acid titration method. *Soil Sci.* **1934**, *37*, 29–38. [CrossRef]
67. Pöhlitz, J.; Rücknagel, J.; Koblenz, B.; Schlüter, S.; Vogel, H.J.; Christen, O. Computed tomography and soil physical measurements of compaction behaviour under strip tillage, mulch tillage and no tillage. *Soil Tillage Res.* **2018**, *175*, 205–216. [CrossRef]
68. Cavalieri, K.M.V.; da Silva, A.P.; Tormena, C.A.; Leão, T.P.; Dexter, A.R.; Håkansson, I. Long-term effects of no-tillage on dynamic soil physical properties in a Rhodic Ferrasol in Paraná, Brazil. *Soil Tillage Res.* **2009**, *103*, 158–164. [CrossRef]
69. López-Garrido, R.; Madejón, E.; León-Camacho, M.; Girón, I.; Moreno, F.; Murillo, J.M. Reduced tillage as an alternative to no-tillage under Mediterranean conditions: A case study. *Soil Tillage Res.* **2014**, *140*, 40–47. [CrossRef]
70. Bogunovic, I.; Pereira, P.; Kisic, I.; Sajko, K.; Sraka, M. Tillage management impacts on soil compaction, erosion and crop yield in Stagnosols (Croatia). *Catena* **2018**, *160*, 376–384. [CrossRef]
71. Fernández-Ugalde, O.; Virto, I.; Bescansa, P.; Imaz, M.J.; Enrique, A.; Karlen, D.L. No-tillage improvement of soil physical quality in calcareous, degradation-prone, semiarid soils. *Soil Tillage Res.* **2009**, *106*, 29–35. [CrossRef]
72. da Silva, J.F.; Neto, M.M.G.; da Silva, G.F.; Borghi, E.; Calonego, J.C. Soil Organic Matter and Aggregate Stability in Soybean, Maize and Urochloa Production Systems in a Very Clayey Soil of the Brazilian Savanna. *Agronomy* **2022**, *12*, 1652. [CrossRef]

73. Joschko, M.; Gebbers, R.; Barkusky, D.; Rogasik, J.; Höhn, W.; Hierold, W.; Fox, C.A.; Timmer, J. Location-dependency of earthworm response to reduced tillage on sandy soil. *Soil Tillage Res.* **2009**, *102*, 55–66. [CrossRef]
74. Wright, A.L.; Hons, F.M.; Lemon, R.G.; McFarland, M.L.; Nichols, R.L. Stratification of nutrients in soil for different tillage regimes and cotton rotations. *Soil Tillage Res.* **2007**, *96*, 19–27. [CrossRef]
75. Tang, X.; Qiu, J.; Xu, Y.; Li, J.; Chen, J.; Li, B.; Lu, Y. Responses of soil aggregate stability to organic C and total N as controlled by land-use type in a region of south China affected by sheet erosion. *Catena* **2022**, *218*, 106543. [CrossRef]
76. Baumgartl, T.; Horn, R. Effect of aggregate stability on soil compaction. *Soil Tillage Res.* **1991**, *19*, 203–213. [CrossRef]
77. Cavalaris, C.; Karamoutis, C.; Papamichail, D.; Gemtos, T.A. Soil tillage effect on weed infestation in a sugar beet crop. In Proceedings of the 4th National Conference of Agricultural Mechanization, Athens, Greece, 6–8 October 2005; pp. 151–158. (In Greek)
78. Cavalaris, C.; Gemtos, T.A. Evaluation of tillage efficiency and energy requirements for five methods of soil preparation in the sugar beet crop. In Proceedings of the 2004 CIGR International Conference, Chinese Academy of Agricultural Mechanization Sciences, Benhing, Beijing China, 11–14 October 2004; pp. 97–101.
79. Cavalaris, C.; Gemtos, T.A.; Georgiou, C. Use of low inputs methods for soil tillage in corn. In Proceedings of the 1st National Conference of Agricultural Mechanization, Athens, Greece, 11–12 December 1998; pp. 377–387. (In Greek)
80. Gemtos, T.A.; Cavalaris, C. Soil tillage effect in the sugar beet crop. In Proceedings of the 1st World Congress on Conservation Agriculture, Madrid, Spain, 1–5 October 2001; pp. 539–543.
81. Cavalaris, C.C.; Gemtos, T.A. Evaluation of four conservation tillage methods in the sugar beet crop. *Agric. Eng. Int. (CIGR) E-J.* **2002**, *IV*.
82. Gemtos, T.A.; Cavalaris, C.; Demis, V.; Pateras, D.; Tsidari, C. Effect of changing tillage practices after four years of continuous reduced tillage. In Proceedings of the 2002 ASABE Annual International Meeting/CIGR World Congress, Chicago, IL, USA, 28–31 July 2002; p. No 021135.
83. Cavalaris, C.; Karamoutis, C.; Aggelopoulou, A.; Gentos, T.A. Effect of changing tillage on soil, plants and yield. In Proceedings of the 5th National Conference of Agricultural Mechanization, Larisa, Greece, 18–20 October 2007; pp. 127–134. (In Greek)
84. Mattila, T.J.; Hagelberg, E.; Söderlund, S.; Joona, J. How farmers approach soil carbon sequestration? Lessons learned from 105 carbon-farming plans. *Soil Tillage Res.* **2022**, *215*, 105204. [CrossRef]

soil systems

MDPI

Article

Selenium and Heavy Metals in Soil–Plant System in a Hydrogeochemical Province with High Selenium Content in Groundwater: A Case Study of the Lower Dniester Valley

Sergey Sheshnitsan [1,*], Nadezhda Golubkina [2], Tatiana Sheshnitsan [1], Otilia Cristina Murariu [3], Alessio Vincenzo Tallarita [4] and Gianluca Caruso [4]

1 Faculty of Forestry, Voronezh State University of Forestry and Technologies, Timiryazeva Str. 8, Voronezh 394087, Russia; tatianasheshnitsan@gmail.com
2 Analytical Laboratory Department, Federal Scientific Vegetable Centre, Selektsionnaya Str. 14, Vnissok, Moscow 143072, Russia; segolubkina45@gmail.com
3 Department of Food Technology, "Ion Ionescu de la Brad" Iasi University of Life Sciences, M. Sadoveanu Alley, 700440 Iasi, Romania; otiliamurariu@uaiasi.ro
4 Department of Agricultural Sciences, University of Naples Federico II, 80055 Naples, Italy; lexvincentall@gmail.com (A.V.T.); gcaruso@unina.it (G.C.)
* Correspondence: sheshnitsan@gmail.com

Abstract: The bioaccumulation of selenium (Se) and heavy metals (HMs) in plants is important because it can affect plant health and human nutrition. Recognizing the factors affecting Se accumulation in plants may have important implications for agricultural practices and human health in selenium-rich regions. The study primarily focused on the interactions between Se and HMs in the soil–plant system of the Lower Dniester Valley. Total concentrations of HMs (Cu, Mn, Zn) were determined by atomic absorption spectrometry, while Se concentrations were determined by a sensitive single-test-tube fluorometric method in solutions and extracts. Water-soluble Se (0.09 ± 0.03 mg·kg^{-1}) in soils was 32.1% of the total Se (0.33 ± 0.13 mg·kg^{-1}) and increased with the total rising Se content (r = 0.845). The results indicated that plants had a greater Zn accumulation capacity than that of the other HMs, suggesting its importance as a trace element for plant requirements. Se also had a high bioaccumulation rate. Se and Zn accumulation varied in different soil types, reflecting differences in bioavailability. In contrast, Mn and Cu showed low bioaccumulation, which varied with soil conditions and anthropogenic Cu pollution. Despite the Cu contamination of the soils in the investigated region, it can be inferred that the hydrogeochemical province with high Se content in groundwater has favorable conditions for Se mobilization in soils. The absence of antagonistic interactions with HMs in the soil–plant system contributes to the enhanced Se accumulation in plants in the Lower Dniester Valley. These results emphasize the complexity of the interactions between Se and HMs in the soil–plant system and their potential impact on agricultural practices.

Keywords: bioaccumulation; soil properties; mobility; pollution; Fluvisols

1. Introduction

Bioaccumulation is the gradual accumulation of certain substances, such as heavy metals and selenium, in plant tissues. These elements are derived from a variety of sources, including natural deposits, industrial activities, and agricultural practices [1–5]. The accumulation of selenium (Se) and heavy metals (HMs) in plants is of great importance due to the potential threat posed to plant health and human consumption. However, Se and HMs such as copper (Cu), manganese (Mn), and zinc (Zn) are also known as essential micronutrients that can be beneficial or detrimental to plants and animals, with a narrow range between beneficial and toxic depending on their concentration and species [6–8]. Although the essentiality of Se has not yet been established for higher plants, it is responsible for a number of beneficial effects in several plant species [9,10] and actively contributes to plant antioxidant defense against all forms of biotic and abiotic stresses [11]. However, it is indispensable for animals and humans [12]. The essentiality of Se for humans provides important protection against viral and cardiovascular diseases and several forms of cancer, and improves immunity, fertility, and mental health among other benefits [13].

The bioaccumulation of HMs and Se in plants and their ecological effects is increasingly being investigated. It was shown that these compounds can either affect plant growth and development or also disrupt the ecological balance of ecosystems [7,8,14–16]. The study of bioaccumulation in hydrogeochemical provinces with high concentrations of Se is of great importance for several reasons. Firstly, such areas are generally at higher risk of Se contamination in both soils and water sources, with direct implications for plant uptake. Understanding the mechanisms and patterns of bioaccumulation in these regions may help to develop effective strategies to manage and mitigate the risks associated with Se accumulation in plants. Secondly, as these territories are usually located in agricultural areas, Se accumulation in crops may have important consequences for food safety and health.

The Moldavian hydrogeochemical province with increased Se content in groundwater and shallow waters was identified in the 1980s due to the presence of a significant source of this trace element in rocks and its high concentration in groundwater (0.n–n·10 $\mu g \cdot L^{-1}$). Sedimentary formations of the Middle Sarmatian are rich in disseminated selenium-containing sulfides, which are the main source of Se in groundwater [17]. According to Hannigan et al. [18], Neogene (Middle Sarmatian) clays contained abundant Se and they were the source of Se in groundwater, the concentrations of which were shown to locally exceed the maximal permitted value [19] by 1.5–24.0 times and varied from 15.0 to 240.0 $\mu g \cdot L^{-1}$. Therefore, a comprehensive investigation is particularly needed to estimate the regional Se distribution because of the possible ecological problems for agricultural practices and the water supply [17]. A more extensive study, which included the Dniester Valley, revealed Se concentrations in the soils ranging from 0.10 to 0.67 $g \cdot kg^{-1}$. The majority of soils analyzed had an optimal Se content. The highest Se content was observed at a depth of 0.4–0.7 m and decreased nearer to the parent rock. High Se concentrations were also found in the local surface waters with values ranging from 0.2 to 6.1 $\mu g \cdot L^{-1}$, indicating the abundant presence of soluble Se that was available to plants [20]. Se accumulation in some components of the regional food chains, including insects [21], bee products [22], and mushrooms [23], was found to be relatively high in the Dniester Valley. Recent studies of Se content in human hair reported a high supply of the trace element in the environment of Moldova [24].

Heavy metal pollution in soils in Moldova was found to have significant impacts on the soil and water environment [25,26]. Moss biomonitoring was used to assess air pollution levels and sources in the Republic of Moldova. The results indicated consistent pollution levels for a wide range of HMs [27]. According to some comprehensive studies [27–29], the main sources of anthropogenic HM pollution in the region are industrial emissions, transport, and agricultural practices.

The aim of this study was to determine the level of Se in soils and its accumulation in plants as well as to identify the factors associated with Se bioaccumulation in the hydrogeochemical province with high Se in groundwater. At the same time, it is expected that HMs (Cu, Mn, and Zn) in the soil–plant system can act antagonistically [30] and may reduce the Se bioaccumulation rate in plants.

The choice of these HMs from a wide range of elements was due to the following reasons. Firstly, Cu, Mn, and Zn are essential for plant, animal, and human nutrition in certain concentrations [30]. Secondly, these elements have a higher affinity to accumulate in living organisms in the geochemical landscapes of the Lower Dniester Valley than their water migration ability. This implies that HMs are more actively involved in biological cycling and consequently highlights the distribution of matter within ecosystems [31].

2. Materials and Methods

2.1. Study Area and Geochemical Characteristics of the Lower Dniester Valley (LDV)

The study area of about 1848 km^2 represents a terraced plain in the south-eastern part of Moldova (46°48′ N 29°38′ E) and occupies mainly the left bank part of the Lower Dniester Valley as well as the small right bank part of the area adjacent to the Bendery (Figure 1).

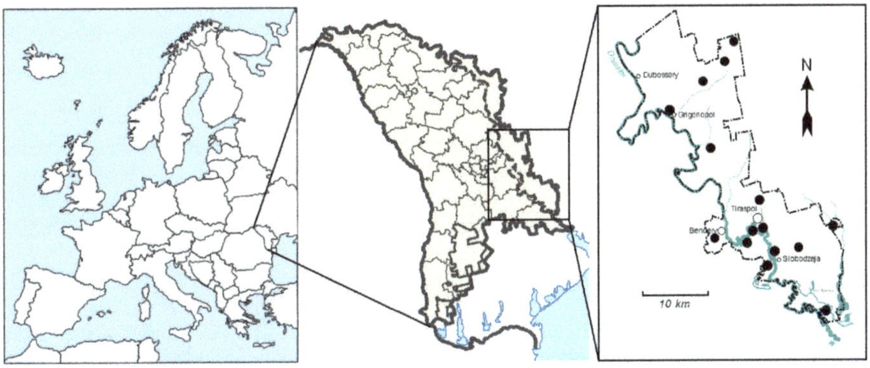

Figure 1. Location of the study area and sampling sites (•) in the Lower Dniester Valley. White circles represent geographic position of towns.

Most of the region consists of Quaternary terraces. The terraces are covered by a thick layer of loess. Towards the north-east, the terraces merge into the root slope of the Dniester Valley, which is covered by a lower layer of deluvial sediments and is formed by sands and clays of the Upper Sarmatian. Floodplains cover over 20% of the region, with predominantly loamy and sandy soils in central and riverine floodplains, accompanied by shallow groundwater. All floodplain areas in the LDV were artificially isolated from flood waters by dyke systems constructed to prevent river flooding of the agricultural lands. In the northeastern part, absolute heights reach 150–200 m, while in the southeastern part they rarely exceed 80 m. The surface of the terrain is somewhat dissected by small dry gullies with flat, uneroded slopes.

The depth of the groundwater table in the Pleistocene terraces of the Dniester River varies significantly, usually ranging from 10 to 15 m. In the Dniester floodplains, the groundwater level is typically set between 0.5 and 8 m, depending on the fluctuations of the Dniester River, and is susceptible to flooding [32].

The climate of the area is temperate continental, i.e., warm and arid. The region has an average annual air temperature of 9.6 °C with a cumulative temperature (above 10 °C) of 3270°, and active vegetation for approximately 200 days. The yearly average precipitation is 420–430 mm, with a corresponding moisture coefficient of 0.50. The study area is dominated by Chernozem soils which cover almost 77% of the territory. The most important types are Luvy-Calcic Chernozems (43.8%), Vorony-Calcic Chernozems (24.5%), and Voronic Chernozems (4.5%). They are followed by Fluvisols which comprise up to 8% of the whole soil cover.

2.2. Field Studies and Sample Preparation

The study was conducted in the second half of the vegetation season during July and August. Soils were collected from the topsoil at a depth of 0 to 30 cm using a 1 m stainless steel auger, taking into account the mesorelief characteristics (watersheds and terraces, slopes, hollows, floodplains). The combined soil sample was the mixture of 7 to 10 samples taken randomly in a given area. Samples of aboveground parts grown in the current year (shoots and leaves) of cultivated (wheat *Triticum aestivum* L., sunflower

Helianthus annuus L.) and wild (couch grass *Elytrigia repens* (L.) Nevski) mature plants most common for the study area were collected at soil sampling sites. Dead plant parts from previous years were not included in the sample.

The obtained soil and plant samples were dried at room temperature and then thoroughly powdered (soil samples were powdered with an agate mortar and after that sieved with a 1.0 mm sieve, plant samples were powdered in an electric mill with steel blades) and homogenized. The larger mass of the average sample was reduced by quartering and then packed in polyethylene bags for subsequent analysis.

Samples were dried in a heating oven at 103 ± 2 °C to a constant weight for conversion to dry weight.

2.3. Chemical Analysis and Extraction

Soil organic carbon content (SOC) was obtained from the dichromate redox titration method using N-phenanthranilic acid as an indicator [33]. The soil pH in water extract was measured using a portable pH-meter WTW pH 3110 SET 2 with a SenTix 41 pH electrode in suspension obtained from stirring of the soil sample with deionized water with the ratio 1:5.

The procedure of alkalimetric determination of carbonates (as $CaCO_3$) was the following. A soil sample (0.5–2 g) was put to a dry 250 mL Erlenmeyer flask, placing the sample near the wall of the flask. An empty porcelain crucible was placed in the flask. Five mL of 2 N HCl was poured into the crucible, using a pipette. Then, a tube without welt (25 × 90 mm) with 5 mL of 0.4 N NaOH was placed in the flask. This tube was leaned against the wall of the flask, and it was immediately closed with a rubber stopper moistened with distilled water. Then, the flask was tilted, overturning the crucible with acid allowed to distribute along the bottom of the flask. After 4–5 h, the flask was opened, the test tube was removed, rinsed with distilled water from the outside, and dried with filter paper. Two drops of a solution of phenolphthalein and about 1 mL of a saturated solution of $BaCl_2$ were added to the test tube. Then, the excess NaOH was titrated with 0.2 N HCl until the disappearing of the pink color.

The carbonate content ($CaCO_3$, $g \cdot kg^{-1}$) was calculated as follows:

$$CaCO_3 = \frac{(V - V_1) \cdot C \cdot 0.022 \cdot 2.273 \cdot 1000}{m \cdot K}, \quad (1)$$

where V is the acid volume used for titration of the NaOH solution in the control experiment, mL; V_1 is the acid volume used for titration of the excess NaOH in the soil analysis, mL; C is the concentration of HCl, mmol (eq)$\cdot mL^{-1}$; m is the mass of the air-dry sample, g; 0.022 is the molar mass of the carbon dioxide equivalent (1/2 CO_2), $g \cdot mol^{-1}$ (eq); 2.273 is the coefficient for conversion from CO_2 to $CaCO_3$ concentration; 1000 is the coefficient for conversion to $g \cdot kg^{-1}$; K is the coefficient for conversion of the analysis result to dry soil.

The sum of exchangeable cations Ca^{2+} + Mg^{2+} was determined in a 1 N KCl extract according to Shaimukhametov [34].

The determination of total HMs (Cu, Mn, Zn) was performed by atomic absorption spectrometry with an Aanalyst800 (Perkin Elmer, Shelton, CT, USA) using a flow-injection system FLAS-400 in aqua regia extracts of soil according to ISO 11047 [35]. Exchangeable forms (EXC) of trace metals were extracted by the acetate-ammonium buffer solution with pH = 4.8 with a soil:solution ratio of 1:10. The Cu acid-soluble forms in an analytical soil sample were determined through soil extraction. This was achieved by suspending the sample in a 1 N nitric acid solution for 24 h. The determination of exchangeable forms of metals and acid-soluble Cu in soil extracts was carried out using atomic absorption spectrophotometry with the use of a SHIMADZU AA-7000 (Shimadzu, Kyoto, Japan).

Total soil Se was determined in solutions obtained from the acid digestion of perchloric and nitric acids at 120 °C for 1 h, 150 °C for 1 h, and 180 °C for 1 h. Water-soluble (WS) Se, a form of exchangeable Se in soil, was extracted using deionized water with a soil to water ratio of 1:5 in a hot-water bath for 1 h [36]. Se was determined in solutions and extracts by a sensitive single-test-tube procedure for the fluorometric determination [37].

2.4. Bioccumulation Assessment and Statistical Analysis

The bioaccumulation factor (BAF) was used to calculate the Se and HMs transfer from soil to the aboveground part of various plants according to the following equation [38]:

$$\mathrm{BAF} = \frac{C_c}{C_s}, \quad (2)$$

where C_c represents the element contents (in dry matter) in the aboveground part of the plant and C_s shows the element concentration in the corresponding soils.

Descriptive statistics and a correlation analysis among various sample groups were carried out using STATISTICA 10.0 software (StatSoft Inc., Tulsa, OK, USA, 2011). Each variable was checked for outliers, which were excluded from further statistical calculations based on the rejection criterion at $p < 0.05$. The statistical significance of the differences between the two variables was assessed according to the distribution patterns of the sample data, using non-parametric methods (Mann–Whitney U-test, Kruskal–Wallis ANOVA). Pearson or Spearman correlation coefficients were calculated, and a regression analysis was performed to examine the tightness and type of relationship between the two variables. All statistical calculations were performed at a significance level of $p < 0.05$.

3. Results

3.1. Total Se and Heavy Metals Content in Soils

3.1.1. Soil Physicochemical Characteristics

The main physicochemical characteristics of soils in the LDV, relevant for assessing Se and HM content and mobility, are shown in Table 1.

Table 1. Soil's physical and chemical properties.

Soil Type	pH	SOC	$CaCO_3$	Exchangeable Cations Ca^{2+} + Mg^{2+}	Texture
		($g \cdot kg^{-1}$)	($g \cdot kg^{-1}$)	(mg (eq)$\cdot$100 g^{-1})	
Fluvisols	8.09 ± 0.10 [ab]	13.9 ± 5.9 [a]	67.2 ± 38.0 [ab]	25.0 ± 8.9 [a]	Sandy loam
Luvy-Calcic Chernozems	8.17 ± 0.11 [a]	13.6 ± 6.3 [a]	43.9 ± 23.8 [b]	32.8 ± 10.0 [a]	Loam
Vorony-Calcic Chernozems	7.88 ± 0.24 [b]	8.4 ± 8.1 [a]	28.4 ± 37.0 [c]	18.2 ± 10.5 [a]	Loam
Voronic Chernozems	7.95 ± 0.10 [b]	27.1 ± 1.5 [b]	12.6 ± 7.9 [c]	46.4 ± 4.8 [b]	Silt loam

SOC: soil organic carbon; $CaCO_3$: carbonate content. Data (n = 34) are presented as the mean values ± SD. Within each column, values with the same letters do not differ statistically according to Mann–Whitney U-test at $p < 0.05$.

Soil pH varied from 7.88 in Vorony-Calcic Chernozems to 8.17 in Luvy-Calcic Chernozems, corresponding to a slightly alkaline soil environment. The ranges of SOC and carbonate content were very wide, from 8.4 to 27.1 $g \cdot kg^{-1}$ and from 12.6 to 67.2 $g \cdot kg^{-1}$, respectively. The highest content of exchangeable cations was observed in Voronic Chernozems with a silt loam texture (46.4 mg (eq)$\cdot$100 g^{-1}), but the lowest content of 18.2 mg (eq)$\cdot$100 g^{-1} was found in Vorony-Calcic Chernozems with a loam texture.

3.1.2. Selenium in Soils

Statistical data for the estimation of the content of total and mobile Se and HMs content in the soils of the Lower Dniester Valley are presented in Table 2.

Table 2. Selenium and heavy metal concentrations in different soil types (0–30 cm) (mg·kg^{-1}).

Soil Type	Se		Mn		Zn		Cu	
	Total	WS	Total	EXC	Total	EXC	Total	EXC
Fluvisols	0.36 ± 0.09 ^{ab}	0.11 ± 0.03 ^{a}	429 ± 173 ^{ab}	96.1 ± 24.0 ^{a}	35.5 ± 25.0 ^{a}	1.91 ± 1.29 ^{a}	27.6 ± 10.9 ^{ab}	0.35 ± 0.22 ^{ab}
Luvy-Calcic Chernozems	0.32 ± 0.16 ^{b}	0.09 ± 0.02 ^{a}	458 ± 122 ^{a}	76.2 ± 25.0 ^{ab}	29.3 ± 16.4 ^{a}	0.97 ± 0.70 ^{ab}	39.1 ± 10.8 ^{a}	0.52 ± 0.51 ^{a}
Vorony-Calcic Chernozems	0.33 ± 0.12 ^{b}	0.08 ± 0.03 ^{a}	374 ± 115 ^{ab}	64.6 ± 12.5 ^{ab}	25.2 ± 10.3 ^{a}	0.46 ± 0.19 ^{b}	13.1 ± 8.5 ^{b}	0.18 ± 0.03 ^{b}
Voronic Chernozems	0.23 ± 0.07 ^{bc}	0.08 ± 0.01 ^{a}	565 ± 17 ^{b}	56.1 ± 6.2 ^{b}	44.0 ± 16.2 ^{a}	0.82 ± 1.25 ^{ab}	26.2 ± 7.7 ^{b}	0.09 ± 0.05 ^{b}
All soil types	0.33 ± 0.13	0.09 ± 0.03	464 ± 129	75.7 ± 24.5	33.8 ± 18.4	1.04 ± 0.90	31.2 ± 12.7	0.41 ± 0.45

WS: water-soluble forms; EXC: exchangeable forms. Data (n = 42 for Se, n = 34 for Mn, Zn, and n = 30 for Cu) are presented as the mean values ± SD. Within each column, values with similar letters do not differ statistically according to Mann–Whitney U-test at $p < 0.05$.

The topsoil layer had an average content of total Se of 0.33 ± 0.13 mg·kg^{-1}. Total Se was the highest in alluvial soils, which were mainly found in floodplain ecosystems, with an average value of 0.36 ± 0.09 mg·kg^{-1}. The mean Se concentration decreased to 0.32 ± 0.16 mg·kg^{-1} and 0.33 ± 0.12 mg·kg^{-1} in Luvy-Calcic Chernozems and Vorony-Calcic Chernozems, respectively, with a rather large variation. Finally, the topsoil of Voronic Chernozems, which are evolutionarily more mature steppe soils, contained the least total Se of 0.23 ± 0.07 mg·kg^{-1}. It should be noted that the maximum concentrations of this element were found in various locations including the Dniester-Turunchuk interfluve soils (0.40 mg·kg^{-1}), the right bank floodplain near Kitskani village (0.44 mg·kg^{-1}), and Copanca village (0.47 mg·kg^{-1}). The highest Se levels were discovered in the soils of the left bank, where concentrations reached up to 0.65 mg·kg^{-1}. However, based on the average concentration of this element, the considered soil types can be ranked in the following order: Fluvisols > Luvy-Calcic Chernozems ≈ Vorony-Calcic Chernozems > Voronic Chernozems.

The average water-soluble Se decreased in a similar sequence as the soil types. Generally, soils have 0.09 ± 0.03 mg·kg^{-1} of water-extractable Se forms, which is 32.1% of their total content. Nonetheless, even though total Se was noticeably higher in alluvial soils, an analysis of statistics revealed no significant differences for mobile Se concentrations in various soil types (Table 2). A regression analysis was conducted to examine the correlation between the total Se content and its water-soluble forms in soil. The results are shown in Figure 2, indicating a high positive linear correlation (r = 0.845, p < 0.00001). It is evident that water-soluble Se in the topsoil layer of the Dniester Valley increased with the rising total Se content.

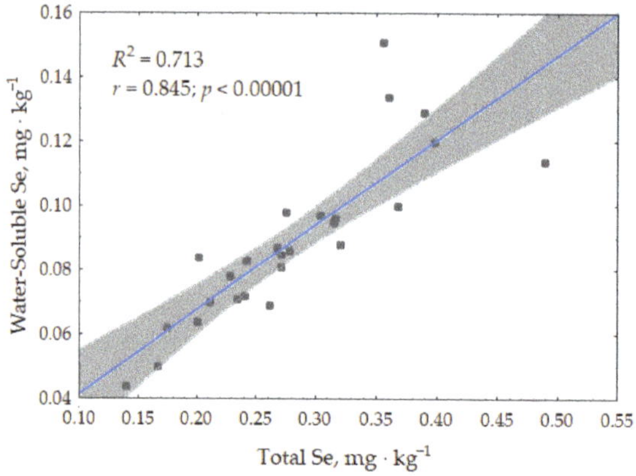

Figure 2. Pearson correlation between total Se in soils and water-soluble Se in soils. The grey shade area represents the 95% confidence interval.

3.1.3. Heavy Metals in Soils

Manganese

The soils of the LDV had different levels of Mn, ranging from 196 to 676 mg·kg^{-1}, with an average of 464 mg·kg^{-1}. Mobile Mn in the upper soil horizons, which is available for plant uptake, averaged 75.7 mg·kg^{-1} or 17.9% of the total Mn (Table 2). Table 2 also displays the significant difference in total Mn content between Voronic Chernozems and Luvy-Calcic Chernozems. This discrepancy can be attributed to two interrelated factors, i.e., the higher content of SOC and exchangeable cations found in Voronic Chernozems, since SOC ($r = 0.715$, $p < 0.05$) and exchangeable cations $Ca^{2+} + Mg^{2+}$ ($r = 0.698$, $p < 0.05$) strongly correlated with the total Mn content (Figure 3). This led to statistically significant differences for exchangeable Mn in Fluvisols and Voronic Chernozems. No significant influence of the investigated physico-chemical properties on the concentration of mobile Mn was observed in the soils. The correlation between the total Mn content and its mobile forms was found to be weak and statistically insignificant.

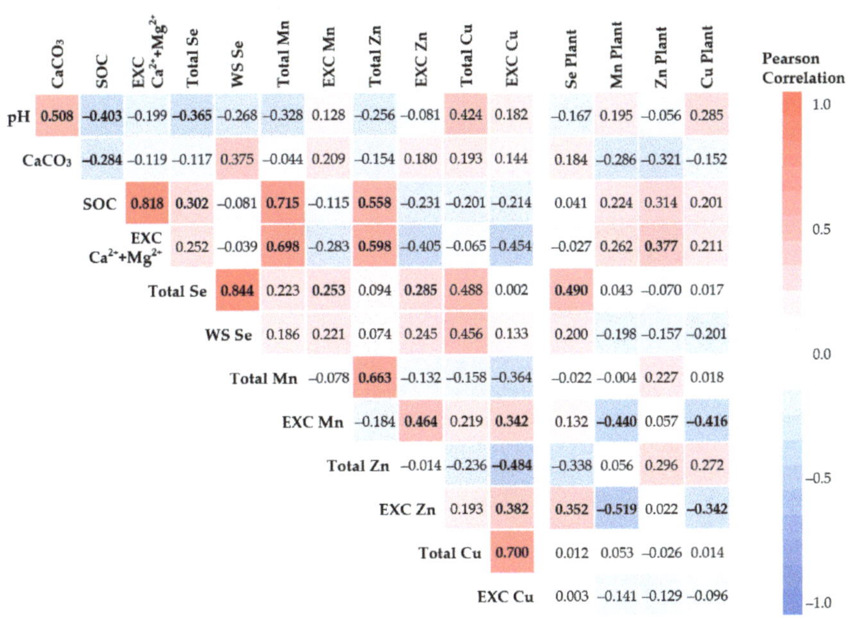

Figure 3. Pearson correlation matrix: correlation coefficients between soil and plant variables. Bold values are statistically significant ($p < 0.05$). The abbreviations are the same as in Tables 1 and 2.

Zinc

The soils of the LDV had relatively low total Zn concentrations ranging from 13.2 to 63.0 mg·kg^{-1}, with an average of 33.8 mg·kg^{-1}. Table 2 shows a decrease in total Zn content in the soil subtypes with no significant differences between them: Voronic Chernozems > Fluvisols > Vorony-Calcic Chernozems > Luvy-Calcic Chernozems. Similar to Mn, analyses of total Zn concentration dependance from soil properties suggested that SOC ($r = 0.558$, $p < 0.05$) and exchangeable cations $Ca^{2+} + Mg^{2+}$ ($r = 0.598$, $p < 0.05$) played a crucial role in the accumulation of Zn in the upper soil horizons (Figure 3). This explains the higher metal content in Voronic Chernozems, which have a silt clay texture and higher organic matter content. The proportion of mobile Zn forms extracted by an ammonium acetate buffer with pH = 4.8 ranged from 0.3 to 15.6% of the total content, with an average of 4.2%. Statistical significance was found only for Vorony-Calcic Chernozems and Fluvisols.

Copper

The analysis of the homogeneity of the samples for both total and mobile Cu revealed anomalous maximum values. These soil concentrations were found locally. For example, floodplain soils near the village of Kitskani, Slobodzeja district, contained 158.3 mg·kg^{-1} of total Cu, 115.8 mg·kg^{-1} of acid-soluble Cu, and 23.7 mg·kg^{-1} of exchangeable Cu. High Cu concentrations were also found in the vicinity of Slobodzeja town, with exchangeable Cu of 5.98 and 3.88 mg·kg^{-1}. These values were excluded from the calculations for Table 2 since they are outliers and do not belong to the general population.

Data were categorized based on the percentage of acid-soluble Cu compared to total content (Table 3). On the basis of this criterion, two categories of soils were defined according to the level of Cu loading: unpolluted and low-polluted soils with an acid soluble Cu content of up to 35%, and highly polluted soils with an acid-soluble Cu content higher than 50% of the total Cu.

Table 3. Copper concentrations (mg·kg^{-1}) in soil samples with different levels of pollution.

Pollution Level		**Cu**		
		Total	**AS**	**EXC**
Unpolluted and low-polluted soils	Mean ± SD	27.3 ± 8.1 ^{a}	6.2 ± 3.5 ^{a}	0.23 ± 0.19 ^{a}
(n = 14)	Min–max	12.1–38.3	2.0–11.5	0.03–0.52
Highly polluted soils	Mean ± SD	73.7 ± 57.7 ^{b}	53.8 ± 43.7 ^{b}	8.00 ± 10.71 ^{b}
(n = 4)	Min–max	28.9–158.3	20.6–115.8	0.85–23.70

EXC: exchangeable forms; AS: acid-soluble forms. Within the pollution levels, values with the same letters do not differ statistically according to Mann–Whitney U-test at $p < 0.05$.

The significant variation in soil Cu concentrations is evident in Table 3. For instance, the total Cu showed a range of two magnitudes, from 12.1 to 158.3 mg·kg^{-1}, while acid-soluble Cu varied within three orders of magnitude, from 2.0 to 115.8 mg·kg^{-1}, whereas the concentrations of mobile Cu varied within four orders of magnitude, from 0.03 to 23.70 mg·kg^{-1}. Hence, the discrepancies in the mean metal content amongst soils with different levels of pollution are dramatically different and statistically significant, both for the total metal content and for the concentrations of acid-soluble and mobile forms. Notably, the values of the acid extraction criterion reached 71–93% for the group of highly polluted soils.

The mean total Cu content in Luvy-Calcic Chernozems was the highest compared to other subtypes of Chernozems and to Fluvisols (Table 2), although statistical calculations allowed us to prove the significance of the differences between Vorony-Calcic Chernozems and Vorony Chernozems. Therefore, the sequence of soil types, according to the average total Cu content in the upper soil horizons, was as follows: Luvy-Calcic Chernozems > Fluvisols > Vorony Chernozems > Vorony-Calcic Chernozems. The content of exchangeable Cu was strongly dependent on the total Cu ($r = 0.700$, $p < 0.01$) (Figure 3), and the percentage of mobile Cu in the total content was 1.41% (excluding outliers). The differences in mobile Cu content between Vorony Chernozems and Luvy-Calcic Chernozems were significant according to the Mann–Whitney U-test at $p < 0.05$.

3.2. Se, Mn, Zn, and Cu Concentrations in Plants and Their Bioaccumulation

The Se content in plants of the Dniester Valley varied from 0.06 to 0.58 mg·kg^{-1}. The differences in Se concentrations in aboveground plant tissues and their bioaccumulation among crops (sunflower, wheat) and wild plant species (couch grass) are shown in Table 4.

Sunflowers had the highest Se accumulation capacity among the studied plants. Their tissues contained from 0.08 to 0.58 mg·kg^{-1} of this element with a mean value of 0.23 mg·kg^{-1}, while wheat and couch grass contained 0.15 and 0.12 mg·kg^{-1}, respectively. A slight positive correlation ($r = 0.494$, $p < 0.001$) was observed between the Se content in plants and the total Se in soils (Figure 4). Conversely, no relationship was found between the water-soluble Se and its concentration in plants (Figure 3).

Table 4. Selenium and heavy metal concentrations in plants ($mg\cdot kg^{-1}$) and their bioaccumulation factors (BAFs).

Plant Species		Se ($mg\cdot kg^{-1}$)	Se BAF	Mn ($mg\cdot kg^{-1}$)	Mn BAF	Zn ($mg\cdot kg^{-1}$)	Zn BAF	Cu ($mg\cdot kg^{-1}$)	Cu BAF
Sunflower	Mean ± SD	0.23 ± 0.17^{a}	0.70 ± 0.41^{a}	94.9 ± 46.7^{a}	0.18 ± 0.08^{a}	21.2 ± 5.5^{b}	0.74 ± 0.46^{a}	5.1 ± 6.4^{a}	0.15 ± 0.13^{a}
(*Helianthus annuus*)	Min–max	0.08–0.58	0.30–1.43	43.6–166.0	0.08–0.30	14.6–27.7	0.36–1.35	1.5–16.4	0.06–0.38
Wheat	Mean ± SD	0.15 ± 0.08^{ab}	0.53 ± 0.26^{ab}	90.4 ± 28.5^{a}	0.20 ± 0.01^{a}	30.1 ± 4.4^{a}	0.75 ± 0.22^{a}	3.1 ± 0.9^{a}	0.11 ± 0.04^{a}
(*Titicum aestivum*)	Min–max	0.07–0.36	0.29–1.05	26.1–126.5	0.19–0.20	22.4–35.6	0.55–1.06	1.9–4.3	0.06–0.15
Couch grass	Mean ± SD	0.12 ± 0.07^{b}	0.43 ± 0.20^{b}	78.2 ± 37.4^{a}	0.20 ± 0.17^{a}	25.4 ± 7.7^{b}	1.03 ± 0.56^{a}	3.3 ± 1.6^{a}	0.14 ± 0.13^{a}
(*Elytrigia repens*)	Min–max	0.06–0.32	0.20–0.86	26.2–152.5	0.05–0.49	13.3–43.4	0.32–2.58	1.5–7.9	0.04–0.49

Data ($n = 12$ for wheat, $n = 10$ for sunflower, and $n = 26$ for couch grass) are presented as the mean values ± SD. For each plant species, mean values with the same letters do not differ statistically according to Mann–Whitney U-test at $p < 0.05$.

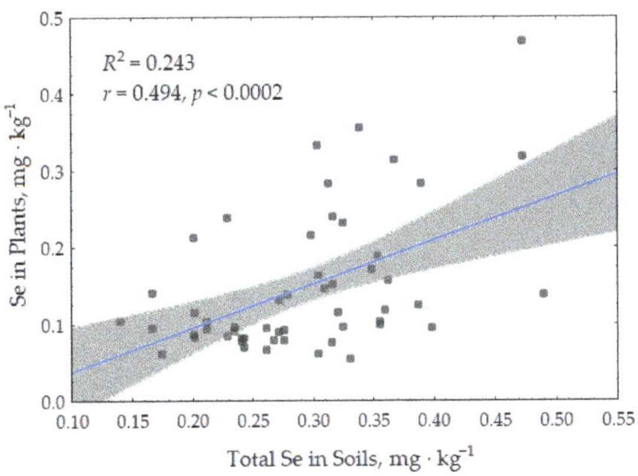

Figure 4. Pearson correlation between total Se in soils and Se concentrations in plant tissues. The grey shade area represents the 95% confidence interval.

The bioaccumulation factors (BAF) showed that Se concentrations decreased in the order sunflower > wheat > couch grass. This proves that sunflower had a higher ability to accumulate Se. Generally, plants accumulated the highest Se concentrations in Fluvisols (0.30 $mg\cdot kg^{-1}$) compared to Luvy-Calcic, Vorony-Calcic, and Vorony Chernozems. Moreover, a statistical analysis using Kruskal–Wallis ANOVA proved that BAFs were significantly higher for plants growing on Fluvisols ($p < 0.05$) compared to those on Chernozem soils (Figure 5). Generally, among plants growing on different Chernozems, only plants growing in Luvy-Calcic and Vorony-Calcic Chernozems showed statistically significant differences.

Similar to Se, higher concentrations of HMs, such as Mn and Cu, were found in the aboveground part of the sunflower but the differences were not statistically significant for the different plant species. Most of the BAFs for these metals were lower than 0.20 and generally did not exceed 0.50. In contrast to Se, Mn, and Cu, the variation in Zn content and accumulation between different plant species was quite different. For example, despite the similarity of HM concentrations in the tissues of different plant species, a significantly higher level of the metal was observed in wheat. In the aboveground parts of sunflower and wheatgrass, its mean value reached up to 25.4 $mg\cdot kg^{-1}$. Zn accumulated more intensively in the plants, and its BAF reached a value of 1.0 and higher. However, the Mann–Whitney U-test did not prove any species-dependent peculiarities in plant Zn bioaccumulation. A single plant species, couch grass, was used to investigate the influence of soil factors on the concentration and accumulation of metals (see Figure 5).

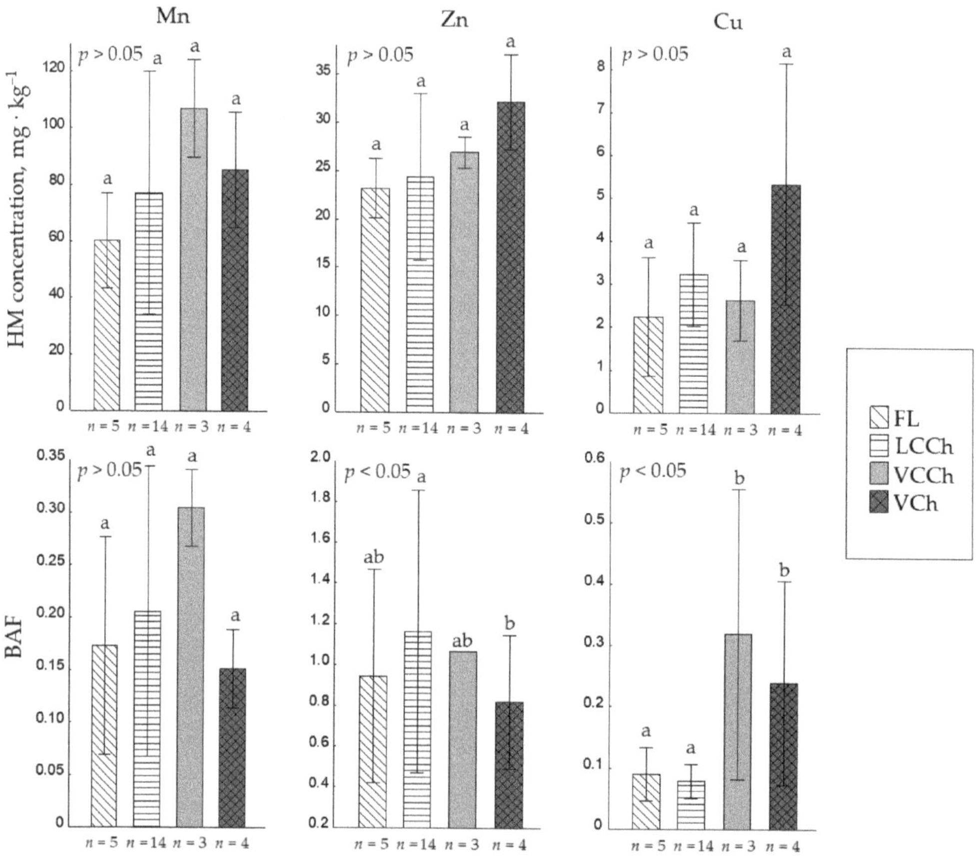

Figure 5. Heavy metal concentrations and their bioaccumulation in couch grass (*E. repens*) in the Lower Dniester Valley. Soil types: FL—Fluvisols, LCCh—Luvy-Calcic Chernozems, VCCh—Vorony-Calcic Chernozems, VCh—Voronic Chernozems. Data are presented as the mean values ± SD ($n = 30$). p-values represent the statistical significance of difference between soil types by Kruskal–Wallis ANOVA. Values with the same letters do not differ statistically at $p < 0.05$.

It is obvious that there was variability in HM accumulation and HM content in plants among different soil types. However, according to Kruskal–Wallis ANOVA, the reliability of these differences ($p < 0.05$) has only been established for the BAFs of Zn and Cu, but not for Mn BAF as well as HM concentrations in plant tissues.

3.3. Se and HMs Interactions in Soil–Plant System

Se and HM interrelations in the soil–plant system in the Lower Dniester Valley are summarized in Figure 6.

As reported above, Se in plants was positively correlated with its total content in soils. Furthermore, a positive correlation was observed between Se in plants and the content of the mobile forms of Zn in soils ($r = 0.352$, $p < 0.05$). No antagonistic effects on Se accumulation by HMs in plants were found. Conversely, negative relationships between metals were more common. Indeed, the increase of EXC Mn in soils negatively affected the content of Mn ($r = -0.440$, $p < 0.01$) and Cu ($r = -0.416$, $p < 0.01$) in plant tissues. Similar relationships were found between EXC Zn and the content of Mn ($r = -0.519$, $p < 0.001$) and Cu ($r = -0.342$, $p < 0.05$) in the aboveground plant biomass.

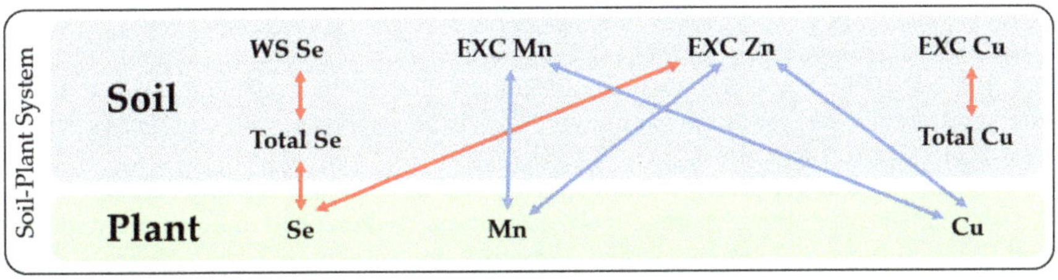

Figure 6. Interactions between Se and HMs concentrations in soil–plant system in the Lower Dniester Valley. Red lines indicate positive correlations, blue lines indicate negative correlations. Only significant correlations, as determined by Pearson's correlation matrix in Figure 3, were shown. Abbreviations are consistent with those presented in Table 2.

4. Discussion

According to the obtained results, Voronic Chernozems were distinguished from other soils by specific conditions predominantly caused by a relatively high content of SOC (up to 27.1 g·kg^{-1}), a higher cation exchange capacity, leaching the upper horizons from carbonates, and the associated lower soil pH. Luvy-Calcic Chernozems were similar to Fluvisols in their physical and chemical properties. Vorony-Calcic Chernozems were characterized by their low carbonate content and pH values comparable to Voronic Chernozems due to leaching from the upper horizon; at the same time, their lighter texture and lower SOC determined the lowest content of exchangeable cations there. Generally, the peculiarities of the soil development process in different soil types under the conditions of the steppe zone of the Dniester Valley have a significant influence on the content of total Se. The top horizons of Fluvisols in floodplains had the highest total Se stocks in comparison with other Chernozems subtypes. Erosion-denudation washout from slopes into floodplains and alluvial-delluvial processes are probably responsible for this phenomenon [20,39]. The mean total Se in soils obtained in this study (0.33 ± 0.13 mg·kg^{-1}) was relatively higher than those in previous studies, 0.143–0.200 mg·kg^{-1} [18] and 0.246 ± 0.073 mg·kg^{-1} [20], but it fell within the range of 0.054–0.711 mg·kg^{-1} [18] and 0.100–0.668 mg·kg^{-1} [20].

The total soil Se is not a good indicator of the Se supply in the food chain [40]. Biogeochemical processes involving the formation of various compounds by the free biogeochemical energy of living organisms occur in soils. Such biogeochemical processes result in the production of mobile forms of the element, which may be available to plants, involving it in biogenic migration [7,41]. The water-soluble Se forms in soils are of the greatest importance to plants, which are a reliable indicator of the element's status in landscapes and ecosystems [30,42,43]. It was found that concentrations of WS Se in the upper layer of soils in the LDV increased with rising total soil Se (r = 0.845). The same trend was found for arable soils in arid regions of China, but the correlation coefficient was lower (r = 0.58), while higher values of the coefficient were characteristic only for soils under natural vegetation [44]. This study confirms previous research indicating that the relative abundances of Se species in soil are dependent on total soil Se [18].

The mobility of trace elements in the soils of Moldova is largely determined by the binding of organic matter and soil-absorbing complexes [45]. Meanwhile, the predominantly slightly alkaline conditions of soil solutions, a low content of soil organic matter, and the cation exchange capacity of soils had no significant effect on the content of water-soluble Se. Furthermore, the total soil Se content was only weakly dependent on the soil humus content and the soil pH. Favorable conditions for Se mobility in soils are indicated by all of the above results. On average, at least 32% of the total Se in the soils of the Dniester Valley can be transferred to the water extract. Bogdevich et al. [18] found that the fraction of soluble and ligand-exchangeable organic and inorganic Se species was up to 24.5% of the total soil Se near Carpineni township, which are similar to the data obtained in the present study. Data related to Se in other regions and biogeochemical provinces were provided for comparison. For example, in different types of soils in China, the WS Se was observed in less amounts from 1.07 to 6.69% of total Se [44]; in the acidic soils of Japan—0.5–7.1% [46]; in the soils of Serbia—0.59–16.35% [47]. In Hungary, most of the soils contained no more than 10% of WS Se, and only in some individual samples did its concentrations reach a similar level of 20–35% of the total amount [48]. Therefore, our results are consistent with those reported in other studies.

To identify the geochemical characteristics of Se and studied HMs in the Lower Dniester Valley, we compared the obtained results with data on element content in soils worldwide, as well as with mean values and concentration ranges in the soils of the study region (see Table 5).

Table 5. Selenium and heavy metal concentrations in soils and their threshold levels (mg·kg^{-1}).

Chemical Elements		World Soils [30]	Soils of Moldova [20,45]		Soils of the LDV (This Study)		Threshold Levels [30]	
		Total	Total	EXC	Total	EXC	MAC	TAV
Se	Mean	0.44	0.25	–	0.33	0.09 *	–	–
	Min–max	–	0.10–0.65	–	0.08–0.65	0.04–0.15 *	–	3–10
Mn	Mean	488	790	2.4	464	75.7	–	–
	Min–max	–	150–2250	0.4–195	196–676	43–136	–	–
Zn	Mean	70	71	1.4	33.8	1.04	–	–
	Min–max	–	10–166	0.1–4.9	13.2–63.0	0.14–3.34	100–130	200–1500
Cu	Mean	38.9	32	1.6	31.2 **	0.41 **	–	–
	Min–max	–	2–400	0.1–60	7.0–158.3	0.01–23.70	60–150	60–500

(*) Data on water-soluble forms; (**) mean value was calculated excluding outliers. Threshold levels: ranges of maximum allowable concentrations (MAC) and trigger action values (TAV) for potentially harmful chemical elements in agricultural soils.

In general, the Se content in LDV soils was found to be relatively higher than reported for a broader area, including the Dniester-Prut interfluve. At the same time, world soils had a higher mean value of total soil Se. The total Cu in the soils of the LDV (excluding contaminated sites) was comparable to that of soils in Moldova, which was slightly lower than the average content of soils in the world. The total Mn in the investigated soils was similar to the mean value of the metal in soils of the world, but 1.7 times lower than the total concentrations in the soils of Moldova. Meanwhile, the LDV had relatively low concentrations of total Zn in the soils. On average, they were more than twice as low as the regional background and similar to those found in soils worldwide. It is important to note that the Se and HMs of soils were typically far below the known thresholds.

The investigation of Mn, Zn, and Cu contents in the soil, which are possible anthropogenic pollutants, confirmed the absence of any systematic contamination with Mn and Zn, but not with Cu [28,45,49]. Agricultural pollution is the main source of high Cu concentrations in the environment of Moldova [29,50]. Unlike total metal concentrations, which reflect general soil contamination, the mobile forms are the main indicator that fully characterizes the degree of adverse effect on plants. Many researchers explain that aggressive extracting agents, such as acids, extract the significant quantity of HMs from the soil. Acidic extracts contain not only the forms available to plants, but also a part of the total, which can be considered as the next reserve capable of being mobilized [51,52]. Acid-soluble Cu varied within three orders of magnitude from 3.0 to 115.8 mg·kg^{-1} soil, then the concentrations of mobile metal forms varied within four orders of magnitude from 0.04 to 23.70 mg·kg^{-1} soil. According to Kiriluk [45], the range for the soils of Moldova was much wider (0.1–60.0 mg·kg^{-1}) with an average value of 1.6 mg·kg^{-1} soil. The acid extract contained significantly large amounts of mobile Cu in polluted soils, potentially available for assimilation by plants. Chemicals have been used extensively in Moldavian agriculture, with approximate estimates of copper-containing preparations being around 6000–8000 tons per year from the early 1950s to the early 1990s [45]. The accumulation of Cu in soils can result in high concentrations of Cu even after the cessation of pesticide use [53,54]. In this respect, the analysis revealed the existence of local Cu contamination in the alluvial soils of the Dniester Valley with relatively low total content and mobility in most of the studied region.

Regarding the average Se and HMs content in plants, there were no significant differences in element concentrations, except for Se in sunflower and Zn in wheat, which were the highest. This variability, in addition to species-dependent accumulation, could imply a different availability of elements in the soil.

The studied plants belonged to the ecological group of non-accumulators according to their ability to accumulate Se in their tissues [55]. Se content in the aboveground part of the studied crops and wild-growing plants was found to be relatively low compared to its content in most soils. Substantially, plants growing on Fluvisols, the soils associated with floodplains, accumulated more Se than plants growing on Chernozems on terraces and slopes in steppe and arable lands. This can be explained by the higher total Se content in floodplain soils, which are subordinate in landscape-geochemical systems compared to Chernozem soils. However, it cannot be ignored that artificially constructed dyke systems to prevent river flooding isolated the agricultural lands in the floodplain from flood waters. This altered soil formation process in alluvial soils may have resulted in a greater influence of groundwater on Se loading in alluvial soils, as shallow groundwater provides additional moisture to the root zone via capillary fluxes [56]. This may have implications for the cultivation of selenium-enriched crops and the production of functional foods. Considering the slightly alkaline nature of soil solutions, the relatively low SOC content, and the significant cation exchange capacity of soils [30,55,57], selenates, selenites, and ligand-exchangeable organoselenium compounds are expected to be readily available for plant uptake. In another study of the soil–plant–groundwater system [58], the Se content in plants was also relatively low.

The use of biogeochemical indicators (BAFs) characterizing interrelations can be used to assess certain cases of the soil–plant system. Biological selectivity allows plants to control their chemical composition within certain boundaries [59,60]. Plants accumulated Zn more intensively, as indicated by BAFs ≥ 1, in order to meet their requirements. Se was also characterized by high BAFs ≥ 1, although it is not an essential trace element for higher plants [11,12,61]. Significant differences in the accumulation of Se and Zn were observed in different soil types of the Dniester Valley. This indicates that the bioavailability of these elements differs depending either on the specific landscape and geochemical conditions or the intensity of their migration. No such patterns were observed for Mn and its BAFs were low. The bioaccumulation of Cu also varied as a function of the soil geochemical conditions and the level of anthropogenic pollution, but it was low too (BAFs < 0.5).

5. Conclusions

The conditions for Se bioaccumulation in plants in the Dniester Valley were found to be favorable. The reasons are connected to the territory location within a hydrogeochemical province with high selenium content in groundwater, the Se optimal level in soils, its high mobility, and the occurrence of favorable physical and chemical conditions of soils for selenium mobilization. Se concentrations in plants varied from 0.06 to 0.58 $mg \cdot kg^{-1}$ depending on either the plant species or the local landscape and geochemical conditions. Plants growing on Fluvisols accumulated more Se than those on Chernozems due to the higher total Se in floodplain soils. This may have implications for producing Se-fortified crops and functional foods. However, the absence of systematic soil contamination caused by Mn and Zn, their relatively low levels in soils, as well as local foci of soil pollution with Cu, determined the absence of antagonistic interactions in the soil–plant system. Further studies are required to identify the reasons for the high mobility of Se and the significant content of its water-soluble forms in soils.

Author Contributions: Conceptualization, S.S. and N.G.; methodology, S.S. and T.S.; software, S.S.; validation, S.S., O.C.M. and G.C.; formal analysis, G.C.; investigation, S.S., T.S. and N.G.; resources, S.S. and T.S.; data curation, S.S. and A.V.T.; writing—original draft preparation, S.S.; writing—review and editing, N.G. and G.C.; visualization, S.S.; supervision, N.G.; project administration, S.S.; funding acquisition, S.S. All authors have read and agreed to the published version of the manuscript.

Funding: This research received no external funding.

Institutional Review Board Statement: Not applicable.

Informed Consent Statement: Not applicable.

Data Availability Statement: Data are contained within the article.

Conflicts of Interest: The authors declare no conflicts of interest.

References

1. Szynkowska, M.I.; Pawlaczyk, A.; Maćkiewicz, E. Bioaccumulation and Biomagnification of Trace Elements in the Environment. In *Recent Advances in Trace Elements*; Chojnacka, K., Saeid, A., Eds.; Wiley: Hoboken, NJ, USA, 2018; pp. 251–276. [CrossRef]
2. Obaid, H.; Ma, L.; Nader, S.E.; Hashimi, M.H.; Sharifi, S.; Kakar, H.; Ni, J.; Ni, C. Heavy Metal Contamination Status of Water, Agricultural Soil, and Plant in the Semiarid Region of Kandahar, Afghanistan. *ACS Earth Space Chem.* **2023**, *7*, 1446–1458. [CrossRef]
3. Massas, I.; Kairis, O.; Gasparatos, D.; Ioannou, D.; Vatougios, D.; Zafeiriou, I. Impaired Soil Health in Agricultural Areas Close to Fe-Ni Mines on Euboea Island, Greece, Caused by Increased Concentrations of Potentially Toxic Elements, and the Associated Impacts on Human Health. *Environments* **2023**, *10*, 150. [CrossRef]
4. Xie, X.; Liu, Y.; Qiu, H.; Yang, X. Quantifying Ecological and Human Health Risks of Heavy Metals from Different Sources in Farmland Soils within a Typical Mining and Smelting Industrial Area. *Environ. Geochem. Health* **2023**, *45*, 5669–5683. [CrossRef] [PubMed]
5. Wang, Y.; Mo, L.; Yu, X.-X.; Shi, H.-D.; Fei, Y. Enrichment Characteristics, Source Apportionment, and Risk Assessment of Heavy Metals in the Industrial and Mining Area of Northern Guangdong Province. *Huan Jing Ke Xue Huanjing Kexue* **2023**, *44*, 1636–1645. [CrossRef] [PubMed]
6. Li, C.; Zhou, K.; Qin, W.; Tian, C.; Qi, M.; Yan, X.; Han, W. A Review on Heavy Metals Contamination in Soil: Effects, Sources, and Remediation Techniques. *Soil Sediment Contam. Int. J.* **2019**, *28*, 380–394. [CrossRef]
7. Natasha; Shahid, M.; Niazi, N.K.; Khalid, S.; Murtaza, B.; Bibi, I.; Rashid, M.I. A Critical Review of Selenium Biogeochemical Behavior in Soil-Plant System with an Inference to Human Health. *Environ. Pollut.* **2018**, *234*, 915–934. [CrossRef] [PubMed]
8. Alengebawy, A.; Abdelkhalek, S.T.; Qureshi, S.R.; Wang, M.-Q. Heavy Metals and Pesticides Toxicity in Agricultural Soil and Plants: Ecological Risks and Human Health Implications. *Toxics* **2021**, *9*, 42. [CrossRef]
9. Jiang, H.; Lin, W.; Jiao, H.; Liu, J.; Chan, L.; Liu, X.; Wang, R.; Chen, T. Uptake, Transport, and Metabolism of Selenium and Its Protective Effects against Toxic Metals in Plants: A Review. *Metallomics* **2021**, *13*, mfab040. [CrossRef]
10. Farman, M.; Nawaz, F.; Majeed, S.; Ahmad, K.S.; Rafeeq, R.; Shehzad, M.A.; Shabbir, R.N.; Usmani, M.M. Interplay between Selenium and Mineral Elements to Improve Plant Growth and Development. In *Handbook of Bioremediation*; Elsevier: Amsterdam, The Netherlands, 2021; pp. 221–236. [CrossRef]
11. Liu, H.; Xiao, C.; Qiu, T.; Deng, J.; Cheng, H.; Cong, X.; Cheng, S.; Rao, S.; Zhang, Y. Selenium Regulates Antioxidant, Photosynthesis, and Cell Permeability in Plants under Various Abiotic Stresses: A Review. *Plants* **2022**, *12*, 44. [CrossRef]
12. Guignardi, Z.; Schiavon, M. Biochemistry of Plant Selenium Uptake and Metabolism. In *Selenium in Plants*; Pilon-Smits, E.A.H., Winkel, L.H.E., Lin, Z.-Q., Eds.; Plant Ecophysiology; Springer International Publishing: Cham, Germany, 2017; Volume 11, pp. 21–34. [CrossRef]
13. Rayman, M.P. The Importance of Selenium to Human Health. *Lancet* **2000**, *356*, 233–241. [CrossRef]
14. Ali, S.; Sami, U.; Hasnain, U.; Arsalan, S.; Sohaib, N.; Zarmina, A.; Hamza, J.M.; Zain, U.A.; Rimsha, Z. Effects of Heavy Metals on Soil Properties and Their Biological Remediation. *Indian J. Pure Appl. Biosci.* **2022**, *10*, 40–46. [CrossRef]
15. Wang, Z.; Huang, W.; Pang, F. Selenium in Soil–Plant-Microbe: A Review. *Bull. Environ. Contam. Toxicol.* **2022**, *108*, 167–181. [CrossRef] [PubMed]
16. Naveed, S.; Oladoye, P.O.; Alli, Y.A. Toxic Heavy Metals: A Bibliographic Review of Risk Assessment, Toxicity, and Phytoremediation Technology. *Sustain. Chem. Environ.* **2023**, *2*, 100018. [CrossRef]
17. Moraru, C. Groundwater Quality in the Republic of Moldova. In *Management of Water Quality in Moldova*; Duca, G., Ed.; Water Science and Technology Library; Springer International Publishing: Cham, Germany, 2014; Volume 69, pp. 177–194. [CrossRef]
18. Hannigan, R.E.; Bogdevich, O.P.; Izmailova, D.N. Selenium in Soils and Groundwater of Moldova. *Environ. Geosci.* **2006**, *13*, 267–279. [CrossRef]
19. World Health Organization. *Guidelines for Drinking-Water Quality: Fourth Edition Incorporating the First Addendum*; World Health Organization: Geneva, Switzerland, 2017.

20. Kapitalchuk, I.; Golubkina, N.; Kapitalchuk, M.; Sheshnitsan, S. Selenium in Soils of Moldova. *J. Environ. Sci. Eng. A* **2014**, *3*, 268–273. [CrossRef]
21. Golubkina, N.; Sheshnitsan, S.; Kapitalchuk, M. Ecological Importance of Insects in Selenium Biogenic Cycling. *Int. J. Ecol.* **2014**, *2014*, 835636.
22. Golubkina, N.A.; Sheshnitsan, S.S.; Kapitalchuk, M.V.; Erdenotsogt, E. Variations of Chemical Element Composition of Bee and Beekeeping Products in Different Taxons of the Biosphere. *Ecol. Indic.* **2016**, *66*, 452–457. [CrossRef]
23. Golubkina, N.A.; Kapitalchuk, M.V.; Sheshnitsan, S.S.; Grishina, T.L.; Kapitalchuk, I.P. Selenium Accumulation by Mushrooms of the Dniester River Valley. *Trace Elem. Med. Mosc.* **2014**, *15*, 19–26.
24. Kapitalchuk, M.V.; Golubkina, N.A.; Kapitalchuk, I.P. Hair Concentrations of Selenium in the Moldovan Population. *Ekol. Cheloveka Hum. Ecol.* **2023**, *30*, 363–373. [CrossRef]
25. Botnaru, V.; Mirlean, N.; Quintana, G.C.R. Informative Eco-Geochemical Assessment of Soil Layer Pollution in Chisinau during the Peak Period of Industrial Activity. *Bull. Inst. Geol. Seismol.* **2022**, *1*, 33–40. [CrossRef]
26. Zubcov, E.; Zubcov, N. The Dynamics of the Content and Migration of Trace Metals in Aquatic Ecosystems of Moldova. *E3S Web Conf.* **2013**, *1*, 32009. [CrossRef]
27. Zinicovscaia, I.; Hramco, C.; Duliu, O.G.; Vergel, K.; Culicov, O.A.; Frontasyeva, M.V.; Duca, G. Air Pollution Study in the Republic of Moldova Using Moss Biomonitoring Technique. *Bull. Environ. Contam. Toxicol.* **2017**, *98*, 262–269. [CrossRef] [PubMed]
28. Zinicovscaia, I.; Duliu, O.; Culicov, O.A.; Frontasyeva, M.; Sturza, R. Major and Trace Elements Distribution in Moldavian Soils. *Romanian Rep. Phys.* **2018**, *70*, 1–10.
29. Zinicovscaia, I.; Sturza, R.; Duliu, O.; Grozdov, D.; Gundorina, S.; Ghendov-Mosanu, A.; Duca, G. Major and Trace Elements in Moldavian Orchard Soil and Fruits: Assessment of Anthropogenic Contamination. *Int. J. Environ. Res. Public Health* **2020**, *17*, 7112. [CrossRef] [PubMed]
30. Kabata-Pendias, A. *Trace Elements in Soils and Plants*, 4th ed.; CRC Press: Boca Raton, FL, USA, 2010. [CrossRef]
31. Antosyak, G. *Atlas of Moldavian SSR*; Main Directorate of Geodesy and Cartography: Moscow, Russia, 1978.
32. Grebenshchikov, V.P.; Grebenshchikova, N.V. Hydrogeological Features of the Territory of Tiraspol. *Sci. Dev. Trends Educ.* **2019**, *56*, 79–82. [CrossRef]
33. Skjemstad, J.; Baldock, J. Total and Organic Carbon. In *Soil Sampling and Methods of Analysis*, 2nd ed.; Carter, M., Gregorich, E., Eds.; CRC Press: Boca Raton, FL, USA, 2007. [CrossRef]
34. Shaimukhametov, M.S. On the Methods of Exchangeable Ca and Mg Determination in Chernozemic Soils. *Eurasian Soil Sci.* **1993**, *12*, 105–111.
35. *ISO 11047:1998*; Soil Quality—Determination of Cadmium, Chromium, Cobalt, Copper, Lead, Manganese, Nickel and Zinc—Flame and Electrothermal Atomic Absorption Spectrometric Methods. German Institute for Standardization: Berlin, Germany, 1998. Available online: https://www.iso.org/standard/24010.html (accessed on 10 September 2023).
36. Gupta, U.; Hettiarachchi, G. Boron, Molybdenum, and Selenium. In *Soil Sampling and Methods of Analysis*, 2nd ed.; Carter, M., Gregorich, E., Eds.; CRC Press: Boca Raton, FL, USA, 2007. [CrossRef]
37. Alfthan, G. A Micromethod for the Determination of Selenium in Tissues and Biological Fluids by Single-Test-Tube Fluorimetry. *Anal. Chim. Acta* **1984**, *165*, 187–194. [CrossRef]
38. Hussain, B.; Abbas, Y.; ur-Rahman, S.; Ali, H.; Zafar, M.; Ali, S.; Ashraf, M.N.; Zehra, Q.; Espinoza, S.T.L.; Valderrama, J.R.D. Metal and Metalloids Speciation, Fractionation, Bioavailability, and Transfer toward Plants. In *Metals Metalloids Soil Plant Water Systems*; Elsevier: Amsterdam, The Netherlands, 2022; pp. 29–50. [CrossRef]
39. Song, T.; Liu, C.; Cui, G.; Tong, S. Research on the Migration and Transformation Behaviors of Soil Selenium in the Flood Irrigation Process. *Arch. Agron. Soil Sci.* **2021**, *67*, 1388–1399. [CrossRef]
40. Jones, G.D.; Winkel, L.H.E. Multi-Scale Factors and Processes Controlling Selenium Distributions in Soils. In *Selenium in Plants*; Pilon-Smits, E.A.H., Winkel, L.H.E., Lin, Z.-Q., Eds.; Plant Ecophysiology; Springer International Publishing: Cham, Germany, 2017; Volume 11, pp. 3–20. [CrossRef]
41. Murphy, D.V.; Stockdale, E.A.; Brookes, P.C.; Goulding, K.W.T. Impact of Microorganisms on Chemical Transformations in Soil. In *Soil Biological Fertility*; Abbott, L.K., Murphy, D.V., Eds.; Springer Netherlands: Dordrecht, The Netherlands, 2004; pp. 37–59. [CrossRef]
42. Jump, R.K.; Sabey, B.R. Soil Test Extractants for Predicting Selenium in Plants. In *SSSA Special Publications*; Jacobs, L.W., Ed.; Soil Science Society of America and American Society of Agronomy: Madison, WI, USA, 2015; pp. 95–105. [CrossRef]
43. Wang, Z.; Gao, Y. Biogeochemical Cycling of Selenium in Chinese Environments. *Appl. Geochem.* **2001**, *16*, 1345–1351. [CrossRef]
44. Tan, J.; Zhu, W.; Wang, W.; Li, R.; Hou, S.; Wang, D.; Yang, L. Selenium in Soil and Endemic Diseases in China. *Sci. Total Environ.* **2002**, *284*, 227–235. [CrossRef]
45. Kiriluk, V.P. *Trace Elements in the Components of the Biosphere of Moldova*; Pontos: Chișinău, Moldova, 2006.
46. Yamada, H.; Hattori, T. Forms of Soluble Selenium in Soil. *Soil Sci. Plant Nutr.* **1989**, *35*, 553–563. [CrossRef]
47. Čuvardić, M.S. Selenium in Soil. *Proc. Nat. Sci.* **2003**, *104*, 23–27.
48. Gondi, F.; Pantó, G.; Fehér, J.; Bogye, G.; Alfthan, G. Selenium in Hungary: The Rock-Soil-Human System. *Biol. Trace Elem. Res.* **1992**, *35*, 299–306. [CrossRef] [PubMed]

49. Kapitalchuk, I.P.; Sheshnitsan, T.L.; Sheshnitsan, S.S.; Kapitalchuk, M.V. Migration of Manganese, Zinc, Copper and Molybdenum in Landscape-Geochemical Catena of the Lower Dniester Valley. *South Russ. Ecol. Dev.* **2018**, *13*, 96–112. [CrossRef]
50. Sapozhnikova, Y.; Zubcov, N.; Hungerford, S.; Roy, L.A.; Boicenco, N.; Zubcov, E.; Schlenk, D. Evaluation of Pesticides and Metals in Fish of the Dniester River, Moldova. *Chemosphere* **2005**, *60*, 196–205. [CrossRef] [PubMed]
51. Udeigwe, T.K.; Eichmann, M.; Eze, P.N.; Ogendi, G.M.; Morris, M.N.; Riley, M.R. Copper Micronutrient Fixation Kinetics and Interactions with Soil Constituents in Semi-Arid Alkaline Soils. *Soil Sci. Plant Nutr.* **2016**, *62*, 289–296. [CrossRef]
52. Yuan, D.; Lian, W.; Lianggang, Z.; Dezhi, C. Expression of Copper and Cadmium Plant Availability in Soils. *Environ. Sci. Technol.* **2010**, *33*, 27–30.
53. Vázquez-Blanco, R.; Nóvoa-Muñoz, J.C.; Arias-Estévez, M.; Fernández-Calviño, D.; Pérez-Rodríguez, P. Changes in Cu Accumulation and Fractionation along Soil Depth in Acid Soils of Vineyards and Abandoned Vineyards (Now Forests). *Agric. Ecosyst. Environ.* **2022**, *339*, 108146. [CrossRef]
54. Li, X.; Zhang, J.; Gong, Y.; Liu, Q.; Yang, S.; Ma, J.; Zhao, L.; Hou, H. Status of Copper Accumulation in Agricultural Soils across China (1985–2016). *Chemosphere* **2020**, *244*, 125516. [CrossRef]
55. White, P.J. Selenium Accumulation by Plants. *Ann. Bot.* **2016**, *117*, 217–235. [CrossRef]
56. Soylu, M.E.; Bras, R.L. Dataset on the Global Distribution of Shallow Groundwater. *Data Brief* **2023**, *47*, 108973. [CrossRef] [PubMed]
57. Wang, S.; Liang, D.; Wang, D.; Wei, W.; Fu, D.; Lin, Z. Selenium Fractionation and Speciation in Agriculture Soils and Accumulation in Corn (*Zea mays* L.) under Field Conditions in Shaanxi Province, China. *Sci. Total Environ.* **2012**, *427–428*, 159–164. [CrossRef] [PubMed]
58. Eliopoulos, G.D.; Eliopoulos, I.-P.D.; Tsioubri, M.; Economou-Eliopoulos, M. Distribution of Selenium in the Soil–Plant–Groundwater System: Factors Controlling Its Bio-Accumulation. *Minerals* **2020**, *10*, 795. [CrossRef]
59. Hasanuzzaman, M.; Hawrylak-Nowak, B.; Islam, T.; Fujita, M. (Eds.) *Biostimulants for Crop Production and Sustainable Agriculture*; CABI: Wallingford, UK, 2022. [CrossRef]
60. Rahman, S.U.; Wang, X.; Shahzad, M.; Bashir, O.; Li, Y.; Cheng, H. A Review of the Influence of Nanoparticles on the Physiological and Biochemical Attributes of Plants with a Focus on the Absorption and Translocation of Toxic Trace Elements. *Environ. Pollut.* **2022**, *310*, 119916. [CrossRef]
61. Ferreira, G.D.S.; De Brito, P.O.B.; Lima, T.D.A.; Aderaldo, F.Í.C.; De Carvalho, G.T.; De Sousa Filho, E.D.N.; Gondim, F.A. Assessment of the Effects of Selenium Application on Leaves or Substrate on the Growth of Sunflower Plants: Avaliação Dos Efeitos Da Aplicação de Selênio Nas Folhas Ou No Substrato Sobre o Crescimento de Plantas de Girassol. *Braz. J. Anim. Environ. Res.* **2022**, *5*, 3972–3982. [CrossRef]

soil systems

MDPI

Article

Impact of Pot Farming on Plant-Parasitic Nematode Control

Silvia Landi [1,*], Beatrice Carletti [2], Francesco Binazzi [1], Sonia Cacini [3], Beatrice Nesi [3], Emilio Resta [4], Pio Federico Roversi [1] and Sauro Simoni [1]

1 Research Centre for Plant Protection and Certification, Council for Agricultural Research and Economics, Via di Lanciola 12A, 51025 Firenze, FI, Italy; francesco.binazzi@crea.gov.it (F.B.); piofederico.roversi@crea.gov.it (P.F.R.); sauro.simoni@crea.gov.it (S.S.)
2 Pro Plantis—Laboratorio di Nematologia Fitopatologica e Diagnostica Fitopatologica su Insetti, Via Nazionale 9, 52018 Castel San Niccolò, AR, Italy; beacarletti@libero.it
3 Research Centre for Vegetables and Ornamental Crops, Council for Agricultural Research and Economics, Via dei Fiori 8, 51017 Pescia, PT, Italy; sonia.cacini@crea.gov.it (S.C.); beatrice.nesi@crea.gov.it (B.N.)
4 Associazione Vivaisti Italiani, Via G. Deledda 30/7, 51100 Pistoia, PT, Italy; dr.emilioresta@gmail.com
* Correspondence: silvia.landi@crea.gov.it

Abstract: In the Pistoia Nursery-Ornamental Rural District (Italy), a leader in Europe in ornamental nurseries covering over 5200 hectares with over 2500 different species of plant, plant-parasitic nematodes represent a serious concern. The potential efficacy of a pot cultivation system using commercial substrates to control plant-parasitic nematodes was assessed. On two different plant species, two different pot cultivation managements, potted plants, and potted plants previously cultivated in natural soil were compared to plants only cultivated in natural soil. The entire soil nematode structure with and without plants was evaluated. The relationship between soil properties and soil nematode community was investigated. All the studied substrates were free from plant-parasitic nematodes. Regarding free-living nematodes, Peat–Pumice showed nematode assemblage established by colonizer and extreme colonizer bacterial feeders, whereas Peat–Perlite included both bacterial and fungal feeders, and, finally, coconut fiber also included omnivores and predators. In farming, the substrates rich in organic matter such as coconut fiber could still play an important role in suppressing plant-parasitic nematodes because of the abundance of free-living nematodes. In fact, they are of crucial importance in both the mineralization of organic matter and the antagonistic control of plant-parasitic nematodes. Potting systems equally reduce virus-vector nematodes and improve the prey/predator ratio favoring natural control.

Keywords: soil nematode community; soil nematode indicators; Pistoia Nursery-Ornamental District; *Acer palmatum*; x *Cupressocyparis leylandii*

1. Introduction

The Pistoia Nursery-Ornamental Rural District (Italy) is a leader in Europe in ornamental nurseries covering over 5200 hectares, with approximately 1000 hectares of potting plants, 1500 companies, and over 5500 employees. In total, the Gross Saleable Product (GSP) is over 700 million Euros of which 360 is related to export (www.cespevi.it; https://group.intesasanpaolo.com/it/research/monitor-dei-distretti-agroalimentari; accessed on 11 January 2024). Over 2500 plant species are cultivated and plant-parasitic nematodes characterized by high polyphagies represent a serious concern. Even though extensive research has been devoted to the investigation of the effect and management of plant-parasitic nematodes in crops, the effect of these pests on the ornamental plant industry remains a relatively understudied field [1]. There are many categories of plant-parasitic nematodes that affect ornamental plants, with the main genera being *Meloidogyne*, *Aphelenchoides*, *Paratylenchus*, *Pratylenchus*, *Helicotylenchus*, *Radopholus*, *Xiphinema*, *Trichodorus*, *Paratrichodorus*, *Rotylenchulus*, and *Longidorus*.

The majority of states in the world have implemented measures to contrast the spread of plant pests by enforcing strict controls on the import/export of plant materials. The new European legislation on protective measures against plant pests (Regulation EU 2016/2031) imposes a "prohibition on the introduction and movement of EU regulated non-quarantine pests on plants for planting". The nurserymen as "professional operators shall not introduce a Union regulated non-quarantine pest into, or move that pest within, the EU territory on the plants for planting through which it is transmitted" (art. 37 Reg. EU 2016/2031). A list of eighteen nematode species belonging to the genera *Ditylenchus, Tylenchulus, Aphelenchoides, Heterodera, Meloidogyne, Pratylenchus, Longidorus*, and *Xiphinema* was established in Annex IV of Regulation (EU) 2019/2072. These plant-parasitic species, already present and widespread in the European Union, represent a phytosanitary risk and potentially determine a severe economic, social, and environmental impact in the EU territory. It is worth remembering that, for these nematodes, the accepted damage tolerance threshold is "zero". It is worth noting also that in the Pistoia Nursery-Ornamental Rural District (Italy), the application of the EU legislation represents a serious threat to the plant industry because, concurrently, the 91/414 EEC Directive also imposed an important reduction in the number of plant protection products suitable for marketing. For this reason, a progressive elimination of many pesticides is still ongoing and, thus, the control of plant-parasitic nematodes is becoming more difficult.

In this context, it is crucial to understand the role of different commercial substrates or natural soil in reducing the risk of plant-parasitic nematode infestation in the nursery. Using healthy commercial substrates or soil is critical for maintaining low infestation levels of plant-parasitic nematodes. Furthermore, this "exclusion approach" contributes to the prevention of accidental pest introduction.

At the same time, soil and commercial substrates may also host many free-living nematodes playing a significant role in controlling plant-parasitic nematodes [2,3]. Several studies have shown that populations of these nematodes decrease with the increase in saprophytic nematodes (bacterial and fungal feeders) [4,5]. Moreover, soil predator nematodes deserve consideration for their potential activity in pest-regulation services [3].

To date, few studies have been conducted to evaluate the impact of pot cultivation systems of ornamental plants on soil plant-parasitic nematodes. The use of solid substrates such as peat, coir, bark, sawdust, compost, rockwool, perlite, pumice, sand, and vermiculite is considered a valid soil management aimed at controlling plant-parasitic nematodes [6]. Short-term trials on tomatoes demonstrated the absence of plant-parasitic nematodes in pots with coconut and peat substrates [2]. However, other authors asserted that soilless culture systems are not effective in the elimination of these noxious organisms [4,7]. Moreover, plant-parasitic nematodes can be introduced by infested propagation materials, from farm appliances, and by irrigation [8,9].

The aim of this study was to assess the potential efficacy of pot cultivation systems using commercial substrates to avoid plant-parasitic nematode infestations and simultaneously increase free-living nematode populations. Two different pot cultivation managements, potted plants, and potted plants previously cultivated in natural soil, were compared to plants only cultivated in natural soil. In detail, the effect of the pot cultivation system was evaluated on (i) the entire soil nematode structure, with and without plants, in *Acer palmatum* and x *Cupressocyparis leylandii* cultivations and (ii) the relationship between soil properties (i.e., physical and chemical) and soil nematode community.

2. Materials and Methods

2.1. Field Site

The experimental site was located at "Vannucci Piante" farm, in the Pistoia district (Central Italy; 43.85096 N, 10.98172 E; 55 m a.s.l.), one of the most important ornamental nursery areas in Europe. The local climate was classified as Cfa (temperate, no dry season, hot summer) by Köppen Climate Types, characterized by a mean air temperature of 15.5 °C and a mean annual precipitation of 1120 mm throughout the two years of the study from 2020 to 2021 (Regione Toscana–Settore Idrologico e Geologico Regionale http://www.sir.toscana.it, accessed on 11 January 2024). The average minimum temperature was 2.5 °C in January 2020, and the average maximum temperatures were 24.2 °C and 26.7 °C in July and August, respectively; the precipitation was concentrated in winter, particularly in December, and it ranged from 261.6 to 326.8 mm in the three years of study.

The plants used in the trial, *A. palmatum* and x *C. leylandii*, were cultivated under (i) open-air conditions in natural soil and (ii) shaded structures in pots (18 liters/potted plant) with the standard coconut-fiber substrate.

As regards plant cultivation in the pot, *A. palmatum* and x *C. leylandii* seedlings or sprouted vegetative propagation material were initially planted in Peat–Perlite (75:25 $v\ v^{-1}$) (for 3–4 months), then transplanted into Peat–Pumice (60:40 $v\ v^{-1}$) (for 18–24 months), and finally into coconut fiber (coir dust–coir fiber 70:30 $v\ v^{-1}$) for each of the following transplants. All these commercial substrates were "added" with controlled released fertilizers. Concerning commercialization, plants cultivated in natural soil were used to be potted and then left in the nursery for a variable time until marketing. Irrigation and fertigation were the same for all plots; the same phytosanitary treatments were applied, even though no chemical was used to control plant-parasitic nematodes. Specifically, for both plant species, Universol Blue ICL fertilizer (18-11-18 for N, P_2O_5, and K_2O, respectively, plus 2.5 MgO and microelements: B 0.01%, Cu 0.01%, Fe 0.1%, Mn 0.04%, Mo 0.001%, Zn 0.01%) were used (1 g/L twice a week). As regards phytosanitary treatments, one treatment with Sulphoxaflor and another with Acetamiprid were used against aphids on *A. palmatum*, whereas 2–3 treatments with copper oxychloride were applied against fungal pathogens on *C. leylandii*.

2.2. Experimental Design and Soil Sampling

First trial: substrates in the absence of plants. In spring 2020, three different substrates, Peat–Perlite (PPE), Peat–Pumice (PPU), and coconut fiber (CF), were compared, before their use, to evaluate soil nematode community structures in the absence of plants. Moreover, as regards coconut fiber (the substrate used for the longest period in the trial), possible changes in the soil nematode community were evaluated at intervals of 0, 10, 30, 60, and 90 days throughout the time in which this material remained in the transplant area of the nursery before usage. To characterize the soil nematode community, eighteen soil samples were collected for each substrate and at each time interval.

Second trial: substrates with a plant. Young plants of *A. palmatum* and x *C. leylandii* were selected for the experiments. The absence of plant-parasitic nematodes was confirmed by previous sampling and analysis of the roots of these plants following the EPPO protocol [10]. For each plant species, potted plants (PP) and potted plants previously cultivated in natural soil (PNS) were compared to plants only cultivated in natural soil (NS). A total of fifty plants per plant species and per treatment was used. In 2020 and 2021, three samples of soil were withdrawn in each season, with the purpose of determining soil nematode structure (i.e., 3 samples/treatment, for each season, for 2 years, for a total of 72 samples). Once surface residues were removed, a hand auger was used to carry out each sampling (5 cm diameter inside) from the 30 cm deep top layer of bare soil. For every sample of soil, six cores were randomly withdrawn and subsequently mixed in order to form a single composite sample.

2.3. Soil Physical and Chemical Analysis

The physical and chemical properties of representative soil profiles and the commercial substrates reported in the field site were determined. Regarding soil, considered parameters were texture, according to the International Society of Soil Science (ISSS) standards, pH and EC (1:2.5 and 1:5 water extraction method, respectively), total Kjeldahl nitrogen, N-NO_3, N-NH_4, P and K content, organic matter percentage, C/N ratio, and active $CaCO_3$. Physical characterization of commercial substrates included the water volume for the determination of available water capacity, the air-filled porosity, the holding water capacity (W −1 kPa), the bulk density (BD), and the total porosity (TP), determined as described by De Boodt et al. [11]. Water tensions at 0, −1, −2, −3, −5, and −10 kPa were used to elaborate the water retention curve. Then, pH was measured according to the EN 13037/1999 method, while EC according to the EN 13038/1999.

2.4. Analysis of Soil Nematode Community

The cotton-wood filter extraction method was used to isolate nematodes from 100 mL of each soil sample. Extraction was performed at room temperature for 48 h at approximately 20 °C. A 25 μm mesh was used to sieve each nematode suspension and nematodes were then counted by a stereomicroscope (50× magnification). Temporary slides were made for all specimens and their identification to genus level was carried out at higher magnification using the keys of Mai and Lyon [12], Bongers [13], and Marinari-Palmisano and Vinciguerra [14]. Taxonomic families were assigned to trophic groups based on Yeates et al. [15] and Okada et al. [16]. Nematode communities were characterized adopting the following criteria: (i) total abundance of individuals; (ii) richness determined by counting the number of taxa; (iii) Maturity (MI) and Plant Parasitic indices (PPI) determined following Bongers [17] and the food web indicators (BI, Basal Index; EI, Enrichment Index; SI, Structure Index; CI, Channel Index) following Ferris et al. [18]; (iv) diversity-weighted abundance (θ) expressed as biomass [19] and categorizing populations of soil nematodes on a functional basis into plant-parasitic nematodes, predators (including omnivores), and detritivores (bacterial and fungal feeders), in accordance with Ferris and Tuomisto [20] for evaluating the eco-system services efficiency; (v) prey-to-predator θ mass ratio aimed at evaluating regulation function for evaluating the ecosystem services efficiency.

2.5. Statistical Analysis

One-way ANOVA was performed to assess the influence of substrate on nematode taxa abundance and indicators of nematode community structure (first trial). Two-way ANOVA was carried out to assess management and plant species effects on nematode taxa abundance and nematode indicators in the second trial. When the F-test was significant at $p < 0.05$, mean treatments were compared using the Student–Newman–Keuls test (SNK) using CoStat Statistical Software 6.4 2021. Moreover, the comparison of nematode communities was performed by the multi-variate methods of the past analysis package [21]. Nematode communities were compared using analysis of similarity (ANOSIM), and SIMPER analysis based on the Bray–Curtis similarity index, nearest neighbor [22]. The nematode abundance data were square root transformed before the analysis. Bonferroni correction was then applied. Canonical Correspondence Analysis (CCA) was performed in order to evaluate the interactions between the communities of nematodes (abundance of nematode taxa and its indicators) and both soil chemical and physical variables (EC, soil pH, and TOC). After the analysis, we selected only the significant environmental axes, which were then graphically represented by vectors. Finally, we assessed the statistical significance of the relationship between environmental and community variables by the permutation test of both the first ordination, axis as well as the combination of both the first and second axes.

3. Results

3.1. Soil Physical and Chemical Properties

Natural soil was classified as loamy sand according to the USDA Soil Taxonomy. The electrical conductivity (EC) values ranged from 448 μS cm^{-1} (1:2 dry material/water extract) in the x *C. leylandii* plot to 2680 μS cm^{-1} in the *A. palmatum* plot, and the soil pH values were 4.7 and 6.8 for *A. palmatum* and x *C. leylandii* areas, respectively. The organic matter content was always less than 2%, which is considered the soil quality critical threshold in temperate regions [23]. Specifically, the natural soils, in which *A. palmatum* and *C. leylandii* were cultivated, had an organic matter content of 1.4% and 0.72%, respectively.

Regarding the substrate's main chemical–physical characterization, data are shown in Table 1, while the retentions curve did not show remarkable trends with respect to the recognized optimal range [24,25].

Table 1. Chemical–physical characterization of the used substrates for potted plant cultivation.

Substrate	EC $\mu S\ cm^{-1}$	pH	BD $g\ cm^{-3}$	TP %	W −1 kPa %	AFP %	AWC %
Peat/Perlite	191.2 ± 10.7 c	5.67 ± 0.12 a	0.12 ± 0.002 b	93.3 ± 0.16 a	58.52 ± 1.07 a	35.73 ± 1.07	28.5 ± 0.27 a
Peat/Pumice	618.3 ± 74.5 b	4.81 ± 0.04 b	0.25 ± 0.002 a	86.8 ± 0.12 c	52.54 ± 1.52 b	34.30 ± 1.51	24.1 ± 0.99 b
Coconut fiber	1545.0 ± 120.5 a	4.00 ± 0.17 c	0.11 ± 0.001 c	92.4 ± 0.09 b	56.23 ± 3.17 ab	36.16 ± 3.16	25.8 ± 2.44 a
ANOVA	***	***	***	***	***	n.s.	**
Acceptable range [1]	<500	5.5–6.5	<0.40	>85	55–70	20–30	20–30

EC: electrical conductivity, BD: bulk density, TP: total porosity, W −1 kPa: water holding capacity, AFP: air-filled porosity, AWC: available water content. Data are reported as mean ± standard deviation (n = 3). ANOVA analysis: n.s. = not significant, *, **, *** = significant $0.01 < p < 0.05$, $0.001 < p < 0.01$ and $p < 0.001$, respectively; different letters for the same parameter indicate significantly different values (STD test, $p < 0.05$). [1] Abad et al., 2001 [24].

3.2. First Trial: Substrates in Absence of Plants

Ten families of free-living nematodes were identified in soil samples from three different commercial substrates. No plant-parasitic nematode family was found. Most families were consistently found in CF and, to a lesser extent, in PPE. Only two families were present in PPU. One-way ANOSIM analysis on nematode abundance showed significant differences in nematode taxa abundance (R 0.61 and $p = 0.0005$). The R values, calculated for substrate CF-PPE and CF-PPU pairwise comparison, were 0.52 ($p = 0.05$) and 0.57 ($p = 0.001$), respectively, indicating that the coconut fiber substrate was different from the other ones. No significant difference was found between PPE-PPU.

The SIMPER analysis showed 70.36% of the overall dissimilarity and family breakdown of similarity indicating that eight families accounted for 95% of this similarity (Table 2). Differences were mainly due to a higher proportion of Neotylenchidae, Aphelenchoidae, and Dorylaimidae. Following the one-way ANOVA analysis, the abundance of individuals belonging to the Dorylaimidae was significantly higher in CF than in other substrates. In general, the fungal feeder nematodes were more abundant in PPE and, albeit to a lesser extent, in CF: specifically, Neotylenchidae and Anguinidae (mainly *Ditylenchus myceliophagus*) were more numerous in PPE than in CF and PPU; and the abundance of Aphelenchoidae (mainly *Aphelenchoides composticola*, *Bursaphelenchus fungivorus*, and *Ektaphelenchoides* sp.) was higher in PPE and CF than PPU. In PPU, only bacterial feeder nematodes were found.

Table 2. Percentage contribution to the Bray–Curtis dissimilarity in family nematode abundance among the different substrates (SIMPER analysis). Mean values and standard errors are reported. Different letters for the same parameter indicate significantly different values (SNK test, $p < 0.05$).

Taxon	Contribution	Cumulative %	Peat–Perlite	Peat–Pumice	Coconut Fiber	*p* Value
Neotylenchidae	8.828	21.13	70.0 ± 0 a	0 b	13.4 ± 7.46 b	0.0008
Aphelenchoidae	6.945	37.75	30.0 ± 0 a	0 b	25.0 ± 5.77 a	0.04
Dorylaimidae	6.225	52.65	0 b	0 b	18.0 ± 2.92 a	0.002
Rhabditidae	5.014	64.64	40.0 ± 2.89 b	100.0 ± 5.77 a	52.7 ± 8.20 b	0.007
Cephalobidae	4.976	76.55	40.0 ± 2.89 b	100.0 ± 5.77 a	50.2 ± 7.67 b	0.004
Anguinidae	4.554	87.45	20.0 ± 0 a	0 c	0 c	0.04
Tylenchidae	3.058	94.77	0 a	0 a	8.9 ± 4.15 a	0.29
Seinuridae	1.127	97.47	0 a	0 a	2.2 ± 1.47 a	0.52
Mononchidae	0.6611	99.05	0 a	0 a	1.4 ± 1.40 a	0.75
Diplogasteridae	0.3972	100	0 a	0 a	0.3 ± 0.30 a	0.75
Total abundance			200 ± 5.00 a	200 ± 0 a	172.2 ± 24.36 a	0.69
Overall average dissimilarity	70.36					

The evaluation of coconut fiber throughout the time in which this material remained in the transplant area (at 10, 30, and 60 days before usage) showed a significant increase in omnivores, predators, and fungal feeder nematodes (Figure 1). Bacterial feeder nematodes remained the dominant trophic group, even though no significant evidence emerged.

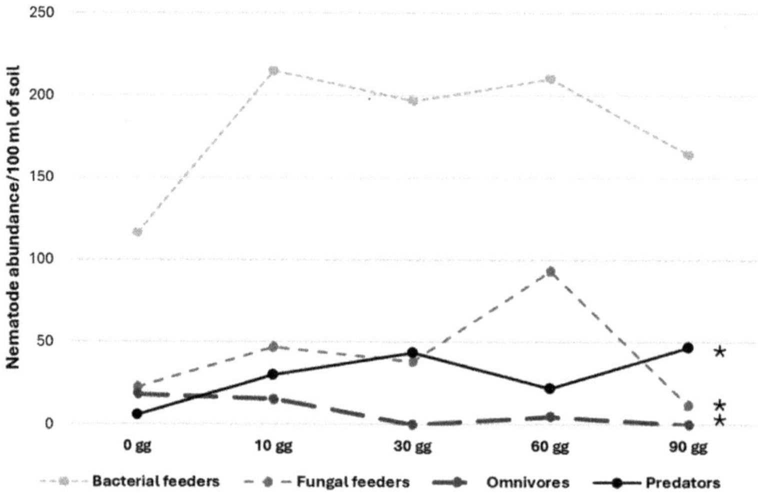

Figure 1. Time trends of different nematode trophic groups in coconut fiber placed in the transplant area before usage. ANOVA analysis * = significant $0.01 < p < 0.05$.

In Table 3, MI values ranged from 1.5 to 1.9, indicating the conspicuous presence of generalist and opportunistic species; the significantly highest values were found in CF and PPE. EI was higher in PPU and CF, while SI was significantly higher in CF than in other substrates. Finally, CI showed values lower than 50 in all types of substrates, and the highest values were found in PPE.

Table 3. Soil nematode indices refer to different commercial substrates. Different letters for the same parameter indicate significantly different values (SNK test, $p < 0.05$).

Indices	Peat–Perlite	Peat–Pumice	Coconut Fiber	*p* Value
Maturity Index (MI)	1.8 ± 0.01 a	1.5 ± 0.03 b	1.9 ± 0.05 a	0.0007
Basal Index (BI)	126.7 ± 1.33	80.0 ± 4.62	78.0 ± 14.27	0.15
Enrichment Index (EI)	50.1 ± 1.56 b	79.7 ± 1.16 a	68.2 ± 3.52 a	0.004
Structure Index (SI)	0 b	0 b	39.4 ± 6.66 a	0.002
Channel Index (CI)	43.0 ± 1.79 a	0 b	17.7 ± 5.17 b	0.005
Richness	5.0 ± 0 a	2.0 ± 0 b	5.4 ± 0.18 a	0.00001

The CCA conducted between nematode taxa abundance and commercial substrate variables (organic matter content (OM), EC, and pH), evidenced that axis 1 was dominated by EC (−0.63) and pH (0.63), while axis 2 was dominated by OM (0.77). Tylenchidae, Diplogasteridae, Seinuridae, Dorylaimidae, and Mononchidae families were positively associated with organic matter content while EC Neotylenchidae and Anguinidae were positively associated with pH. The bacterial feeder families of Rhabditidae and Cephalobidae were poorly affected by soil parameters (Figure 2A).

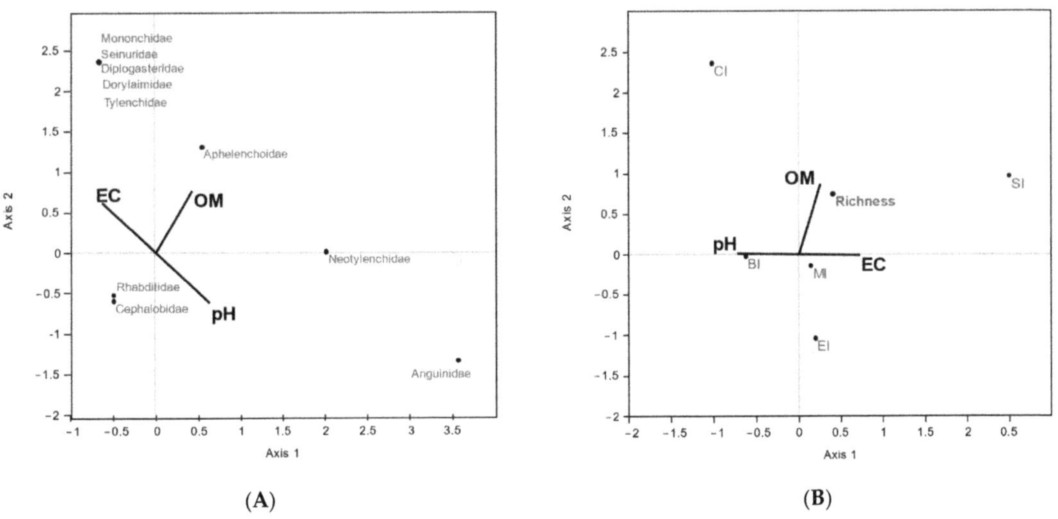

Figure 2. Scatter plot of CCA ordination showing relationships between soil properties and nematode taxa abundance (**A**) and soil nematode indicators (**B**). (**A**) Percentages of variance were 69.27% ($p < 0.002$) for axis 1 and 30.73% ($p < 0.001$) for axis 2; (**B**) percentages of variance were 62.65% ($p < 0.01$) for axis 1 and 37.35% ($p < 0.002$) for axis 2.

The biplot of CCA including soil nematode indicators and the OM, EC, and pH environmental variables showed axis 1 dominated by EC (0.73) and pH (−0.73) and axis 2 by OM (0.87). SI was positively influenced by EC and organic matter content, and, instead, CI was positively correlated with organic matter and to a lesser extent with pH (Figure 2B).

3.3. Second Trial: Substrates with Plants

Seventeen plant-parasitic and free-living nematode families were identified in soil samples collected. In detail, 16 and 15 families were found on A. palmatum and x *C. leylandii*, respectively. Regarding management, consistent differences in the number of taxa were found: the lowest number of families was in PP (7 families), especially for plant-parasitic nematodes (only one family), compared to PNS (16 families) and NS (15 families). In general, ANOSIM analysis on nematode abundance confirmed the low difference between the two plant species (R 0.23, $p = 0.0001$) and higher differences per soil management (R 0.36, $p = 0.0001$). Pot conditions were different from cultivation in natural soil: R values in pairwise comparison were for the management PP-PNS 0.44 ($p = 0.002$); for PP-NS, 0.53 ($p = 0.0003$); and for PNS-NS 0.31 ($p = 0.001$). The analysis of the contribution of family abundance to the average Bray–Curtis dissimilarity, among managements using SIMPER, was 57.27% of the overall dissimilarity; the family breakdown of similarity showed that 15 families accounted for 95% of this similarity. Differences were mainly due to a higher proportion of Rhabditidae, Cephalobidae, and Dorylaimidae in PNS than in the PP and NS.

Two-way ANOVA for plant factor further evidenced that Diphterophoridae, Telotylenchidae (mainly *Trophurus* spp.), Heteroderidae, and Hoplolaimidae (mainly Rotylenchus spp. and to a lesser extent *Helicotylenchus* spp. and *Scutellonema* spp.) were significantly higher in *A. palmatum* than x *C. leylandii*, while Aphelenchoidae, Tylenchidae, Psilenchidae, Pratylenchidae, and Trichodoridae were significantly more abundant in x *C. leylandii* than in *A. palmatum*. As regards management, Seinuridae, Tylenchidae, Psilenchidae, and Pratylenchidae (mainly *Pratylenchus vulnus*) were significantly higher in PNS than others (Table 4). The virus-vector plant-parasitic nematodes Trichodoridae and Longidoridae showed the highest abundance in NS. Moreover, significant interactions between plant and management factors were found.

Table 4. Effects of different managements in *A. palmatum* and x *C. leylandii* on the abundance of nematode taxa (mean number ± SE of nematodes/100 mL soil). Samples were collected from PP, potted plants; PNS, potted plants previously cultivated in natural soil; NS, plants only cultivated in natural soil. Different letters for the same parameter indicate significantly different values (SNK test, $p < 0.05$).

	Management			Plant		Significant Effects (p Value)		
	PP	PNS	NS	*Acer palmatum*	*Cupress. leylandii*	M	P	M + P
Rhabditidae	4.2 ± 1.42	47.0 ± 11.11	7.4 ± 1.94	31.4 ± 14.55	43.3 ± 7.67	0.15	0.70	0.69
Cephalobidae	5.2 ± 2.98	43.8 ± 8.79	9.8 ± 2.09	27.0 ± 11.05	44.2 ± 7.20	0.11	0.35	0.55
Aphelenchidae	0	1.3 ± 0.44	2.8 ± 0.95	1.2 ± 0.36	1.7 ± 0.71	0.19	0.69	0.01
Aphelenchoidae	0	1.3 ± 0.38	0	0.15 ± 0.08 b	1.9 ± 0.61	0.23	0.003	0.16
Anguinidae	3.5 ± 0.50	2.4 ± 0.79	0	3.2 ± 1.04	0.9 ± 0.40	0.31	0.08	0.41
Diphterophoridae	0	1.5 ± 0.72	2.5 ± 1.44	2.6 ± 1.05 a	0.4 ± 0.26 b	0.32	0.03	0.06
Dorylaimidae	1.7 ± 1.12	16.6 ± 4.46	4.6 ± 1.08	11.9 ± 5.67	15.2 ± 3.64	0.29	0.84	0.50
Mononchidae	0	4.1 ± 1.02	4.6 ± 1.08	3.8 ± 0.84	3.8 ± 1.46	0.32	0.66	0.61
Seinuridae	2.3 ± 1.17 ab	7.2 ± 0.99 a	0 b	6.5 ± 1.06	4.6 ± 1.37	0.001	0.08	0.42
Tylenchidae	5.3 ± 0.33 b	23.9 ± 2.26 a	12.9 ± 1.44 b	10.8 ± 0.86 a	32.1 ± 2.86 b	0.00001	0.00001	0.00001
Telotylenchidae	2.5 ± 1.23 b	0 c	7.1 ± 2.92 a	2.6 ± 1.02 a	0 b	0.00001	0.0006	0.00001
Psilenchidae	0 b	3.2 ± 0.77 a	0 b	0 b	5.3 ± 1.15 a	0.006	0.00001	0.02
Heteroderidae	0	0.0 ± 0.07	0.2 ± 0.17	0.2 ± 0.11 a	0 b	0.51	0.05	0.72
Pratylenchidae	0 b	17.3 ± 2.68 a	1.4 ± 0.84 b	3.2 ± 0.94 b	24.9 ± 3.76 a	0.002	0.00001	0.01
Hoplolaimidae	0	32.2 ± 18.12	8.6 ± 3.90	44.4 ± 24.92	3.4 ± 0.75	0.49	0.09	0.60
Trichodoridae	0 b	0.1 ± 0.07 b	3.9 ± 1.62 a	0.2 ± 0.11 b	1.4 ± 0.67 a	0.00001	0.01	0.00001
Longidoridae	0 b	0.06 ± 0.04 b	4.3 ± 1.70 a	1.1 ± 0.57	0.3 ± 0.17	0.00001	0.06	0.00001
Total abundance	24.7 ± 4.92	202.1 ± 41.08	69.7 ± 10.59	150.1 ± 55.25	183.1 ± 24.30	0.15	0.88	0.60
Richness	5.5 ± 0.22 b	6.9 ± 0.17 a	7.3 ± 0.58 a	6.8 ± 0.24	6.9 ± 0.23	0.01	0.68	0.00001

The soil nematode indicators showed few differences between the two plant species: BI was significantly higher in *A. palmatum* than x *C. leylandii*, while an opposite result was recorded for SI Table 5. On the contrary, many differences were found per management. Soil nematode communities in natural soil showed the highest MI, PPI, and CI values, while BI and EI values were significantly higher in PNS. As expected, the plants only cultivated in pots using coconut fiber substrate exhibited the lowest MI, PPI, and BI values.

Table 5. Effects of different managements in *A. palmatum* and *C. leylandii* on soil nematode indices (±SE). PP, potted plants; PNS, potted plants previously cultivated in natural soil; NS, plants only cultivated in natural soil. Different letters for the same parameter indicate significantly different values (SNK test, $p < 0.05$).

	Management			Plant		Significant Effects (p Value)		
	PP	PNS	NS	*Acer palmatum*	*Cupress. leylandii*	M	P	M + P
MI	1.9 ± 0.12 b	2.0 ± 0.03 b	2.2 ± 0.10 a	2.0 ± 0.05	1.9 ± 0.03	0.05	0.13	0.04
PPI	1.5 ± 0.67 c	2.6 ± 0.14 b	3.6 ± 0.11 a	2.7 ± 0.19	2.7 ± 0.19	0.0003	0.49	0.27
BI	10.5 ± 1.94 b	58.7 ± 7.51 a	21.7 ± 2.63 b	34.8 ± 8.75 b	64.7 ± 7.40 a	0.02	0.03	0.11
EI	51.5 ± 5.10 ab	62.6 ± 1.93 a	46.1 ± 4.63 b	59.4 ± 2.12	58.4 ± 3.22	0.001	0.50	0.33
SI	40.7 ± 7.06	45.2 ± 2.82	45.2 ± 7.33	50.1 ± 2.96 a	38.6 ± 3.97 b	0.47	0.01	0.18
CI	39.7 ± 7.54 ab	25.6 ± 2.49 b	45.5 ± 5.55 a	30.5 ± 2.95	29.7 ± 3.84	0.003	0.84	0.40

The average values of the diversity-weighted abundance (θ) index are reported in Figure 3. Significant differences were found only in the plant-parasitic channel: the highest value was reported in NS. The regulation functions of opportunistic and plant-parasitic nematodes by predation were greater in PP and to a lesser extent in PNS; they ranged from 1 to 1.1 and from 1 to 1.47 for PP and PNS, respectively. By contrast in NS, the high θ value of the plant-parasitic channel together with the low θ value in the predator channel, caused a predator/prey ratio of 1:5.74 indicating an insufficient regulation. No differences were found between the two different plant species: the predator/prey ratio ranged from 1:1.63 to 1:1.69 for *A. palmatum* and x *C. leylandii*, respectively.

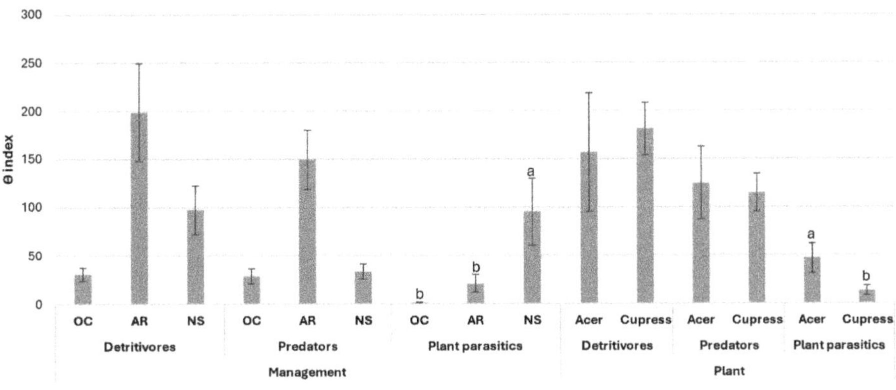

Figure 3. Diversity-weighted abundance (θ) index for functional classes of soil nematode assemblage. PP, potted plants; PNS, potted plants previously cultivated in natural soil; NS, plants only cultivated in natural soil. Standard errors are reported. Different letters are significant differences at $p < 0.05$.

The CCA, conducted between nematode taxa abundance from PP and NS and soil variables (organic content, soil pH, and EC) showed that axis 1 was dominated by EC (0.85) and pH (−0.89), while axis 2 was by OM (0.89). The families of Seinuridae and Anguinidae were positively associated with OM; instead, the families of Trichodoridae and Pratylenchidae were related to pH. The biplot of CCA conducted between soil nematode indicators and the same soil variable showed that axis 2 was dominated by OM (0.48 and pH −0.46); the PPI, which plotted furthest from the origin and so varied the most within this environmental gradient, was positively associated with the pH and inversely with OM (Figure 4A).

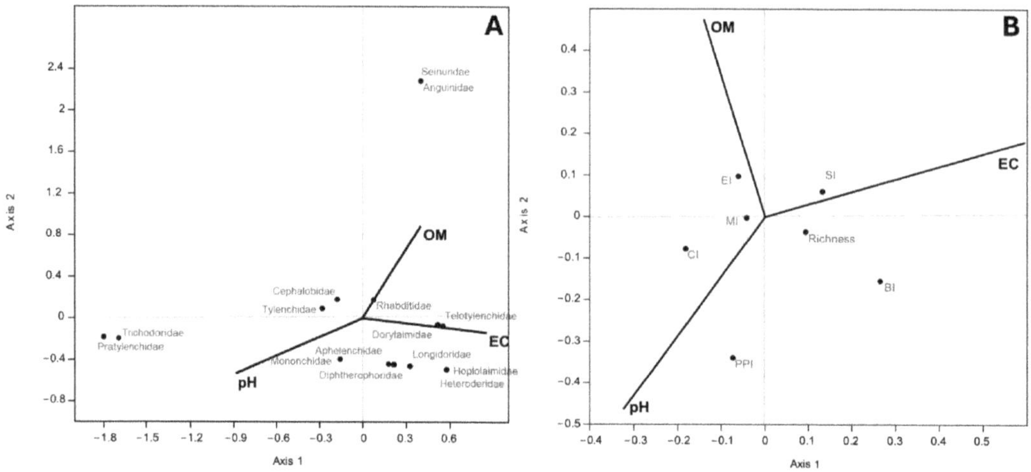

Figure 4. Scatter plot of CCA ordination showing relationships between soil properties and nematode taxa abundance (**A**) and soil nematode indicators (**B**). (**A**) Percentages of variance were 54.96% ($p < 0.001$) for axis 1 and 45.04% ($p < 0.001$) for axis 2; (**B**) percentages of variance were 30.08% ($p < 0.016$) for axis 2 and no significant axis 1.

4. Discussion

This study investigated the response of the soil nematode community (i.e., both plant-parasitic and free-living nematodes) to artificial substrates both before and during agricultural practices. It is worth noting that it is relatively difficult to obtain a comprehensive picture of the general problem, as ornamental nurseries represent a complex network formed by a large variety of plants, with their own cultivation, nutritional, and phytosanitary requirements. In this study, the context was simplified by choosing only two different plant species. Specifically, X *Cupressocyparis leylandii* and *A. palmatum* were selected as they are representative of conifers and broad-leaved trees and because they are economically relevant cultivated species in the Pistoia district.

4.1. Effect of Pot Cultivation on Artificial Substrate Soil Nematode Community Structure

As previously reported by several authors, the use of artificial substrates, such as perlite, pumice, and coconut fiber, represents a valid practice aimed at reducing the introduction of plant-parasitic nematodes in nurseries [2,6]. This study provided a reasonable explanation of this trend. These substrates created an unsuitable environment for plant-parasitic nematodes due to the absence of plants and the abundance of free-living nematodes showing a composition typical of organic materials (i.e., compost) [26]. It is well known that as free-living nematodes increase, plant-parasitic nematodes decrease [4,27]. The dominant trophic groups were bacterial and fungal feeders, both involved in the organic matter decomposition. As reported by Ferris & Matute [26] and Georgieva et al. [28], after the application of compost, it is possible to identify a food web succession driving the mineralization process: colonizer and extreme colonizer bacterial feeders start the activity of decomposition, while fungal feeders may subsequently develop. The three analyzed substrates were presumably different phases of this ecological succession: PPU was characterized by the presence of only colonizer and extreme colonizer bacterial feeders, PPE included both bacterial and fungal feeders, and, finally, CF also included omnivores and predators. These differences might be attributed to different content in organic matter, higher in CF than others, and the different characteristics of inert materials such as pumice and perlite. Moreover, fungal feeders, omnivores, and predators increased throughout the period in which CF remained in the transplant area exposed to the air. Substrate aeration probably stimulated microorganism and nematode growth and consequently allowed predator development. The soil nematode indicators confirmed this trend: CF showed the highest values of MI and SI, indicating both the presence of colonizers and a better soil nematode community structure than PPE and PPU. On the contrary, CF showed the lowest EI value. CI indicated that the composition channel was driven by bacteria, even if in PPE the fungal channel was also relevant.

Pot cultivation changed the composition of the nematode community, as shown by similarity analysis. In general, the soilless practice conducted only in pot conditions produced a decrease in the total number of taxa and abundance of nematodes when compared with natural soil conditions. The potted plants, previously cultivated in natural soil, showed an opposite trend. In general, free-living nematodes were moderately affected by pot cultivation, and, in all cases, the dominant trophic group was represented by bacterial feeders followed by fungal feeders, omnivores, and predators. The high presence of free-living nematodes involved in the detritus food web and the graze on bacteria and fungi in the soil may regulate decomposition and nitrogen mineralization in soil ecosystems, as well as in pot cultivations [2,29]. On the contrary, plant-parasitic nematodes were the trophic group mainly affected by variations due to plant species and management. Plant-parasitic nematode communities were peculiar to each examined plant species. Palomares-Rius et al. [30] found that nematode community populations in the rhizosphere of cultivated olives differed according to plant genotype. Although the families were the same, Telotylenchidae, Hoplolaimidae, and Longidoridae were dominant in *A. palmatum*, while Pratylenchidae, Psilenchidae, and Trichodoridae were more numerous in x *C. leylandii*. A key role could be played by host–parasite interaction, a factor scarcely investigated yet.

Soilless farming in pots reduced the number of plant-parasitic nematodes in *A. palmatum* and x *C. leylandii* plants more than in cultivation in natural soil. In this regard, only a few individuals belonging to the Telotylenchidae family were found. Because coconut fiber substrates were nematode-free, it can be assumed that these plant-parasitic nematodes were accidentally introduced [8,9]. As reported by Hug and Malan [31], the risk of contamination with plant-parasitic nematodes was low only when capped boreholes were the source of irrigation water. According to the same authors, the plant-parasitic nematodes of economic importance found in irrigation water belonged mainly to the genera *Meloidogyne*, *Xiphinema*, *Tylenchulus*, *Trichodorus*, *Criconemoides*, and *Pratylenchus*. On the contrary, in potted plants previously cultivated in natural soil, plant-parasitic nematodes found environments more suitable for growth. The abundance of the families Telotylenchidae, Pratylenchidae, and Hoplolaimidae increased, while virus-vector nematodes belonging to Longidoridae and Trichodoridae decreased. The more homogeneous fertilization and irrigation in pot probably favored root development and consequently enhanced plant-parasitic nematodes, leading to their exponential growth. On the contrary, the constrained space negatively affected the development of virus-vector nematodes, as evidenced by Ali et al. [32] and Landi et al. [33] reporting that the intensification of agricultural practices disfavoured these nematodes.

The nematode indicators confirmed that differences were mainly due to management rather than to plant species. Pot cultivation reduced PPI because of the decrease in k-strategist plant-parasitic nematodes such as virus vectors from natural soil. Moreover, soilless conditions also reduced MI due to the increment of colonizer species, especially in PP [3]. BI and EI were higher after repotting, suggesting that the fungal feeders benefited from these practices, probably due to the high organic matter content supplied by the coconut fiber substrate. CI evidenced that the channel decomposition was driven by bacteria. In terms of ecosystem services, the application of diversity-weighted abundance, expressed as biomass, evidenced that free-living nematodes implicated in nutrient mineralization and plant-parasitic nematodes appeared more regulated in pot farming than in natural soil. However, PP and PNL showed optimal (prey/predator ratio 1:1) and suboptimal regulation values, respectively.

4.2. Soil Factor Influencing Soil Nematode Structure

In general, the commercial substrates were rich in organic matter, especially CF. As reported by Landi et al. [3,5], many families were positively affected by organic carbon, especially the predators Mononchidae and the omnivores Dorylaimidae, while the Anguinidae family was favored by soil pH. On the whole, the soil nematode indicators confirmed that organic matter improved the soil nematode community structure. SI and CI were mainly positively correlated with organic matter. In accordance with several authors, during farming, the presence of roots in soil confirmed that organic matter was the relevant factor for the development of free-living nematodes and the decrease in plant-parasitic nematodes [3,4]. The families Seinuridae (predators) and Anguinidae (fungal feeder nematodes) were always favored by organic matter, whereas the families Pratylenchidae and Trichodoridae (plant-parasitic nematodes) were mostly influenced by soil pH. Regarding soil nematode indicators, the studied indices were only moderately influenced by the explored environmental variables. Only PPI, and to an even lesser extent BI, were weakly influenced by soil pH and CE, respectively.

5. Conclusions

This work shows that substrates rich in organic matter such as coconut fiber, even though unable to prevent accidental introduction during cultivation, could still play an important role in suppressing plant-parasitic nematodes. This is due to the abundance of free-living nematodes, which are of crucial importance in both the mineralization of organic matter and antagonistic control of plant-parasitic nematodes. Potting systems, also, reduce virus-vector nematodes and improve the prey/predator ratio favoring natural control. However, it is worth noting that the population of some plant-parasitic nematode species, such as *P. vulnus* found in *C. leylandii*, showed a strong increase in abundance. This pest, listed in Annex IV, Teg. (EU) 2072/2019 among regulated non-quarantine pests, was the most numerous species representing more than 70% of the entire plant-parasitic nematode population. Further studies are required to confirm these findings. Additional research is recommended to investigate the mechanisms determining differences in plant-parasitic nematodes between plant species and to explore the effectiveness of combining pot cultivation with other control methods.

Author Contributions: Conceptualization, S.L. and E.R.; methodology, B.C., S.C. and B.N.; investigation, B.C., F.B., S.C. and B.N.; data curation, S.L. and F.B.; writing—original draft preparation, S.L.; writing—review and editing, F.B., B.C., S.C., B.N., E.R. and S.S.; visualization, S.L. and F.B.; supervision, S.L., P.F.R. and S.S.; funding acquisition, P.F.R. All authors have read and agreed to the published version of the manuscript.

Funding: This research was funded by the Rural Development Program, Tuscany Region, Italy, UE, grant number 2014IT6RDRP010 under the project AUTOFITOVIV "Good practices for a self-controlled and sustainable management in ornamental nurseries". The APC was not due as this paper was presented upon invitation, and was therefore published free of charge.

Institutional Review Board Statement: No ethical approval was required for the sample types collected in this study.

Informed Consent Statement: Not applicable.

Data Availability Statement: The data presented in this study are available upon request from the corresponding author.

Acknowledgments: We wish to thank "Azienda Agricola Vannucci Piante di Vannino Vannucci" for their support and field logistics.

Conflicts of Interest: The authors declare no conflicts of interest.

References

1. Howland, A.D. Plant-Parasitic Nematodes and their Effects on Ornamental Plants: A Review. *J. Nematol.* **2023**, *55*, 20230007. [CrossRef]

2. Sas Paszt, L.; Trzciński, P.; Bakalarska, M.; Holownicki, R.; Konopacki, P.; Treder, W. The influence of heated soil in crop of "Tamaris" tomato plants on the biological activity of the rhizosphere soil. *Adv. Microbiol.* **2014**, *4*, 191–201. [CrossRef]
3. Landi, S.; Valboa, G.; Vignozzi, N.; d'Errico, G.; Pellegrini, S.; Simoncini, S.; Torrini, G.; Roversi, P.F.; Priori, S. Response of nematode community structure to different restoration practices in two vineyard soils in Tuscany (Italy). *Biol. Agric. Hortic.* **2023**, *39*, 149–169. [CrossRef]
4. Barker, K.R.; Koenning, S.R. Developing sustainable systems for nematode management. *Annu. Rev. Phytopathol.* **1998**, *36*, 165–205. [CrossRef]
5. Landi, S.; Papini, R.; d'Errico, G.; Brandi, G.; Rocchini, A.; Roversi, P.F.; Bazzoffi, P.; Mocali, S. Effect of different set-aside management systems on soil nematode community and soil fertility in North, Central and South Italy. *Agric. Ecosyst. Environ.* **2018**, *261*, 251–260. [CrossRef]
6. Phani, V.; Khan, M.R.; Dutta, T.K. Plant-parasitic nematodes as a potential threat to protected agriculture: Current status and management options. *Crop Prot.* **2021**, *144*, 105573. [CrossRef]
7. Hallmann, J.; Hänisch, D.; Braunsmann, J.; Klenner, M. Plant-parasitic nematodes in soil-less culture systems. *Nematology* **2005**, *7*, 1–4. [CrossRef]
8. Sabir, N.; Walia, R.K. Management of nematodes in protected cultivation with shortnotes on key pests. In *All Indian Coordinated Research Project on Nematodes in Cropping Systems, ICAR*; Indian Agricultural Research Institute: New Delhi, India, 2017.
9. Noling, J.W.; Rich, J.R. Greenhouse Nematode Management. University of Florida, IFAS Extension. Available online: https://hortintl.cals.nesu.udu/articles/greenhouse-nematode-management (accessed on 30 August 2020).
10. Eppo. Diagnostic. PM 7/119 (1) Nematode extraction. *Bull. OEPP/EPPO Bull.* **2013**, *43*, 471–495. [CrossRef]
11. De Boodt, M.F.; Verdonck, O.F.; Cappaert, I.M. Method for measuring the water release curve of organic substrates. *Acta Hortic.* **1974**, *37*, 2054–2062. [CrossRef]
12. Mai, W.F.; Lyon, H.H. *Pictorial Key to Genera of Plant Parasitic Nematodes*; Plates Reproduced by Art Craft of Ithaca, Inc.: Ithaca, NY, USA, 1962.
13. Bongers, T. *De Nematoden van Nederland*; KNNV: Utrecht, The Nederlands, 1988.
14. Marinari-Palmisano, A.; Vinciguerra, M. Classificazione dei nematodi. In *Nematologia Agraria Generale e Applicata*; Ambrogioni, L., d'Errico, F.P., Greco, N.A., Marinari-Palmisano, A., Roversi, P.F., Eds.; Società Italiana di Nematologia: Bari, Italy, 2014; pp. 23–42.
15. Yeates, G.W.; Bongers, T.; De Goede, R.; Freckman, D.; Georgieva, S. Feeding habits in soil nematode families and genera in outline for soil ecologists. *J. Nematol.* **1993**, *25*, 315.
16. Okada, H.; Harada, H.; Kadota, I. Fungal-feeding habits of six nematode isolates in the genus *Filenchus*. *Soil Biol. Biochem.* **2005**, *37*, 1113–1120. [CrossRef]
17. Bongers, T. The maturity index: An ecological measure of environmental disturbance based on nematode species composition. *Oecologia* **1990**, *83*, 14–19. [CrossRef] [PubMed]
18. Ferris, H.; Bongers, T.; De Goede, R. A framework for soil food web diagnostics: Extension of the nematode faunal analysis concept. *Appl. Soil Ecol.* **2001**, *18*, 13–29. [CrossRef]
19. Ferris, H. Form and function: Metabolic footprints of nematodes in the soil food web. *Eur. J. Soil Biol.* **2010**, *46*, 97–104. [CrossRef]
20. Ferris, H.; Tuomisto, H. Unearthing the role of biological diversity in soil health. *Soil Biol. Biochem.* **2015**, *85*, 101–109. [CrossRef]
21. Hammer, O.; Harper, D.A.T.; Ryan, P.D. Past version 1.95: Paleontological Statistical Software Package for Education and Data Analysis. *Paleontol. Electron.* **2001**, *4*, 9.
22. Clarke, K.R. Non-parametric multivariate analysis of changes in community structure. *Aust. J. Ecol.* **1993**, *18*, 117–143. [CrossRef]
23. Loveland, P.; Legendre, L. Is there a critical level of organic production practices on soil quality indicators. *J. Environ. Qual.* **1998**, *28*, 1601–1609.
24. Abad, M.; Noguera, P.; Bures, S. National inventory of organic wastes for use as growing media for ornamental potted plant production: Case study in Spain. *Bioresour. Technol.* **2001**, *77*, 197–200. [CrossRef]
25. Carlile, W.R.; Raviv, M.; Prasad, M. Organic soilless media components. In *Soilless Culture: Theory and Practice Theory and Practice*; Raviv, M., Lieth, J.H., Bar-Tal, A., Eds.; Elsevier: Amsterdam, The Netherlands, 2019; pp. 303–378. [CrossRef]
26. Ferris, H.; Matute, M.M. Structural snf functional succession in the nematode fauna of a soil food web. *Appl. Soil Ecol.* **2003**, *23*, 93–110. [CrossRef]
27. Rahman, L.; Chan, K.Y.; Heenan, D.P. Impact of tillage, stubble management and crop rotation on nematode populations in a long-term field experiment. *Soil Till Res.* **2007**, *95*, 110–119. [CrossRef]
28. Georgieva, S.; Christensen, S.; Stevnbak, K. Nematode succession and microfauna-microorganism interactions during root residue decomposition. *Soil Biol. Biochem.* **2005**, *37*, 1763–1774. [CrossRef]
29. Sohlenius, B.; Bostrom, S.; Sandor, A. Carbon and nitrogen budgets of nematodes in arable soil. *Biol. Fertil. Soils* **1988**, *6*, 1–8. [CrossRef]
30. Palomares-Rius, J.E.; Castillo, P.; Montes-Borrego, M.; Müller, H.; Landa, B.B. Nematode community populations in the rhizosphere of cultivated olive differs according to the plant genotype. *Soil Biol. Biochem.* **2012**, *45*, 168–171. [CrossRef]
31. Hugo, H.J.; Malan, A.P. Occurrence and Control of Plant-parasitic Nematodes in Irrigation Water—A Review. *S. Afr. J. Enol. Vitic* **2010**, *31*, 169–180. [CrossRef]

32. Ali, N.; Tavoillot, J.; Besnard, G.; Khadari, B.; Dmowska, E.; Winiszewska, G.; Fottati-Gaschignard, O.; Ater, M.; Hamza, M.A.; El Mousadik, A.; et al. How anthropogenic changes may affect soil-borne parasite diversity? Plant-parasitic nematode communities associated with olive trees in Morocco as a case study. *BMC Ecol.* **2017**, *17*, 4. [CrossRef]
33. Landi, S.; d'Errico, G.; Papini, R.; Cutino, I.; Simoncini, S.; Rocchini, A.; Brandi, G.; Rizzo, R.; Gugliuzza, G.; Germinara, G.S.; et al. Impact of Super-High Density Olive Orchard Management System on Soil Free-Living and Plant-Parasitic Nematodes in Central and South Italy. *Animals* **2022**, *12*, 1551. [CrossRef]

soil systems

MDPI

Article

The Effect of Soil Heterogeneity on the Content of Macronutrients and Micronutrients in the Chickpea (*Cicer arietinum* L.)

Antonín Kintl [1], Ján Šmeringai [1], Tomáš Lošák [2,3], Igor Huňady [1], Julie Sobotková [1], Tadeáš Hrušovský [2], Ladislav Varga [4], Karel Vejražka [1] and Jakub Elbl [1,5,*]

1 Agricultural Research, Ltd., Zahradní 1, 664 41 Troubsko, Czech Republic; kintl@vupt.cz (A.K.); smeringai@vupt.cz (J.Š.); hunady@vupt.cz (I.H.); sobotkova@vupt.cz (J.S.); vejrazka@vupt.cz (K.V.)
2 Department of Environmentalistics and Natural Resources, Faculty of Regional Development and International Studies, Mendel University in Brno, Zemědělská 1, 613 00 Brno, Czech Republic; tomas.losak@mendelu.cz (T.L.); tadeas.hrusovsky@mendelu.cz (T.H.)
3 Department of Agroecosystems, Faculty of Agriculture and Technology, University of South Bohemia in Ceske Budejovice, Branišovská 1645/31A, 370 05 Ceske Budejovice, Czech Republic
4 Institute of Agronomic Sciences, Faculty of Agrobiology and Food Resources, Slovak University of Agriculture in Nitra, Tr. A. Hlinku 2, 949 76 Nitra, Slovakia; ladislav.varga@uniag.sk
5 Department of Agrosystems and Bioclimatology, Faculty of AgriSciences, Mendel University in Brno, Zemědělská 1, 613 00 Brno, Czech Republic
* Correspondence: jakub.elbl@mendelu.cz; Tel.: +420-545-133-081

Abstract: Chickpea (*Cicer arietinum* L.) is one of the most important legumes currently grown. It is an important source of proteins and nutrients, such as calcium, potassium and iron. As a result, precise crop management is necessary for maximizing its production. The presented study deals with the effect of soil heterogeneity caused by variable contents of macro- and micronutrients on the uptake of nutrients by chickpea. The values measured (contents of macro- and micronutrients in plant samples) indicate that soil heterogeneity is an important factor for the contents of nutrients and soil reactions, which strongly affect the growth of chickpea. We investigated the soil heterogeneity in a chickpea field. Two zones (A and B) with different stand development were found in the model plot. Zone A showed a healthy (green) growth, while Zone B exhibited a yellow-coloured growth, indicating deficits in nutrient uptake. The contents of selected nutrients (P, K, Ca, Mg, Fe, Cu, Zn and Mn) in the soil and in the plant biomass (i.e., stems, leaves, pods and seeds) were analyzed. In the zone with the yellow-coloured biomass, the results showed significantly ($p < 0.05$) reduced contents of N, P, K, Mg, Fe, Mn, Cu and Zn in the leaves; higher values of soil reaction (pH); and higher contents of calcium and calcium carbonate in the soil. The uptake of nutrients by the plants and their translocation were affected by the above-mentioned soil parameters and by their mutual interactions. Therefore, it is possible to state that soil heterogeneity (caused by variable contents of nutrients in soil) should be taken into account in the precise crop management of chickpeas.

Keywords: chickpea; macronutrients; micronutrients; management practices; soil heterogeneity

1. Introduction

Nowadays, mineral fertilizers have to be used to obtain proper crop yields [1], which has led to a rising demand for fertilizers [2]. Plants are able to absorb only about 50% of applied mineral fertilizers, while the rest escape into the environment with negative impacts on ecosystems [3,4]. The sustainability of ecological systems and the minimization of impacts on the environment, on the one hand, and sufficient food production, on the other hand, should be the main goals of modern agriculture [5]. The high costs of mineral fertilizers (especially nitrogenous ones), caused by the recent increase in the price of natural gas, which is necessary for their production, has made farmers look for partial solutions, such as growing legumes [6].

One of the possible solutions to reduce the use of nitrogenous mineral fertilizers is the integration of legumes, which are able to assimilate atmospheric nitrogen through symbiosis with *Rhizobium* bacteria, into cropping systems [7]. In ideal conditions, nitrogen fixation can produce more than 100 kg/ha of N during one year, which is 85% of the overall nitrogen demand of cicer [8]. According to Flowers et al. [9], cicer can fix 140 kg/ha in one year, decreasing the occurrence of plant diseases and improving soil structure and the availability of K and P in soil [10]. According to Carlsson and Huss-Danell [11], this symbiosis enriches soil with nitrogen, which leads to reduced consumption of mineral fertilizers. Worldwide, this alternative reduces the overall fertilization of agricultural soils by 13% [7], which could lead to higher crop yields and reduced N losses into the environment.

Chickpea or cicer (*Cicer arietinum* L.), belonging to the *Fabaceae*, is a legume from southeast Turkey and Syria [12–14]. Worldwide, it is grown on approximately 17.8 million hectares, with an annual production of 17.2 million tons [15]; the main producers are India (65%), Pakistan (10%), Iran (8%) and Turkey (5.5%) [16–18]. Despite the fact that cicer is a legume grown in temperate zones and is most tolerant to high temperatures and droughts, these climatic factors can inflict 40–45% of losses in yield worldwide [19]. One reason for considering the introduction of new procedures (e.g., monitoring of soil heterogeneity) in growing chickpeas is the crop's significance for nutrition. Cicer is a high-quality source of proteins for the human population and for livestock [15,18,20]. Apart from proteins (whose concentration is twice as high as in cereals), cicer is very rich in fibre and minerals such as calcium, potassium, iron, phosphorus, magnesium, selenium and zinc [14,21].

Plot heterogeneity in terms of nutrient contents in soil and soil reactions may be reflected in the chemical composition of plants and their organs, yield and quality. There also may be some visual changes in the colour of leaves, etc. This heterogeneity can be caused by a number of biotic and abiotic factors, which may be difficult to determine. A very frequent cause of heterogeneity can be a lack of nutrients in different parts of a plot, which can be due to various factors, including a deficit of soil nutrients due to the absence of fertilization and liming and uneven application of fertilizers [22]. Many farmers try to prevent this by using technologies of precise agriculture [23]. Based on information about the contents of soil nutrients (soil sampling) and spectral analysis of growth, these technologies allow the application of optimum doses of nutrients or the identification of problematic sites in stands [24]. From the viewpoint of natural ecosystems, spatial heterogeneity in the availability of soil nutrients affects species diversity [25] and directly affects the yields of crops in agroecosystems [26,27] through the dynamics of nutrients [26,28]. Different crops have naturally different nutrient requirements (N, P, K, Ca, Mg, etc.), but there are generally valid principles which affect the uptake of nutrients in all plants [27,28]. The most important of them include mutual interactions of nutrients; for example, surplus Ca^{2+} cations in soil tend to bind P in calcium compounds [29]. In a model case, a heterogeneous plot with different contents of Ca^{2+} will have different levels of P available to crops grown in different parts of the plot. This can be resolved by variable application of fertilizers, i.e., by precision agriculture technology. According to Habib-ur-Rahman [27], the effectiveness of resources can be increased by precision agriculture when management procedures are adapted to the heterogeneity of plant growth conditions.

The goal of this study was to explain the potential influence of soil heterogeneity in terms of nutrient contents on differences in the chemical composition of individual parts of chickpea plants (stems, leaves, pods and seeds).

2. Material and Methods

2.1. Description of the Experimental Location

The health condition of chickpea plants was monitored in 2021 in Horní Moštenice, near Přerov in the Olomouc Region, Moravia, Czech Republic (Figure 1). The basic meteorological parameters of area of our interest are shown in Table 1. The area belongs to the sugar beet-growing region. The location is characterized by Luvisol chernozem soil on sandy–loamy sediments. The soil type of the area of interest is shown in Figure A1 (Appendix A). Two sites were selected, which exhibited different conditions for the growth of plants over the long term (Figure 1). Basic agrochemical parameters of the experimental plot (Table 2) were identified in 2019 by regular basal monitoring that is carried out in the area every three years.

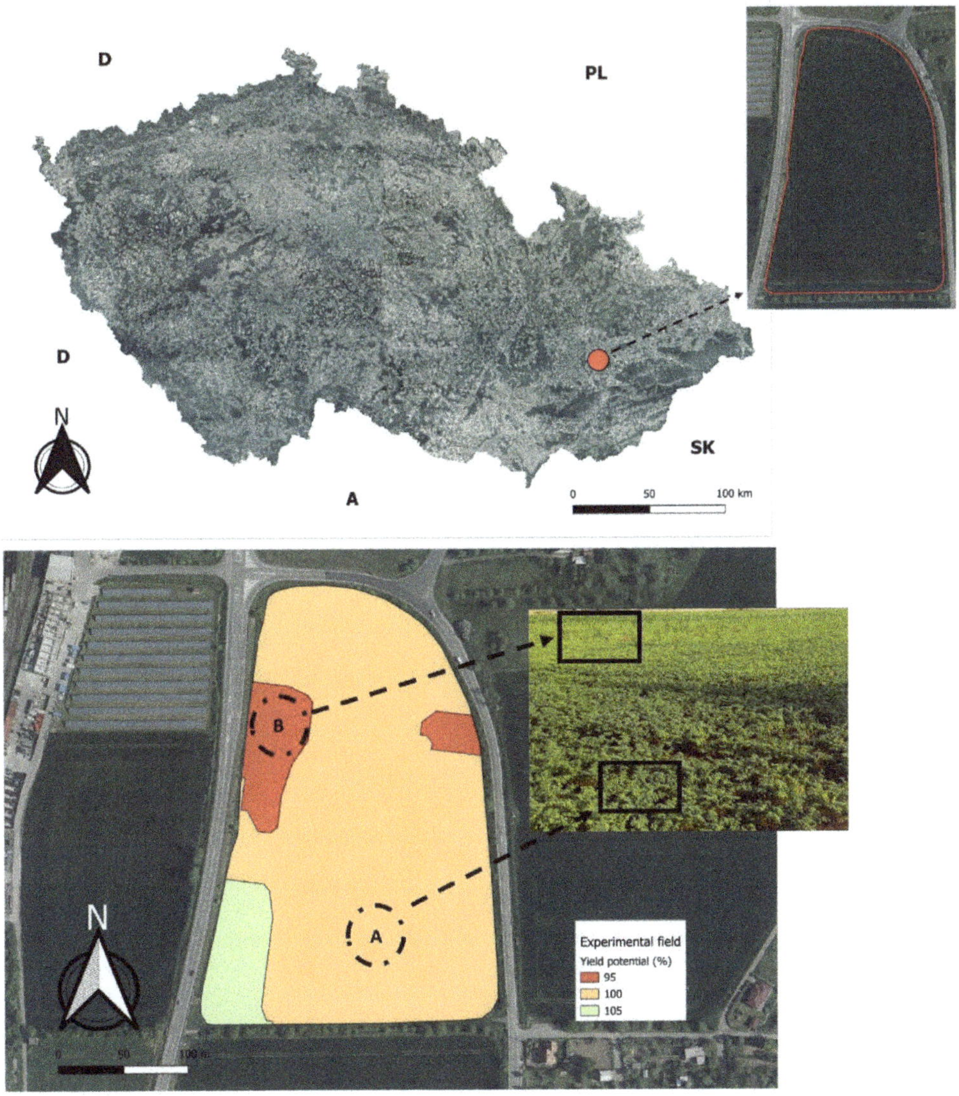

Figure 1. Map of the position of the experimental field within Central Europe and the Czech Republic (A = Austria, D = Deutschland, PL = Poland, SK = Slovakia) and the yield potential, with the zones (A and B) for the collection of soil and plant samples indicated. Source of base maps: www.cuzk.cz (accessed on 10 March 2024).

Table 1. Meteorological and climatological parameters.

Year	Mean Annual Temperature (°C)	Mean Annual Precipitation (mm)
2021	10.1	559
Long-term standard (1991–2020)	7.8	708

Comments: Meteorological data were measured using the DAVIS Vantage Pro2 weather station (Davis Instruments, Hayward, CA, USA), which was located in Horní Moštěnice (250 m a.s.l.). Data for the long-term standard (1991–2020) are for Olomouc Region and were prepared based on data available from the Czech Hydrometeorological Institute (http://portal.chmi.cz/historicka-data/ (accessed on 10 March 2024)).

Table 2. Basic agrochemical parameters of the experimental field—content of nutrients available to plants and soil reaction.

pH KCl	P ± SD	K ± SD	Ca ± SD	Mg ± SD
	mg/kg	mg/kg	mg/kg	mg/kg
6.85	46 ± 6.44	283 ± 10.75	3958 ± 384.47	220 ± 13.46

2.2. Design of the Field Experiment

The experiment was carried out with the chickpea variety Orion (*Cicer arietinum* L.), which was sown on 24 April 2021 (120 kg.ha^{-1}). The field was treated with the NP fertilizer AMMOPHOS (BelFert, Gomel, Russia; 120 kg of fertilizer/ha) 2 weeks before sowing, and 5 weeks after sowing it was treated with the nitrogen fertilizer LOVOFERT LAD 27 (80 kg of fertilizer/ha; Lovochemie Ltd., Lovosice, Czech Republic).

Plant samples (stems, leaves, pods and seeds) and soil samples were collected from two sites (Zones A and B, Figure 1), which exhibited visibly different conditions for plant growth and development, supporting (A) healthy-looking plants and (B) yellowish plants. The zones for sampling were selected based on the yield potential map. The zone with the average yield potential (=100%) represented an area (A) where the growth of chickpeas occurred without visually conspicuous changes. The zone with the lower yield potential (≤95%) represented an area (B) where, evidently, there were problems with plant development that were indicated by the change in leaf colour (yellowing). The map of the yield potential (Figure 1) was prepared based on an analysis of multispectral images of the area taken over the last 8 years, and the potential was calculated by the Laboratory of Precise Agriculture PrezemLab (Assoc. Prof. Vojtěch Lukas, Mendel University in Brno), according to Lukas et al. [30].

2.3. Plant and Soil Analyses

Three mixed samples of plants and soil were collected from each zone. The mixed soil samples were collected in line with ISO 10381-6 [31] from three sampling points regularly distributed in each zone. A final mixed soil sample of 500 g (min.) was obtained from the specific sampling points after three collections from the 0–20 cm layer with the use of a sampling probe. Thus, there were three mixed soil samples collected for each variant.

The sampling of plant biomass proceeded as follows: one mixed sample contained five plants collected from three points in each repetition. There were, altogether, 3 mixed samples of plant biomass collected from each zone.

In the plant samples, contents of N, P, K, Ca, Mg, Cu, Zn, Fe and Mn were determined. All these elements (with the exception of N and P) were established with the use of atomic absorption spectrometry (AAS; Agilent 55B AA; Agilent Technologies, Santa Clara, CA, USA), according to Jones [32]. P content was measured spectrophotometrically using the Onda VIS V-10 Plus spectrophotometer (Giorgio Bormac, Carpi, Italy), according to Olsen and Summers [33]. Kjeldahl's method was used to determine the total N content in the biomass samples.

In addition to basic nutrients (macroelements) and microelements (micronutrients) in the plants, contents of macroelements (P, K, Ca and Mg) and microelements (Fe, Mn, Cu, Zn) in the soil were determined. The individual elements were established via Mehlich 3 extraction [34]. Soil reaction pH/$CaCl_2$ was determined in 0.01 M pH/$CaCl_2$ using the ion-selective electrode Radelkis OP 211 (Radelkis Electrochemical Instruments, Budapest, Hungary). The K, Mg and Ca contents of plant-available nutrients in the Mehlich 3 extract were determined using AAS (Agilent 55B AA; Agilent Technologies, Santa Clara, CA, USA), according to Sarojam [35]. P contents were determined colourimetrically, according to Olsen and Summers [33].

2.4. Statistical Analysis

For statistical analysis of the acquired data, the software STATISTICA version 13.5.0.17 (TIBCO Software Inc., Palo Alto, CA, USA) was used. The results presented in this study are means of at least 3 repetitions for each presented parameter. Statistically significant differences in the contents of selected elements in the soil, chickpea above-ground organs, and seeds of the A and B zones were obtained by *t*-tests and Tukey's post hoc HSD tests. Correlation analysis was used to establish Spearman's correlation coefficients (R) between the contents of selected elements in the soil and in the chickpea above-ground organs and seeds. The level of significance chosen for all implemented statistical analyses was $p < 0.05$. Map documents were prepared in the QGIS 3.28 programme (QGIS Development Team; General Public License), with WMS data of CUZK (State Administration of Land Surveying and Cadastre of the Czech Republic) used as underlying layers.

3. Results and Discussion

The presented study deals with the monitoring of micro- and macronutrients in soil and plant samples in plots with assumed growth differences in an experimental field. For greater clarity, the measured values are presented in two subsections.

3.1. Contents of Macro- and Micronutrients in the Soil

In general, the contents of micronutrients in the soil of the experimental variants were average to high. The highest values of microelements were measured for Mn, followed by Fe, Zn and Cu, in the two experimental variants (Table 3). There were no significant differences found between the contents of Mn and Fe within the A variants ($p > 0.05$), while a significant difference was recorded in the B variants ($p < 0.05$). In both variants, a demonstrably higher Zn content was recorded compared with that of Cu. In terms of the contents of macronutrients, a similar trend could be observed in both variants, i.e., average contents of P, K and Mg compared with high contents of Ca (Table 4).

Table 3. Contents of micronutrients in the soil.

Variants	Fe		Mn		Cu		Zn	
	mg/kg	±SD	mg/kg	±SD	mg/kg	±SD	mg/kg	±SD
Zone A	189 ± 10.44 ^{A}		205 ± 1.07 ^{B}		5.1 ± 0.006 ^{A}		9.01 ± 0.48 ^{A}	
Classification	Medium		High		High		High	
Zone B	190 ± 3.09 ^{A}		216 ± 2.92 ^{A}		5.2 ± 0.04 ^{A}		6.27 ± 0.44 ^{B}	
Classification	Medium		High		High		High	

Different letters indicate significant differences in the measured values between Zone A and Zone B at a significance level of $p < 0.05$.

The observed statistically significant differences in the contents of the selected elements in soil samples from locations A and B point to a certain degree of soil heterogeneity with respect to the presence or availability of these elements. Zone B was deficient in P, K and Zn, whereas it was enriched with Mn (Tables 3 and 4). Such soil heterogeneity can be caused by various factors and could be related to an uneven distribution of basic soil sources [36,37].

Some of the factors are of natural origin (differences in bedrock, calcium carbonate content, etc.), and others are of anthropogenic origin (level of organic and mineral fertilization, precise dosing in the individual field parts, etc.) [37–41]. Habib-ur-Rahman [27] suggested that available water capacity and slope elevation significantly affect soil heterogeneity, the latter factor being the most significant. The analysis of Shukla et al. [42] showed that soil pH is significantly correlated with concentrations of extractable Zn, Cu, Mn and Fe. In our results, statistically significant differences in soil pH were observed (Figure 2). While the value of exchange or potential soil reaction pH/$CaCl_2$ in Zone A was 6.52, i.e., slightly acid, the pH/$CaCl_2$ in Zone B was 6.84, i.e., neutral (Figure 2). The actual pH (H_2O) copied the trend of the potential pH, and it only reached higher values. The higher pH value in Zone B was related both to the higher content of soil calcium (4 504 mg/kg) and the higher content of calcium carbonate ($CaCO_3$; 0.78%) as compared with Zone B (Table 4).

Table 4. Contents of macronutrients in the soil.

Variants	$CaCO_3$		P		K (mg/kg)		Ca (mg/kg)		Mg (mg/kg)	
	%	±SD	mg/kg	±SD	mg/kg	±SD	mg/kg	±SD	mg/kg	±SD
Zone A	0.59	–	111 ± 8.06 ^{A}		190 ± 6.51 ^{A}		3 421 ± 203.05 ^{A}		180 ± 13.01 ^{A}	
Classification	Medium		Good		Good		High		Good	
Zone B	0.78	–	95 ± 0.57 ^{B}		173 ± 4.72 ^{B}		4 537 ± 33.86 ^{B}		190 ± 4.08 ^{A}	
Classification	Medium		Good		Sufficient		High		Good	

Different letters indicate significant differences in the measured values between Zone A and Zone B at a significance level of $p < 0.05$.

Furthermore, the greatest difference in our results between Zone A and Zone B was observed in the lower soil Zn content in Zone B (Table 3). In contrast, the content of Mn was higher in Zone B compared with Zone A. Such a decrease can be connected to a higher $CaCO_3$ concentration [42], which was the case in this study (0.59% $CaCO_3$ in Zone A versus 0.78% $CaCO_3$ in Zone B). Cicer improves soil zinc availability [12], which positively affects the development of symbiotic nodules and nitrogen fixation [43,44].

The contents of soil macronutrients exhibited a significant decrease in P and K in Zone B (Tables A1, A2 and 4). Conversely, the Ca content in Zone B was higher (4504 mg/kg) than in Zone A (3455 mg/kg). Contents of macronutrients in the soil (established via Mehlich 3 extraction) can be evaluated verbally, according to Joines and Hardy [45], as "low–sufficient–good–high–very high." When the content of a particular nutrient is high or very high, fertilization with the nutrient is not necessary. In our case, only high contents of calcium in the soil were found in both zones. The contents of the other macroelements (or basic nutrients) were "good" or only "sufficient" in the case of potassium in Zone B (Table 4). The low content of a nutrient in soil indicates a low content of the nutrient in plants, which was demonstrated in the case of N in the chickpea leaves in Zone B (Figure 3). Basic soil parameters measured in the two zones did not show any extreme values, and their contents in the soil were likely affected by the soil management system and the soil type in the given region.

The observed fluctuations in the contents of the macronutrients (or plant-available nutrients) P, K and Ca between the variants (zones) were probably caused by plot heterogeneity based on different soil conditions (soil nutrient contents) and water regimes (field water capacities, Table 5) [36,38,46]. According to Liu et al. [37], soil heterogeneity has two components: qualitative and configuration components. The qualitative component defines differences in the contents of specific parameters (e.g., nutrients in specific areas), and the configuration component defines the size of these areas. In the presented study, the area was not defined in terms of its size and precise location. The goal was to find out whether real differences existed between two qualitatively different zones (according to indications of growth conditions and different yield potentials) in terms of contents of plant and soil nutrients (Figure 1 and Table 5). Based on the measured values of the contents of micro- and macronutrients, it was possible to state that a difference existed between Zone A and Zone B with respect to their suitability for growing chickpeas.

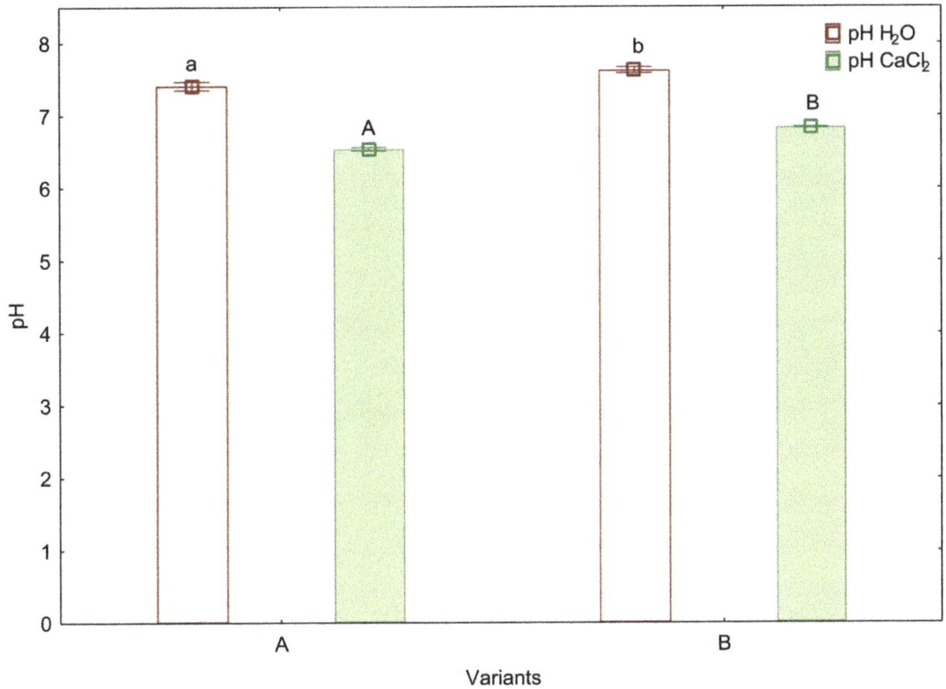

Figure 2. Mean values of actual (pH H_2O) and potential (pH $CaCl_2$) soil reaction (n = 3) ± SDs. Different lowercase letters indicate significant differences in pH H_2O ($p < 0.05$; ANOVA Tukey's post hoc HSD test); different capital letters indicate significant differences in pH $CaCl_2$.

Table 5. Differences in contents of macronutrients between individual variants of the experiment and initial states.

Differences in Contents of Plant-Available Nutrients	P (mg/kg)	K (mg/kg)	Ca (mg/kg)	Mg (mg/kg)
Zone A	65 *	−93 *	−537	−40
Zone B	49 *	−110 *	579	−30

The * symbol indicates a significant difference ($p < 0.05$, t-test) between the individual variants with respect to one nutrient and the initial state in 2019.

3.2. Contents of Macro- and Micronutrients in Plant Samples

The plant materials were analyzed separately for contents of micronutrients (Fe, Mn, Cu and Zn—Figure 3) and macronutrients (N—Figure 4; P, K, Ca and Mg—Figure 5) in roots, stems, leaves and seeds. What was particularly interesting in our observations was the reduced contents of all microelements (Figure 3) in the leaves in Zone B as compared to Zone A and there being no change in their contents in the seeds between the two zones (Figure 3). This reduction in the contents of microelements in chickpea leaves (at medium to high contents in the soil) could be attributed to inappropriate soil properties, particularly alkaline soil reactions, a high content of soil calcium and a high content of calcium carbonate in the soil of Zone B. Moreover, mutually negative interactions are likely to exist in the uptake of nutrients by roots in the form of ion antagonism.

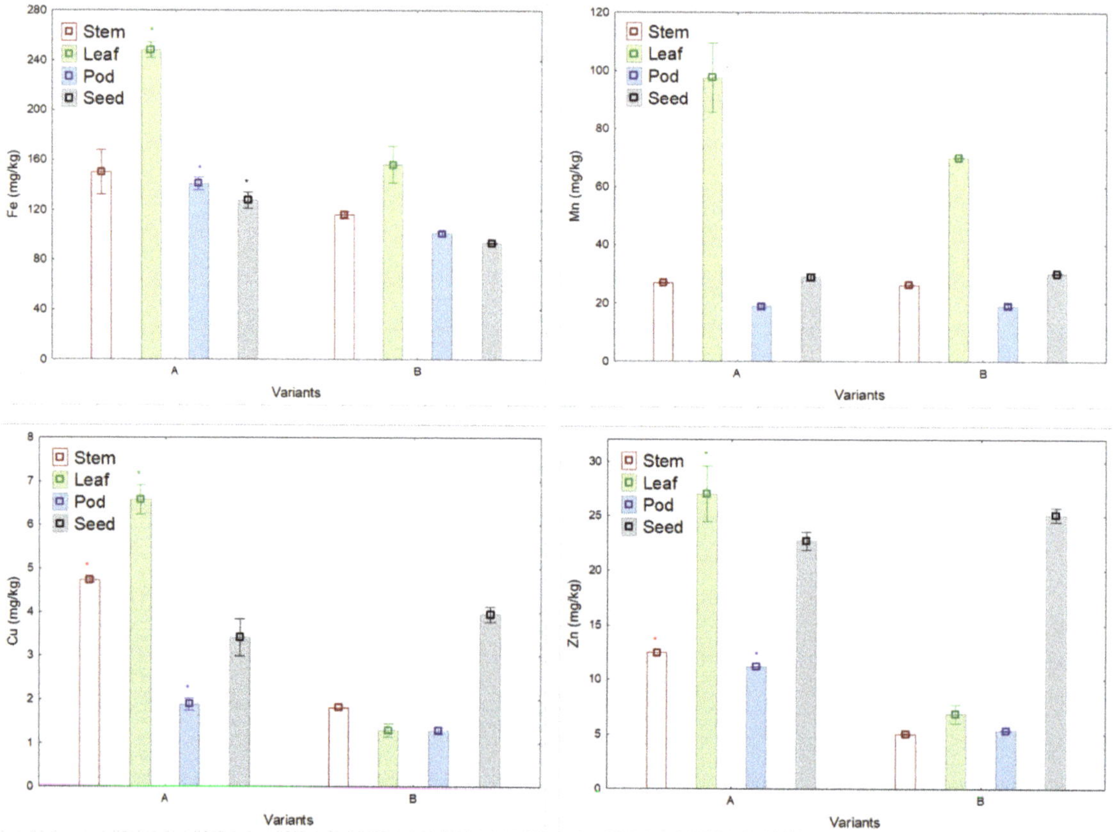

Figure 3. Contents of micronutrients—Fe (A), Mn (B), Cu (C) and Zn (D)—in selected plant organs (stem, leaves and pods) and seeds. Columns represent average values of the contents of elements (n = 3) ± SDs. The * symbol indicates a significant difference ($p < 0.05$, *t*-test) between the individual variants with respect to one nutrient and a specific plant organ.

The analysis of micronutrient (Figure 3, Table A3) contents in plant organs revealed that, in the case of Fe, its presence was significantly reduced by 44.22% in leaf tissues and by 31.98% in pods. In the case of Mn contents, no significant differences were observed in any of the selected organs or seeds. The contents of Cu in Zone B were significantly reduced in the stems, leaves and pods by 62.1, 83.43 and 40.01%, respectively. The Zn contents in Zone B were also significantly lower in the stems, leaves and pods by 58.95, 79.6 and 51.93%, respectively. The recorded contents of micronutrients in seeds differed significantly between the A and B plants only in the case of Fe.

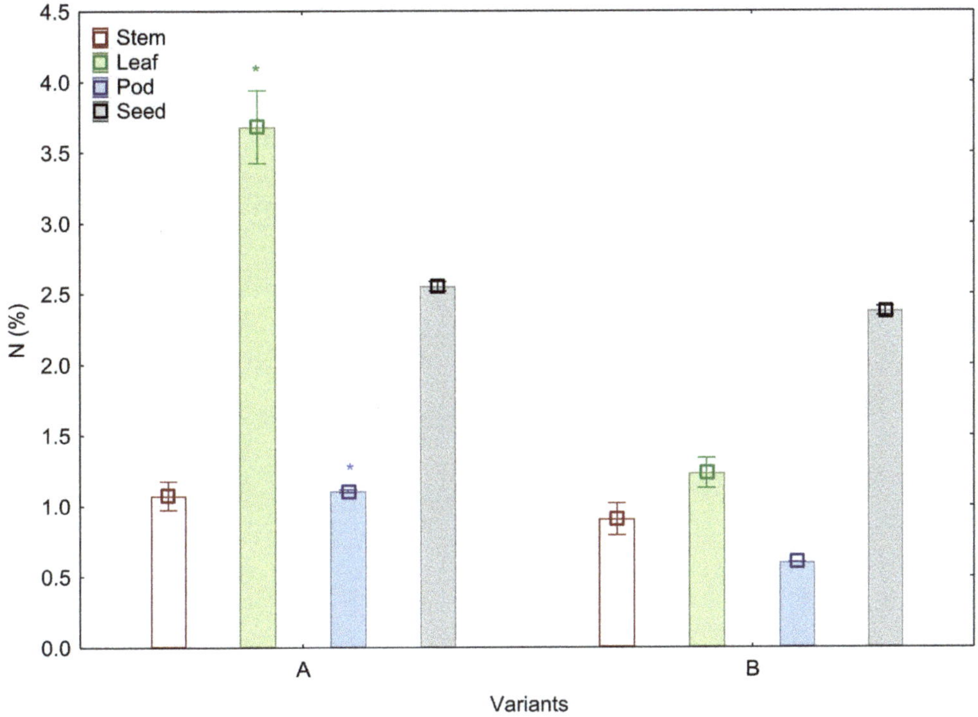

Figure 4. N contents in selected plant organs (stems, leaves and pods) and seeds. Columns represent average values for N contents (n = 3) ± SDs. The * symbol indicates a significant difference ($p < 0.05$, *t*-test) between the individual variants with respect to a specific plant organ.

The B-variant chickpea plants were recognizable by yellowish leaves, which can point to an imbalance in the availability of micro- and macronutrients. This particularly relates to the content of Fe, which was demonstrably lower in the leaves of the B plants. The analysis of the presence of the selected elements revealed deficiencies in those that are responsible for the sufficient production of chlorophyll and the proper functioning of photosynthesis. Iron (Fe), which is an important element in the biosynthesis of chlorophyll [47], is a vital component of various enzymes [48]. It is also present in various protein complexes involved in the processes of photosynthesis [49]. In a study by Mahmoudi et al. [50], chickpeas with iron deficiency suffered from yellowing of young leaves, a large decrease in chlorophyll concentration and a significant decline in plant biomass. However, a decrease in iron content is more damaging in roots than in shoots [51]. When compared to other legumes, chickpea shows stronger resistance to Fe deficiency. This resistance could be explained by the higher seed iron reserves in chickpea [51]. In our analysis, the Fe content in the leaves of the B plants was reduced to 156.49 mg/kg as compared with 248.40 mg/kg recorded in the leaves of the A plants (Figure 3).

The main factors responsible for reduced cicer yields include a lack of nutrients, namely, zinc (Zn), and low soil fertility [52,53]. Zinc is important for the proper development of plants, especially pollen, and can negatively affect their reproduction [44]. The content of Zn in leaves was markedly reduced to 6.86 mg/kg in Zone B, while in Zone A it reached 27.02 mg/kg. The differences in Zn contents recorded in the individual parts of plants between Zone A and B, namely, in the stems, leaves and pods, were some of the most distinct for all the micronutrients assessed (Figure 3).

Copper (Cu) is an important micronutrient, and it is necessary for proper growth of the plant body. In our study, the Cu contents in the B variants were significantly reduced; for example, the Cu content in the leaves in Zone A was 6.58 mg/kg, and in Zone B it was 1.29 mg/kg. Chickpea can increase the bioavailable content of Cu in the soil. In mixed cropping systems, chickpea significantly increased the content of Cu in the roots of *Eucalyptus globulus* [54]. According to Kambhampati et al. [55], chickpea is a cost-effective and environmentally friendly accumulator of Cu. However, the addition of Ethylenediaminetetraacetic acid (EDTA) is necessary for the acceleration of Cu absorption. Cu deficiency results in smaller and chlorotic leaves as well as reduced contents of nitrogen, starch and sugars [56].

Manganese (Mn) is involved in a number of enzymatic processes in plants. Its content in the leaves was 97.68 mg/kg in Zone A and 70.13 mg/kg in Zone B. However, the difference was not statistically significant ($p > 0.05$).

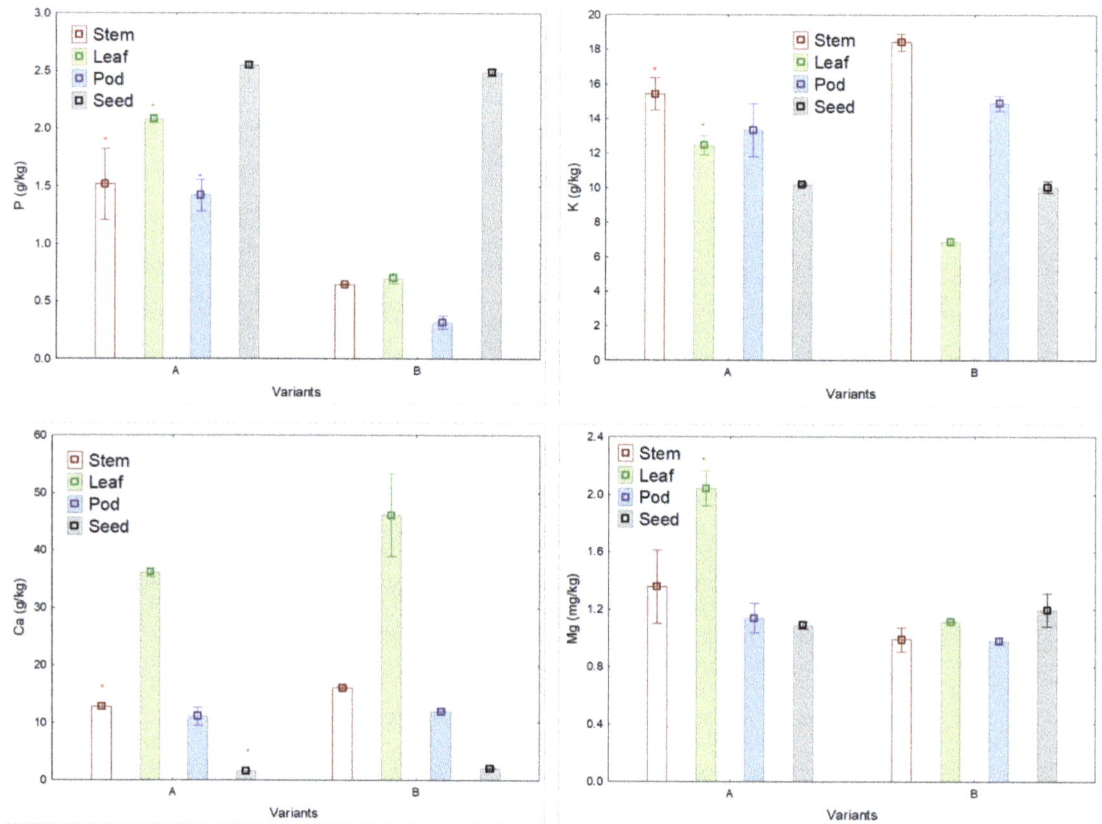

Figure 5. Contents of P, K, Ca and Mg macronutrients in selected plant organs (stems, leaves and pods) and seeds. Columns represent average values of the contents of elements (n = 3) ± SDs. The * symbol indicates a significant difference ($p < 0.05$, *t*-test) between the individual variants with respect to one nutrient and a specific plant organ.

In many instances, the contents of macronutrients (N—Figure 4; P, K, Ca and Mg—Figure 5) in plant organs and seeds showed a similar trend to those of microelements. Particularly interesting was the significant decrease in N, P, K and Mg in the leaves of plants growing in Zone B compared with Zone A and an increased content of Ca in Zone B as compared with Zone A. However, the differences were not significant (Figure 5, Tables A4 and A5). The reduced contents of macronutrinetss in the chickpea leaves were caused mainly by alkaline soil reactions, the high Ca content in the soil, and the higher content of $CaCO_3$ in the soil of Zone B compared to Zone A. Antagonism between Ca, K and Mg resulted in reduced K and Mg contents in the leaves (Figure 5).

Plants of the B variant (location) were less able to take up nitrogen, and its content (%) was significantly decreased in the chickpea leaves and pods. The largest decrease was detected in leaves, where the N content was lower by 83.43% (Figure 4). A noticeable depletion of P in Zone B was recorded, the content of which was significantly reduced in the chickpea stems, leaves and pods by 64.93%, 69.22% and 76.03%, respectively. There were no statistically significant differences observed in the seeds of the A and B chickpea variants. The contents of K in the stems and leaves varied significantly between Zones A and B. The content of K in the stems of the A-variant plants increased by 9.08%, whereas in the leaves it decreased by 43.74%. In Zone B, contents of Ca in the stems, leaves and seeds of the chickpea plants significantly increased by 17.71%, 30.66% and 16.24%, respectively. In the case of Mg, a statistically significant difference was recorded only in the leaves, where Mg was decreased by 48.08% (Figure 5). No significant differences were observed in the stems, pods and seeds of chickpeas grown in the A and B zones (Figure 5). Mg is also important in the primary productivity of plants due to its crucial role in the structure of chlorophyll [57].

Another important macronutrient, P, was significantly decreased in the stems, leaves and pods in Zone B (Figure 5). A lack of P is detrimental to the overall fitness of chickpea [58]. Yahiya et al. [59] investigated the effect of P on the nodulation and N fixation of chickpea, and the results showed that P had no direct effect on the nodules. However, the inoculation of chickpea with phosphate-solubilizing bacteria increased the fitness of chickpeas [60]. P is a basic macronutrient, and legumes which bind atmospheric N have higher P requirements than legumes fertilized with mineral N. Therefore, P deficiency results in lower activity with respect to the symbiotic fixation of N, as well as growth retardation and lower subsequent P accumulation in plant biomass [61].

Multiple studies [61–63] have researched the effect of excessive salt content (Mg^{2+}, K^+, Na^+, Cl^-, etc.) in soil on the overall fitness of chickpeas. In our results, only the content of K in the soil was determined to be an indicator of soil salinity (Table 2—initial state from 2019 and Table 4—situation during the field experiment in 2021). This is very important, because Gul and Ullah [62] found that the sodium cation (Na^+) content in chickpeas was significantly affected by salinity. High concentrations of chlorine anions (Cl^{-1}) in chickpea leaves were tolerated, while the increased presence of Na^+ caused growth impairment in multiple phenotypes. Saxena and Rewari [63] found that Na^+ affected the nodulation ability of chickpea, and nodule and shoot dry weights were reduced to 55% and 58%, respectively, in the control. The presence of elevated Na^+ content could have also decreased the content of K^+ [64]. Different results were obtained by Turner et al. [65], where chickpea genotypes more susceptible to salt stress exhibited higher concentrations of Na^+ and K^+ (106 and 364 μmol.g^{-1} DW, respectively) under salt stress. The excessive accumulation of Na^+ in the leaf mesophyll cells resulted in structural damage to chloroplasts. The resistance of some of the studied genotypes was caused by the ability to exclude excessive Na^+ from the photosynthetically active mesophyll cells [66]. The ability of chickpea to create nodules under salt stress is mediated by the presence of Zn and phosphates [63]. In our results, the Zn and P contents in the above-ground chickpea parts were significantly decreased in Zone B (Figures 3 and 5).

The regression analysis showed that the dependence of the concentrations of selected elements in plant organs on their contents in the soil was lowest in seeds, where only the Ca content depended on the presence of Ca in the soil (Table 6). In stems, the presence of Ca, Cu, Zn and Mg correlated with their presence in the soil. In leaves, the contents of Cu, K, and Zn correlated with their contents in the soil. Finally, Cu, Zn and Mg contents in pods depended on the presence of these selected elements in the soil. The content of Mn in the selected chickpea organs was not correlated with the Mn content in the soil (Table 6).

Table 6. Simple linear regression analysis results of the relation between the contents of selected macronutrients (Ca, K and Mg) and micronutrients (Fe, Mn, Cu and Zn) in the chickpea plant organs (stems, leaves, pods and seeds) and in the soil.

Organ	Element	Regression Coefficient	p Value	SE of Estimation	F
Stem	Ca	0.9872	0.0002 *	0.3397	153.1783
	P	0.3729	0.4666	0.6085	0.6461
	Cu	0.8896	0.0176	0.8156	15.1795
	Fe	0.7721	0.0720	19.3931	5.9057
	K	0.8356	0.0383	1.2312	9.2553
	Zn	0.8973	0.0153	1.9978	16.5316
	Mn	0.3623	0.4804	0.6615	0.6043
	Mg	0.9771	0.0008 *	0.0844	84.3579
Leaf	Ca	0.5832	0.2243	8.8070	2.0622
	P	0.8054	0.0531	0.5065	7.3876
	Cu	0.9116	0.0114	1.3431	19.6772
	Fe	0.2896	0.5778	57.1325	0.3662
	K	0.9616	0.0022 *	0.9558	49.0872
	Zn	0.9732	0.0011 *	2.9401	71.7409
	Mn	0.7297	0.0997	15.2389	4.5561
	Mg	0.5980	0.2099	0.4700	2.2271
Pod	Ca	0.4926	0.3208	1.7234	1.2818
	P	0.6901	0.1292	0.5077	3.6377
	Cu	0.9025	0.0138	0.1796	17.5632
	Fe	0.3396	0.5101	23.8670	0.5216
	K	0.6277	0.1821	1.7115	2.6011
	Zn	0.8998	0.0146	1.5591	0.7518
	Mn	0.3106	0.5491	1.0739	0.4270
	Mg	0.9645	0.0019 *	0.0425	53.3639
Seed	Ca	0.8664	0.0256	0.1572	12.0407
	P	0.1919	0.7158	0.06	0.1529
	Cu	0.2580	0.6216	0.6332	0.2852
	Fe	0.4121	0.4169	20.3876	0.8181
	K	0.5086	0.3029	0.3605	1.3957
	Zn	0.3977	0.4348	1.7868	0.7518
	Mn	0.3194	0.5372	1.5507	0.4544
	Mg	0.0446	0.9331	0.1578	0.0080

Results of a simple linear regression analysis of the relation between the contents of selected elements in the soil and in selected chickpea plant organs and seeds are shown. Statistical significant correlation at level of $p < 0.05$ is illustrated with red colour. The * symbol indicates that the difference was significant, also at a significance level of $p < 0.01$.

To obtain certain elements, especially micronutrients, plants need to control several steps during the journey from soil to seed, such as uptake, transport, remobilization and storage [67]. Apart from internal factors, the environment also influences the rate of micronutrient absorption [68–70]. The presence of phosphorus increases the contents of Ca, Mg, Fe, Mn and Zn in wheat, while it decreases the contents of Ca, Mg, Fe and Zn in chickpea [68]. In our results, we could observe a decreased content of phosphorus in the soil of Zone B and decreased contents of Mg, Fe and Zn in the chickpea leaves. Another external factor that affects the uptake of micronutrients is arbuscular mycorrhiza [69].

In our results, we did not observe a correlation between Fe contents in any of the observed chickpea organs and seeds. Contrary to this result, Mahmoudi et al. [51] revealed that the Fe content in plant tissues was strongly dependent on the Fe content in the soil.

4. Conclusions

The measured values confirm that, to reach the maximum effectiveness in producing important crops such as chickpea (*Cicer arietinum* L.), field and soil heterogeneity must be considered. In our study, we revealed the effect of the heterogeneity of certain elements (nutrients) and soil reactions on the ability of chickpea to uptake and translocate these elements into plant organs and seeds, thus proving that soil heterogeneity strongly affects the overall fitness of chickpea. The experimental plot we used was situated in a flatland with a relatively homogeneous chernozem soil type. The measured data indicated that extreme soil heterogeneity could be detected, even on the site which did not otherwise show it, at a level that affects the development of plants. The heterogeneity in the presented study consisted in the variable contents of carbonates in the soil and related changes in soil reactions, which were demonstrated by changes in plant uptake of nutrients and their translocation within the plant. In many cases, farmers can influence detected plot heterogeneity by taking appropriate measures (mineral and organic fertilization, liming, etc.) using a system of precision agriculture—in other words, a system of targeted farming. In this case, a crucial measure appears to be reduced input of calcium fertilizers in the parts of a plot that exhibit increased contents of carbonates in the soil.

Author Contributions: Conceptualization, A.K., J.Š., I.H., J.S., T.H., L.V. and J.E.; methodology, A.K., J.Š., I.H. and J.E.; validation, A.K. and T.L.; formal analysis, A.K., J.Š., I.H. and J.E.; investigation, A.K., J.Š., I.H. and K.V.; resources, A.K.; data curation, A.K., J.Š. and I.H.; writing—original draft preparation, A.K., J.Š., T.L. and J.E.; writing—review and editing, A.K. and J.E.; visualization, A.K. and J.E.; supervision, A.K.; project administration, A.K.; funding acquisition, A.K. All authors have read and agreed to the published version of the manuscript.

Funding: The result was obtained within the framework of institutional support MZE-RO1724.

Institutional Review Board Statement: Not applicable.

Informed Consent Statement: Not applicable.

Data Availability Statement: Data are available upon request from the corresponding author (jakub.elbl@mendelu.cz).

Conflicts of Interest: The authors declare no conflict of interest. Authors Antonín Kint, Ján Šmeringai, Igor Huňady, Julie Sobotková, Karel Vejražka and Jakub Elbl were employed by Agricultural Research, Ltd. The remaining authors declare that the research was conducted in the absence of any commercial or financial relationships that could be construed as a potential conflict of interest.

Appendix A

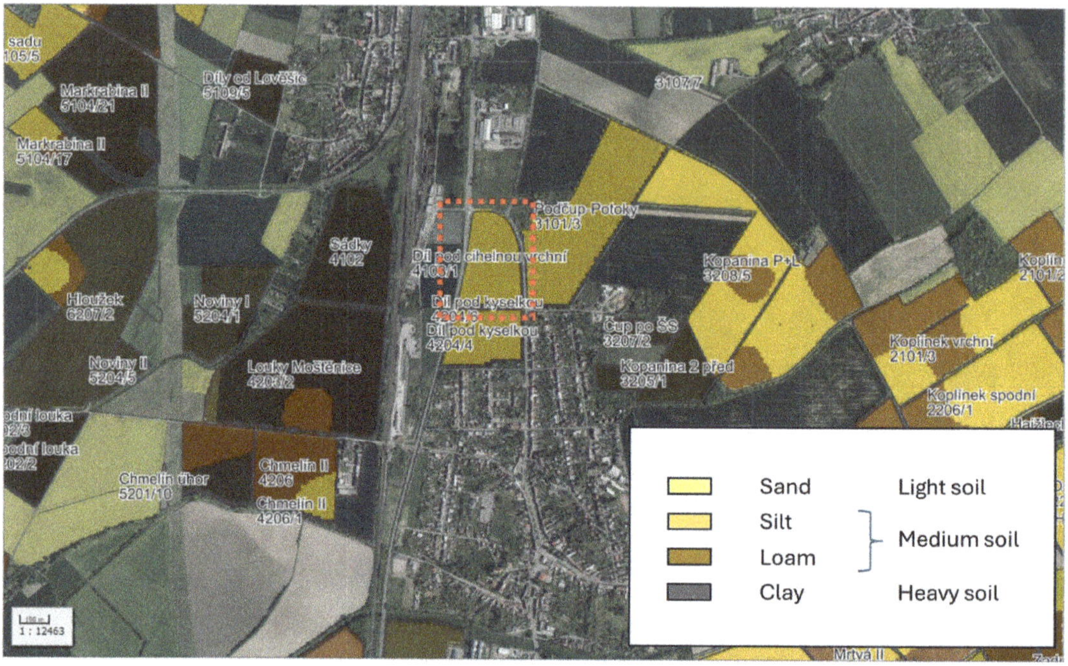

Figure A1. Soil type in the area of interest and the experimental field (marked in red). Source of data: PREFARM© system (MJM, Ltd., Litovel, Czech Republic).

Appendix B

Table A1. Results of *t*-test statistical analysis for independent samples according to groups—contents of macroelements in the soil.

Variable	Average A	Average B	Value		*p*	Number of A Values	Number of B Values	±SD A	±SD B	F	*p*
P	113.578	95.233	3.60617	4	0.022634	3	3	8.7538	1.00167	76.37357	0.025849
K	190.333	173.000	2.87122	4	0.045413	3	3	6.5064	8.18535	1.58268	0.774390
Mg	180.333	190.333	−1.12827	4	0.322297	3	3	13.3167	7.63763	3.04000	0.495050
Ca	3421.500	4537.000	−7.06657	3	0.005826	2	3	287.7925	58.66004	24.06989	0.078248

Statistical significant differences between individual variants of experiment in content of macro(nutrients)elements ($p < 0.05$) is illustrated with red color.

Table A2. Results of *t*-test statistical analysis for independent samples according to groups—contents of microelements in the soil.

Variable (mg/kg)	Average A	Average B	Value		*p*	Number of A Values	Number of B Values	±SD A	±SD B	F	*p*
Fe	188.1700	190.5440	−0.21800	4	0.838095	3	3	18.08849	5.344287	11.45578	0.160568
Mn	204.7267	215.6800	−3.51752	4	0.024505	3	3	1.85133	5.065807	7.48735	0.235645
Cu	5.1200	5.0500	1.91703	4	0.127708	3	3	0.01000	0.062450	39.00000	0.050000
Zn	9.0267	6.2700	4.27192	4	0.012930	3	3	0.82470	0.754387	1.19510	0.911119

Statistical significant differences between individual variants of experiment in content of micro(nutrients)elements ($p < 0.05$) is illustrated with red color.

Table A3. Results of *t*-test statistical analysis for independent samples according to groups—contents of microelements in individual plant parts.

Variable	Average A	Average B	Value		*p*	Number of A Values	Number of B Values	±SD A	±SD B	F	*p*
					Parameter: Fe (mg/kg)						
Stem	150.1387	116.1050	1.869704	4	0.134877	3	3	31.10846	5.12638	36.8244	0.052876
Leaf	248.4022	156.4934	5.662339	4	0.004796	3	3	11.15334	25.80693	5.3538	0.314772
Pod	141.0395	101.0936	7.249932	4	0.001922	3	3	9.03939	3.06005	8.7261	0.205632
Seed	127.6415	93.4381	5.321621	4	0.005999	3	3	11.11641	0.59497	349.0868	0.005713
					Parameter: Mn (mg/kg)						
Stem	27.00713	26.19582	1.960562	4	0.121476	3	3	0.40521	0.591214	2.129	0.639239
Leaf	97.68317	70.12653	2.318110	4	0.081305	3	3	20.58257	0.547139	1415.155	0.001412
Pod	19.00216	19.18008	−0.193791	4	0.855783	3	3	1.21092	1.030787	1.380	0.840323
Seed	29.04051	30.19522	−0.958327	4	0.392167	3	3	2.07546	0.218916	89.882	0.022007
					Parameter: Cu (mg/kg)						
Stem	4.728312	1.813914	54.20307	4	0.000001	3	3	0.093062	0.003547	688.5556	0.002900
Leaf	6.581654	1.297449	14.22110	4	0.000142	3	3	0.582956	0.272701	4.5698	0.359080
Pod	1.891200	1.291513	3.71838	4	0.020503	3	3	0.245971	0.132395	3.4516	0.449274
Seed	3.422010	3.957789	−1.15651	4	0.311839	3	3	0.742713	0.303711	5.9803	0.286522
					Parameter: Zn (mg/kg)						
Stem	12.44880	5.05823	131.9236	4	0.000000	3	3	0.010227	0.096492	89.02685	0.022216
Leaf	27.02489	6.86004	7.3801	4	0.001797	3	3	4.493783	1.484147	9.16791	0.196697
Pod	11.16978	5.33553	93.8513	4	0.000000	3	3	0.059551	0.089706	2.26919	0.611773
Seed	22.71773	25.11203	−2.2877	4	0.084068	3	3	1.412992	1.135567	1.54830	0.784838

Statistical significant differences between individual variants of experiment in content of Fe, Mn, Cu and Zn ($p < 0.05$) is illustrated with red color.

Table A4. Results of *t*-test statistical analysis for independent samples according to groups—contents of macroelements in individual plant parts.

Variable	Average A	Average B	Value		*p*	Number of A Values	Number of B Values	±SD A	±SD B	F	*p*
					Parameter: N (%)						
Stem	1.519458	0.645628	2.82313	4	0.047676	3	3	0.535946	0.013353	1611.008	0.001241
Leaf	2.089035	0.698369	23.83607	4	0.000018	3	3	0.062298	0.079565	1.631	0.760119
Pod	1.423570	0.316445	7.52089	4	0.001673	3	3	0.234417	0.100289	5.463	0.309431
Seed	2.552791	2.490321	1.60241	4	0.184325	3	3	0.048473	0.047010	1.063	0.969376
					Parameter: P (g/kg)						
Stem	1.519458	0.645628	2.82313	4	0.047676	3	3	0.535946	0.013353	1611.008	0.001241
Leaf	2.089035	0.698369	23.83607	4	0.000018	3	3	0.062298	0.079565	1.631	0.760119
Pod	1.423570	0.316445	7.52089	4	0.001673	3	3	0.234417	0.100289	5.463	0.309431
Seed	2.552791	2.490321	1.60241	4	0.184325	3	3	0.048473	0.047010	1.063	0.969376
					Parameter: K (g/kg)						
Stem	15.43472	18.43853	−2.87274	4	0.045343	3	3	1.598768	0.850849	3.53074	0.441429
Leaf	12.45889	6.90129	9.25038	4	0.000759	3	3	0.977703	0.356329	7.52856	0.234506
Pod	13.35620	14.92493	−0.97146	4	0.386330	3	3	2.695461	0.746606	13.03417	0.142509
Seed	10.19347	10.05765	0.40544	4	0.705904	3	3	0.116292	0.568471	23.89572	0.080335
					Parameter: Ca (g/kg)						
Stem	12.82150	16.06492	−5.17691	4	0.006620	3	3	1.018377	0.37480	7.3825	0.238591
Leaf	36.17069	46.16988	−1.36870	4	0.242925	3	3	1.308391	12.58589	92.5322	0.021383
Pod	11.08363	11.99436	−0.58697	4	0.588754	3	3	2.681937	0.17110	245.7086	0.008107
Seed	1.56556	2.01582	−3.62568	4	0.022245	3	3	0.042804	0.21080	24.2534	0.079197
					Parameter: Mg (g/kg)						
Stem	1.357471	0.991020	1.373126	4	0.241656	3	3	0.440246	0.140879	9.7655	0.185778
Leaf	2.042080	1.116067	7.580073	4	0.001624	3	3	0.211358	0.010004	446.3230	0.004471
Pod	1.139188	0.980919	1.508301	4	0.205970	3	3	0.176385	0.043826	16.1981	0.116292
Seed	1.088609	1.196372	−0.919453	4	0.409888	3	3	0.039462	0.199129	25.4636	0.075575

Statistical significant differences between individual variants of experiment in content of N, P, K, Ca and Mg ($p < 0.05$) is illustrated with red color.

Table A5. Results of statistical analysis via Tukey's post hoc HSD test—contents of macroelements in individual plant parts.

				Parameter: P (mg/kg)				
Variants	**A—Stem**	**A—Leaf**	**A—Pod**	**A—Seed**	**B—Leaf**	**B—Pod**	**B—Seed**	**B—Stem**
A—Stem		0.073303	0.999106	0.000592	0.004728	0.000238	0.001018	0.002699
A—Leaf	0.073303		0.026185	0.207612	0.000181	0.000175	0.353142	0.000178
A—Pod	0.999106	0.026185		0.000321	0.013572	0.000360	0.000463	0.007574
A—Seed	0.000592	0.207612	0.000321		0.000175	0.000175	0.999947	0.000175
B—Leaf	0.004728	0.000181	0.013572	0.000175		0.409081	0.000175	0.999983
B—Pod	0.000238	0.000175	0.000360	0.000175	0.409081		0.000175	0.580214
B—Seed	0.001018	0.353142	0.000463	0.999947	0.000175	0.000175		0.000175
B—Stem	0.002699	0.000178	0.007574	0.000175	0.999983	0.580214	0.000175	
				Parameter: K (mg/kg)				
Variants	**A—Stem**	**A—Leaf**	**A—Pod**	**A—Seed**	**B—Leaf**	**B—Pod**	**B—Seed**	**B—Stem**
A—Stem		0.134553	0.490919	0.002107	0.000177	0.999512	0.001652	0.128371
A—Leaf	0.134553		0.984106	0.391229	0.001219	0.298115	0.326429	0.000638
A—Pod	0.490919	0.984106		0.097769	0.000364	0.778691	0.077000	0.002800
A—Seed	0.002107	0.391229	0.097769		0.077875	0.005323	1.000000	0.000179
B—Leaf	0.000177	0.001219	0.000364	0.077875		0.000182	0.098860	0.000175
B—Pod	0.999512	0.298115	0.778691	0.005323	0.000182		0.004137	0.052254
B—Seed	0.001652	0.326429	0.077000	1.000000	0.098860	0.004137		0.000178
B—Stem	0.128371	0.000638	0.002800	0.000179	0.000175	0.052254	0.000178	
				Parameter: Mg (mg/kg)				
Variants	**A—Stem**	**A—Leaf**	**A—Pod**	**A—Seed**	**B—Leaf**	**B—Pod**	**B—Seed**	**B—Stem**
A—Stem		0.014492	0.881953	0.735994	0.821678	0.369054	0.972995	0.399907
A—Leaf	0.014492		0.001256	0.000769	0.000999	0.000345	0.002315	0.000366
A—Pod	0.881953	0.001256		0.999982	1.000000	0.975447	0.999959	0.982905
A—Seed	0.735994	0.000769	0.999982		1.000000	0.997426	0.997415	0.998631
B—Leaf	0.821678	0.000999	1.000000	1.000000		0.989867	0.999609	0.993576
B—Pod	0.369054	0.000345	0.975447	0.997426	0.989867		0.888437	1.000000
B—Seed	0.972995	0.002315	0.999959	0.997415	0.999609	0.888437		0.909912
B—Stem	0.399907	0.000366	0.982905	0.998631	0.993576	1.000000	0.909912	
				Parameter: Ca (mg/kg)				
Variants	**A—Stem**	**A—Leaf**	**A—Pod**	**A—Seed**	**B—Leaf**	**B—Pod**	**B—Seed**	**B—Stem**
A—Stem		0.000399	0.999698	0.115134	0.000175	0.999998	0.141609	0.985386
A—Leaf	0.000399		0.000264	0.000175	0.202119	0.000322	0.000175	0.001387
A—Pod	0.999698	0.000264		0.247113	0.000175	0.999996	0.295686	0.874817
A—Seed	0.115134	0.000175	0.247113		0.000175	0.167709	1.000000	0.023143
B—Leaf	0.000175	0.202119	0.000175	0.000175		0.000175	0.000175	0.000180
B—Pod	0.999998	0.000322	0.999996	0.167709	0.000175		0.203907	0.950893
B—Seed	0.141609	0.000175	0.295686	1.000000	0.000175	0.203907		0.029132
B—Stem	0.985386	0.001387	0.874817	0.023143	0.000180	0.950893	0.029132	

Statistical significant differences between individual plant organs in content of P, K, Mg and Ca (Tukey's post hoc HSD test, $p < 0.05$) is illustrated with red color.

References

1. Hirsch, P.R.; Mauchline, T.H. The Importance of the microbial N cycle in soil for crop plant nutrition. *Adv. Appl. Microbiol.* **2015**, *93*, 45–71. [CrossRef] [PubMed]
2. Mabrouk, Y.; Belhadj, O. Enhancing the biological nitrogen fixation of leguminous crops grown under stressed environments. *Afr. J. Biotechnol.* **2012**, *11*, 10809–10815. [CrossRef]
3. Adams, M.A.; Buchmann, N.; Sprent, J.; Buckley, T.N.; Turnbull, T.L. Crops, Nitrogen, Water: Are Legumes Friend, Foe, or Misunderstood Ally? *Trends Plant Sci.* **2018**, *23*, 539–550. [CrossRef] [PubMed]
4. Fustec, J.; Lesuffleur, F.; Mahieu, S.; Cliquet, J.-B. Nitrogen rhizodeposition of legumes. A review. *Agron. Sustain. Dev.* **2010**, *30*, 57–66. [CrossRef]
5. Jiao, X.; Lyu, Y.; Wu, X.; Li, H.; Cheng, L.; Zhang, C.; Yuan, L.; Jiang, R.; Jiang, B.; Rengel, Z.; et al. Grain production versus resource and environmental costs: Towards increasing sustainability of nutrient use in China. *J. Exp. Bot.* **2016**, *67*, 4935–4949. [CrossRef] [PubMed]
6. Luce, M.C.; Grant, C.A.; Zebarth, B.J.; Ziadi, N.; O'Donovan, J.T.; Blackshaw, R.E.; Harker, K.N.; Johnson, E.N.; Gan, Y.; Lafond; et al. Legumes can reduce economic optimum nitrogen rates and increase yields in a wheat–canola cropping sequence in western Canada. *Field Crop. Res.* **2015**, *179*, 12–25. [CrossRef]
7. Anglade, J.; Billen, G.; Garnier, J. Relationships for estimating N2 fixation in legumes: Incidence for N balance of legume-based cropping systems in Europe. *Ecosphere* **2015**, *6*, 37. [CrossRef]
8. Gan, Y.T.; Warkentin, T.D.; McDonald, C.L.; Zentner, R.P.; Vandenberg, A. Seed Yield and Yield Stability of Chickpea in Response to Cropping Systems and Soil Fertility in Northern Latitudes. *Agron. J.* **2009**, *101*, 1113–1122. [CrossRef]
9. Flowers, T.J.; Gaur, P.M.; Gowda, C.L.L.; Krishnamurthy, L.; Samineni, S.; Siddique, K.H.M.; Turner, N.C.; Vadez, V.; Varshney, R.K.; Colmer, T.D. Salt sensitivity in chickpea. *Plant Cell Environ.* **2010**, *33*, 490–509. [CrossRef]
10. Korbu, L.; Tafes, B.; Kassa, G.; Mola, T.; Fikre, A. Unlocking the genetic potential of chickpea through improved crop management practices in Ethiopia. A Review. *Agron. Sustain. Dev.* **2020**, *40*, 13. [CrossRef]
11. Carlsson, G.; Huss-Danell, K. Nitrogen fixation in perennial forage legumes in the field. *Plant Soil* **2003**, *253*, 353–372. [CrossRef]
12. Ullah, A.; Farooq, M.; Rehman, A.; Hussain, M.; Siddique, K.H.M. Zinc nutrition in chickpea (*Cicer arietinum*): A review. *Crop Pasture Sci.* **2020**, *71*, 199. [CrossRef]
13. Grewal, S.K.; Sharma, K.P.; Bharadwaj, R.D.; Hegde, V.; Tripathi, S.; Singh, S.; Kumar Jain, P.; Kumar Agrawal, P.; Mondal, B. Understanding genotypic variation and identification of promising genotypes for iron and zinc content in chickpea (*Cicer arietinum* L.). *J. Food Compos. Anal.* **2020**, *88*, 1034548. [CrossRef]
14. Singh, M.; Bhardwaj, C.; Singh, S.; Panatu, S.; Chaturvedi, S.K.; Rana, J.C.; Rizvi, A.H.; Kumar, N.; Sarker, A. Chickpea genetic resources and its utilization in India: Current status and future prospects. *Indian J. Genet. Plant Breed.* **2016**, *76*, 515–529. [CrossRef]
15. Maphosa, L.; Richards, M.F.; Norton, S.L.; Nguyen, G.N. Breeding for Abiotic Stress Adaptation in Chickpea (*Cicer arietinum* L.): A Comprehensive Review. *Crop Breed. Genet. Genom.* **2020**, *2*, e200015. [CrossRef]
16. Galili, S.; Ran, H.; Dor, E.; Hershenhorn, J.; Harel, A.; Amir-Segev, O.; Bellalou, A.; Badani, H.; Smirnov, E.; Achdari, G. The history of chickpea cultivation and breeding in Israel. *Isr. J. Plant Sci.* **2018**, *65*, 186–194. [CrossRef]
17. Devasirvatham, V.; Tan, D.K.Y.; Gaur, P.M.; Raju, T.N.; Trethowan, R.M. High temperature tolerance in chickpea and its implications for plant improvement. *Crop Pasture Sci.* **2012**, *63*, 419–428. [CrossRef]
18. Pande, S.; Siddique, K.H.M.; Kishore, G.K.; Bayaa, B.; Gaur, P.M.; Gowda, C.L.L.; Bretag, T.W.; Crouch, J.H. Ascochyta blight of chickpea (*Cicer arietinum* L.): A review of biology, pathogenicity, and disease management. *Aust. J. Agric. Res.* **2005**, *56*, 317–332. [CrossRef]
19. Esfahani, M.N.; Sulieman, S.; Schulze, J.; Yamaguchi-Shinozaki, K.; Shinozaki, K.; Tran, L.-S.P. Mechanisms of physiological adjustment of N_2 fixation in *Cicer arietinum* L. (chickpea) during early stages of water deficit: Single or multi-factor controls. *Plant J.* **2014**, *79*, 964–980. [CrossRef]
20. Kumar, M.; Yusuf, M.A.; Nigam, M.; Kumar, M. An Update on Genetic Modification of Chickpea for Increased Yield and Stress Tolerance. *Mol. Biotechnol.* **2018**, *60*, 651–663. [CrossRef]
21. Rachwa-Rosiak, D.; Nebesny, E.; Budryn, G. Chickpeas—Composition, Nutritional Value, Health Benefits, Application to Bread and Snacks: A Review. *Crit. Rev. Food Sci. Nutr.* **2015**, *55*, 1137–1145. [CrossRef] [PubMed]
22. Kren, J.; Klem, K.; Svobodova, I.; Misa, P.; Lukas, V. Influence of Sowing, Nitrogen Nutrition and Weather Conditions on Stand Structure and Yield of Spring Barley. *Cereal Res. Commun.* **2015**, *43*, 326–335. [CrossRef]
23. Rodriguez-Moreno, F.; Lukas, V.; Neudert, L.; Dryšlová, T. Spatial interpretation of plant parameters in winter wheat. *Precis. Agric.* **2014**, *15*, 447–465. [CrossRef]
24. Mezera, J.; Lukas, V.; Horniaček, I.; Smutný, V.; Elbl, J. Comparison of Proximal and Remote Sensing for the Diagnosis of Crop Status in Site-Specific Crop Management. *Sensors* **2022**, *22*, 19. [CrossRef]
25. Xue, W.; Bezemer, T.M.; Berendse, F. Soil heterogeneity and plant species diversity in experimental grassland communities: Contrasting effects of soil nutrients and pH at different spatial scales. *Plant Soil* **2019**, *442*, 497–509. [CrossRef]
26. Tittonell, P.; Vanlauwe, B.; Corbeels, M.; Giller, K.E. Yield gaps, nutrient use efficiencies and response to fertilisers by maize across heterogeneous smallholder farms of western Kenya. *Plant Soil* **2008**, *313*, 19–37. [CrossRef]

27. Habib-ur-Rahman, M.; Raza, A.; Ahrends, H.E.; Hüging, H.; Gaiser, T. Impact of in-field soil heterogeneity on biomass and yield of winter triticale in an intensively cropped hummocky landscape under temperate climate conditions. *Precis. Agric.* **2021**, *23*, 912–938. [CrossRef]
28. Adomako, M.O.; Roiloa, S.; Yu, F.-H. Potential roles of soil microorganisms in regulating the effect of soil nutrient heterogeneity on plant performance. *Microorganisms* **2022**, *10*, 2399. [CrossRef]
29. Jakobsen, S.T. Interaction between phosphate and calcium in nutrient uptake by plant roots. *Commun. Soil Sci. Plant Anal.* **2008**, *10*, 141–152. [CrossRef]
30. Lukas, V.; Neudert, L.; Duffková, R.; Mezera, J.; Horniaček, I.; Širůček, P.; Krček, V. *Mapa Relativního Výnosového Potenciálu pro Pozemky AGRA Řisuty*; Mendelova Univerzita: Brno, Czech Republic, 2020; ISBN 978-80-7509-746-0.
31. *ISO 10381-6*; Soil Quality—Selection and Application of Sampling Techniques. International Organization for Standardization (ISO): Geneva, Switzerland, 2017.
32. Jones, J.B. *Laboratory Guide for Conducting Soil Tests and Plant Analysis*; CRC Press: Boca Raton, FL, USA, 2001; 384p, ISBN 9780429132117.
33. Olsen, S.R.; Sommers, L.E. Phosphorus. In *Methods of Soil Analysis*; Agronomy Monographs; American Society of Agronomy, Soil Science Society of America: Madison, WI, USA, 1983; pp. 403–430. [CrossRef]
34. Mehlich, A. Mehlich 3 soil test extractant: A modification of Mehlich 2 extractant. *Commun. Soil Sci. Plant Anal.* **1984**, *15*, 1409–1416. [CrossRef]
35. Sarojam, P. *Analysis of Micronutrients in Soil by Using AA 800 Atomic Absorption Spectrophotometer—Application Note*; PerkinElmer, Inc.: Waltham, MA, USA, 2009; 5p. Available online: https://resources.perkinelmer.com/labsolutions/resources/docs/app_micronutrientsinsoilbyaa.pdf (accessed on 28 October 2023).
36. Berhe, A.A. Drivers of soil change. In *Developments in Soil Science*; Busse, M., Giardina, P.C., Morris, D.M., Page-Dumroese, D.S., Eds.; Elsevier: Amsterdam, The Netherlands, 2019; Chapter 3; Volume 36, pp. 27–42. ISBN 9780444639981. [CrossRef]
37. Liu, Y.; Li, G.; Wang, M.; Yan, W.; Hou, F. Effects of three-dimensional soil heterogeneity and species composition on plant biomass and biomass allocation of grass-mixtures. *AoB Plants* **2021**, *13*, plab033. [CrossRef]
38. Elbl, J.; Kintl, A.; Brtnický, M.; Širůček, P.; Mezera, J.; Smutný, V.; Vopravil, J.; Holátko, J.; Huňady, I.; Lukas, V. Assessment of the effect of optimised field plot size on the crop yield. *Plant Soil Environ.* **2023**, *69*, 447–462. [CrossRef]
39. Tekin, A.; Gunal, H.; Sindir, K.; Balcı, Y. Spatial structure of available micronutrient contents and their relationships with other soil characteristics and corn yield. Fresenius Environmental Bulletin. *Fresenius Environ. Bull.* **2011**, *20*, 783–792.
40. García-Palacios, P.; Maestre, F.T.; Gallardo, A. Soil nutrient heterogeneity modulates ecosystem responses to changes in the identity and richness of plant functional groups. *J. Ecol.* **2011**, *99*, 551–562. [CrossRef]
41. Lukas, V.; Hunady, I.; Kintl, A.; Mezera, J.; Hammerschmiedt, T.; Sobotková, J.; Brtnický, M.; Elbl, J. Using UAV to Identify the Optimal Vegetation Index for Yield Prediction of Oil Seed Rape (*Brassica napus* L.) at the Flowering Stage. *Remote Sens.* **2022**, *4*, 4953. [CrossRef]
42. Shukla, A.K.; Behera, S.K.; Lenka, N.K.; Tiwari, P.K.; Prakash, C.; Malik, R.S.; Sinha, N.K.; Singh, V.K.; Patra, A.K.; Chaudhary, S.K. Spatial variability of soil micronutrients in the intensively cultivated Trans-Gangetic Plains of India. *Soil Tillage Res.* **2016**, *163*, 282–289. [CrossRef]
43. Khaitov, B.; Kurbonov, A.; Abdiev, A.; Adilov, M. Effect of chickpea in association with Rhizobium to crop productivity and soil fertility. *Eurasian J. Soil Sci.* **2016**, *5*, 105–112. [CrossRef]
44. Pal, V.; Singh, G.; Dhaliwal, S.S. Symbiotic Parameters, Growth, Productivity and Profitability of Chickpea as Influenced by Zinc Sulphate and Urea Application. *J. Soil Sci. Plant Nutr.* **2020**, *20*, 738–750. [CrossRef]
45. Joines, D.K.; Hardy, D.H. Acetate and Mehlich-3 Extractable Sulfate-Sulfur. In *Soil Test Methods from the Southeastern United States*; Sikora, F.J., Moore, K.P., Eds.; Southern Cooperative Series Bulletin No. 419; Southern Cooperative: Clemson, SC, USA, 2014; pp. 124–130. ISBN 1-58161-419-5.
46. Li, G.; Wang, M.; Ma, C.; Tao, R.; Hou, F.; Liu, Y. Effects of Soil Heterogeneity and Species on Plant Interactions. *Front. Ecol. Evol.* **2021**, *25*, 756344. [CrossRef]
47. Pushnik, J.C.; Miller, G.W.; Manwaring, J.H. The role of iron in higher plant chlorophyll biosynthesis, maintenance and chloroplast biogenesis. *J. Plant Nutr.* **2008**, *7*, 733–758. [CrossRef]
48. Vigani, G.; Murgia, I. Iron-Requiring Enzymes in the Spotlight of Oxygen. *Trends Plant Sci.* **2018**, *23*, 874–882. [CrossRef] [PubMed]
49. Higuchi, K.; Saito, A. Elucidation of efficient photosynthesis in plants with limited iron. *Soil Sci. Plant Nutr.* **2022**, *68*, 505–513. [CrossRef]
50. Mahmoudi, H.; Ksouri, R.; Gharsalli, M.; Lachaâl, M. Differences in responses to iron deficiency between two legumes: Lentil (*Lens culinaris*) and chickpea (*Cicer arietinum*). *J. Plant Physiol.* **2005**, *162*, 1237–1245. [CrossRef] [PubMed]
51. Mahmoudi, H.; Labidi, N.; Ksouri, R.; Gharsalli, M.; Abdelly, C. Differential tolerance to iron deficiency of chickpea varieties and Fe resupply effects. *C. R. Biol.* **2007**, *330*, 237–246. [CrossRef] [PubMed]
52. Ullah, A.; Farooq, M.; Qadeer, A.; Sanaullah, M. Impact of zinc and plant growth-promoting bacteria on soil health as well as aboveground biomass of *desi* and *kabuli* chickpea under arid conditions. *J. Sci. Food Agric.* **2022**, *102*, 2262–2269. [CrossRef]
53. Hidoto, L.; Worku, W.; Mohammed, H.; Bunyamin, T. Effects of zinc application strategy on zinc content and productivity of chickpea grown under zinc deficient soils. *J. Soil Sci. Plant Nutr.* **2017**, *17*, 112–126. [CrossRef]

54. Luo, J.; Qi, S.; Peng, L.; Xie, X. Enhanced phytoremediation capacity of a mixed-species plantation of *Eucalyptus globulus* and *Chickpeas*. *J. Geochem. Explor.* **2017**, *182*, 201–205. [CrossRef]
55. Kambhampati, M.S.; Vu, V.T. EDTA Enhanced Phytoremediation of Copper Contaminated Soils Using Chickpea (*Cicer aeritinum* L.). *Bull. Environ. Contam. Toxicol.* **2013**, *91*, 310–313. [CrossRef] [PubMed]
56. Bhakuni, G.; Dube, B.K.; Sinha, P.; Chatterjee, C. Copper Stress Affects Metabolism and Reproductive Yield of Chickpea. *J. Plant Nutr.* **2009**, *32*, 703–711. [CrossRef]
57. Beale, S.I. Enzymes of chlorophyll biosynthesis. *Photosynth. Res.* **1999**, *60*, 43–73. [CrossRef]
58. Yalçın Gülüt, K.; Özdemir, O. Phosphorus tolerance levels of different chickpea genotypes. *Saudi J. Biol. Sci.* **2021**, *28*, 5386–5390. [CrossRef] [PubMed]
59. Yahiya, M.; Samiullah; Fatma, A. Influence of phosphorus on nitrogen fixation in chickpea cultivars. *J. Plant Nutr.* **1995**, *18*, 719–727. [CrossRef]
60. Imen, H.; Neila, A.; Adnane, B.; Manel, B.; Mabrouk, Y.; Saidi, M.; Bouaziz, S. Inoculation with Phosphate Solubilizing Mesorhizobium Strains Improves the Performance of Chickpea (*Cicer aritenium* L.) Under Phosphorus Deficiency. *J. Plant Nutr.* **2015**, *38*, 1656–1671. [CrossRef]
61. Dokwal, D.; Romshdal, T.B.; Kunz, A.D.; Alonso, A.P.; Dickstein, R. Phosphorus deprivation affects composition and spatial distribution of membrane lipids in legume nodules. *Plant Physiol.* **2021**, *185*, 1847–1859. [CrossRef]
62. Gul, J.; Ullah, M. Biochemical, physiological, and growth evaluation of different chickpea genotypes under varying salinity regimes. *Braz. J. Biol.* **2022**, *82*, e268350. [CrossRef]
63. Saxena, A.K.; Rewari, R.B. Influence of phosphate and zinc on growth, nodulation and mineral composition of chickpea (*Cicer arietinum* L.) under salt stress. *World J. Microbiol. Biotechnol.* **1991**, *7*, 202–205. [CrossRef]
64. Khan, H.A.; Siddique, K.H.M.; Colmer, T.D. Salt sensitivity in chickpea is determined by sodium toxicity. *Planta* **2016**, *244*, 623–637. [CrossRef] [PubMed]
65. Turner, N.C.; Colmer, T.D.; Quealy, J.; Pushpavalli, R.; Krishnamurthy, L.; Kaur, J.; Singh, G.; Siddique, K.H.M.; Vadez, V. Salinity tolerance and ion accumulation in chickpea (*Cicer arietinum* L.) subjected to salt stress. *Plant Soil* **2013**, *365*, 347–361. [CrossRef]
66. Kotula, L.; Clode, P.L.; Jimenez, J.D.L.C.; Colmer, T.D. Salinity tolerance in chickpea is associated with the ability to 'exclude' Na from leaf mesophyll cells. *J. Exp. Bot.* **2019**, *70*, 4991–5002. [CrossRef]
67. Sperotto, R.A.; Ricachenevsky, F.K.; Williams, L.E.; Vasconcelos, M.W.; Menguer, P.K. From soil to seed: Micronutrient movement into and within the plant. *Front. Plant Sci.* **2014**, *5*, 468. [CrossRef]
68. Li, L.; Tang, C.; Rengel, Z.; Zhang, F.S. Calcium, magnesium and microelement uptake as affected by phosphorus sources and interspecific root interactions between wheat and chickpea. *Plant Soil* **2004**, *261*, 29–37. [CrossRef]
69. Farzaneh, M.; Vierheilig, H.; Lössl, A.; Kaul, H.P. Arbuscular mycorrhiza enhances nutrient uptake in chickpea. *Plant Soil Environ.* **2011**, *57*, 465–470. [CrossRef]
70. Marschner, P.; Rengel, Z. Nutrient availability in soils. In *Marschner's Mineral Nutrition of Plants*; Academic Press: Cambridge, MA, USA, 2023; pp. 499–522. [CrossRef]

soil systems

MDPI

Article

Impact of Crop Residue, Nutrients, and Soil Moisture on Methane Emissions from Soil under Long-Term Conservation Tillage

Rajesh Choudhary [1], Sangeeta Lenka [2,*], Dinesh Kumar Yadav [2], Narendra Kumar Lenka [2,*], Rameshwar S. Kanwar [3], Abhijit Sarkar [2], Madhumonti Saha [2], Dharmendra Singh [2] and Tapan Adhikari [2]

1 Department of Soil Science and Agriculture Chemistry, College of Agriculture, Gwalior 474011, India; choudharyrajesh721@gmail.com
2 ICAR—Indian Institute of Soil Science, Nabibagh, Bhopal 462038, India; yadavdinu501@gmail.com (D.K.Y.); abhijitiiwm@gmail.com (A.S.); madhumonti2609@gmail.com (M.S.); dharamstomar17@gmail.com (D.S.); tapan_12000@rediffmail.com (T.A.)
3 Department of Agricultural and Biosystems Engineering, Iowa State University, Ames, IA 50011, USA; rskanwar@iastate.edu
* Correspondence: sangeeta.lenka@icar.gov.in (S.L.); nklenka@rediffmail.com (N.K.L.)

Abstract: Greenhouse gas emissions from agricultural production systems are a major area of concern in mitigating climate change. Therefore, a study was conducted to investigate the effects of crop residue, nutrient management, and soil moisture on methane (CH_4) emissions from maize, rice, soybean, and wheat production systems. In this study, incubation experiments were conducted with four residue types (maize, rice, soybean, wheat), seven nutrient management treatments {N0P0K0 (no nutrients), N0PK, N100PK, N150PK, N100PK + manure@ 5 Mg ha^{-1}, N100PK + biochar@ 5 Mg ha^{-1}, N150PK+ biochar@ 5 Mg ha^{-1}}, and two soil moisture levels (80% FC, and 60% FC). The results of this study indicated that interactive effects of residue type, nutrient management, and soil moisture significantly affected methane (CH_4) fluxes. After 87 days of incubation, the treatment receiving rice residue with N100PK at 60% FC had the highest cumulative CH_4 mitigation of −19.4 μg C kg^{-1} soil, and the highest emission of CH_4 was observed in wheat residue application with N0PK at 80% FC (+12.93 μg C kg^{-1} soil). Nutrient management had mixed effects on CH_4 emissions across residue and soil moisture levels in the following order: N150PK > N0PK > N150PK + biochar > N0P0K0 > N100PK + manure > N100PK + biochar > N100PK. Decreasing soil moisture from 80% FC to 60% FC reduced methane emissions across all residue types and nutrient treatments. Wheat and maize residues exhibited the highest carbon mineralization rates, followed by rice and soybean residues. Nutrient inputs generally decreased residue carbon mineralization. The regression analysis indicated that soil moisture and residue C mineralization were the two dominant predictor variables that estimated 31% of soil methane fluxes in Vertisols. The results of this study show the complexity of methane dynamics and emphasize the importance of integrated crop, nutrient, and soil moisture (irrigation) management strategies that need to be developed to minimize methane emissions from agricultural production systems to mitigate climate change.

Keywords: methane; mitigation; crop residue; soil moisture; nutrient; residue mineralization

1. Introduction

Methane (CH_4) is a potent greenhouse gas (GHG) that significantly contributes to global warming. Thus, it is crucial to consider the role of CH_4 fluxes in the global carbon cycle. Methane, in particular, has seen a significant increase in atmospheric concentration, reaching 1.5 times the levels observed in pre-industrial times [1]. Methane contributes 18% of the global warming potential, making it the second-highest contributor to long-lived GHGs. Agriculture, forestry, and other land use (AFLOU) are responsible for approximately 22%

of global net anthropogenic emissions, with AFLOU-CH_4 accounting for almost 41% of the total net anthropogenic CH_4 emissions, with agriculture accounting for 88% of the AFOLU component [2]. When it comes to agricultural activities like managing residue and applying nutrients, it is important to comprehend how different types of residue, nutrient management, and soil moisture interact and influence methane emissions. That is because these factors have varying effects on the release of methane [3–7]. Different residue types, including maize, rice, soybean, and wheat, possess unique chemical compositions and decomposition rates, which can affect the soil's methane production and consumption processes [8–11]. Nutrient management practices, such as fertilization with nitrogen (N), phosphorus (P), and potassium (K), as well as the use of biochar and organic manure, can alter soil microbial activity and nutrient availability, influencing methane emissions [9,12–17]. Soil moisture content is another critical factor that regulates methane production and consumption, as it affects the availability of oxygen required for methane oxidation [4,18–20]. This study aims to develop effective mitigation strategies by examining the effect of crop residue type, nutrient management, and soil moisture with the underlying mechanisms on CH_4 emission.

Crop residues, commonly used to enhance soil fertility and soil health [21], can serve as both sources and sinks of atmospheric CO_2 and CH_4 [8]. Adding crop residues provides carbon substrates and nutrients that promote methanogenesis, increasing CH_4 production and consumption [16,22–24]. Residue incorporation can create anaerobic microsites, increasing soil moisture and favoring methanogenesis and CH_4 emissions [25]. On the other hand, improved aeration due to residue addition enhances CH_4 oxidation by promoting methanotroph activity [26,27]. Methanogenesis, carried out by methanogenic archaea in anaerobic environments, is stimulated by carbon-rich crop residues, leading to increased CH_4 emissions [28]. Conversely, methane oxidation, performed by methane-oxidizing bacteria in aerobic conditions, can be influenced by residue addition through changes in soil properties and oxygen availability [29]. Additionally, crop residues can indirectly influence methane oxidation by altering soil properties, such as oxygen availability, pH, and nutrient availability, which affect the activity and abundance of methanotrophs [29–31].

The effect of crop residue on CH_4 emissions is also driven by soil moisture and nutrient concentration. Soil moisture content is critical in regulating CH_4 emissions by influencing soil water and oxygen availability for microbial activity, carbon and nitrogen mineralization, and CO_2 respiration [18,20,32]. Decreased soil moisture levels can enhance CH_4 uptake under semi-arid conditions due to increased oxygen diffusivity, stimulating soil CH_4 oxidation [32–34]. Excess soil moisture through flood irrigation with straw incorporation resulted in the highest average CH_4 fluxes, leading to a total CH_4 emission of -0.94 kg ha^{-1}. In the wheat–maize cropping system, straw incorporation (ca. straw removal) reduced CH_4 emission by 17.1% with surface drip irrigation and 14.0% with partial root-zone irrigation [26]. Du et al. [35] reported that limited irrigation and nitrogen management resulted in a relatively higher cumulative CH_4 uptake in the wheat season in the wheat–maize cropping system and reduced greenhouse gas intensity without additional cost. In a laboratory experiment, Korkiakoski et al. [36] demonstrated that excess soil moisture with fresh carbon input reduced the CH_4 oxidation potential of soil. A literature review reported a shift in the balance between methanogens and methanotrophs' activities and abundance influencing either an increase/decrease in soil CH_4 emission under different soil moisture content driven by soil organic input and nitrate nitrogen concentration [37].

Soil nutrient concentrations, particularly nitrogen and phosphorus, affect CH_4 emissions or oxidation, with excess nitrogen promoting methanogenesis and nutrient limitation enhancing methane oxidation [12,27]. Additionally, adding organic residues has increased CH_4 oxidation in clay soil [38]. Shaukat et al. [12] reported that incorporating a biochar amendment at a rate of 2% in conjunction with a nitrogen (N) application of 140 kg N per hectare can be a promising approach to mitigate CH_4 emissions from paddy rice cultivation in an Alfisol soil. Sainju et al. [7] reported that N rates did not affect soil CH_4 uptake but increased soil CO_2 fluxes in the northern Great Plains, USA. A meta-analysis concluded that CH_4 emissions were stimulated at low N application rates (<100 kg N ha^{-1}) but inhibited at high N rates (>200 kg N ha^{-1}) as compared to no N fertilizer (control) [39]. Applying chemical NPK fertilizer (240 kg urea-N ha^{-1}, 90 kg P_2O_5 ha^{-1}, and 120 kg K_2O ha^{-1}), manure, and their combination increased seasonal mean CH_4 emissions by 67.4%, 20.4%, and 101.2%, respectively, compared with PK (90 kg P_2O_5 ha^{-1} and 120 kg K_2O ha^{-1}) treatment without N fertilizer input in rice paddies of China [16]. Nutrient addition alters the soil elemental stoichiometry (C:N:P ratio) with residue C input, resulting in varied responses of GHG emission from residue return soils [1]. There is growing evidence of biochar as an amendment for soil carbon sequestration [40–42]; however, previous researchers have reported positive [43], negative [40], and uncertain [44] effects of biochar on mitigating CH_4 emissions from agricultural soils. Several environmental [44], soil [40,44], and management [45,46] factors regulate the effectiveness of biochar, including the rate of N fertilization and crop residue type. Therefore, in this study, we aimed to assess the integrated application of synthetic fertilizers with manure/biochar as a nutrient management strategy for evaluating the responses to CH_4 emissions.

By considering these factors (residue types, nutrient application, and soil moisture) and investigating the associated mechanisms, we can develop effective strategies to mitigate methane emissions while maintaining crop productivity. Furthermore, quantifying the impact of residue, nutrients, and soil moisture on residue carbon mineralization can improve our understanding of greenhouse gas inventories and enhance predictive models for assessing climate change impacts. In this study, we hypothesize the following: (1) residue types and nutrient management will significantly impact methane emissions, with variations observed among different residue types and nutrient treatments; (2) soil moisture content will interact with residue and nutrient management, leading to distinct methane flux patterns under different moisture conditions; and (3) specific soil properties and microbial processes, such as labile organic carbon, nutrient availability, and residue C mineralization, will significantly mediate methane emissions in response to residue, nutrient, and moisture conditions.

2. Materials and Methods

2.1. Study Site

The experiment was conducted in the ICAR-Indian Institute of Soil Science laboratory in Bhopal, India. The place is at 23°15′ N latitude and 77°25′ E longitude, with an elevation of 427 m above sea level and a humid subtropical climate. The soil is deep Vertisols (IsohyperthermicTypicHaplustert) with a clayey texture (54% clay). Its bulk density is 1.34 Mg m^{-3} at 0.27 g g^{-1} soil water content and has 0.99% total soil organic carbon content (0–15 cm depth). The soil is neutral to alkaline (pH—7.85) with an electrical conductivity of 0.3 ds m^{-1}, and Ca^{2+} is the main exchangeable cation in the Ap horizon. The soil sample for incubation was collected from the top 0–15 cm of soil after harvesting wheat in 2020. It came from a 12-year conservation tillage experiment in a soybean–wheat system. The production method involved reduced tillage with 30% residue return plus fertilizer (30:60:30 kg N–P_2O_5–K_2O ha^{-1} for soybean and 100:60:30 kg N–P_2O_5–K_2O ha^{-1} for wheat).

2.2. Incubation Experimental Detail

The soil sample for incubation was collected from the top 0–15 cm of soil after harvesting wheat in 2020. It came from a 12-year conservation tillage experiment in a soybean–wheat system. The production method involved reduced tillage with 30% residue return plus fertilizer (30:60:30 kg N–P_2O_5–K_2O ha^{-1} for soybean and 100:60:30 kg N–P_2O_5–K_2O ha^{-1} for wheat). The soil samples were sieved to remove big fragments and stored at 4 °C until further study. Various crop residues like rice, maize, soybean, and wheat were air-dried, then milled and sieved to 2 mm. A subsample of crop residues was dried for water content assessment, while others were analyzed chemically. An elemental analyzer (NC analyzer, Thermofisher Scientific, Rodano, Italy, Flash 2000 model) and the acid detergent fiber method were used to determine C and N concentrations and lignin and cellulose contents, respectively. The total carbon/nitrogen (TC:TN) ratio of the organic amendments was 8:1 in manure, 45:1 (biochar), 76:1 (wheat straw), 50:1 (rice straw), 61:1 (soybean straw), and 65:1 (maize straw), respectively. The mean lignin content was 13.1% (*w*/*w*), 13.5% (*w*/*w*), 15.4% (*w*/*w*), and 9.0% (*w*/*w*) for wheat, rice, soybean, and maize straw, respectively. The mean cellulose content was 56.0% (*w*/*w*), 29.0% (*w*/*w*), 42.3% (*w*/*w*), and 49.7% (*w*/*w*) for wheat, rice, soybean, and maize straw, respectively. Reference is made to our previous work (Lenka et al. [5] and Raul et al. [47]) for more details on the properties of biochar and crop residues.

The soil had been pre-incubated for ten days at 70% of the two moisture levels (80% FC and 60% FC) and room temperature to kickstart microbial activity. Following pre-incubation, the crop residues (<2 mm) were completely mixed with soil (<2 mm) for incubation. A factorial experiment was set up with three replications to investigate the impact of crop residue type, nutrient levels, and soil moisture on CH_4 emission and carbon mineralization. The experiment consisted of five different levels of crop residue (wheat straw, maize straw, soybean straw, rice straw, and no residue), two levels of soil moisture content (80% FC and 60% FC), and seven nutrient treatments (N0P0K0, no nutrients; N0PK; N100PK; N150PK; N100PK + manure@ 5 Mg ha^{-1}; N100PK + biochar@ 5 Mg ha^{-1}; and N150PK+ biochar@ 5 Mg ha^{-1}). The various degrees of nutrient management have been used to simulate the impact of synthetic fertilizer or the combined use of synthetic fertilizers and organic amendment (manure/biochar) on CH_4 emissions. Two treatments were set up in 460 mL glass jars: (a) 20 g soils (dry weight basis) combined with wheat, maize, soybean, or rice straw residues at a rate of 2.23 mg g^{-1} soil, equaling 5 Mg ha^{-1} residue incorporation; and (b) 20 g soil (dry weight basis) without crop residue (control). Seven nutrient treatments were applied to both the control soil and residue-incorporated soil: N0P0K0 (no nutrients), N0PK, N100PK, N150PK, N100PK + manure@ 5 Mg ha^{-1}, N100PK + biochar@ 5 Mg ha^{-1}, and N150PK+ biochar@ 5 Mg ha^{-1}. The treatments N100 and N150 represented N application rates of 100 kg N ha^{-1} and 150 kg N ha^{-1}, respectively, using AR-grade ammonium nitrate. Besides the no-nutrient (N0P0K0) treatment, the same concentrations of phosphorus (P)@ 22 kg ha^{-1} and potassium (K)@ 21 kg ha^{-1} were added to the rest of the six nutrient treatments to assess the effect of increasing N levels. The application of phosphorus and potassium was made through AR-grade potassium dihydrogen phosphate to maintain a nutrient ratio of N: P_2O_5: K_2O of 4:2:1, equivalent to 100 kg N ha^{-1}. Nutrients were given in a solution made with distilled water containing NH_4NO_3 + KH_2PO_4 with pH adjusted to 7 using 1 M NaOH while maintaining incubation moisture levels at 80% FC and 60% FC. Field capacity was measured at matric potentials of −33 kPa using sieved (< 2 mm) soil samples in pressure plate extractors from Soil Moisture Equipment Corp., Santa Barbara, CA, with FC moisture content at 0.27 m^3 m^{-3}. The two soil moisture levels, 80% FC and 60% FC, were selected for the incubation study to represent optimal and deficit moisture conditions, respectively. The 80% FC provides an environment conducive to microbial activity, supporting organic matter decomposition and nutrient cycling processes. This level mimics near-optimal conditions where microbial communities are most active. In contrast, 60% FC represents a moderate moisture deficit, which helps study how reduced water availability impacts microbial processes, particularly those involved in decomposing organisms and methane production and consumption. That allows for assessing microbial responses and greenhouse gas emissions under varying moisture regimes. A blank glass jar without soil or residue was included to consider atmospheric CO_2 and CH_4 concentration in the headspace of incubation jars for determining evolved gasses from the treatments. Evolved gasses (CO_2/CH_4) from the treatments were calculated alongside a blank jar without soil or residue, accounting for atmospheric CO_2/CH_4 concentration at an incubation temperature of 30 °C based on the region's long-term average temperature over an incubation period spanning 87 days where soil moisture was maintained through regular weighing and water addition to make up for evaporation losses during gas sampling intervals.

2.3. Greenhouse Gas Sampling and Measurements

Headspace gasses were sampled with a syringe before being promptly transferred to an evacuated glass vial at set intervals on specific days and analyzed using gas chromatography (Agilent Technologies model 7890A, Santa Clara, CA, USA). These days included 0, 1, 4, 10, 17, 26, 33, 40, 47, 57, 67, 77, and 87 days of incubation. The purpose of these uneven intervals was to capture the usual asymptotic decrease observed in incubation experiments. All jars remained open for half an hour to replenish headspace oxygen and CO_2 to the normal concentration before being tightly sealed with aluminum caps on each sampling day. The flux rate (CH_4/CO_2) was determined by calculating the change in headspace concentration (µg C or mg C) per kg of soil (dry wt. equivalent) over a unit time of incubation using the ideal gas law. Cumulative CO_2 and CH_4 emissions were computed by integrating the fluxes from each measurement time. Apparent residue C mineralization was evaluated as the difference in CO_2 emission between soil amended with residue and control soil at the corresponding nutrient level [5].

2.4. Post Incubation Soil Analysis

During the incubation period of 87 days, soil samples were collected to analyze soil mineral N components—NO_3, NO_2, and NH_4. Dehydrogenase (DHA) and labile SOC were also included in the analysis. The moisture content was determined gravimetrically using the oven dry method. For soil mineral nitrogen extraction, 2 M KCl was used, and the subsequent analysis employed standard methods [48]. Dehydrogenase activity assessment involved tracking the production rate of triphenylformazon (TPF) [49]. Labile SOC calculations utilized the potassium permanganate oxidation method [50,51]. More information on the incubation experiment and soil analysis can be found in our prior research work reference [5].

2.5. Statistical Analysis

Data underwent testing for normality and homogeneity of variance. In cases where a significant improvement was observed in normality variance, log transformation was utilized. Given the factorial design of our experiment setup, the statistical analysis was conducted using SPSS software (version 21.0, SPSS Inc., Chicago, IL, USA) through an analysis of variance under the generalized linear model. This analysis explored potential differences in the response variable (CH_4/residue C mineralization) regarding residue types, nutrients, and soil moisture treatments. The significance level was established at $\alpha = 0.05$. Tukey's HSD multiple comparisons were employed to compare the main factor and interaction means and derive homogenous subsets. To identify predictor variables of soil CH_4 emissions, Pearson correlation (two-tailed significance) and a stepwise multiple regression analysis were conducted.

3. Results

3.1. Methane (CH_4) Fluxes

Soil CH_4 fluxes were significantly influenced by the main factor effect of crop residue type, nutrient, and soil moisture, and the interactive effects of residue × nutrient, residue × moisture, nutrient × moisture, and residue × nutrient × moisture (Table 1). The results of two-way and three-way interactions are presented in this section because, while a three-way interaction indicates that the relationship between any two factors depends on the third, discussing two-way interactions helps clarify how these relationships behave in simpler contexts. That can guide the interpretation of the three-way interaction.

Table 1. Analysis of variance (significance *p* value) for methane (CH_4) emissions to study the interactive effect of residue type and nutrient management at 80 and 60% FC soil moisture content after 87 days of incubation.

Source of Variation	Degrees of Freedom	CH_4 Emission	Degrees of Freedom	Residue C Mineralization
Residue	4	<0.001	3	<0.001
Nutrient	6	0.002	6	<0.001
Moisture	1	<0.001	1	<0.001
Residue × Nutrient	24	<0.001	18	0.066
Residue × Moisture	4	<0.001	3	0.002
Nutrient × Moisture	6	0.003	6	0.001
Residue × Nutrient × Moisture	24	<0.001	18	0.124
Error	140		112	
Total	210		168	
Corrected Total	209		167	

3.1.1. At 80% FC Interactive Influence of Residue × Nutrient

The magnitude and trend of temporal dynamics of CH_4 fluxes differed with treatments (Figures S1 and S2). For 60% FC, the average CH_4 fluxes ranged from −8.25 µg C kg^{-1} soil day^{-1} (maize + N0P0K0) to 5.41 µg C kg^{-1} soil day^{-1} (soil without residue + N100PK + biochar@ 5 Mg/ha). Similarly, at 80% FC, the average CH_4 fluxes ranged from −1.76 µg C kg^{-1} soil day^{-1} (soybean + N100PK) to 6.91 µg C kg^{-1} soil day^{-1} (soybean + N0PK). Irrespective of nutrient management, the cumulative mean CH_4 flux was negative and the lowest in soil amended with maize residue (−1.87 µg C kg^{-1} soil). The cumulative CH_4 fluxes (µg C kg^{-1} soil) during 87 days of incubation followed the order maize (−1.87) >rice (−0.33) ≈ soybean (2.21) > wheat (6.88) >control soil (7.18) (Figure 1). Averaged across residue types, the trend of nutrient management on CH_4 fluxes (µg C kg^{-1} soil) was N100PK (0.78) < N100PK + biochar (1.28) < N150PK + biochar (1.43) < N0P0K0 (2.76) < N100PK + manure (4.10) < N150PK (4.27) < N0PK (5.08). Mixed responses of nutrient application were observed on CH_4 fluxes; e.g., N100PK + manure, N150PK, and N0PK nutrient treatments increased CH_4 emission by 48–84% compared with minus nutrients, N0P0K0. On the contrary, N100PK, N100PK + biochar, N150PK+ biochar decreased CH_4 emission by 48 to 72% across residue. Further, the effect of nutrient management varied significantly with different residue treatments and soil without residue. In soil without residue, the nutrient effect followed the order N100PK + biochar < N0P0K0 < N150PK < N0PK < N100PK < N100PK + manure < N150PK + biochar. However, in soil amended with rice residue, the nutrient effect varied in the order N100PK + biochar < N100PK < N150PK < N0P0K0 < N100PK + manure < N150PK + biochar.

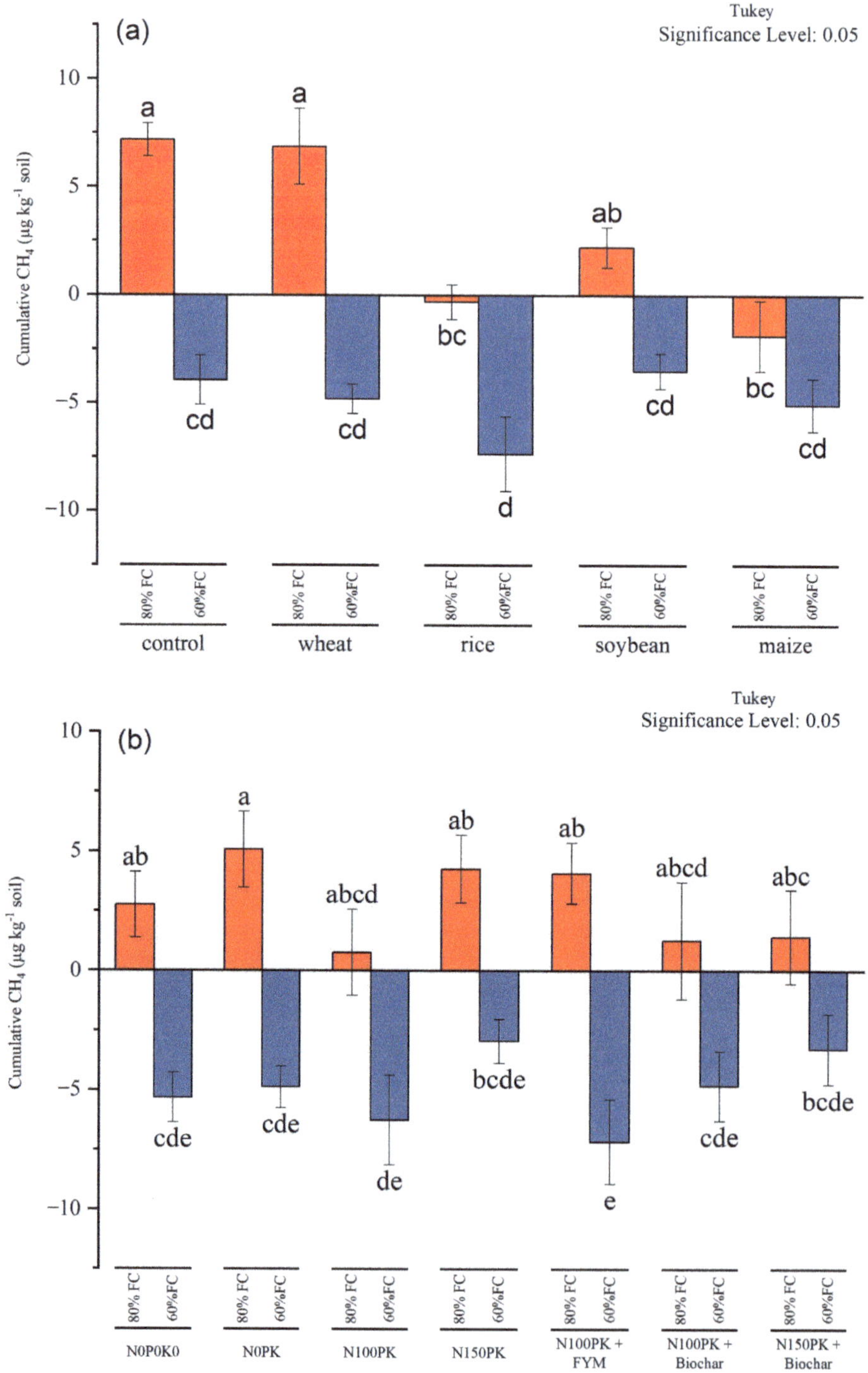

Figure 1. *Cont.*

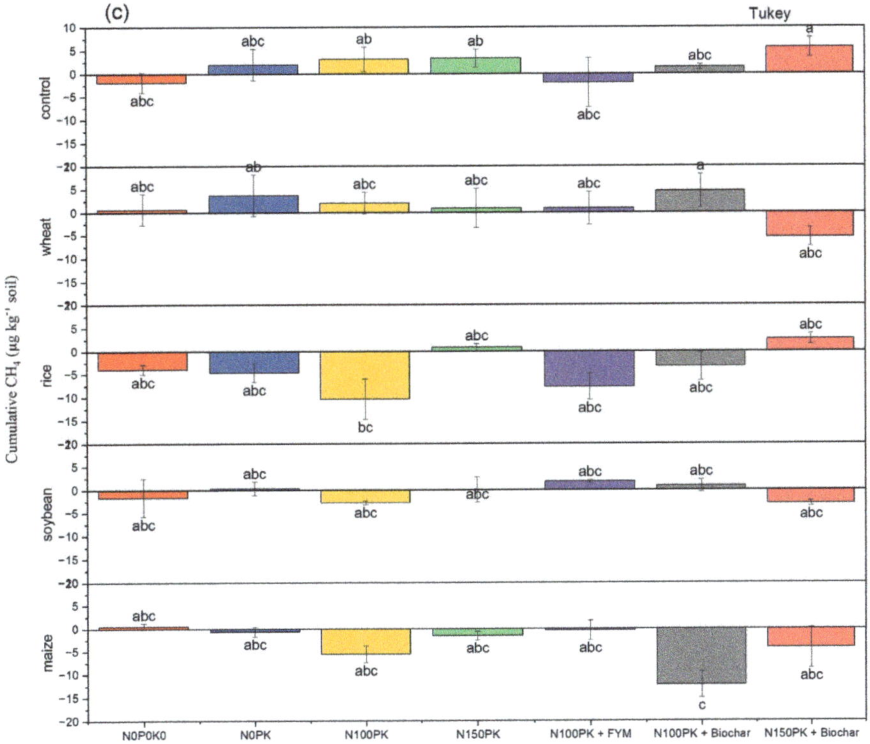

Figure 1. Cumulative soil methane (CH_4) flux (μg-C kg^{-1} soil) (**a**) effect of residue types and soil moisture across nutrient management, (**b**) effect of nutrient management and soil moisture across residue treatment, and (**c**) effect of nutrient management and residue types across soil moisture treatment. Vertical bars represent mean ± standard error (n = 3). Different lower-case letters indicate significant differences among treatments at $\alpha < 0.05$.

3.1.2. At 60% FC Interactive Influence of Residue × Nutrient

Soil CH_4 flux was significantly influenced by soil moisture ($p < 0.001$) (Table 1). The mean cumulative CH_4, flux was negative indicating methane oxidation after the end of the 87-day incubation period in all treatments at 60% FC. Similar to 80% FC, the interaction effect of the residue and nutrient was found to be significant ($p < 0.001$) (Table 1). The highest impact of methane oxidation was observed in soil amended with maize residue + N100PK + biochar (−13.87 μg C kg^{-1} soil), and the lowest in wheat + N0PK (+12.93 μg C kg^{-1} soil) (Figure 1). Irrespective of nutrient management, the residue treatments followed the order of rice < maize < wheat < soil without residue < soybean. In rice, maize, and wheat residue-amended soils, the methane oxidation increased by 86, 29, and 22%, respectively, compared to soil without residue. In contrast, the soybean residue decreased methane oxidation by 11% across nutrient management. The nutrient management had an inconsistent effect on methane oxidation from soils amended with and without residue. Overall, across soils with and without residue, the nutrient effect on CH_4 fluxes (μg C kg^{-1} soil) followed the order N100PK + manure < N100PK < N0P0K0 < N0PK < N100PK + biochar < N150PK + biochar < N150PK. The treatments N100PK + manure, N100PK, N0PK, N100PK + biochar, N150PK + biochar, and N150PK increased the cumulative mean CH_4 consumption by 35.5%, 17.7%, −8.3%, −9.3%, −38.4%, and −44.7%, respectively, compared with the control treatment (N0P0K0), across with and without residue treatment. Similar to 80% FC, the effect of nutrient management varied with residue type and control soil without residue.

3.1.3. Interaction Effect

A significant influence of residue type, nutrient management, and soil moisture and their interactions on CH_4 fluxes was observed in the present study (Tables 1 and S1). The cumulative mean CH_4 flux was the highest in the treatment receiving wheat residue +80% FC + N0PK (+12.93 μg C kg^{-1} soil) and the lowest in rice residue + 60% FC + N100PK (−19.24 μg C kg^{-1} soil). On decreasing the soil moisture from 80% FC to 60% FC, the CH_4 fluxes decreased by −1.76 times, indicating CH_4 consumption across residue types and nutrient management. Across nutrients and soil moisture, residue application decreased the methane fluxes (μg C kg^{-1} soil) in the order soil without residue (+1.61) > wheat (1.03) > soybean (−0.65) > maize (−3.48) > rice (−3.84). The increasing effect of different nutrient management across residue and soil moisture on CH_4 fluxes was N150PK > N0PK > N150PK + biochar > N0P0K0 > N100PK + manure > N100PK + biochar > N100PK. Overall, the nutrient treatments N100PK + manure, N100PK + biochar, and N100PK decreased methane fluxes by 20%, 39%, and 115%, respectively, compared to those without nutrients (N0P0K0), and N150PK, N0PK, and N150PK + biochar increased emissions by 152%, 108%, and 28%, across soil moisture and residue types. The results showed that the effects of nutrient inputs on CH_4 emission varied significantly with different nitrogen application rates, integrated use of nutrients, crop residue types, and soil moisture.

3.2. *Apparent Residue C Mineralization*

Apparent residue C mineralization (% residue C yr^{-1}) was significantly influenced by residue type, nutrient management, soil moisture, and residue × nutrient × moisture interaction ($p < 0.01$) (Table 1). The mineralization of total residue C was the highest in wheat (39–86%) and maize (40–94%), followed by rice (32–74%) and soybean (14–47%) residue, at 80% FC (Figure 2). The cumulative residue C mineralization was three times ($p < 0.001$) higher in 80% FC than 60% FC soil moisture, suggesting that soil moisture affected the mineralization of residue C. Nutrient input decreased the mineralization of residue C, and the decreasing order observed was N0P0K0 > N150PK > N0PK > N100PK + biochar > N100PK + manure > N150PK + biochar > N100PK. Regardless of soil moisture, the residue C mineralization decreased by 8% (N0PK), 37% (N100PK), 4% (N150PK), 18% (N100PK + manure), 8% (N100PK + biochar), and 31% (N150PK + biochar) over N0P0K0 treatment. The treatment combination maize residue + N0P0K0 + 80% FC recorded the highest residue C mineralization (94.08%) compared with soybean + N100PK + 60% FC being the lowest (10.46%).

3.3. *Correlation and Regression between CH_4 Emission, Residue C Mineralization, and Measured Variables*

The partial correlation tests showed that the cumulative residue C mineralization was significantly correlated with soil and residue characteristics (Table 2). Among the soil properties, the correlation was positively robust for NH_4-N, NO_3-N, labile C, dehydrogenase activity, and soil moisture ($p < 0.01$), and among the properties of residue, negative for lignin/TC but positive for cellulose/TC ($p < 0.05$). Factors significantly influencing the residue C mineralization were chosen by the stepwise regression analysis, which showed that soil labile C, lignin, soil moisture content, and residue TC exerted powerful effects. The constant and each coefficient of variables were significant ($p < 0.001$), including the R^2 (0.391) and adjusted R^2 (0.376) in Equation (1).

$$\text{Residue C mineralization (\% residue C yr}^{-1}) = -7.895 + 0.128 \text{ labile C (mg/kg)} - 2.457 \text{ lignin (\%)} + 1.155 \text{ soil moisture (\% FC)} - 1.878 \text{ residue TC (\%)} \quad (1)$$

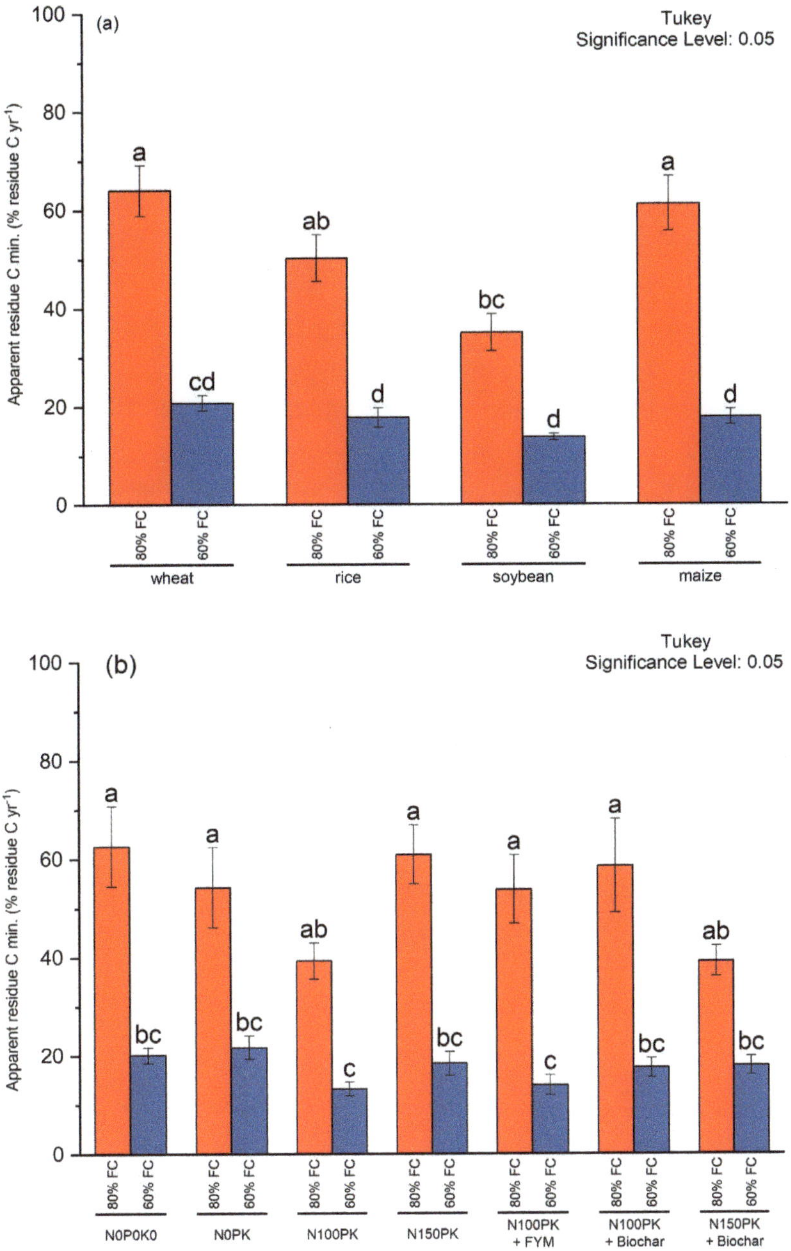

Figure 2. The apparent residue C mineralization (% residue C yr^{-1}) (**a**) effect of residue types and soil moisture across nutrient management, (**b**) effect of nutrient management and soil moisture across residue treatment, and the effect of nutrient management and residue types across soil moisture treatment were found to be nonsignificant; therefore, the figure is given in the Supplementary File as Figure S3. Vertical bars represent the mean ± standard error (n = 3). Different lower-case letters indicate significant differences among treatments at $\alpha < 0.05$.

Table 2. Pearson correlation coefficients (r) between measured variables during soil incubations and residue characteristics.

	CH_4	NO_3	NO_2	NH_4	DHA	Labile C	CO_2	TC: TN	Soil Moisture	Lignin/TN	Cellulose/lignin	Cellulose/TC	Res C min	Res. TC	Res. TN	Lignin	Cellulose
CH_4	1	0.083	0.160 *	0.330 **	0.143	0.222 **	0.450 **	0.270 **	0.511 **	0.274 **	−0.004	0.148	0.387 **	0.174 *	−0.254 **	0.131	0.186 *
NO_3	0.083	1	0.511 **	0.270 **	0.642 **	0.276 **	0.199 **	−0.038	0.158 *	−0.029	0.047	−0.079	0.058	0.276 **	0.076	0.127	−0.026
NO_2	0.160 *	0.511 **	1	0.269 **	0.599 **	0.368 **	0.437 **	0.110	0.369 **	0.028	0.016	0.171 *	0.332 **	0.182 *	−0.106	−0.077	0.190 *
NH4	0.330 **	0.270 **	0.269 **	1	0.370 **	0.722 **	0.739 **	−0.072	0.373 **	−0.054	−0.038	−0.072	0.601 **	−0.022	0.075	0.015	−0.073
DHA	0.143	0.642 **	0.599 **	0.370 **	1	0.380 **	0.410 **	0.083	0.270 **	0.098	0.005	0.039	0.298 **	−0.043	−0.088	0.028	0.035
Labile C	0.222 **	0.276 **	0.368 **	0.722 **	0.380 **	1	0.687 **	0.052	0.336 **	0.045	−0.101	0.046	0.501 **	−0.010	−0.056	−0.009	0.044
CO_2	0.450 **	0.199 **	0.437 **	0.739 **	0.410 **	0.687 **	1	0.105	0.488 **	0.054	−0.041	0.156 *	0.866 **	−0.049	−0.125	−0.110	0.138
TC: TN	0.270 **	−0.038	0.110	−0.072	0.083	0.052	0.105	1	0.472 **	0.903 **	0.048	0.747 **	0.096	0.455 **	−0.989 **	0.169 *	0.828 **
Soil moisture	0.511 **	0.158 *	0.369 **	0.373 **	0.270 **	0.336 **	0.488 **	0.472 **	1	0.517 **	−0.057	0.182 *	0.332 **	0.450 **	−0.419 **	0.368 **	0.281 **
Lignin/TN	0.274 **	−0.029	0.028	−0.054	0.098	0.045	0.054	0.903 **	0.517 **	1	−0.012	0.396 **	0.031	0.425 **	−0.856 **	0.544 **	0.508 **
Cellulose/lignin	−0.004	0.047	0.016	−0.038	0.005	−0.101	−0.041	0.048	−0.057	−0.012	1	0.123	0.004	0.002	−0.062	−0.116	0.112
Cellulose/TC	0.148	−0.079	0.171 *	−0.072	0.039	0.046	0.156 *	0.747 **	0.182 *	0.396 **	0.123	1	0.186 *	0.202 **	−0.813 **	−0.517 **	0.979 **
Res C min	0.387 **	0.058	0.332 **	0.601 **	0.298 **	0.501 **	0.866 **	0.096	0.332 **	0.031	0.004	0.186 *	1	−0.155 *	−0.132	−0.185 *	0.146
Res. TC	0.174 *	0.276 **	0.182 *	−0.022	−0.043	−0.010	−0.049	0.455 **	0.450 **	0.425 **	0.002	0.202 **	−0.155 *	1	−0.346 **	0.438 **	0.390 **
Res. TN	−0.254 **	0.076	−0.106	0.075	−0.088	−0.056	−0.125	−0.989 **	−0.419 **	−0.856 **	−0.062	−0.813 **	−0.132	−0.346 **	1	−0.043	−0.865 **
Lignin	0.131	0.127	−0.077	0.015	0.028	−0.009	−0.110	0.169 *	0.368 **	0.544 **	−0.116	−0.517 **	−0.185 *	0.438 **	−0.043	1	−0.358 **
Cellulose	0.186 *	−0.026	0.190 *	−0.073	0.035	0.044	0.138	0.828 **	0.281 **	0.508 **	0.112	0.979 **	0.146	0.390 **	−0.865 **	−0.358 **	1

*: Correlation is significant at the 0.05 level (2-tailed); **: Correlation is significant at the 0.01 level (2-tailed). CH_4: methane; NO_3: nitrate N; NO_2: nitrite N; NH_4: ammoniacal N; DHA: dehydrogenase activity; CO_2: carbon dioxide; Res C min: residue C mineralization; Res. TC: residue total carbon; Res. TN: residue total nitrogen.

The cumulative CH_4 emission was significant ($p < 0.01$) and positively correlated with NH_4-N, labile SOC, residue C mineralization, C: N ratio of crop residue, and lignin/N ratio of crop residue, and negatively correlated with plant total nitrogen (Table 2). The quantification of the variation partitioning analysis of the effects of studied soil and crop properties on CH_4 emissions was performed by the stepwise multiple regression analysis. Equation (2) describes CH_4 emissions as a function of studied soil and crop residue properties. The constant and each coefficient of variables were significant ($p < 0.001$), including the R^2 (0.314) and adjusted R^2 (0.306) in Equation (2).

$$CH_4\ (\mu g\ C\ kg^{-1}\ soil) = -27.037 + 0.352\ soil\ moisture\ (\%\ FC) + 0.069\ residue\ C\ mineralized\ (mg\ C\ kg^{-1}\ soil) \quad (2)$$

The regression analysis showed that the two dominant predictor variables within the studied variables were soil moisture and residue C mineralization, which estimated 31% of soil methane fluxes in Vertisols.

4. Discussion

4.1. Apparent Residue C Mineralization

The results indicate that apparent residue carbon (C) mineralization was significantly influenced by several factors, including residue type, nutrient management, soil moisture, and their interactions. The mineralization of total residue C varied among different crop residues, with wheat and maize residues showing the highest mineralization rates, followed by rice and soybean residues. Different crop residues have varying biochemical compositions, with wheat and maize residues often containing higher amounts of labile organic matter that decomposes faster than rice and soybean residues [52–54]. As a result, the higher mineralization rates observed in wheat and maize residues are attributed to their higher content of easily decomposable carbon compounds [55]. On the other hand, soybean residues typically have higher lignin content (cf. wheat, maize, and rice), which makes them more resistant to microbial degradation and, consequently, leads to slower carbon mineralization [5]. This study also found that soil moisture played a crucial role in residue C mineralization, with higher moisture levels (80% FC) leading to three times higher mineralization than lower moisture levels (60% FC). The role of soil moisture in regulating residue C mineralization is well established in the literature. Adequate soil moisture levels promote microbial activity and enhance the decomposition of organic matter, including crop residues [5,18,56,57]. Under conditions of waterlogged soils or high soil moisture, anaerobic conditions prevail, leading to a slowdown in residue decomposition and lower carbon mineralization rates [8]. The residue C mineralization was weakly correlated with residue TC: TN and total nitrogen (TN); however, it was negatively correlated to lignin/total carbon (TC) and positively with cellulose/TC at $p < 0.05$ (Table 2). In our study, the application of nutrients (NPK/NPK + organic amendments) eliminated the nutrient stoichiometry imbalance from crop residue incorporation (cf. N0P0K0), which explains why residue TC: TN ratio/TN probably did not significantly affect the residue C mineralization. Furthermore, previous studies have indicated that the mechanism of residue C mineralization involves soluble residue C as the crucial factor during the initial stage, while lignin, cellulose, and hemicellulose are the primary drivers in the later stages [53]. Factors significantly influencing the residue C mineralization were chosen by the stepwise regression analysis, which showed that soil labile C, lignin, soil moisture content, and residue TC exerted powerful effects ($p < 0.001$; Equation (1)).

Additionally, averaged across residue types and soil moisture, nutrient inputs decreased the mineralization of residue C, with the lowest mineralization observed in N100PK treatment (cf. N0P0K0). The findings align with previous studies that have reported the inconsistent influence of nutrient management on residue decomposition and carbon mineralization in agricultural soils [16,55,58,59]. The inverse relationship between nutrient inputs and residue C mineralization may be attributed to the priming effect [60]. Nutrient application, particularly nitrogen (N), can stimulate microbial activity and increase crop residues and soil organic matter decomposition. However, in the presence of an abundant external carbon source (crop residue in this case), the microbes preferentially utilize the labile carbon from the residue, leading to a reduced decomposition of native soil organic matter [60]. Nutrient (NPK alone or in combination with organic amendments) application would retard the mineralization of crop residues when soil nutrient enrichment satisfies the microbial N and nutrient demand and thereby decreases the need for microbes to decompose crop residue for obtaining N and nutrients [61,62].

4.2. Methane Fluxes

The results of this study clearly indicate a significant variation in soil methane fluxes based on residue types and nutrient management, aligning with prior research that has underscored the importance of these factors in greenhouse gas emissions from soils [14,15]. The negative cumulative mean CH_4 flux in soil amended with maize residue (-1.87 μg C kg^{-1} soil) is particularly noteworthy, as it suggests a methane oxidation process, turning the soil into a CH_4 sink rather than a source. This finding is consistent with previous research that has shown that different crop residues can have varying effects on methane emissions from soils [4,7,9,10,30]. Studies have reported that the type of crop residue added to the soil can influence methane emissions due to differences in the composition and decomposition rates of the residues. For example, rice residues are known to be a significant source of methane emissions due to their high carbon content and the presence of easily decomposable organic matter, which facilitates methanogenesis in flooded paddy soils [1]. Rice straw application significantly increased seasonal CH_4 flux by an average of 28–122% over no straw [13]. On the other hand, maize residues typically have a higher lignin content, resulting in slower decomposition and lower methane production [53,63]. Further crop residues provide a source of carbon for methanogenic microbes, which are microorganisms that produce methane. The availability of fresh organic carbon from crop residues can stimulate the growth and activity of these methanogenic microbes, leading to increased methane production [13,22,24], thereby accelerating soil C and N cycling. In contrast, wheat straw return and soil warming in northern China Plain increased the CH_4 uptake due to reduced decomposition and mineralization from soil warming [22].

The mixed responses of nutrient application on CH_4 fluxes highlight the complexity of these interactions, necessitating a nuanced approach to nutrient management in agroecosystems. For instance, the increase in CH_4 emissions with N100PK + manure, N150PK, and N0PK treatments could be due to the enhanced availability of labile organic carbon, stimulating methanogenic microbes [30]. In contrast, the decrease in CH_4 emissions with N100PK, N100PK + biochar, and N150PK + biochar treatments could be attributed to the stimulation of methane-oxidizing bacteria or changes in soil physicochemical properties that suppress methane production [12,16,64]. On the other hand, the nutrient management treatments show a clear trend, with N0PK inducing the highest CH_4 emissions, which could be associated with its potential to stimulate methanogenic microbial activities or suppress methane-oxidizing bacteria [8,55,65]. For instance, the addition of biochar has been reported to increase soil porosity, enhance soil microbial activity, and thus potentially augment CH_4 oxidation [21,43,66,67]. In contrast, high nitrogen rates have often been linked to higher CH_4 emissions, as excessive nitrogen can inhibit methane oxidation. Further, nitrogen (N) fertilization, in particular, has been shown to stimulate methane production in soil. Nitrogen fertilizer application can enhance microbial activity and organic matter decomposition, promote anoxic conditions, and be favorable for methanogens, leading to higher methane emissions [14]. However, the effect of other nutrients like phosphorus (P) and potassium (K) on methane emissions is less consistent and may vary depending on soil conditions and microbial activity [30,68]. Similar to our results, Yang et al. [55] reported that N fertilization significantly increased cumulative CH_4 emissions from maize straw incorporation during the spring season, and cumulative CH_4 absorption decreased with a higher N fertilization rate in autumn in dual maize cropping in China. The integration of biochar with N100PK significantly decreased CH_4 emissions, which could be due to enhanced porosity and nutrient availability, fostering methanotrophic activity. This reinforces the argument that biochar application can significantly enhance methane oxidation in agricultural soils [15,24,30,67]. In contrast, the application of manure with N100PK increased CH_4 emissions, which might be due to increased organic matter providing substrates for methanogenesis [31,38].

The distinct influence of soil moisture on CH_4 flux is in line with the existing literature, as soil moisture levels are well known to significantly impact methane emissions and oxidation in soils [18,32,36,69–72]. The observed decrease in CH_4 flux during the incubation period suggested enhanced methane oxidation at lower soil moisture levels. This aligns with previous studies that have reported increased CH_4 oxidation in drier soil conditions [18,32,73–75]. The significant impact of residue and nutrient interaction at 80% and 60% FC further highlights the complexity of these relationships. The highest cumulative methane consumption in soil amended with rice residue + N100PK at 60% FC (-19.24 μg C kg^{-1} soil) was comparable to maize residue + N100PK + biochar (-13.87 μg C kg^{-1} soil) and is a promising result, indicating the potential of these combinations in mitigating CH_4 emissions from soils. This result highlights the effectiveness of these combinations in mitigating methane emissions from soils, demonstrating a promising strategy for reducing greenhouse gas emissions in agricultural practices. Soil moisture significantly influences methane flux optimal moisture levels (e.g., 60% FC) with the addition of nutrients (N100PK)/N100PK + biochar along with maize and rice residue possibly having supported methanotrophic bacteria while preventing conditions that favor methanogenesis [30]. In contrast, the highest cumulative CH_4 fluxes (production) were observed in treatment receiving wheat residue + N0PK at 80% FC (+12.93 μg C kg^{-1} soil), suggesting that not all residue types contribute positively to CH_4 mitigation. The inconsistent effects of nutrient management on methane oxidation in soils amended with and without residue highlight the need for site-specific and residue-specific nutrient management strategies to mitigate CH_4 emissions effectively.

This study explored the relationship between methane (CH_4) emissions and various soil and crop properties. The results revealed significant correlations between cumulative CH_4 emissions and specific parameters. Notably, CH_4 emissions were positively correlated with NH_4-N (ammoniacal nitrogen), labile SOC (labile soil organic carbon), residue C mineralization, and the TC: TN ratio of crop residue. On the other hand, CH_4 emissions were negatively correlated with plant total nitrogen (N). The positive correlation between CH_4 emissions and NH4-N and labile SOC is consistent with previous research. Ammonium nitrogen is a precursor for methanogenesis, and its availability in the soil positively influences methane production by promoting the growth and activity of methanogenic microorganisms [55]. Similarly, labile SOC provides a readily available carbon source for methanogens, enhancing methane production in the soil [1]. The negative correlation between CH_4 emissions and plant total nitrogen is likely due to competition for nitrogen between methane-producing microbes and other heterotrophic microorganisms [1]. The positive correlation between CH_4 emissions and the TC: TN ratio of crop residue suggests that crop residues with higher TC: TN ratios (e.g., wheat) may contribute more to methane emissions than residues with low TC: TN ratios (e.g., rice). Residues with lower TC: TN ratios decompose more rapidly, releasing labile carbon that supports higher methane oxidation [30]. The regression analysis using Equation (2) provides insights into the relative contributions of soil moisture and residue C mineralization to CH_4 emissions. Soil moisture and residue C mineralization were identified as the dominant predictor variables, explaining 31% of the variation in soil methane fluxes.

While this study provides valuable insights into methane flux dynamics, the relationships between methane emissions, and various soil and crop properties, it is essential to acknowledge some limitations. This study was conducted under controlled laboratory conditions, and the results may not fully represent the complexities of methane emissions in actual field environments. Additionally, this study focused on short-term incubation experiments, and long-term field studies are needed to confirm the findings and assess the sustainability of the observed effects. However, the findings underscore the importance of adopting sustainable soil management practices to mitigate greenhouse gas emissions and enhance soil carbon sequestration.

5. Conclusions

This research shows the importance of relationships between methane (CH_4) fluxes, residue C mineralization, and agricultural management practices (nutrient and irrigation practices) to mitigate climate change. Key findings of this study emphasize that CH_4 emissions are significantly affected by residue type, nutrient management, and soil moisture levels, with rice and maize residue exhibiting the lowest CH_4 flux under specific nutrient practices. The effect of N100PK with biochar was found to be the best strategy for mitigating CH_4 emission. Additionally, soil moisture plays a pivotal role, and at lower soil moisture levels (60% FC), methane oxidation becomes evident across treatments. Residue C mineralization was significantly influenced by nutrient management and soil moisture levels. Nutrient inputs, particularly nitrogen, decreased residue C mineralization. The results of this study elucidate the intricate relationships between soil and residue characteristics, C mineralization, and methane emissions in agricultural fields. They underscore the significance of considering both intrinsic soil properties and residue quality in understanding and predicting organic matter decomposition and greenhouse gas emissions from soils. Two predictors, soil moisture and residue C mineralization, were identified in this study, and their relationships with CH_4 emissions are significant findings of this study, which will help develop more accurate models for mitigating greenhouse gas emissions from agricultural soils.

Supplementary Materials: The following supporting information can be downloaded at: https://www.mdpi.com/article/10.3390/soilsystems8030088/s1, Figure S1: Effect of crop residue types and nutrient management on temporal dynamics of soil methane (CH_4) flux (μg-C kg^{-1} soil day^{-1}) at 80% FC during the incubation period of 87 days; Figure S2. Effect of crop residue types and nutrient management on temporal dynamics of soil methane (CH_4) flux (μg-C kg^{-1} soil day^{-1}) at 60% FC during the incubation period of 87 days: Figure S3. Apparent residue C mineralization (% residue C yr^{-1} (c) effect of nutrient management and residue types across soil moisture treatment. Vertical bars represent the mean ± standard error (n = 3). Different lower-case letters indicate significant differences among treatments at $\alpha < 0.05$: Table S1: Effect of crop residue type, nutrient management, and soil moisture on soil cumulative CH_4 flux (μg C/kg soil) over 87 days of incubation.

Author Contributions: Conceptualization, S.L. and N.K.L.; methodology, R.C., S.L., N.K.L. and M.S.; software, D.S., N.K.L. and A.S.; validation, N.K.L. and D.K.Y.; formal analysis, R.C., S.L., D.S. and M.S.; investigation, R.C., S.L., A.S. and D.S.; resources, S.L.; data curation, R.C., S.L., D.K.Y. and R.S.K.; writing—original draft preparation, R.C. and S.L.; writing—review and editing, N.K.L., R.S.K. and T.A.; visualization, S.L. and R.S.K.; supervision, S.L., A.S. and T.A.; project administration, S.L. and N.K.L.; funding acquisition, S.L. All authors have read and agreed to the published version of the manuscript.

Funding: This research received external funding from National Agricultural Science Fund of the Indian Council of Agricultural Research, New Delhi (Grant No. NASF/CA-7019/2018–2019).

Institutional Review Board Statement: Not applicable.

Informed Consent Statement: Not applicable.

Data Availability Statement: The data presented in this study are available upon request from the corresponding author.

Acknowledgments: We gratefully acknowledge the Science and Engineering Research Board POWER Fellowship granted to the corresponding author (S.L.), and the invaluable contributions of the editor and reviewers in shaping this research.

Conflicts of Interest: The authors declare no conflicts of interest.

References

1. Tian, H.; Chen, G.; Lu, C.; Xu, X.; Ren, W.; Zhang, B.; Banger, K.; Tao, B.; Pan, S.; Liu, M.; et al. Global methane and nitrous oxide emissions from terrestrial ecosystems due to multiple environmental changes. *Ecosyst. Health Sustain.* **2015**, *1*, 11878978. [CrossRef]
2. Nabuurs, G.-J.; Mrabet, R.; Abu Hatab, A.; Bustamante, M.; Clark, H.; Havlík, P.; House, J.; Mbow, C.; Ninan, K.; Popp, A.; et al. *IPCC Sixth Assessment Report. Mitigation of Climate Change, Chapter 7: Agriculture, Forestry and Other Land Uses*; Cambridge University Press: Cambridge, UK; New York, NY, USA, 2022; ISBN 9781009157926. [CrossRef]
3. Korres, N.E.; Singh, A.; Prasad, S. *Agricultural Residues Management: Life Cycle Assessment Implications for Sustainable Agricultural Practices and Reduction of Greenhouse Gases Emissions*, 1st ed.; Elsevier: Amsterdam, The Netherlands, 2023.
4. Lenka, S.; Lenka, N.K.; Rao, A.S.; Raghuwanshi, J.; Singh, B.; Saha, J.K.; Patra, A.K. Tillage and nutrient management influence net global warming potential and greenhouse gas intensity in soybean-wheat cropping system. *Indian J. Exp. Biol.* **2022**, *60*, 207–214.
5. Lenka, S.; Choudhary, R.; Lenka, N.K.; Saha, J.K.; Amat, D.; Patra, A.K.; Gami, V.; Singh, D. Nutrient Management Drives the Direction and Magnitude of Nitrous Oxide Flux in Crop Residue-Returned Soil Under Different Soil Moisture. *Front. Environ. Sci.* **2022**, *10*, 857233. [CrossRef]
6. Singh, D.; Lenka, S.; Lenka, N.K.; Yadav, D.K.; Yadav, S.S.; Kanwar, R.S.; Sarkar, A.; Kushwaha, J. Residue Management and Nutrient Stoichiometry Control Greenhouse Gas and Global Warming Potential Responses in Alfisols. *Sustainability* **2024**, *16*, 3997. [CrossRef]
7. Sainju, U.M.; Ghimire, R.; Dangi, S. Soil carbon dioxide and methane emissions and carbon balance with crop rotation and nitrogen fertilization. *Sci. Total Environ.* **2021**, *775*, 145902. [CrossRef]
8. Wang, N.; Yu, J.G.; Zhao, Y.H.; Chang, Z.Z.; Shi, X.X.; Ma, L.Q.; Li, H.B. Straw enhanced CO_2 and CH_4 but decreased N2O emissions from flooded paddy soils: Changes in microbial community compositions. *Atmos. Environ.* **2018**, *174*, 171–179. [CrossRef]
9. Battaglia, M.L.; Thomason, W.E.; Fike, J.H.; Evanylo, G.K.; Stewart, R.D.; Gross, C.D.; Seleiman, M.; Babur, E.; Sadeghpour, A.; Harrison, M.T. Corn and Wheat Residue Management Effects on Greenhouse Emissions in the Mid-Atlantic USA. *Land* **2022**, *11*, 846. [CrossRef]
10. Akiyama, H.; Yamamoto, A.; Uchida, Y.; Hoshino, Y.T.; Tago, K.; Wang, Y.; Hayatsu, M. Effect of low C/N crop residue input on N2O, NO, and CH4 fluxes from Andosol and Fluvisol fields. *Sci. Total Environ.* **2020**, *713*, 136677. [CrossRef]
11. Weller, S.; Kraus, D.; Ayag, K.R.P.; Wassmann, R.; Alberto, M.C.R.; Butterbach-Bahl, K.; Kiese, R. Methane and nitrous oxide emissions from rice and maize production in diversified rice cropping systems. *Nutr. Cycl. Agroecosyst.* **2015**, *101*, 37–53. [CrossRef]
12. Shaukat, M.; Samoy-Pascual, K.; Maas, E.D.v.L.; Ahmad, A. Simultaneous effects of biochar and nitrogen fertilization on nitrous oxide and methane emissions from paddy rice. *J. Environ. Manag.* **2019**, *248*, 109242. [CrossRef]
13. Song, H.J.; Lee, J.H.; Jeong, H.C.; Choi, E.J.; Oh, T.K.; Hong, C.O.; Kim, P.J. Effect of straw incorporation on methane emission in rice paddy: Conversion factor and smart straw management. *Appl. Biol. Chem.* **2019**, *62*, 70. [CrossRef]
14. Banger, K.; Tian, H.; Lu, C. Do nitrogen fertilizers stimulate or inhibit methane emissions from rice fields? *Glob. Change Biol.* **2012**, *18*, 3259–3267. [CrossRef]
15. Anderson, C.R.; Condron, L.M.; Clough, T.J.; Fiers, M.; Stewart, A.; Hill, R.A.; Sherlock, R.R. Biochar induced soil microbial community change: Implications for biogeochemical cycling of carbon, nitrogen and phosphorus. *Pedobiologia* **2011**, *54*, 309–320. [CrossRef]
16. Kong, D.; Li, S.; Jin, Y.; Wu, S.; Chen, J.; Hu, T.; Wang, H.; Liu, S.; Zou, J. Linking methane emissions to methanogenic and methanotrophic communities under different fertilization strategies in rice paddies. *Geoderma* **2019**, *347*, 233–243. [CrossRef]
17. Lenka, S.; Malviya, S.K.; Lenka, N.K.; Sahoo, S.; Bhattacharjya, S.; Jain, R.C.; Saha, J.K.; Patra, A.K. Manure addition influences the effect of tillage on soil aggregation and aggregate associated carbon in a Vertisol of central India. *J. Environ. Biol.* **2021**, *41*, 1585–1593. [CrossRef]
18. Zhou, X.; Smaill, S.J.; Gu, X.; Clinton, P.W. Manipulation of soil methane oxidation under drought stress. *Sci. Total Environ.* **2021**, *757*, 144089. [CrossRef]
19. Lenka, S.; Lenka, N.K.; Singh, A.B.; Singh, B.; Raghuwanshi, J. Global warming potential and greenhouse gas emission under different soil nutrient management practices in soybean–wheat system of central India. *Environ. Sci. Pollut. Res.* **2017**, *24*, 4603–4612. [CrossRef] [PubMed]
20. Zhou, X.; Zhang, M.; Krause, S.M.B.; Bu, X.; Gu, X.; Guo, Z.; Jia, Z.; Zhou, X.; Wang, X.; Chen, X.; et al. Soil aeration rather than methanotrophic community drives methane uptake under drought in a subtropical forest. *Sci. Total Environ.* **2021**, *792*, 148292. [CrossRef] [PubMed]
21. Al-Kaisi, M.M.; Kwaw-Mensah, D.; Ci, E. Effect of nitrogen fertilizer application on corn residue decomposition in Iowa. *Agron. J.* **2017**, *109*, 2415–2427. [CrossRef]
22. Wu, G.; Ling, J.; Xu, Y.P.; Zhao, D.Q.; Liu, Z.X.; Wen, Y.; Zhou, S.L. Effects of soil warming and straw return on soil organic matter and greenhouse gas fluxes in winter wheat seasons in the North China Plain. *J. Clean. Prod.* **2022**, *356*, 131810. [CrossRef]
23. Wang, X.-g.; Luo, Y. Crop residue incorporation and nitrogen fertilizer effects on greenhouse gas emissions from a subtropical rice system in Southwest China. *J. Mt. Sci.* **2018**, *15*, 1972–1986. [CrossRef]

24. Wang, C.; Shen, J.; Liu, J.; Qin, H.; Yuan, Q.; Fan, F.; Hu, Y.; Wang, J.; Wei, W.; Li, Y.; et al. Microbial mechanisms in the reduction of CH4 emission from double rice cropping system amended by biochar: A four-year study. *Soil Biol. Biochem.* **2019**, *135*, 251–263. [CrossRef]
25. Oertel, C.; Matschullat, J.; Zurba, K.; Zimmermann, F.; Erasmi, S. Greenhouse gas emissions from soils—A review. *Chemie der Erde* **2016**, *76*, 327–352. [CrossRef]
26. Zhang, H.; Liang, Q.; Peng, Z.; Zhao, Y.; Tan, Y.; Zhang, X.; Bol, R. Response of greenhouse gases emissions and yields to irrigation and straw practices in wheat-maize cropping system. *Agric. Water Manag.* **2023**, *282*, 108281. [CrossRef]
27. Jiang, Y.; Qian, H.; Huang, S.; Zhang, X.; Wang, L.; Zhang, L.; Shen, M.; Xiao, X.; Chen, F.; Zhang, H.; et al. Acclimation of methane emissions from rice paddy fields to straw addition. *Sci. Adv.* **2019**, *5*, eaau9038. [CrossRef] [PubMed]
28. Nguyen, B.T.; Trinh, N.N.; Bach, Q.V. Methane emissions and associated microbial activities from paddy salt-affected soil as influenced by biochar and cow manure addition. *Appl. Soil Ecol.* **2020**, *152*, 103531. [CrossRef]
29. Jin, Z.; Shah, T.; Zhang, L.; Liu, H.; Peng, S.; Nie, L. Effect of straw returning on soil organic carbon in rice–wheat rotation system: A review. *Food Energy Secur.* **2020**, *9*, e200. [CrossRef]
30. Shakoor, A.; Shakoor, S.; Rehman, A.; Ashraf, F.; Abdullah, M.; Shahzad, S.M.; Farooq, T.H.; Ashraf, M.; Manzoor, M.A.; Altaf, M.M.; et al. Effect of animal manure, crop type, climate zone, and soil attributes on greenhouse gas emissions from agricultural soils—A global meta-analysis. *J. Clean. Prod.* **2021**, *278*, 124019. [CrossRef]
31. Kaleeem Abbasi, M.; Mahmood Tahir, M.; Sabir, N.; Khurshid, M. Impact of the addition of different plant residues on nitrogen mineralization-immobilization turnover and carbon content of a soil incubated under laboratory conditions. *Solid Earth* **2015**, *6*, 197–205. [CrossRef]
32. Liu, L.; Estiarte, M.; Peñuelas, J. Soil moisture as the key factor of atmospheric CH4 uptake in forest soils under environmental change. *Geoderma* **2019**, *355*, 113920. [CrossRef]
33. Anandakumar, S.; Bakhoum, N.; Chinnadurai, C.; Malarkodi, M.; Arulmozhiselvan, K.; Karthikeyan, S.; Balachandar, D. Impact of long-term nutrient management on sequestration and dynamics of soil organic carbon in a semi-arid tropical Alfisol of India. *Appl. Soil Ecol.* **2022**, *177*, 104549. [CrossRef]
34. Ruf, T.; Emmerling, C. The effects of periodically stagnant soil water conditions on biomass and methane yields of Silphium perfoliatum. *Biomass Bioenergy* **2022**, *160*, 106438. [CrossRef]
35. Du, C.; Liu, Y.; Guo, J.; Zhang, W.; Xu, R.; Zhou, B.; Xiao, X.; Zhang, Z.; Gao, Z.; Zhang, Y.; et al. Novel annual nitrogen management strategy improves crop yield and reduces greenhouse gas emissions in wheat-maize rotation systems under limited irrigation. *J. Environ. Manag.* **2024**, *353*, 120236. [CrossRef] [PubMed]
36. Korkiakoski, M.; Määttä, T.; Peltoniemi, K.; Penttilä, T.; Lohila, A. Excess soil moisture and fresh carbon input are prerequisites for methane production in podzolic soil. *Biogeosciences* **2022**, *19*, 2025–2041. [CrossRef]
37. Nwokolo, N.L.; Enebe, M.C. Methane production and oxidation—A review on the pmoA and mcrA genes abundance for understanding the functional potentials of the agricultural soil. *Pedosphere* **2024**, *in press*. [CrossRef]
38. Brenzinger, K.; Drost, S.M.; Korthals, G.; Bodelier, P.L.E. Organic residue amendments to modulate greenhouse gas emissions from agricultural soils. *Front. Microbiol.* **2018**, *9*, 3035. [CrossRef] [PubMed]
39. Sun, B.F.; Zhao, H.; Lü, Y.Z.; Lu, F.; Wang, X.K. The effects of nitrogen fertilizer application on methane and nitrous oxide emission/uptake in Chinese croplands. *J. Integr. Agric.* **2016**, *15*, 440–450. [CrossRef]
40. Qi, L.; Pokharel, P.; Chang, S.X.; Zhou, P.; Niu, H.; He, X.; Wang, Z.; Gao, M. Biochar application increased methane emission, soil carbon storage and net ecosystem carbon budget in a 2-year vegetable–rice rotation. *Agric. Ecosyst. Environ.* **2020**, *292*, 106831. [CrossRef]
41. Barrow, C.J. Biochar: Potential for countering land degradation and for improving agriculture. *Appl. Geogr.* **2012**, *34*, 21–28. [CrossRef]
42. Han, J.; Zhang, A.; Kang, Y.; Han, J.; Yang, B.; Hussain, Q.; Wang, X.; Zhang, M.; Khan, M.A. Biochar promotes soil organic carbon sequestration and reduces net global warming potential in apple orchard: A two-year study in the Loess Plateau of China. *Sci. Total Environ.* **2022**, *803*, 150035. [CrossRef]
43. Han, X.; Sun, X.; Wang, C.; Wu, M.; Dong, D.; Zhong, T.; Thies, J.E.; Wu, W. Mitigating methane emission from paddy soil with rice-straw biochar amendment under projected climate change. *Sci. Rep.* **2016**, *6*, 24731. [CrossRef]
44. Nan, Q.; Xin, L.; Qin, Y.; Waqas, M.; Wu, W. Exploring long-term effects of biochar on mitigating methane emissions from paddy soil: A review. *Biochar* **2021**, *3*, 125–134. [CrossRef]
45. Bhoi, T.K.; Samal, I.; Saraswat, A.; Hombegowda, H.C.; Samal, S.K.; Dash, A.K.; Sharma, S.; Lawate, P.; Vyas, V.; Raza, M.B. Biochar as a soil amendment: Effects on microbial communities and soil health. In *Biochar Production for Green Economy*; Academic Press: Cambridge, MA, USA, 2024; ISBN 9780443155062.
46. Li, H.; Lin, L.; Peng, Y.; Hao, Y.; Li, Z.; Li, J.; Yu, M.; Li, X.; Lu, Y.; Gu, W.; et al. Biochar's dual role in greenhouse gas emissions: Nitrogen fertilization dependency and mitigation potential. *Sci. Total Environ.* **2024**, *917*, 170293. [CrossRef]
47. Raul, C.; Bharti, V.S.; Dar Jaffer, Y.; Lenka, S.; Krishna, G. Sugarcane bagasse biochar: Suitable amendment for inland aquaculture soils. *Aquac. Res.* **2021**, *52*, 643–654. [CrossRef]
48. Kempers, A.J. Determination of sub-microquantities of ammonium and nitrates in soils with phenol, sodium nitroprusside and hypochlorite. *Geoderma* **1974**, *12*, 201–206. [CrossRef]

49. Klein, D.A.; Loh, T.C.; Goulding, R.L. A rapid procedure to evaluate the dehydrogenase activity of soils low in organic matter. *Soil Biol. Biochem.* **1971**, *3*, 385–387. [CrossRef]
50. Blair, G.J.; Lefroy, R.D.; Lisle, L. Soil carbon fractions based on their degree of oxidation, and the development of a carbon management index for agricultural systems. *Aust. J. Agric. Res.* **1995**, *46*, 1459–1466. [CrossRef]
51. Islam, K.R.; Stine, M.A.; Gruver, J.B.; Samson-Liebig, S.E.; Weil, R.R. Estimating active carbon for soil quality assessment: A simplified method for laboratory and field use. *Am. J. Altern. Agric.* **2003**, *18*, 3–17. [CrossRef]
52. Hadas, A.; Kautsky, L.; Goek, M.; Kara, E.E. Rates of decomposition of plant residues and available nitrogen in soil, related to residue composition through simulation of carbon and nitrogen turnover. *Soil Biol. Biochem.* **2004**, *36*, 255–266. [CrossRef]
53. Trinsoutrot, I.; Recous, S.; Bentz, B.; Linères, M.; Chèneby, D.; Nicolardot, B. Biochemical Quality of Crop Residues and Carbon and Nitrogen Mineralization Kinetics under Nonlimiting Nitrogen Conditions. *Soil Sci. Soc. Am. J.* **2000**, *64*, 918–926. [CrossRef]
54. Malhi, S.S.; Lemke, R.; Wang, Z.; Chhabra, B.S. Tillage, nitrogen and crop residue effects on crop yield, nutrient uptake, soil quality, and greenhouse gas emissions. *Soil Tillage Res.* **2006**, *90*, 171–183. [CrossRef]
55. Yang, L.; Muhammad, I.; Chi, Y.X.; Liu, Y.X.; Wang, G.Y.; Wang, Y.; Zhou, X.B. Straw return and nitrogen fertilization regulate soil greenhouse gas emissions and global warming potential in dual maize cropping system. *Sci. Total Environ.* **2022**, *853*, 158370. [CrossRef]
56. Guntiñas, M.E.; Gil-Sotres, F.; Leirós, M.C.; Trasar-Cepeda, C. Sensitivity of soil respiration to moisture and temperature. *J. Soil Sci. Plant Nutr.* **2013**, *13*, 445–461. [CrossRef]
57. Zhou, W.; Hui, D.; Shen, W. Effects of soil moisture on the temperature sensitivity of soil heterotrophic respiration: A laboratory incubation study. *PLoS ONE* **2014**, *9*, e92531. [CrossRef] [PubMed]
58. Liu, Y.; Zang, H.; Ge, T.; Bai, J.; Lu, S.; Zhou, P.; Peng, P.; Shibistova, O.; Zhu, Z.; Wu, J.; et al. Intensive fertilization (N, P, K, Ca, and S) decreases organic matter decomposition in paddy soil. *Appl. Soil Ecol.* **2018**, *127*, 51–57. [CrossRef]
59. Muhammad, W.; Vaughan, S.M.; Dalal, R.C.; Menzies, N.W. Crop residues and fertilizer nitrogen influence residue decomposition and nitrous oxide emission from a Vertisol. *Biol. Fertil. Soils* **2011**, *47*, 15–23. [CrossRef]
60. Fang, Y.; Nazaries, L.; Singh, B.K.; Singh, B.P. Microbial mechanisms of carbon priming effects revealed during the interaction of crop residue and nutrient inputs in contrasting soils. *Glob. Change Biol.* **2018**, *24*, 2775–2790. [CrossRef] [PubMed]
61. Nottingham, A.T.; Turner, B.L.; Stott, A.W.; Tanner, E.V.J. Nitrogen and phosphorus constrain labile and stable carbon turnover in lowland tropical forest soils. *Soil Biol. Biochem.* **2015**, *80*, 26–33. [CrossRef]
62. Ji, D.; Ding, F.; Dijkstra, F.A.; Jia, Z.; Li, S.; Wang, J. Crop residue decomposition and nutrient release are independently affected by nitrogen fertilization, plastic film mulching, and residue type. *Eur. J. Agron.* **2022**, *138*, 126535. [CrossRef]
63. Liang, X.; Yuan, J.; Yang, E.; Meng, J. Responses of soil organic carbon decomposition and microbial community to the addition of plant residues with different C:N ratio. *Eur. J. Soil Biol.* **2017**, *82*, 50–55. [CrossRef]
64. Yao, Z.; Zheng, X.; Wang, R.; Xie, B.; Butterbach-Bahl, K.; Zhu, J. Nitrous oxide and methane fluxes from a rice-wheat crop rotation under wheat residue incorporation and no-tillage practices. *Atmos. Environ.* **2013**, *79*, 641–649. [CrossRef]
65. Yin, X.; Peñuelas, J.; Sardans, J.; Xu, X.; Chen, Y.; Fang, Y.; Wu, L.; Singh, B.P.; Tavakkoli, E.; Wang, W. Effects of nitrogen-enriched biochar on rice growth and yield, iron dynamics, and soil carbon storage and emissions: A tool to improve sustainable rice cultivation. *Environ. Pollut.* **2021**, *287*, 117565. [CrossRef]
66. Feng, J.; Zhu, B. Global patterns and associated drivers of priming effect in response to nutrient addition. *Soil Biol. Biochem.* **2021**, *153*, 108118. [CrossRef]
67. Abhishek, K.; Shrivastava, A.; Vimal, V.; Gupta, A.K.; Bhujbal, S.K.; Biswas, J.K.; Singh, L.; Ghosh, P.; Pandey, A.; Sharma, P.; et al. Biochar application for greenhouse gas mitigation, contaminants immobilization and soil fertility enhancement: A state-of-the-art review. *Sci. Total Environ.* **2022**, *853*, 158562. [CrossRef] [PubMed]
68. Omonode, R.A.; Vyn, T.J.; Smith, D.R.; Hegymegi, P.; Gál, A. Soil carbon dioxide and methane fluxes from long-term tillage systems in continuous corn and corn-soybean rotations. *Soil Tillage Res.* **2007**, *95*, 182–195. [CrossRef]
69. Jin, X.; Wu, F.; Wu, Q.; Heděnec, P.; Peng, Y.; Wang, Z.; Yue, K. Effects of drying-rewetting cycles on the fluxes of soil greenhouse gases. *Heliyon* **2023**, *9*, e12984. [CrossRef] [PubMed]
70. da Silva Cardoso, A.; Junqueira, J.B.; Reis, R.A.; Ruggieri, A.C. How do greenhouse gas emissions vary with biofertilizer type and soil temperature and moisture in a tropical grassland? *Pedosphere* **2020**, *30*, 607–617. [CrossRef]
71. Yue, P.; Zuo, X.; Li, K.; Li, X.; Wang, S.; Misselbrook, T. Precipitation changes regulate the annual methane uptake in a temperate desert steppe. *Sci. Total Environ.* **2022**, *804*, 150172. [CrossRef] [PubMed]
72. Le Mer, J.; Roger, P. Production, oxidation, emission and consumption of methane by soils: A review. *Eur. J. Soil Biol.* **2001**, *37*, 25–50. [CrossRef]
73. Xu, X.; Xia, Z.; Liu, Y.; Liu, E.; Müller, K.; Wang, H.; Luo, J.; Wu, X.; Beiyuan, J.; Fang, Z.; et al. Interactions between methanotrophs and ammonia oxidizers modulate the response of in situ methane emissions to simulated climate change and its legacy in an acidic soil. *Sci. Total Environ.* **2021**, *752*, 142225. [CrossRef]

74. Chai, L.L.; Hernandez-Ramirez, G.; Hik, D.S.; Barrio, I.C.; Frost, C.M.; Chinchilla Soto, C.; Esquivel-Hernández, G. A methane sink in the Central American high elevation páramo: Topographic, soil moisture and vegetation effects. *Geoderma* **2020**, *362*, 114092. [CrossRef]
75. Liu, W.; Yuan, W.; Xu, S.; Shao, C.; Hou, L.; Xu, W.; Shi, H.; Pan, Q.; Li, L.; Kardol, P. Spatiotemporal patterns and drivers of methane uptake across a climate transect in Inner Mongolia Steppe. *Sci. Total Environ.* **2021**, *757*, 143768. [CrossRef] [PubMed]

soil systems

MDPI

Article

The Effect of Manure Application Rates on the Vertical Distribution of Antibiotic Resistance Genes in Farmland Soil

Yuqian Wang [1,2], Liqiong Yang [1], Weipeng Liu [1,*] and Jie Zhuang [3]

1 Key Laboratory of Pollution Ecology and Environmental Engineering, Institute of Applied Ecology, Chinese Academy of Sciences, Shenyang 110016, China; wangyuqian20@mails.ucas.ac.cn (Y.W.)
2 University of Chinese Academy of Sciences, Beijing 100039, China
3 Department of Biosystems Engineering and Soil Science, Institute for a Secure and Sustainable Environment, The University of Tennessee, Knoxville, TN 37996, USA
* Correspondence: liuwp@iae.ac.cn

Abstract: Manure application is the primary input route for antibiotic resistance genes (ARGs) in farmland soil. This study investigated the effects of varying the rates of five chicken manure applications on the accumulation and distribution of ARGs across different soil depths (0–20, 20–40, and 40–60 cm) using metagenomic sequencing. The results revealed that the distribution of ARGs in farmland soil was closely linked to soil depth and influenced to some extent by the fertilizer quantity after 30 days of fertilization. ARGs were predominantly concentrated in the surface soil and exhibited a significant decrease in type and abundance with an increased soil depth. Compared with soil treated with chemical fertilizers alone, chicken manure-treated surface soil presented a higher diversity and abundance of ARGs. However, the diversity and abundance of ARGs did not increase proportionally with the increasing ratios of chicken manure application (0, 25, 50, 75, and 100%). ARGs in soil primarily conferred resistance to host bacteria through antibiotic efflux pumps (~33%), antibiotic target alteration (~31%), antibiotic inactivation (~20%), and antibiotic target protection (~8%). Correlation analysis involving ARGs and soil microorganisms revealed widespread multidrug resistance among soil microorganisms. Furthermore, two genera of human pathogenic bacteria (*Pseudomonas* sp. and *Listeria* sp.) were identified as potential microbial hosts of ARGs in all treatments. Correlation analysis involving ARGs and environmental factors indicated that soil ARGs are predominantly influenced by heavy metals and microorganisms. This paper offers valuable insights for environmental risk assessments regarding the utilization of livestock manure resources. Additionally, it furnishes a scientific foundation for farmland application strategies pertaining to livestock manure.

Keywords: chicken manure; antibiotic resistance genes (ARGs); farmland soil; microorganisms; soil depths

1. Introduction

Livestock manure is rich in nutrients (e.g., nitrogen, phosphorus, and potassium) that promote crop growth [1–5]. The application of manure to farmlands has been proven to be an effective strategy for mitigating soil constraints, augmenting soil organic matter, enhancing soil health, and boosting crop yields [6–9]. Prolonged manure application not only improves soil microbial characteristics but also enhances soil chemical properties, and thus, it plays a pivotal role in sustaining agricultural productivity and ecosystem services [10]. Concurrently, pollutants commonly found in manure, including heavy metals, antibiotics, antibiotic resistance genes (ARGs), and pathogens, may accumulate in soil with manure application, leading to soil contamination and posing risks to human health.

Although the careful administration of veterinary antibiotics can prevent animal illnesses and promote growth, significant portions of these antibiotics are not fully absorbed and utilized within the animal gut. With the rapid development of the livestock and poultry industries, the consumption of veterinary antibiotics has gradually increased. The emissions of 80 veterinary antibiotics ranged from 23,110 tonnes/year to 40,850 tonnes/year in 2100–2020 [11]. Tetracyclines, sulfonamides, chloramphenicols, and quinolones account for 94% [11]. Approximately 75% to 95% of antibiotics, along with ARGs induced by the prolonged selective pressure of antibiotics, are excreted in feces and urine [12–14], and most of these excreted compounds retain their potency in soil [15]. Numerous studies have shown that the application of manure can elevate antibiotic concentrations in soil (from µg/kg to mg/kg levels) [16–18], increasing the number of drug-resistant bacteria and the abundance of ARGs [19,20].

The application of manure significantly enriches soil with carbon-containing substrates and nutrients, stimulating the growth and activity of soil microbial populations [21,22]. This enrichment increases soil microbial biomass and diversity [23,24], subsequently influencing the maintenance and distribution of ARGs. For example, numerous nutrient-rich bacterial communities have emerged in carbon-rich fertilizer soils [25]. Soil fertilized with manure exhibits a higher microbial biomass compared with soil that uses only chemical fertilizers [26]. Peng et al. [27] demonstrated that 3 years of continuous manure application could alter the structures of microbial communities and antibiotic resistance in soil. Moreover, Li et al. [20] revealed that manure application significantly enhanced the abundance and α diversity of ARGs in soil. The effect of manure application on soil ARGs is attributable not only to nutrient increase but also to antibiotics, heavy metals, and other factors. Gabini et al. [28] observed that 20 mg/kg of sulfamethoxazole caused only preliminary and short-term changes in soil bacterial community composition while no significant effect was noted on fungal communities. However, higher concentrations (100–300 mg/kg) of sulfamethoxazole induced substantial and persistent changes in the β diversity of both bacteria and fungi [29]. Additionally, the presence of heavy metals in livestock manure leads to the compound pollution of antibiotics in soil [30,31]. These heavy metals can act as selective agents for the proliferation of ARGs and promote their persistence in the environment [32]. For example, Kuppusamy et al. [33] discovered a positive correlation between soil ARGs and residues of tetracyclines, sulfonamides, quinolones, copper, and cadmium. However, most studies have, to date, been focused on the surface soil or on a few types of ARGs. Although these studies are very useful for understanding the environmental risks of ARG accumulation in farmland soils, there is little information on the vertical distribution of ARGs along the soil profile. In addition, insufficient attention has been paid to actual production conditions in studies. Soil depth leads to heterogeneity in the soil environment, and within farmland, the vertical variation of soil properties with depth is much greater than the horizontal spatial variability. Research on the vertical distribution of ARGs in farmland soils after manure application is helpful for improving the understanding of ARG environmental risks. The co-application of organic and chemical fertilizers is a common fertilization strategy in agricultural production. Understanding the effect of the ratio of manure to chemical fertilizer application on the accumulation of ARGs in farmland can help us formulate ARG environmental risk management strategies that better align with production needs. We believe that a reasonable application ratio of manure and chemical fertilizers can alleviate the accumulation of ARGs caused by the resource utilization of livestock manure in agricultural production to a certain extent and control the diffusion of ARGs into deep soil. This study aimed to investigate the influence of different manure application rates on the vertical distribution of ARGs in agricultural soils, assess the relationship between ARGs and microbial communities, and identify the principal environmental factors that influence ARG retention. The findings of this research can provide theoretical guidance for ecological risk assessment pertaining to the sustainable utilization of livestock manure.

2. Materials and Methods

2.1. Sample Collection

The soil samples utilized in this study were collected from corn farmland in Changtu County (123°58′0.340″ E, 42°48′14.083″ N, Liaoning Province, China). Corn cultivation follows rotary tillage practices. The fertilizers consist of chemical fertilizers (self-made corn-specific fertilizers) and organic fertilizer (chicken manure after high-temperature composting). Various proportions of chemical and organic fertilizers were combined and applied to the farmland soil in three repeat regions. The chicken manure was labeled JF in figure and the application rates of organic fertilizers and sample numbers for different treatments are provided in Table 1. Sampling was performed at depths of 0–20 cm (Layer A), 20–40 cm (Layer B), and 40–60 cm (Layer C). Each sample comprised a blend of soils collected from five sampling points after fertilization for 30 days. A portion of the soil samples was stored at −80 °C in Whirl-Pak bags for the subsequent analysis of microorganisms, ARGs, and antibiotics. The remaining portion of samples was cleared of plant roots, large gravel particles, leaves, and other debris. After indoor drying, it was sifted through a 2 mm standard sieve and homogenized for the assessment of soil physicochemical properties and heavy metal content.

Table 1. Application rate of organic fertilizer (chick manure) under different treatments.

Soil Layers	Control Group (100% Chemical Fertilizer)	Treatment 1 (25% Organic Fertilizer Replaced Chemical Fertilizer)	Treatment 2 (50% Organic Fertilizer Replaced Chemical Fertilizer)	Treatment 3 (75% Organic Fertilizer Replaced Chemical Fertilizer)	Treatment 4 (100% Organic Fertilizer)
A (0–20 cm)	A0	A1	A2	A3	A4
B (20–40 cm)	B0	B1	B2	B3	B4
C (40–60 cm)	C0	C1	C2	C3	C4

2.2. Antibiotic and Heavy Metal Determination

Eight antibiotics that were commonly used in livestock and occurring in poultry manure (oxytetracycline, tetracycline, doxycycline, tilmicosin, tylosin, sulfamonomethoxine, sulfamethazine, and sulfadiazine) were determined in this study [34–36]. The recoveries of the eight antibiotics measured via the external standard method [37] in the manure samples ranged from 70% to 120%. In particular, the antibiotics were extracted from 5.0 g of the manure samples by following this procedure: 20 mL of mixed extractant (0.4 g Na_2EDTA + acetonitrile mixed with phosphate buffer at the ratio of 1:1, v/v) was added to the samples and shaken vigorously for 3 min and then centrifuged at 21,500× g for 5 min. The supernatant was transferred into a conical flask, adding 10 mL of mixed extractant. Then the extraction was repeated again. Finally, the pH of the extracting solution was adjusted to 2–2.5 and extraction was performed using HLB solid-phase extraction columns. The HLB columns was pre-activated with 6 mL methanol and 6 mL water. After extraction, each sample was vacuumed for 5 min and then 6 mL of methanol was used to elute the HLB columns. The eluate was collected and blown to near-dryness at 40 °C under nitrogen atmosphere, dissolved in 2 mL of 1:1 methanol: water v/v) with 0.2% formic acid, and centrifuged at 32,250× g for 10 min. Liquid chromatographic conditions for antibiotics analysis were set as follows: Poroshell 120 EC-C18 column (2.1 × 50 mm, 1.9 µm), flow rate of 0.3 mL/min, column temperature of 40 °C, injection volume of 2 µL, mobile phase A with 0.1% formic acid in methanol, and mobile phase B with 2.5 mmol/ammonium acetate (containing 0.1% formic acid) in water.

The concentrations of Cu, Zn, Cd, As, Pb, and Hg in the digested soil and manure were determined using inductively coupled plasma mass spectrometry (ICP-MS) (NexION 350, Perkin Elmer, Waltham, MA, USA). About 0.1 g soil/manure sample was placed in a digestion flask, and 5 mL HCl, 10 mL HNO_3, 2 mL HF, and 1 mL HClO were added, then the samples were digested using a graphite furnace digestion apparatus. The accuracy of the experimental data was verified using reagent blank samples and duplicate soil samples. The recovery rate of the standard samples was 90–110% and the relative standard deviation (RSD) of the data was less than 5%.

2.3. Soil Physical and Chemical Indicators

All soil samples were air-dried at 25–30 °C in the laboratory, ground and sieved through a 0.25 mm mesh, and then analyzed for physicochemical properties. The pH was measured using a pH analyzer (Sartorius PB-10, Shanghai, China). The cation exchange capacity (CEC) was determined using atomic absorption spectrometry (AA700, Perkin Elmer, Waltham, MA, USA) [38]. The contents of total carbon (TC), total nitrogen (TN), and total phosphorus (TP) were measured with an elemental analyzer (Vario MICRO cube, Hanau, Germany). The concentration of potassium (K) was determined using an inductively coupled plasma mass spectrometer (ICP-MS) (NexION 350, Perkin Elmer, Waltham, MA, USA).

2.4. DNA Extraction and Metagenomic Sequencing

Manure/soil DNA was extracted using HiPure Soil DNA Kit B. Integrity and purity of extracted DNA were assessed using 1% agarose gel electrophoresis and its concentration and purity were determined using a NanoDrop 2000 ultra-micro spectrophotometer (Thermo Fisher, Wilmington, DE, USA) and a QuantiFluor fluorometer (Promega, Madison, WI, USA). PE150 bipartite sequencing was performed using the Illumina HiSeq/Illumina Novaseq/MGI2000 platform. Sequencing raw data (pass filter data) were removed from splices and low-quality sequences using the second-generation sequencing data quality statistics software, Cutadapt (v1.9.1), with primer and splicer sequences removed. The clean data for subsequent information analysis were obtained by removing bases with quality values lower than 20 at both ends, excluding sequences with N-base content greater than 10%, and retaining the minimum read length of 75 bp. Clean reads were assembled into overlapping clusters for each sample using MEGAHIT [39] and coding genes were predicted using the Prodigal software (v3.02) [40]. Then, the gene sequences of all the samples were integrated and further de-redundated by sequence clustering software MMseq2 (v11-e1a1c), which defaulted to 95% identity and 95% coverage for clustering to obtain a non-redundant gene set of unigene sequences. The representative sequences in the redundant gene set were compared with those in the NCBI NR database to obtain bacterial annotation and taxonomic information. Identification and analysis of ARGs were conducted based on the Comprehensive Antibiotic Research Database (CARD), employing a threshold set above 90.

2.5. Statistical Analysis

Figures were plotted using TB tools (heat map) and Origin 2021 (Origin Lab, Northampton, MA, USA). The Duncan test was performed using SPSS (27 IBM, Armonk, NY, USA) to compare the differences among treatments at a probability level of less than 0.05.

3. Results and Discussion

3.1. Types and Abundance of ARGs in Manure

The identification and quantification of ARGs in manure are essential for evaluating the environmental ramifications of manure utilization in agriculture and formulating efficient fertilization methodologies [41–44]. A total of 462 ARG subtypes were identified in chicken manure. Among them, lincosamide (eighteen subtypes), aminoglycoside (fourteen subtypes), macrolide (eleven subtypes), chloramphenicol (nine subtypes), and glycopeptide (nine subtypes) were the primary types of ARGs (Figure 1A). ARGs that belonged to the lincomycin, macrolide, and aminoglycoside classes exhibited higher abundance, with 3187, 2651, and 1614 counts, respectively (Figure 1B). The major ARG subtypes in chicken manure were *tetA(58), saur_walk_dap, msbA, bcrA, Ecol_fabG_TRC, macB*, and *novA*. Among these, *tetA (58)* and *bcrA* showed higher abundance (Figure 2). *tetA(58)* is a common gene that confers resistance to tetracyclines while *bcrA* confers resistance to peptides. Both genes are widely distributed in chicken manure and soil [45–47]. Multidrug resistance genes are defined as ARGs that confer resistance to three or more antibiotics [48], facilitating the spread of ARGs among different bacteria [49]. The classification and analysis of ARGs in chicken manure revealed that subtypes of ARGs that are resistant to one-class drugs (40) >two-class drugs (23) > multidrug-resistant ARGs (19) (Figure 3). In terms of abundance, ARGs resistant to one-class drugs were still the largest (4032), but the abundance of multidrug-resistant ARGs (2653) > that of two-class drugs (1172). The major resistance mechanisms of ARGs detected in chicken manure were antibiotic target alteration (33.33%), antibiotic efflux pump (27.49%), antibiotic inactivation (17.75%), and antibiotic target protection (10.17%) (Figure S1). Antibiotics affect fundamental bacterial functions, including protein synthesis, RNA polymerase transcription, chromosome segregation, and folate metabolism. Antibiotic target alteration directly affects associated proteins or indirectly triggers the activation or inhibition of regulatory proteins that oversee these modifications, diminishing drug affinity and bolstering bacterial resistance [50]. Antibiotic efflux pumps are transporter proteins that expel toxic substrates out of cells and into the external environment [51]. Characterized by the expulsion of intracellular antibiotics from the cell via transporters, they assume a pivotal role in bacterial multidrug resistance and the development of numerous drug-resistant phenotypes [52].

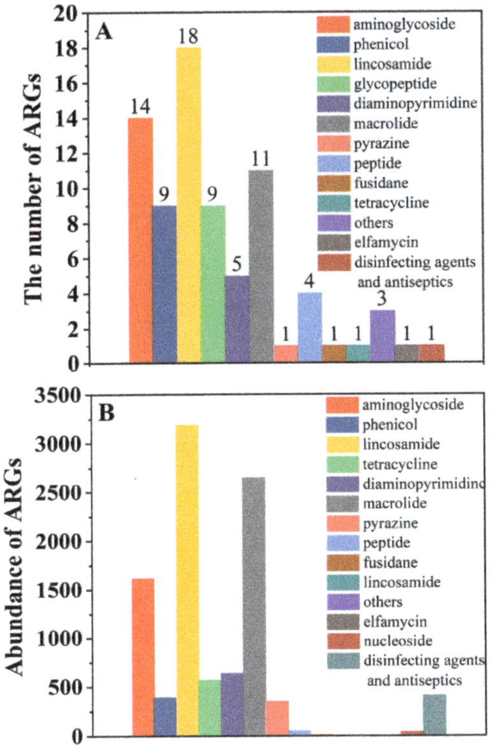

Figure 1. Quantities (**A**) and abundance (**B**) of ARGs in chicken manure.

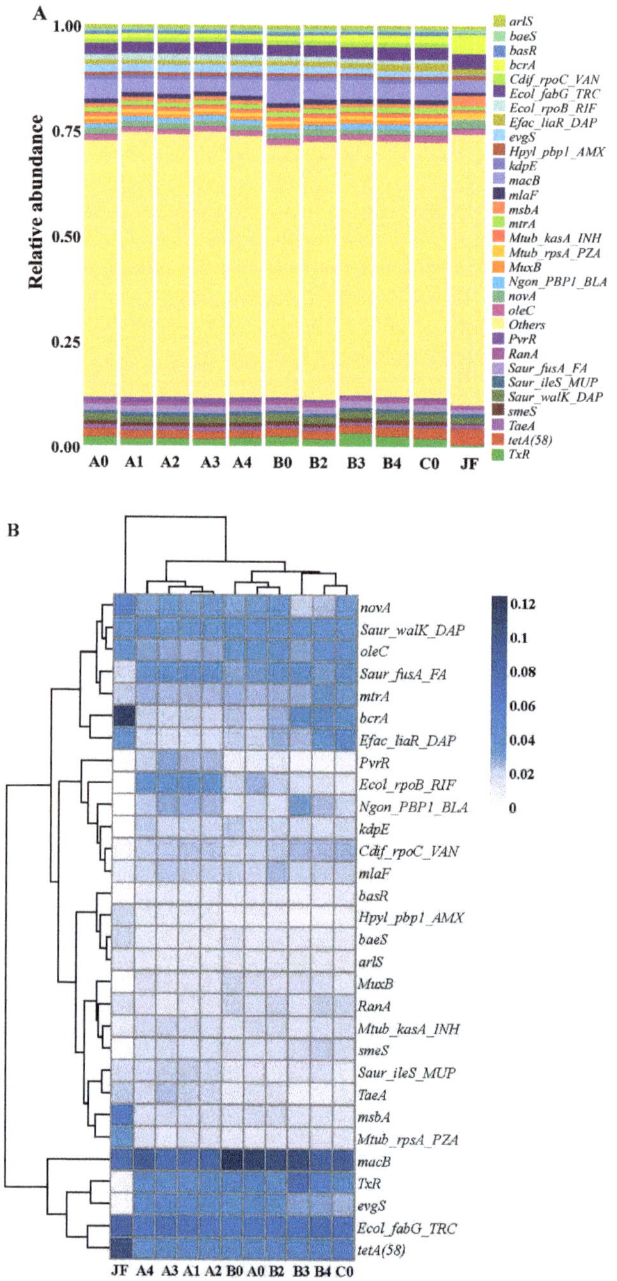

Figure 2. Changes in the relative abundance of major ARGs in soil and heatmap: subtypes and abundance of ARGs under different treatments (**A**) and the heatmap of ARGs under different treatments (**B**). A, B, and C represent the 0–20 cm, 20–40 cm, and 40–60 cm soil layers, respectively. The numbers 0, 1, 2, 3, and 4 correspond to the application of 0, 25, 50, 75, and 100% manure in the soil, respectively. JF: organic fertilizer (chicken manure after high-temperature composting).

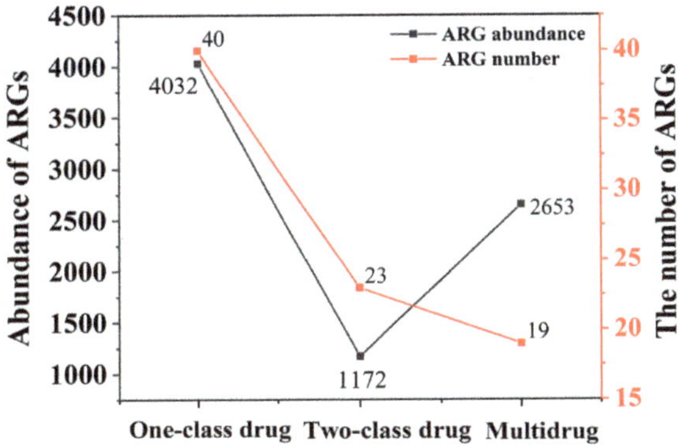

Figure 3. Abundance and quantities of ARGs in chicken manure in terms of resistance to drug classes.

3.2. Diversity and Abundance of ARGs in Fertilized Farmland Soil

After the application of chicken manure to soil, various factors such as the diffusion of ARGs from manure, the alteration of environmental conditions, nutrient dynamics, and antibiotic stress can influence the relative abundance of ARGs in soil [53–55]. Although no significant positive correlation was observed between the quantity of applied chicken manure and the diversity of ARGs in soil, the soil treated with 100% chicken manure exhibited the highest diversity of ARG types (A4). Notably, the application of chicken manure contributed to an increase in the diversity of ARGs in agricultural soil. ARGs were still detected in soils treated only with chemical fertilizers, indicating that the application of organic fertilizers had led to the long-term presence of ARGs in farmland soils. A total of 652 ARG subtypes were identified in soil, with peptide, glycopeptide, fluoroquinolone, and elfamycin as the primary types of ARGs. The dominant ARGs detected in soil from each treatment included *macB*, *TRC*, *tetA (58)*, *Ecol_fabG_TxR*, *evgS*, *Saur_walK_DAP*, and *Saur_fusA_FA* (Figure 2A). A comparison between ARGs detected in chicken manure and those in soil post-application revealed similarities in major ARG subtypes, albeit with significant variations in their abundance (Figure 4). Variations in the abundance of resistance genes are intricately linked to microbial communities within the environment [56]. In particular, the relative abundance of *bcrA* in soil ranged from 58.66% to 77.83%, lower than that in chicken manure, while the relative abundance of *TxR* ranged from 73.94% to 86.56%, higher than that in chicken manure. The dominant ARGs observed in Layer A (0–20 cm) and Layer B (20–40 cm) were similar, with *macB* as the most prevalent ARG in both layers (Figures S2 and S3). *macB* confers resistance to macrolides in bacteria through the mechanism of antibiotic efflux pumps [57]. ARGs detected in soil predominantly exhibit four mechanisms of antibiotic resistance (Figure S1), including antibiotic target protection (~8%), antibiotic target alteration (~31%), antibiotic inactivation (~20%), and antibiotic efflux pumps (~33%) [58,59].

The distribution of ARGs in farmland soil was closely associated with soil depth and influenced by the application rate of manure. As shown in Figure 4A, the surface layer (Layer A, 0–20 cm) presented a rich diversity of ARGs. As the soil depth increased, the diversity of ARGs in each soil treatment decreased to varying extents (e.g., 41 subtypes of ARGs in A4, 16 subtypes of ARGs in B4, and none in C4). Notably, in the deepest soil layer (Layer C, 40–60 cm), no ARGs were detected in soils treated with chicken manure. The surface soil, as the primary recipient of manure, retained a more diverse array of ARGs. In terms of gene abundance, the ARG abundance in surface soil (Layer A) was significantly higher than that in deeper soils (Figure 5A). For instance, in soils treated with 100% chicken manure, the highest ARG abundance in Layer A was 347 (A4), which decreased to 109 in B4 and was absent in C4. However, the distribution of individual ARG types varied across soil layers. The abundance of fluoroquinolone ARGs, which was dominant in A4, decreased by 85.45% in B4 and was entirely absent in C4. The abundance of peptide ARGs in B2 was 59, which was 47.9% higher than that in A2. The variation in individual ARG types distribution across soil layers is linked to microbial communities, antibiotics, and mobile genetic elements [20]. Conversely, the abundance of ARGs in the surface soil did not exhibit a significant increase with the rise in chicken manure application (e.g., the abundance values of A1, A2, A3, and A4 were 207, 173, 223, and 347, respectively). This result is attributed to the application of chemical fertilizers also influencing the maintenance of ARGs in soil [60]. The application of manure increased the diversity and abundance of ARGs in surface soil, with the most notable effect observed in soil treated only with manure. The accumulation of ARGs can foster the emergence of multidrug-resistant bacteria in the environment [61].

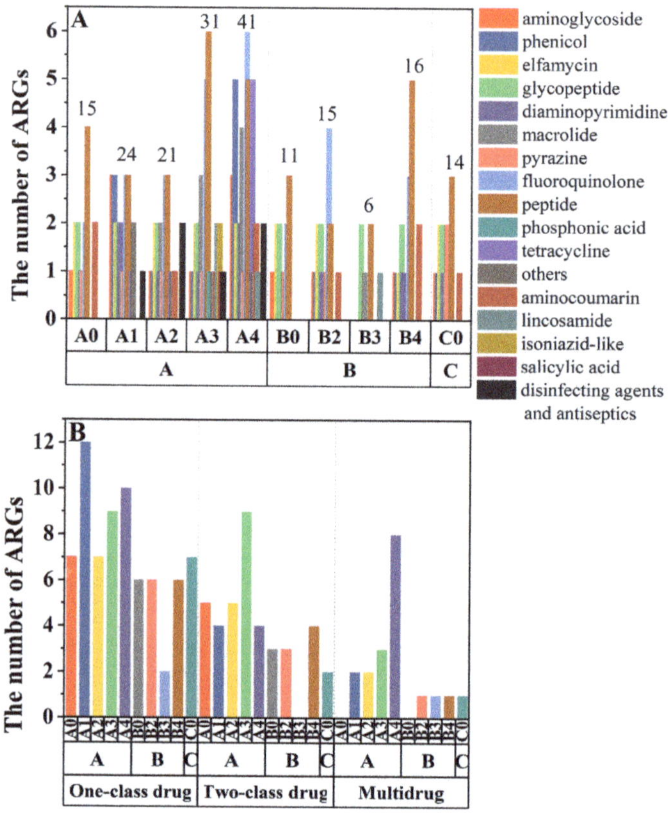

Figure 4. Quantities of ARGs (**A**) and drug resistance classes (**B**) in different soil layers (A: 0–20 cm, B: 20–40 cm, C: 40–60 cm). The numbers 0, 1, 2, 3, and 4 correspond to the application of 0, 25, 50, 75, and 100% manure in the soil, respectively.

According to the analysis of ARG resistance classes in soil (Figure 4B), ARGs that were resistant to only one-class drugs comprised the highest number (648) while the abundance of multidrug-resistant ARGs was the lowest (137) (Figure 5B). This distribution was aligned with the proportion of ARG resistance types observed in chicken manure. All types of ARG resistance were predominantly concentrated in the surface soil (Layer A), with abundances decreasing significantly with an increased soil depth. For example, in samples treated with 100% chicken manure, the numbers of ARGs that were resistant to one class of drugs were ten (A4), six (B4), and zero (C4), while the numbers of ARGs that were resistant to two classes of drugs were four (A4), four (B4), and zero (C4). No significant correlation was observed between ARG resistance types in soil and the application amount of chicken manure, which was also true for abundance (Figure 5A). The abundance of multidrug-resistant ARGs was the highest in topsoil treated with 100% organic fertilizer while the abundance of ARGs that were resistant to one and two classes of drugs was higher in surface soil treated with 75% organic fertilizer.

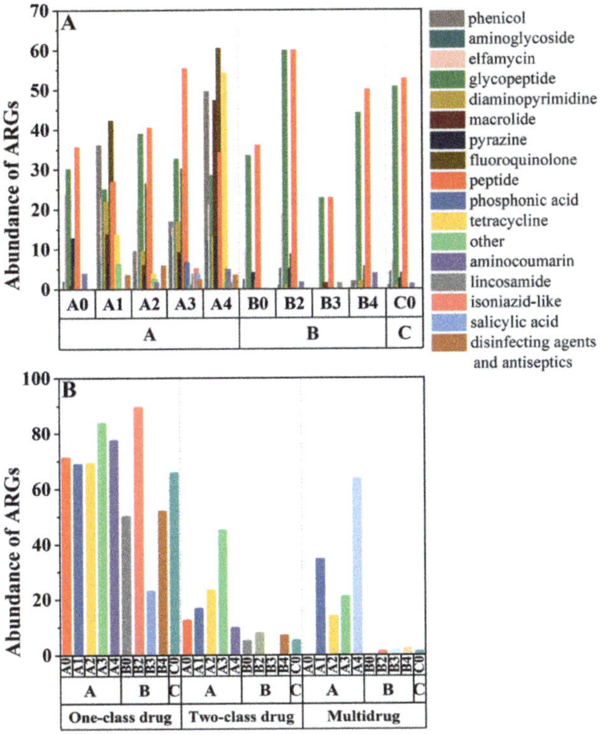

Figure 5. Abundance of ARGs (**A**) and drug resistance classes (**B**) in different soil layers (A: 0–20 cm, B: 20–40 cm, C: 40–60 cm). The numbers 0, 1, 2, 3, and 4 correspond to the application of 0, 25, 50, 75, and 100% manure in the soil, respectively.

3.3. Soil Microbial Community in Fertilized Farmland Soil

Nutrients, microorganisms, ARGs, antibiotics, and heavy metals present in manure can directly or indirectly influence the accumulation and maintenance of ARGs in soil [62–65]. The contents of carbon, antibiotics, and heavy metals in soil treated with chicken manure increased to varying degrees (Figure S4). These components exhibit the potential to affect the structure of the microbial community in soil, thereby influencing the maintenance of ARGs. The predominant bacterial phyla identified in chicken manure were Actinobacteria, Firmicutes, and unclassified with relative abundance values of 46.95%, 40.95%, and 10.82%, respectively (Figure 6). Actinobacteria and Firmicutes were recognized as major bacterial hosts of ARGs [66–69]. Changes in the relative abundance of species at the phylum level in soil from different treatments are depicted in Figure 6A, revealing that Chloroflexi (1.03–7.67%), Verrucomicrobiota (0.84–10.61%), Firmicutes (0.49–1.17%), Gemmatimonadetes (2.96–8.47%), unclassified (11.42–14.29%), Actinobacteria (7.06–20.46%), Acidobacteria (12.2–30.02%), and Proteobacteria (14.86–46.93%) were the predominant bacterial phyla in all treated samples.

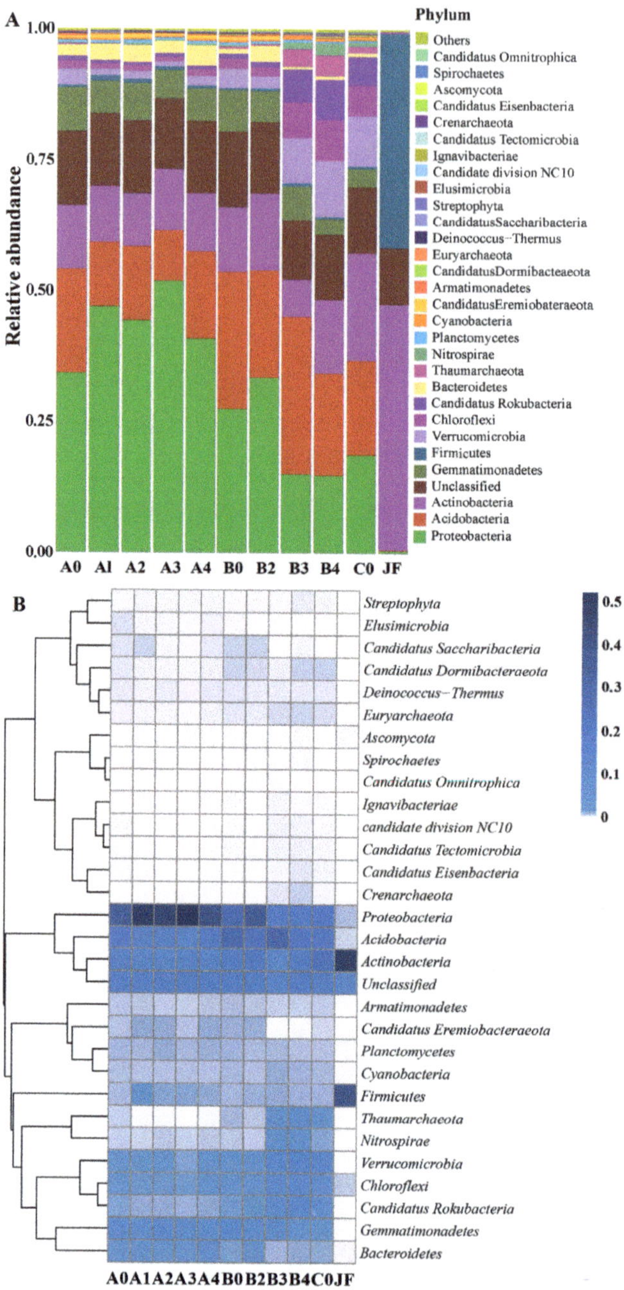

Figure 6. Changes in the relative abundance of microbial species in different soil layers: the relative abundance of species at the phylum level under different treatments (**A**) and heatmaps of species at the phylum level under different treatments (**B**) (others represent all phyla or genera with a relative abundance of less than 1%). A, B, and C represent the 0–20 cm, 20–40 cm, and 40–60 cm soil layers, respectively. The numbers 0, 1, 2, 3, and 4 correspond to the application of 0, 25, 50, 75, and 100% manure in the soil, respectively. JF: organic fertilizer (chicken manure after high-temperature composting).

There were significant differences in the structures of bacterial communities in soils at different depths (Figures S5 and S6). In Layer A soil, the dominant phyla were Proteobacteria, Acidobacteria, Actinobacteria, unclassified, and Gemmatimonadetes. In Layer B soil, the relative abundance of Proteobacteria and Gemmatimonadetes decreased by 48.06% and 14.19%, respectively, while the relative abundance of Acidobacteria and Actinobacteria increased by 66.48% and 7.69%, respectively. Nutrients in manure can promote the growth and reproduction of soil bacteria, enhancing bacterial diversity [70]. However, heavy metals exert a suppressive effect on microbial activities [71]. Therefore, changes in nutrient composition and heavy metal content in soil collectively influence the structure of the microbial community.

Compared with the control treatment that only used chemical fertilizers, the application of chicken manure exerted varied effects on the bacterial community structure in soil. At the phylum level, Proteobacteria exhibited an increase in Layer A soil (levels in A1, A2, A3, and A4 were 37.18%, 27.47%, 51.42%, and 19.41% higher than the control group A0) (Figure 6A). Proteobacteria, which consist of Gram-negative bacteria, prefer environments with a high nutrient content and are significantly affected by the type of fertilizer used [72,73]. Consequently, Proteobacteria decreased by 58.45% in Layer B soil compared to layer A soil. Acidobacteria also declined in layer A soil (it decreased by 38.32%, 29.22%, 51.57%, and 16.43% in A1, A2, A3, and A4 compared to control A0). Acidobacteria, which are oligotrophic bacteria, can thrive in complex environments and under oligotrophic conditions [74]. Hence, Acidobacteria exhibited higher relative abundance in deeper Layer B soil. Due to the complex conditions of field soil, changes in the bacterial community structure were only correlated with the application rate of chicken manure in surface soil.

Figure S7 shows the KEGG cluster and functional analyses of microbial community genomes in chicken manure and soil. The enzymes primarily involved in genes were glycosyltransferases and glycoside hydrolases. The major functional pathways were related to metabolism, genetic information processing, and environmental information processing.

The differences in microbial function and classification composition among different layers of soil were analyzed (Figure 7). According to the analysis of similarities (ANOSIM), the differences in microbial function (Figure 7A) and classification (Figure 7B) between groups in Layers A and B were greater than those within groups ($R > 0$, $p < 0.05$). The principal component analysis (PCA) based on the Bray–Curtis distance matrix (Figure 7) indicated that the first two principal components (PCs) explained 62.04% of the variance in microbial communities. PC1 was a principal component associated with soil layers, accounting for 40.91% of the variance, while PC2 was a principal component associated with treatments within groups, accounting for 21.13% of the variance (Figure 7C). The correlation between the soil layers and microbial communities was greater than that between the treatments within the groups and microbial communities. Soil properties, such as chemical composition and physical structure, change from surface to deeper layers; hence, significant changes also occur in community composition and functional characteristics with depth in the soil microbiome [75]. The compositions of microbial communities in soil samples with different fertilization treatments exhibited differences on the PC axis (Figure 7D). On the PC1 axis, each treatment group was relatively scattered, with chicken manure (JF) distributed in the positive direction, while the other treatment groups were mostly distributed on the negative axis of PC1. On the PC2 axis, JF and C5 were distributed on the positive axis while Groups A and B were distributed on both the positive and negative axes, with group A mainly in the negative direction. The results indicated that PC1 was a principal component related to the characteristics of chicken manure, and it could explain 50.42% of the variance. PC2 was a principal component related to soil depth, accounting for 18.46% of the variance. The microbial community structures in A1, A2, A3, and A4 were similar while those in JF and soil samples with different fertilization rates showed significant differences.

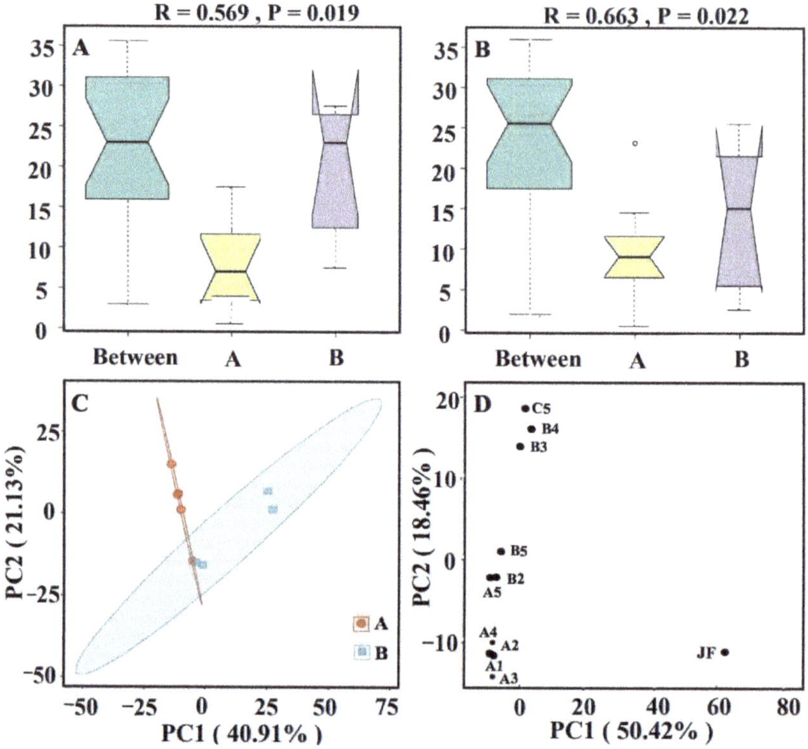

Figure 7. Differences in microbial function and classification composition at different soil layers: differences in microbial function (**A**) and classification (**B**) among ANOSIM groups; PCA of microbial community distribution characteristics in different soil layers (**C**) and fertilization rates (**D**) based on the Bray–Curtis distance matrix.

3.4. Correlation Analysis of ARGs with Microorganisms in Soil

The correlation between the major ARGs and microorganisms in soil and their antibiotic resistance types were analyzed. Figure 8 shows that multidrug-resistant ARGs are widely found in soil microorganisms, and microorganisms can harbor multiple ARGs. The presence of multiple ARGs in the same potential microbial host increases the risk of ARGs spreading among pathogens [76–78]. Upon comparing the potential microbial hosts of ARGs in soil with pathogenic bacteria in the NCBI database (Table S1), the potential microbial hosts of ARGs in all the samples encompassed three species of human pathogenic bacteria: *Pseudomonas* sp. (*Pseudomonas aeruginosa LESB58* and *P. aeruginosa PAO1*) and *Listeria* sp. (*Listeria monocytogenes EGD-e*). The three pathogenic genera simultaneously contain multiple ARGs and nearly every pathogen has developed resistance to at least one antibiotic. The antibiotic resistance mechanism of *Pseudomonas* involved antibiotic efflux and the alteration of antibiotic targets while that of *Listeria* was antibiotic inactivation. *Listeria,* which consists of a series of Gram-positive bacteria, can obtain resistance genes from plasmids and associated transposons possessing various resistance mechanisms [79]. The dissemination of ARGs resulting from manure application and their accumulation in pathogenic bacteria significantly increase risks to human health.

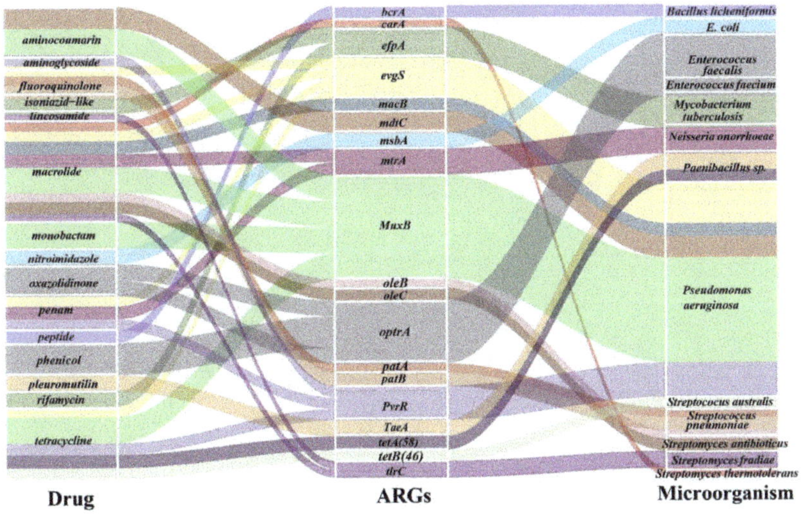

Figure 8. Mulberry plot of microbes, ARGs, and antibiotic resistance types in soil.

3.5. Correlation Analysis of ARGs in Soil with Environmental Factors

The effects of environmental factors on ARGs in soil were assessed with an aggregation boosting tree (ABT). The analysis of Bray–Curtis distance indicated that factors that influenced the abundance of ARGs included heavy metals, microorganisms, antibiotics, and nutrients, with relative effects of 26.9%, 23.7%, 16.5%, and 17.4%, respectively (Figure 9A). The redundancy analysis (RDA) of ARG abundance and environmental factors further suggested that the selected microorganisms accounted for 71.3% of abundance changes in ARGs, with RDA1 and RDA2 accounting for 41.5% and 29.8%, respectively. Among them, *Enterococcus faecalis* exerted the most significant effect on ARGs. The abundance of *E. faecalis* was positively correlated with *macB*, *TxR*, *olec_walk_DAP*, *saur_walk_DAP*, and *evgS*, but negatively correlated with *tetA(58)*, *Ecol_fabG_TRC*, *Saur_fusA_FA*, and *novA*. Meanwhile, the abundance of *E. faecalis* was positively correlated with *P. aeruginosa*, *Streptomyces fradiae*, and *Neisseria gonorrhoeae* [80]. *E. faecalis* is an opportunistic pathogen of animals and humans. It not only acts as the primary host of multidrug-resistant bacteria but also possesses unique virulence factors that facilitate the transfer of ARGs and virulence genes. It is also an indicator of food and manure contamination [45,81].

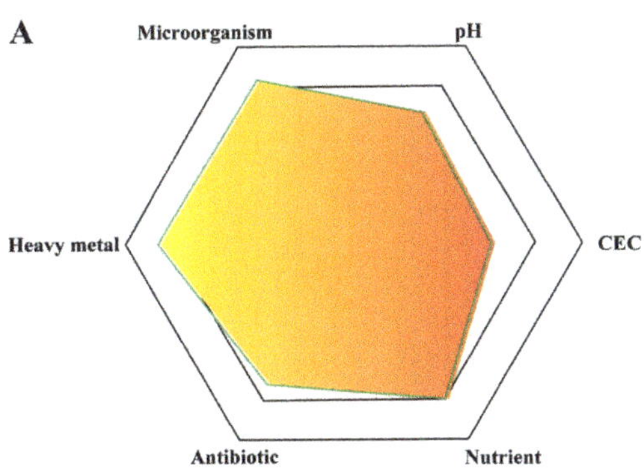

Figure 9. *Cont.*

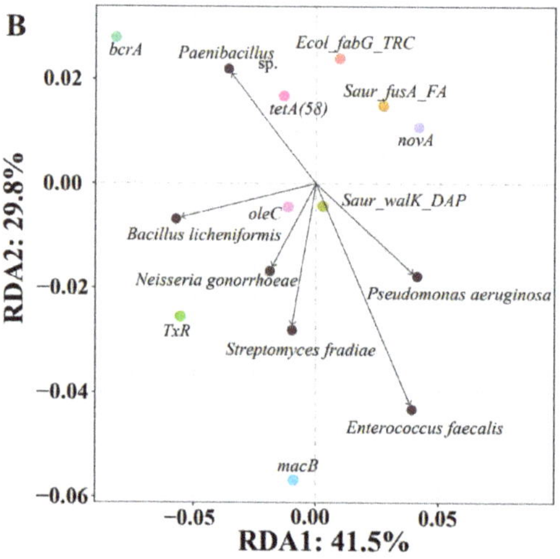

Figure 9. ABT assessment of the effect of environmental factors on ARGs (based on the Bray–Curtis distance) (**A**) and RDA based on ARGs and microbial abundance (**B**).

4. Conclusions

A total of 462 ARG subtypes, which are predominantly resistant to lincomycin, aminoglycoside, macrolides, chloramphenicol, and tetracyclines, were detected in chicken manure. After manure application, 652 ARG subtypes were detected in soil after 30 days. Among them, peptide, glycopeptide, fluoroquinolone, and elfamycin were the primary types of ARGs. Variations in environmental conditions resulted in disparities in the composition of ARGs. The distribution of ARGs in farmland soil was closely related to soil depth and influenced by the amount of manure applied. Although the types and abundance of ARGs were significantly higher in surface soil treated with chicken manure compared with the control samples treated only with chemical fertilizers, they decreased significantly with an increased soil depth. However, ARGs did not exhibit a gradual change with increasing fertilizer application. The subtype compositions of ARGs in chicken manure and soil were similar, but differences in abundance were observed. Resistance to one class of drugs was identified as the major type of drug resistance among ARGs in soil and chicken manure.

The analysis of soil microbial communities revealed that Chloroflexi, Verrucomicrobia, Firmicutes, Gemmatimonadetes, an unclassified phylum/phyla, Actinobacteria, Acidobacteria, and Proteobacteria were the dominant bacterial phyla in all the treated samples. The structure of soil bacterial communities was affected by the amount of chicken manure applied. ARGs in soil endowed host bacteria with resistance through antibiotic efflux pumps (~33%), antibiotic target protection (~8%), antibiotic target alteration (~31%), and antibiotic inactivation (~20%). Further analysis showed that multidrug-resistant ARGs were widespread among soil microorganisms, with the potential microbial hosts of ARGs including two human pathogenic genera (*Pseudomonas* sp. and *Listeria* sp.). The transfer of ARGs due to manure application considerably increases human health risks. Heavy metals (26.9%) and microorganisms (23.7%) exerted relatively greater effects on the accumulation and a maintenance of ARGs in soil. Therefore, the harmless treatment of livestock manure and the reasonable manure application strategy are crucial for reducing the accumulation of pollutants and avoiding the occurrence of compound pollution in farmland soil.

Supplementary Materials: The following supporting information can be downloaded at: https://www.mdpi.com/article/10.3390/soilsystems8030089/s1, Figure S1: The composition of the detected ARGs drug resistance mechanism in different soil layers. A, B, and C represent the 0–20 cm, 20–40 cm, and 40–60 cm soil layers, respectively. The numbers 0, 1, 2, 3, and 4 correspond to the application of 0, 25, 50, 75, and 100% manure in the soil, respectively. JF: organic fertilizer (chicken manure after high-temperature composting). Figure S2: Relative abundance of ARGs in different soil layers. A: 0–20 cm soil layer, B: 20–40 cm soil layer. Figure S3: Heatmap of relative abundance of ARGs in different soil layers. A: 0–20 cm soil layer, B: 20–40 cm soil layer. Figure S4: Changes in pH (A), CEC (B), heavy metals (C), antibiotics (D), carbon (E), and nutrients (F) in different soil layers. A, B, and C represent the 0–20 cm, 20–40 cm, and 40–60 cm soil layers, respectively. The numbers 0, 1, 2, 3, and 4 correspond to the application of 0, 25, 50, 75, and 100% manure in the soil, respectively. JF: organic fertilizer (chicken manure after high-temperature composting). Figure S5: The relative abundance of microbial phylum in different soil layers (others represent all phyla with relative abundance less than 1%). A: 0–20 cm soil layer, B: 20–40 cm soil layer. Figure S6: Heatmap of microbial phylum in different soil layers. A: 0–20 cm soil layer, B: 20–40 cm soil layer. Figure S7: KEGG cluster analysis and functional analysis of microbial community genomes in chicken manure and different soil layers: Enzyme activity annotation of microbial communities in chicken manure and different soil layers (A); KEGG clustering of microbial community genomes in chicken manure and different soil layers (B). A, B, and C represent the 0–20 cm, 20–40 cm, and 40–60 cm soil layers, respectively. The numbers 0, 1, 2, 3, and 4 correspond to the application of 0, 25, 50, 75, and 100% manure in the soil, respectively. JF: organic fertilizer (chicken manure after high-temperature composting). Table S1: Pathogenic bacteria with ARGs and resistance mechanisms in soil.

Author Contributions: Conceptualization, Y.W. and J.Z.; methodology, Y.W.; software, Y.W.; validation, Y.W. and W.L.; formal analysis, Y.W.; investigation, Y.W. and W.L.; resources, W.L.; data curation, Y.W.; writing—original draft preparation, Y.W.; writing—review and editing, W.L., L.Y. and J.Z.; visualization, Y.W.; supervision, L.Y.; project administration, L.Y.; funding acquisition, L.Y. All authors have read and agreed to the published version of the manuscript.

Funding: This research was funded by National Natural Science Foundation of China grant number 52379051.

Institutional Review Board Statement: Not applicable.

Informed Consent Statement: Not applicable.

Data Availability Statement: Dataset available on request from the authors.

Conflicts of Interest: The authors declare no conflict of interest.

References

1. Fatondji, D.; Ibrahim, A. Improving nutrient use efficiency from decomposing manure and millet yield under Plinthosols in Niger. *Nutr. Cycl. Agroecosys.* **2018**, *110*, 485–499. [CrossRef]
2. Mažeika, R.; Arbačiauskas, J.; Masevičienė, A.; Narutytė, I.; Šumskis, D.; Žičkienė, L.; Rainys, K.; Drapanauskaite, D.; Staugaitis, G.; Baltrusaitis, J. Nutrient Dynamics and Plant Response in Soil to Organic Chicken Manure-Based Fertilizers. *Waste Biomass Valori.* **2021**, *12*, 371–382. [CrossRef]
3. Baxter, A.E.; Leytem, A.B.; Dungan, R.S.; Bjorneberg, D. Potential of winter double crops and tillage for managing manure-based nutrient loading. *Plant Soil* **2023**, *11*, 06072. [CrossRef]
4. Keskinen, R.; Suojala-Ahlfors, T.; Sarvi, M.; Hagner, M.; Kaseva, J.; Salo, T.; Uusitalo, R.; Rasa, K. Granulated broiler manure based organic fertilizers as sources of plant available nitrogen. *Environ. Technol. Innov.* **2020**, *18*, 100734. [CrossRef]
5. Samara, E.; Matsi, T.; Barbayiannis, N.; Lithourgidis, A. Liquid Cattle Manure Effect on Corn Yield and Nutrients' Uptake and Soil Fertility, in Comparison to the Common and Recommended Inorganic Fertilization. *J. Soil Sci. Plant Nutr.* **2020**, *20*, 2283–2293. [CrossRef]
6. Lehmann, J.; Hansel, C.M.; Kaiser, C.; Kleber, M.; Maher, K.; Manzoni, S.; Nunan, N.; Reichstein, M.; Schimel, J.P.; Torn, M.S.; et al. Persistence of soil organic carbon caused by functional complexity. *Nat. Geosci.* **2020**, *13*, 529–534. [CrossRef]
7. Zhai, L.; Wang, Z.; Zhai, Y.; Zhang, L.; Zheng, M.; Yao, H.; Lv, L.; Shen, H.; Zhang, J.; Yao, Y.; et al. Partial substitution of chemical fertilizer by organic fertilizer benefits grain yield, water use efficiency, and economic return of summer maize. *Soil Till. Res.* **2022**, *217*, 105287. [CrossRef]
8. Hu, B.; Ni, H.; Xie, M.; Li, H.; Wen, Y.; Chen, S.; Zhou, Y.; Teng, H.; Bourennane, H.; Shi, Z. Mapping soil organic matter and identifying potential controls in the farmland of Southern China: Integration of multi-source data, machine learning and geostatistics. *Land Degrad. Dev.* **2023**, *34*, 5468–5485. [CrossRef]
9. Zhao, Z.; Zhang, C.; Wang, H.; Li, F.; Pan, H.; Yang, Q.; Li, J.; Zhang, J. The Effects of Natural Humus Material Amendment on Soil Organic Matter and Integrated Fertility in the Black Soil of Northeast China: Preliminary Results. *Agronomy* **2023**, *13*, 794. [CrossRef]

10. Usharani, K.V.; Roopashree, K.M.; Naik, D. Role of soil physical, chemical and biological properties for soil health improvement and sustainable agriculture. *J. Pharmacog. Phytoch.* **2019**, *8*, 1256–1267.
11. Li, S.; Zhu, Y.; Zhong, G.; Huang, Y.; Jones, K.C. Comprehensive Assessment of Environmental Emissions, Fate, and Risks of Veterinary Antibiotics in China: An Environmental Fate Modeling Approach. *Environ. Sci. Technol.* **2024**, *58*, 5534–5547. [CrossRef] [PubMed]
12. Chee-Sanford, J.C.; Mackie, R.I.; Koike, S.; Krapac, I.G.; Lin, Y.F.; Yannarell, A.C.; Maxwell, S.; Aminov, R.I. Fate and Transport of Antibiotic Residues and Antibiotic Resistance Genes following Land Application of Manure Waste. *J. Environ. Qual.* **2009**, *38*, 1086–1108. [CrossRef] [PubMed]
13. Gutiérrez, I.R.; Watanabe, N.; Harter, T.; Glaser, B.; Radke, M. Effect of sulfonamide antibiotics on microbial diversity and activity in a Californian *Mollic Haploxeralf*. *J. Soil Sediment.* **2010**, *10*, 537–544. [CrossRef]
14. Munk, P.; Knudsen, B.E.; Lukjancenko, O.; Duarte, A.S.R.; Van Gompel, L.; Luiken, R.E.C.; Smit, L.A.M.; Schmitt, H.; Garcia, A.D.; Hansen, R.B.; et al. Abundance and diversity of the faecal resistome in slaughter pigs and broilers in nine European countries. *Nat. Microbiol.* **2018**, *3*, 898–908. [CrossRef]
15. Gaballah, M.S.; Guo, J.; Sun, H.; Aboagye, D.; Sobhi, M.; Muhmood, A.; Dong, R. A review targeting veterinary antibiotics removal from livestock manure management systems and future outlook. *Bioresour. Technol.* **2021**, *333*, 125069. [CrossRef] [PubMed]
16. Al-Wabel, M.I.; Ahmad, M.; Ahmad, J.; Lubis, N.M.A.; Usman, A.R.A.; Al-Farraj, A.S.F. Assessing the prevalence of veterinary antibiotics and associated potential ecological risk in dryland soil, manure, and compost: A case study from Saudi Arabia. *J. King Saud Univ. Sci.* **2021**, *33*, 101558. [CrossRef]
17. Pan, Z.; Yang, S.; Zhao, L.; Li, X.; Weng, L.; Sun, Y.; Li, Y. Temporal and spatial variability of antibiotics in agricultural soils from Huang-Huai-Hai Plain, northern China. *Chemosphere* **2021**, *272*, 129803. [CrossRef]
18. Gu, J.; Chen, C.; Huang, X.; Mo, J.; Xie, Q.; Zeng, Q. Occurrence and risk assessment of tetracycline antibiotics in soils and vegetables from vegetable fields in Pearl River Delta, South China. *Sci. Total Environ.* **2021**, *776*, 145959. [CrossRef]
19. Abbas, F.; Thomas, P.; Cully-Duse, B.; Andronicos, N.M.; Winter, G. Cattle-compost-soil: The transfer of antibiotic resistance in livestock agriculture. *Microbiologyopen* **2023**, *12*, e1375. [CrossRef]
20. Li, Y.; Kong, F.; Li, S.; Wang, J.; Hu, J.; Chen, S.; Chen, Q.; Li, Y.; Ha, X.; Sun, W. Insights into the driving factors of vertical distribution of antibiotic resistance genes in long-term fertilized soils. *J. Hazard. Mater.* **2023**, *456*, 131706. [CrossRef]
21. Zhang, L.; Chen, X.; Xu, Y.; Jin, M.; Thompson, M.L. Soil labile organic carbon fractions and soil enzyme activities after 10 years of continuous fertilization and wheat residue incorporation. *Sci. Rep.* **2020**, *10*, 11318. [CrossRef] [PubMed]
22. Wang, H.; He, X.; Zhang, Z.; Li, M.; Zhang, Q.; Zhu, H.; Xu, S.; Yang, P. Eight years of manure fertilization favor copiotrophic traits in paddy soil microbiomes. *Eur. J. Soil Biol.* **2021**, *106*, 103352. [CrossRef]
23. Caban, J.R.; Kuppusamy, S.; Kim, J.H.; Yoon, Y.E.; Kim, S.Y.; Lee, Y.B. Green manure amendment enhances microbial activity and diversity in antibiotic-contaminated soil. *Appl. Soil Ecol.* **2018**, *129*, 72–76. [CrossRef]
24. Sayre, J.M.; Wang, D.; Lin, J.; Danielson, R.E.; Scow, K.M.; Rodrigues, J.L.M. Repeated manure inputs to a forage production soil increase microbial biomass and diversity and select for lower abundance genera. *Agr. Ecosyst. Environ.* **2023**, *354*, 108567. [CrossRef]
25. Gautam, A.; Sekaran, U.; Guzman, J.; Kovács, P.; Hernandez, J.L.G.; Kumar, S. Responses of soil microbial community structure and enzymatic activities to long-term application of mineral fertilizer and beef manure. *Environ. Sustain. Indic.* **2020**, *8*, 100073. [CrossRef]
26. Ren, F.; Sun, N.; Xu, M.; Zhang, X.; Wu, L.; Xu, M. Changes in soil microbial biomass with manure application in cropping systems: A meta-analysis. *Soil Till. Res.* **2019**, *194*, 104291. [CrossRef]
27. Peng, S.; Wang, Y.; Chen, R.; Lin, X. Chicken Manure and Mushroom Residues Affect Soil Bacterial Community Structure but Not the Bacterial Resistome When Applied at the Same Rate of Nitrogen for 3 Years. *Front. Microbiol.* **2021**, *12*, 618693. [CrossRef] [PubMed]
28. Garbini, G.L.; Grenni, P.; Rauseo, J.; Patrolecco, L.; Pescatore, T.; Spataro, F.; Barra Caracciolo, A. Insights into structure and functioning of a soil microbial community amended with cattle manure digestate and sulfamethoxazole. *J. Soil Sediment* **2022**, *22*, 2158–2173. [CrossRef]
29. Cheng, S.; Shi, M.; Xing, L.; Wang, X.; Gao, H.; Sun, Y. Sulfamethoxazole affects the microbial composition and antibiotic resistance gene abundance in soil and accumulates in lettuce. *Environ Sci Pollut.* **2020**, *27*, 29257–29265. [CrossRef]
30. Seiler, C.; Berendonk, T.U. Heavy metal driven co-selection of antibiotic resistance in soil and water bodies impacted by agriculture and aquaculture. *Front. Microbiol.* **2012**, *3*, 399. [CrossRef]
31. Lin, Y.; Yu, J.; Wu, L. The status and ecological effects of antibiotics and heavy metals combined pollution:A review. *Appl. Chem. Ind.* **2023**, *52*, 504–510.
32. Wang, X.; Lan, B.; Fei, H.; Wang, S.; Zhu, G. Heavy metal could drive co-selection of antibiotic resistance in terrestrial subsurface soils. *J. Hazard. Mater.* **2021**, *411*, 124848. [CrossRef] [PubMed]
33. Kuppusamy, S.; Venkateswarlu, K.; Megharaj, M.; Sellappa, K.; Lee, Y.B. Contamination of long-term manure-fertilized Indian paddy soils with veterinary antibiotics: Impact on bacterial communities and antibiotics resistance genes. *Appl. Soil Ecol.* **2023**, *192*, 105106. [CrossRef]
34. De Oliveira Paranhos, A.G.; Pereira, A.R.; da Fonseca, I.C.; Sanson, A.L.; de Cássia Franco Afonso, R.J.; de Aquino, S.F. Analysis of tylosin in poultry litter by HPLC-UV and HPLC-MS/MS after LTPE. *Int. J. Environ. Anal. Chem.* **2020**, *101*, 2568–2585. [CrossRef]

35. Acaroz, U.; Kucukkurt, I.; Ince, S.; Arslan-Acaroz, D.; Gurler, Z.; Eryavuz, A. Assessment for the passage of tylosin into the milk of Anatolian buffaloes. *J. Hell. Vet. Med. Soc.* **2021**, *72*, 3127–3132. [CrossRef]
36. Zhang, Y.; Cheng, D.; Xie, J.; Zhang, Y.; Wan, Y.; Zhang, Y.; Shi, X. Impacts of farmland application of antibiotic-contaminated manures on the occurrence of antibiotic residues and antibiotic resistance genes in soil: A meta-analysis study. *Chemosphere* **2022**, *300*, 134529. [CrossRef] [PubMed]
37. Li, C.; Chen, J.; Wang, J.; Ma, Z.; Han, P.; Luan, Y.; Lu, A. Occurrence of antibiotics in soils and manures from greenhouse vegetable production bases of Beijing, China and an associated risk assessment. *Sci. Total Environ.* **2015**, *521–522*, 101–107. [CrossRef] [PubMed]
38. Kaiser, M.; Ellerbrock, R.H.; Gerke, H.H. Cation Exchange Capacity and Composition of Soluble Soil Organic Matter Fractions. *Soil Sci. Soc. Am. J.* **2008**, *72*, 1278–1285. [CrossRef]
39. Li, D.; Liu, C.; Luo, R.; Sadakane, K.; Lam, T.W. MEGAHIT: An ultra-fast single-node solution for large and complex metagenomics assembly via succinct de Bruijn graph. *Bioinformatics* **2015**, *31*, 1674–1676. [CrossRef]
40. Hyatt, D.; Chen, G.; LoCascio, P.F.; Land, M.L.; Larimer, F.W.; Hause, L.J. Prodigal: Prokaryotic gene recognition and translation initiation site identification. *BMC Bioinform.* **2010**, *11*, 119. [CrossRef]
41. Qian, X.; Gu, J.; Sun, W.; Wang, X.; Su, J.; Stedfeld, R. Diversity, abundance, and persistence of antibiotic resistance genes in various types of animal manure following industrial composting. *J. Hazard. Mater.* **2018**, *344*, 716–722. [CrossRef] [PubMed]
42. Lu, Y.; Pang, L.; Chatzisymeon, E.; Liu, X.; Xu, K.; Yang, P.; Gou, M. Copper in different forms and tetracycline affect behavior and risk of antibiotic resistome in thermophilic anaerobic digestion of cattle manure. *Environ. Sci. Pollut. R.* **2023**, *30*, 108162–108175. [CrossRef] [PubMed]
43. Xie, W.; Yuan, Y.; Wang, Y.; Liu, D.; Shen, Q.; Zhao, F. Hazard reduction and persistence of risk of antibiotic resistance during thermophilic composting of animal waste. *J. Environ. Manag.* **2023**, *330*, 117249. [CrossRef]
44. Wu, S.; Cui, L.; Han, Y.; Lin, F.; Huang, J.; Song, M.; Lan, Z.; Sun, S. Characteristics, Whole-Genome Sequencing and Pathogenicity Analysis of *Escherichia coli* from a White Feather Broiler Farm. *Microorganisms* **2023**, *11*, 2939. [CrossRef] [PubMed]
45. Fatoba, D.O.; Amoako, D.G.; Akebe, A.L.K.; Ismail, A.; Essack, S.Y. Genomic analysis of antibiotic-resistant *Enterococcus spp.* reveals novel enterococci strains and the spread of plasmid-borne *Tet(M)*, *Tet(L)* and *Erm(B)* genes from chicken litter to agricultural soil in South Africa. *J. Environ. Manag.* **2022**, *302*, 114101. [CrossRef] [PubMed]
46. Chi, S.; Xu, W.; Han, Y. ARGs distribution and high-risk ARGs identification based on continuous application of manure in purple soil. *Sci. Total Environ.* **2022**, *853*, 158667. [CrossRef] [PubMed]
47. Awasthi, M.K.; Chen, H.; Awasthi, S.K.; Duan, Y.; Liu, T.; Pandey, A.; Varjani, S.; Zhang, Z. Application of metagenomic analysis for detection of the reduction in the antibiotic resistance genes (ARGs) by the addition of clay during poultry manure composting. *Chemosphere* **2019**, *220*, 137–145. [CrossRef]
48. Magiorakos, A.P.; Srinivasan, A.; Carey, R.B.; Carmeli, Y.; Falagas, M.E.; Giske, C.G.; Harbarth, S.; Hindler, J.F.; Kahlmeter, G.; Olsson-Liljequist, B.; et al. Multidrug-resistant, extensively drug-resistant and pandrug-resistant bacteria: An international expert proposal for interim standard definitions for acquired resistance. *Clin. Microbiol. Infect.* **2012**, *18*, 268–281. [CrossRef] [PubMed]
49. Blazejewska, A.; Zalewska, M.; Grudniak, A.; Popowska, M. A Comprehensive Study of the Microbiome, Resistome, and Physical and Chemical Characteristics of Chicken Waste from Intensive Farms. *Biomolecules* **2022**, *12*, 1132. [CrossRef]
50. Hadjadj, L.; Baron, S.A.; Diene, S.M.; Rolain, J.M. How to discover new antibiotic resistance genes? *Expert Rev. Mol. Diagn.* **2019**, *9*, 349–362. [CrossRef]
51. Webber, M.A.; Piddock, L.J.V. The importance of efflux pumps in bacterial antibiotic resistance. *J. Antimicrob. Chemother.* **2003**, *51*, 9–11. [CrossRef] [PubMed]
52. Lambert, P. Bacterial resistance to antibiotics: Modified target sites. *Adv. Drug Deliv. Rev.* **2005**, *57*, 1471–1485. [CrossRef] [PubMed]
53. Chen, Y.; Yang, K.; Ye, Y.; Zhang, Y.; Mi, H.; Li, C.; Li, Z.; Pei, Z.; Chen, F.; Yan, J.; et al. Reductive soil disinfestation attenuates antibiotic resistance genes in greenhouse vegetable soils. *J. Hazard. Mater.* **2021**, *420*, 126632. [CrossRef]
54. Wang, B.; Song, L.; Li, W.; Hou, L.; Li, J.; Xu, X.; Sheng, G. Distribution and migration of antibiotic resistance genes, as well as their correlation with microbial communities in swine farm and its surrounding environments. *Environ. Pollut.* **2023**, *316*, 120618. [CrossRef]
55. Chen, Q.; An, X.; Li, H.; Zhu, Y.; Su, J.; Cui, L. Do manure-borne or indigenous soil microorganisms influence the spread of antibiotic resistance genes in manured soil? *Soil Biol. Biochem.* **2017**, *114*, 229–237. [CrossRef]
56. Wang, W.; Shen, P.; Lu, Z.; Mo, F.; Liao, Y.; Wen, X. Metagenomics reveals the abundance and accumulation trend of antibiotic resistance gene profile under long-term no tillage in a rainfed agroecosystem. *Front Microbiol.* **2023**, *14*, 1238708. [CrossRef] [PubMed]
57. Amarasekara, N.R.; Mafiz, A.I.; Qian, X.; Tiedje, J.M.; Hao, W.; Zhang, Y. Exploring the co-occurrence of antibiotic, metal, and biocide resistance genes in the urban agricultural environment. *J. Agric. Food Res.* **2023**, *11*, 100474. [CrossRef]
58. Chen, C.; Pankow, C.A.; Oh, M.; Heath, L.S.; Zhang, L.; Du, P.; Xia, K.; Pruden, A. Effect of antibiotic use and composting on antibiotic resistance gene abundance and resistome risks of soils receiving manure-derived amendments. *Environ Int.* **2019**, *128*, 233–243. [CrossRef]
59. Khalid, M.; Liu, X.; Zheng, B.; Su, L.; Kotze, D.J.; Setälä, H.; Ali, M.; Rehman, A.; Saeed-ur-Rahman; Hui, N. Distinct climatic regions drive antibiotic resistance genes dynamics across public parks and pristine soil ecosystems. *J. Clean. Prod.* **2023**, *409*, 137275. [CrossRef]

60. Wang, F.; Xu, M.; Stedtfeld, R.D.; Sheng, H.; Fan, J.; Liu, M.; Chai, B.; de Carvalho, T.S.; Li, H.; Li, Z.; et al. Long-term effect of different fertilization and croping systems on the Soil Antibiotic Resistome. *Environ. Sci. Technol.* 2018; 52, 13037–13046. [CrossRef]
61. Kang, M.; Yang, J.; Kim, S.; Park, J.; Kim, M.; Park, W. Occurrence of antibiotic resistance genes and multidrug-resistant bacteria during wastewater treatment processes. *Sci. Total Environ.* **2022**, *811*, 152331. [CrossRef]
62. Deng, W.; Zhang, A.; Chen, S.; He, X.; Jin, L.; Yu, X.; Yang, S.; Li, B.; Fan, L.; Ji, L.; et al. Heavy metals, antibiotics and nutrients affect the bacterial community and resistance genes in chicken manure composting and fertilized soil. *J. Environ. Manag.* **2020**, *257*, 109980. [CrossRef]
63. Shen, C.; He, M.; Zhang, J.; Liu, J.; Su, J.; Dai, J. Effects of the coexistence of antibiotics and heavy metals on the fate of antibiotic resistance genes in chicken manure and surrounding soils. *Ecotox. Environ. Safe* **2023**, *263*, 115367. [CrossRef] [PubMed]
64. Xue, J.; Wu, J.; Hu, Y.; Sha, C.; Yao, S.; Li, P.; Lin, K.; Cui, C. Occurrence of heavy metals, antibiotics, and antibiotic resistance genes in different kinds of land-applied manure in China. *Environ. Sci. Pollut. Res.* **2021**, *28*, 40011–40021. [CrossRef] [PubMed]
65. Zhao, Z.; Wang, J.; Han, Y.; Chen, J.; Liu, G.; Lu, H.; Yan, B.; Chen, S. Nutrients, heavy metals and microbial communities co-driven distribution of antibiotic resistance genes in adjacent environment ofmariculture. *Environ. Pollut.* **2017**, *220 (B)*, 909–918. [CrossRef]
66. Peng, H.; Gu, J.; Wang, X.; Wang, Q.; Sun, W.; Hu, T.; Guo, H.; Ma, J.; Bao, J. Insight into the fate of antibiotic resistance genes and bacterial community in co-composting green tea residues with swine manure. *J. Environ. Manag.* **2020**, *266*, 110581. [CrossRef]
67. Liu, H.; Ye, X.; Chen, S.; Sun, A.; Duan, X.; Zhang, Y.; Zou, H.; Zhang, Y. Chitosan as additive affects the bacterial community, accelerates the removals of antibiotics and related resistance genes during chicken manure composting. *Sci. Total Environ.* **2021**, *792*, 148381. [CrossRef]
68. Zhang, H.; Dong, M.; Zhou, Y.; Sun, J.; Chang, M.; Zhai, Z. Animal Manure Fertilization Promotes Antibiotic Resistance Gene Dissemination Among Manure, Soil, and Vegetables. *Environ. Sci.* **2021**, *42*, 2080–2088. [CrossRef]
69. Kang, J.; Liu, Y.; Chen, X.; Xu, F.; Xiong, W.; Li, X. Shifts of Antibiotic Resistomes in Soil Following Amendments of Antibiotics-Contained Dairy Manure. *Int. J. Environ. Res. Public Health* **2022**, *19*, 10804. [CrossRef]
70. Calleja-Cervantes, M.E.; Menéndez, S.; Fernández-González, A.J.; Irigoyen, I.; Cibriáin-Sabalza, J.F.; Toro, N.; Aparicio-Tejo, P.M.; Fernández-López, M. Changes in soil nutrient content and bacterial community after 12 years of organic amendment application to a vineyard. *Eur. J. Soil Sci.* **2015**, *66*, 802–812. [CrossRef]
71. Zhang, X.; Meng, H.; Shen, Y.S.; Li, J.; Song, L.S. Survey on heavy metal concentrations and maturity indices of organic fertilizer in China. *Int. J. Agric. Biol. Eng.* **2018**, *11*, 172–179. [CrossRef]
72. Byss, M.; Elhottová, D.; Tříska, J.; Baldrian, P. Fungal bioremediation of the creosote-contaminated soil: Influence of *Pleurotus ostreatus* and *Irpex lacteus* on polycyclic aromatic hydrocarbons removal and soil microbial community composition in the laboratory-scale study. *Chemosphere* **2008**, *73*, 1518–1523. [CrossRef]
73. Lazcano, C.; Gómez-Brandón, M.; Revilla, P.; Domínguez, J. Short-term effects of organic and inorganic fertilizers on soil microbial community structure and function. *Biol. Fertil. Soils* **2012**, *49*, 723–733. [CrossRef]
74. Kielak, A.M.; George, C.C.; Kowalchuk, G.A.; Veen, J.A.V.; Kuramae, E.E. The Ecology of Acidobacteria: Moving beyond Genes and Genomes. *Front. Microbiol.* **2016**, *7*, 171794. [CrossRef]
75. Naylor, D.; McClure, R.; Jansson, J. Trends in Microbial Community Composition and Function by Soil Depth. *Microorganisms* **2022**, *10*, 540. [CrossRef]
76. Wei, R.; Ge, F.; Zhang, L.; Hou, X.; Cao, Y.; Gong, L.; Chen, M.; Wang, R.; Bao, E. Occurrence of 13 veterinary drugs in animal manure-amended soils in Eastern China. *Chemosphere* **2016**, *144*, 2377–2383. [CrossRef] [PubMed]
77. Fang, H.; Han, L.; Zhang, H.; Long, Z.; Cai, L.; Yu, Y. Dissemination of antibiotic resistance genes and human pathogenic bacteria from a pig feedlot to the surrounding stream and agricultural soils. *J. Hazard. Mater.* **2018**, *357*, 53–62. [CrossRef]
78. Zhao, B.; Xu, J.; Zhang, G.; Lu, S.; Liu, X.; Li, L.; Li, M. Occurrence of antibiotics and antibiotic resistance genes in the Fuxian Lake and antibiotic source analysis based on principal component analysis-multiple linear regression model. *Chemosphere* **2021**, *262*, 127741. [CrossRef] [PubMed]
79. Baquero, F.; Lanza, V.F.; Duval, M.; Coque, T.M. Ecogenetics of antibiotic resistance in Listeria monocytogenes. *Mol. Microbiol.* **2020**, *113*, 570–579. [CrossRef]
80. Agga, G.E.; Galloway, H.O.; Appala, K.; Mahmoudi, F.; Kasumba, J.; Loughrin, J.H.; Conte, E. Effect of continuous in-feed administration of tylosin to feedlot cattle on macrolide and tetracycline resistant enterococci in a randomized field trial. *Prev. Vet. Med.* **2023**, *215*, 105930. [CrossRef]
81. Tyson, G.H.; Nyirabahizi, E.; Crarey, E.; Kabera, C.; Lam, C.; Rice-Trujillo, C.; McDermott, P.F.; Tate, H. Prevalence and antimicrobial resistance of enterococci isolated from retail meats in the United States, 2002 to 2014. *Appl. Environ. Microb.* **2018**, *84*, e01902–e01917. [CrossRef]

soil systems

MDPI

Article

Mixed Grazing Increases Abundance of Arbuscular Mycorrhizal Fungi in Upland Welsh Grasslands

Annie Lesley Buckle [1,2], Felicity Victoria Crotty [3] and Philip L. Staddon [1,4,*]

[1] Countryside and Community Research Institute, University of Gloucestershire, Cheltenham GL50 4AZ, UK; anniebuckle@connect.glos.ac.uk
[2] School of Agriculture, Food and Environment, Royal Agricultural University, Cirencester GL7 6JS, UK
[3] Agriculture and Land Team, Ricardo, The Gemini Building, Fermi Avenue, Harwell, Didcot OX11 0QR, UK; felicity.crotty@ricardo.com
[4] Staddon Environmental Consulting, 3 Galileo Gardens, Cheltenham GL51 0GA, UK
* Correspondence: pstaddon1@glos.ac.uk

Abstract: Grasslands play a crucial role in exchanges between global ecosystems and the atmosphere and form an integral part of the agricultural industry. Arbuscular mycorrhizal fungi (AMF) are mutualistic symbionts of most grassland plant species and thereby influence the functional capacity of grassland systems. Agricultural grasslands are primarily used for livestock farming and are subjected to various management practices designed to increase production, but which also alter both plant and soil communities in the process. This research investigated the effects of a selection of management practices and environmental factors on the presence and abundance of AMF in upland Welsh grasslands. The aim was to identify how these management practices affected the abundance of AMF, assessed through microscopic observations of four AMF structures: spores, hyphae, vesicles and arbuscules. The results suggest grazing sheep and cattle together had the highest overall influence on AMF abundance compared to grazing sheep or cattle separately. High plant diversity correlated with high arbuscule and vesicle abundance, but conversely, the application of lime reduced vesicle abundance. These findings offer new insights into the effects of management practices on AMF. Mixing livestock, increasing plant diversity and reducing lime applications are shown here to improve the abundance of AMF and could, therefore, help to inform sustainable farm management decisions in the future.

Keywords: arbuscular mycorrhizal fungi; effects of management practices; mixed grazing

1. Introduction

Grasslands are a globally important habitat with the potential to offer substantial resources and extensive ecosystem services, e.g., flood prevention, biologically diverse habitats and carbon storage [1–3]. Current research into the importance and value of these habitats focuses predominantly on their uses within the agricultural industry, a key area of which is the impact of livestock production, through direct grazing, silage or hay-making [1,2]. Multiple avenues of research are now seeking to understand how grasslands can be managed more appropriately by studying the complex interactions that occur both above and below ground [4–9].

Recent analysis suggests 70% of all agricultural grasslands are now used for livestock production, but these practices are also the main threat to many grassland habitats, with inappropriate management cited as the leading cause of grassland degradation and soil erosion [4,10–14]. However, if managed appropriately, livestock can be an essential tool in sustainable agricultural grasslands by increasing plant diversity and improving soil structure and function, to the extent that many conservation organisations now advocate the use of grazing animals in restoration projects [9,15–17].

Soils play a key role in healthy grasslands, containing a rich and varied diversity of life, including meso and macrofauna, as well as archaeal, bacterial and fungal communities [5,9]. The permanent vegetative canopy above buffers surface temperatures and evaporation, helping to regulate water filtration, decomposition rates and microbial activity in the soil below [5,9]. Soil health underpins many aspects of grassland health but is highly influenced by environmental and anthropogenic changes, both short-term and long-term. This research focuses on the soil microbial level, specifically arbuscular mycorrhizal fungi (AMF), and the influence of a range of management practices widely utilised within the livestock farming community [18,19].

1.1. The Role of Mycorrhizal Fungi within Healthy Grassland Soils

Mycorrhizal fungi are a major group of symbiotic soil fungi, predominantly from the phylum *Glomeromycetes*, but have also evolved independently in the phyla *Ascomycota* and *Basidiomycota* [20,21]. Approximately 80% of all terrestrial plant species form symbiotic relationships with mycorrhizal fungi, a phenomenon which is thought to have evolved around 410 Mya ago and is one of the reasons the plant kingdom has been so successful in colonising terrestrial environments [22–24]. The central benefit of these relationships is the exchange of nutrients, especially phosphorus to the plant and carbohydrates to the fungi, although continuing research is finding more varied and complex exchanges which are still not yet fully understood [25–27].

Arbuscular mycorrhizal fungi (AMF) are one of the most notable species in this group of symbiotic fungi and are unique in their appearance, due to the distinctive, highly branched 'tree-like' structure, the arbuscule, which forms inside plant root cells [24]. AMF are also the only group which form balloon-like storage structures, called vesicles [24]. The ability of the hyphal network to extend into the surrounding soil in order to absorb nutrients surpasses that of the host plant's roots [21,25,28,29]. Both the plant and fungal partner typically produce and obtain more carbon and nutrients together than they require individually, further supporting evidence that these associations are mutually beneficial [26,28,29]. The mycorrhizal network formed around the plant's root system has been strongly linked to improved soil aggregation through the release of glomalin, a glue-like deposit released by the hyphae [25]. Additionally, this hyphal network also creates a microscopic habitat for surrounding microbes which, in turn, release further micro-nutrients for the AMF to absorb, increasing the mutual benefits [30,31].

Although some species of AMF have been commercially produced as soil inoculants, most AMF cannot be synthetically cultivated [32]. This makes enhancing their abundance within agricultural soils highly dependent on appropriate management techniques as opposed to reliance on artificial enrichment [33,34]. This study focuses on how agricultural management practices change the abundance of four mycorrhizal structures (spores, hyphae, vesicles and arbuscules).

1.2. Impact of Livestock and Agricultural Management on Grasslands

Cattle and sheep are globally important animals with estimated numbers of 1.5 billion and 1.2 billion respectively [35]. They provide an important food source in places where the land and soil quality are not sufficient to support arable crops and, therefore, provide vital economic benefits [36]. However, their increasing numbers and influence on grassland ecosystems are now the focus of much attention, especially in light of global food security, climate change, environmental degradation and shifting dietary preferences.

Grassland plants have evolved to tolerate a degree of herbivory without long-term damage, and research has shown that the action of non-intensive grazing can promote biodiversity and stimulate plant growth through compensatory vegetative production [16,37]. Large grazing mammals can, therefore, play an important role in maintaining open grassland habitats, by reducing the encroachment of trees and scrub, varying feeding preferences to create a mosaic of plant species, contributing to the nutrient cycle with deposits of dung and urine, trampling the ground to open niche microhabitats and moving seeds both in and on their bodies [38].

However, overgrazing has been the most cited cause of grassland decline, and it is estimated that between 25 and 30% of global grasslands have been degraded by inappropriate levels of livestock grazing [4,39,40]. Multiple studies offer evidence that high-intensity grazing can negatively impact grasslands by excessive vegetation removal, adversely altering species composition and damaging the soils either directly or indirectly [41–44]. To combat these issues and maintain or even increase the productivity of agricultural grasslands, applications of fertilisers and other agrochemicals are used to artificially elevate grassland yields and livestock productivity [45,46]. However, the consequences of these agrochemicals are now emerging, as increasing application rates are required to compensate for diminishing soil health, so understanding how to readdress the natural balance of grassland soils is of vital importance to ensuring a sustainable future [46].

1.3. Agricultural Management

This research focuses on four agricultural practices within the livestock sector which are each controlled to a greater or lesser degree by management decisions of farmers. The aim of each practice is to increase agricultural output, but this study will investigate their effects on AMF.

Livestock type varies largely across the globe depending on the environmental, social and economic context and area of interest [47]. This study focuses on farmland in Wales, which covers 88% of the land area of this region [48]. Livestock production is central to Welsh farming, accounting for around 75% of total agricultural output [49]. It supplies 40% of the UK's sheep and cattle demand, despite only accounting for 10% of the agricultural land area for the whole of the UK. Currently, there is no published research specifically related to different livestock types and their effects on AMF, so this study aims to investigate this knowledge gap.

Grazing livestock on grasslands is one of the dominant agricultural features of Wales, as the low soil quality and the steep topography of the land is typically unsuitable for high arable output [48]. Almost 80% of Welsh agricultural land is used for either permanent of rough grazing, and Powys has one of the lowest arable land uses in Wales [48]. Only around 8% of farmers house their livestock year-round, with most choosing a mix of housing and grazing, highlighting the importance of healthy grasslands [49]. Currently, most research focuses on the effects of grazing intensity and AMF, commonly linking high-intensity grazing with reduced AMF abundance [43,44].

Lime is applied to soils in order to counteract the negative effects of soil acidification. The soils in Wales are predominantly acidic as a result of natural geological formation and, therefore, typically low in organic content [50,51]. Research has shown that different AMF taxa colonise soils of different pH; therefore, any alteration in the soil's pH by the application of lime could cause a disruption to or decline in the AMF community [52,53].

Plant diversity within livestock grasslands is commonly altered by farmers to maximise forage quantity and quality and compliment the application of agrochemicals, such as fertiliser [54,55]. Ryegrass and clover often dominate agricultural grasslands due to the quality and quantity of their forage, fast establishment and persistence within the field [55]. However, this reduces biodiversity and drought resilience. Research has shown that increased plant diversity can improve soil health and overall grassland resilience [55,56]. Plant diversity has also been linked with AMF, but more research is required as there are many compounding factors that influence this relationship, such as environment and soil composition.

Grasslands and their soils are of great importance to global food systems and environmental functions. Understanding the complexities with which these systems interact is vital for a sustainable and viable agricultural future. This research, therefore, aims to investigate the impact of widely used agricultural management practices upon the aforementioned four structures of AMF, a well-known but still relatively poorly understood element of healthy grassland soils. It is hypothesised that each of the management practices outlined in Section 1.3. will impact one or more of the AMF structures to a greater or lesser degree. It is predicted the practices which align more closely with sustainable practices, e.g., reduced grazing and inputs and increased plant diversity, will result in greater AMF abundance.

2. Materials and Methods

2.1. Site and Field Selection

This study focuses on farmland in Wales, with study sites chosen from a self-selected group of farmers consisting of 11 farms within the lower Wye Valley area of Powys. This area is dominated by livestock, consisting mostly of sheep and cattle and is, therefore, representative of the Welsh livestock industry [49]. The farms ranged in size between 12 hectares and 200 hectares and had varied management approaches with a mix of livestock including sheep and/or cattle.

The farmers were asked to identify two fields on their farm that had been under grass for at least five years and were currently used in a grazing rotation using a low or moderate grazing intensity. Information regarding management practices, such as type of livestock used in the grazing rotation (sheep or cows) and the application of lime, was also collected and recorded (Table 1). Each of the fields were sampled across three temporal periods to ensure representative data across the growing season: early spring (end of March 2022), late spring (late May 2022), and mid-summer (mid July 2022). During each temporal sample period, it was noted whether the fields were being actively grazed by livestock (Table 2). All the farmers were given an identification code and completed a consent form prior to research commencing.

Table 1. Field codes and their corresponding management practices.

Field Code	Livestock Type	Lime	Plant Diversity
A.1	Cattle	No lime	Medium
A.2	Sheep and Cattle	No lime	Medium
B.1	Sheep and Cattle	No lime	Medium
B.2	Sheep and Cattle	No lime	Medium
C.1	Cattle	Lime	High
C.2	Cattle	Lime	High
D.1	Sheep	Lime	Medium
D.2	Sheep	Lime	Low
E.1	Sheep	No lime	High
E.2	Sheep	No lime	High
F.1	Sheep and Cattle	No lime	High
F.2	Sheep and Cattle	No lime	High
G.1	Sheep and Cattle	Lime	Medium
G.2	Sheep and Cattle	Lime	Medium
H.1	Sheep and Cattle	Lime	Low
H.2	Sheep and Cattle	Lime	Low
J.1	Sheep and Cattle	Lime	Low
J.2	Sheep and Cattle	Lime	Low
K.1	Sheep	No lime	Medium
K.2	Sheep	No lime	Medium
M.1	Sheep	Lime	Low
M.2	Sheep	Lime	Low

Table 2. Field codes and their corresponding grazing practices per temporal period.

Field Code	Early Spring	Late Spring	Summer
A.1	No grazing	Active grazing	No grazing
A.2	Active grazing	No grazing	Active grazing
B.1	No grazing	No grazing	No grazing
B.2	No grazing	No grazing	No grazing

Table 2. *Cont.*

Field Code	Early Spring	Late Spring	Summer
C.1	No grazing	No grazing	No grazing
C.2	No grazing	No grazing	No grazing
D.1	No grazing	No grazing	No grazing
D.2	No grazing	No grazing	No grazing
E.1	No grazing	No grazing	No grazing
E.2	No grazing	No grazing	No grazing
F.1	No grazing	Active grazing	No grazing
F.2	Active grazing	Active grazing	No grazing
G.1	No grazing	Active grazing	No grazing
G.2	No grazing	No grazing	No grazing
H.1	No grazing	No grazing	Active grazing
H.2	No grazing	No grazing	Active grazing
J.1	No grazing	Active grazing	Active grazing
J.2	Active grazing	No grazing	No grazing
K.1	No grazing	No grazing	No grazing
K.2	No grazing	Active grazing	No grazing
M.1	No grazing	Active grazing	No grazing
M.2	No grazing	No grazing	No grazing

2.2. *Plant Diversity*

A 50 × 50 cm quadrat sample was taken at random from the selected fields, avoiding gateways, individual features such as trees, water troughs or footpaths and not within 20 m of hedgerows to ensure a representative sample (Table 1). The plants were identified using multiple keys to ensure accuracy [57–59].

The diversity was then ranked in one of three categories: low (indicating 6 or fewer species per quadrat), medium (indicating between 7 and 10 species per quadrat), and high (indicating over 10 species per quadrat).

2.3. *Soil Samples*

2.3.1. Collection and Storage

Three soil cores were taken from each of the selected fields per temporal period using a standardised auger, 5 cm in diameter and 15 cm deep. The cores were taken at random within the fields, but no cores were collected within 20 m of hedgerows, gateways or obvious access paths. At each sampling point, the three cores were mixed into a single field sample and any large vegetative matter was removed. In total, 66 samples were taken. The samples were kept cool and stored within 48 h in a freezer at a constant −18 °C.

2.3.2. Staining

Root staining preparations were adapted from the protocols of Wu et al. (2012) and Penn State, (2022) [60,61]. Blue Parker Quink™ ink was selected as the stain, due to its low toxicity and optimal staining performance comparable to other staining options which were deemed too noxious/carcinogenic [60,62].

Approximately half (600 g) of each field sample was removed from the freezer and defrosted thoroughly for a minimum of 24 h. These subsamples were then weighed and fractioned through 6 mm and 2 mm sieves to isolate the roots, which were separated and rinsed in a fine mesh strainer to remove the remaining soil. The isolated roots were then reweighed.

The roots were then cut into ~1 cm fragments and any large, thick or dead roots removed. The fragments were placed in a 10% potassium hydroxide mixture (KOH) and heated to 80 °C for 30 min. The fragments were rinsed three times in distilled water, placed in blue Parker Quink™ ink and clear white vinegar mix and left to stain for 15 min at room temperature, in alignment with the adapted protocols of Wu et al. (2012) and Penn State, (2022). The fragments were rinsed with distilled water and then submersed in distilled water with two drops of vinegar overnight at room temperature to de-stain further. A total of 660 root fragments were selected at random, mounted on plates with distilled water and observed using a compound microscope at ×250 and ×400 magnification.

2.3.3. Identification of Mycorrhizal Structures

The four AMF structures—spores, hyphae, arbuscules and vesicles—were identified using features standardised by Willis, Rodrigues and Harris, 2013 [25], Dixon et al., 2014 [63], and Walker et al., 2018 [64] (Figure 1).

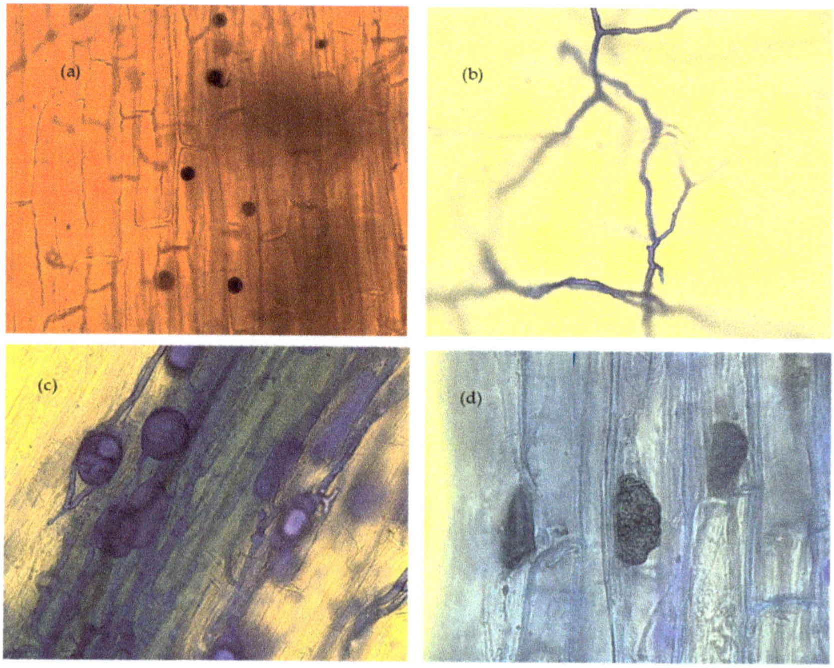

Figure 1. Microscopy images of AMF structures after staining and cleaning process for (**a**) spores, (**b**) hyphae, (**c**) arbuscules, (**d**) vesicles (Copyright Annie Buckle 2022).

For each root fragment selected, the spores, arbuscules and vesicles were assigned an abundance value, shown here in brackets, based on the amount observed through the microscope: none (0), between 1 and 4 (2), between 5 and 9 (5), and 10 or more (10). These scores were used due to the large differences between samples (and to reduce inaccuracy when counting large numbers on microscope slides). Hyphae were noted for their presence (1) or absence (0).

2.4. Data Analysis

All data were analysed in GenStat Version 20.1.2.24528 by VSN International Ltd., Rothamsted Research, St Albans, UK. The data were verified for normality and homoscedasticity prior to analysis and a transformation applied (log10 + 1) if necessary. Two-way sample *T*-tests were used to test for significance in pairwise comparisons, and Analysis of Variance (ANOVA) were used to determine significance for multiple pairwise comparisons. Graphs are presented with means of the abundance values, observed for each AMF structure, to provide an abundance assessment.

3. Results

Overall, all soil samples were found to contain root fragments which had at least one AMF structure, and 78% of the individual root fragments taken from the soil samples were found to contain more than one AMF structure. Significant results are denoted on the graphs by the use of an asterisk (*).

3.1. Livestock Type

There was a significant association between the type of livestock and the relative abundance of arbuscules and the presence of hyphae. When cattle and sheep were grazed together, there was a 43% higher abundance of arbuscules compared to when cattle or sheep were grazed separately ($p = 0.036$, Figure 2a). Similarly, when cattle and sheep were grazed together, there was a 30% higher rate of hyphae presence compared to when cattle or sheep were grazed separately ($p = 0.001$, Figure 2b). No significant differences were found in vesicle or spore abundance.

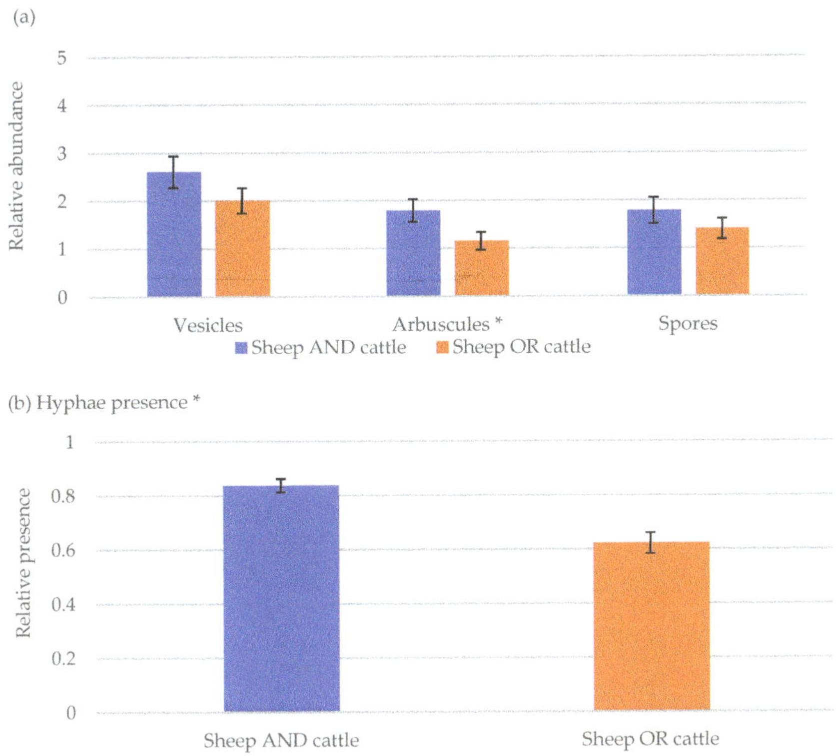

Figure 2. The abundance of (**a**) arbuscules and the presence of (**b**) hyphae were both significantly higher when both sheep and cattle grazed together compared to field systems where sheep or cattle grazed separately. No significant differences were found between vesicles or spores.

3.2. Active Grazing

The fields which were sampled whilst being actively grazed showed significantly lower hyphae presence than those which were not being actively grazed ($p = 0.001$, Figure 3b). No significant differences were found in the abundance of vesicles, arbuscules or spores (Figure 3a).

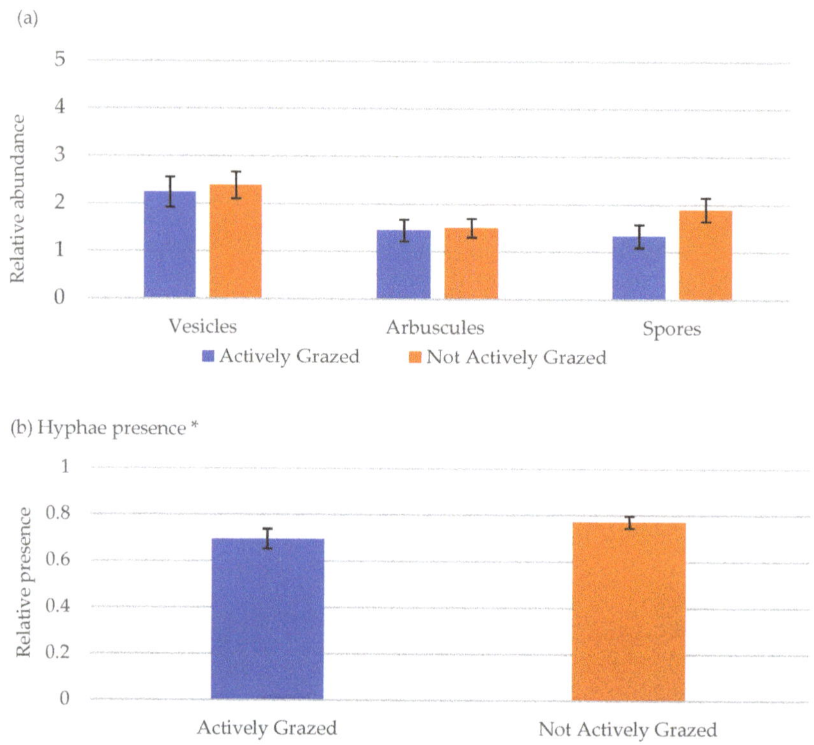

Figure 3. No significant differences were found for the abundance of (**a**) vesicles, arbuscules or spores, but (**b**) the presence of hyphae was significantly lower in fields which had active grazing compared to those which were not being actively grazed.

3.3. Application of Lime

There was a significantly lower abundance of vesicles when lime was applied to the sampled fields, compared to when lime was not applied ($p = 0.014$, Figure 4a). No significant differences were found between the abundance of arbuscules, spores or the presence of hyphae (Figure 4a,b).

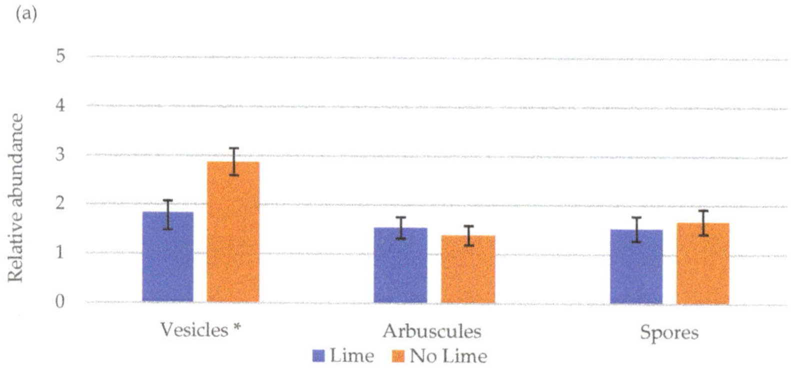

Figure 4. *Cont.*

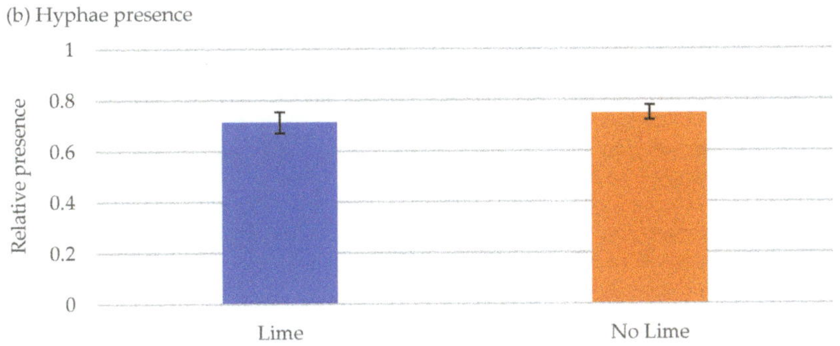

Figure 4. The abundance of (**a**) vesicles was significantly lower in fields which had received an application of lime within the management period of that field but no significant differences were found in the abundance of arbuscules, spores or (**b**) the presence of hyphae.

3.4. Plant Diversity

There was a significantly lower abundance of arbuscules with high plant diversity, compared to that of medium or low diversity ($p = 0.001$, Figure 5a). Conversely, there was a significantly lower abundance of vesicles with low plant diversity, compared to that of medium or high diversity ($p = 0.044$, Figure 5a). No significant differences were found in the abundance of spores or the presence of hyphae (Figure 5a,b).

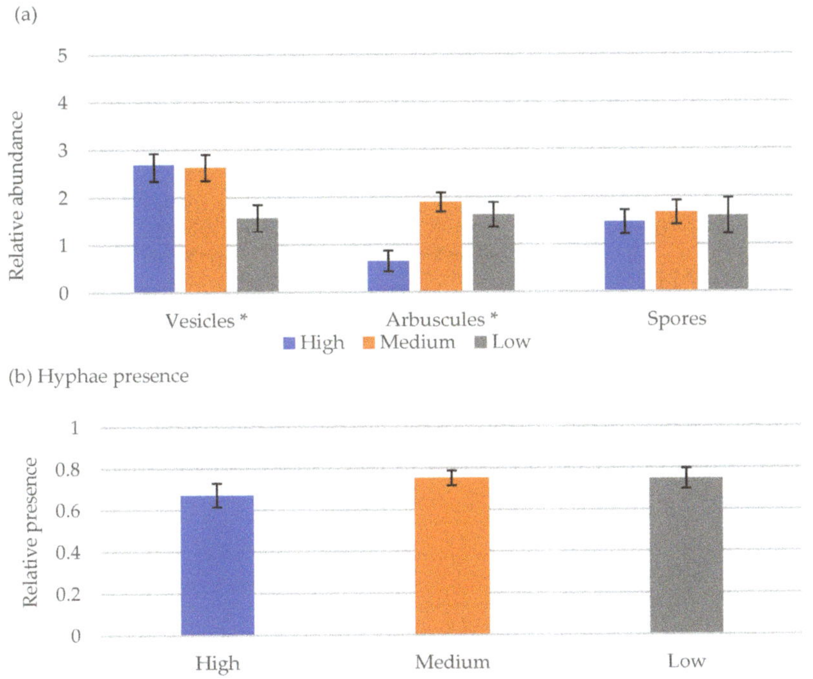

Figure 5. The abundance of (**a**) vesicles was significantly lower in fields with low plant diversity, and the abundance of arbuscules was significantly lower in fields with high plant diversity. No significance was found in the abundance of spores or (**b**) the presence of hyphae.

3.5. Farm Differences

The overall abundance of vesicles, arbuscules and spores, and the presence of hyphae were tested for significance across all eleven farms sampled (Figure 6). The abundance of vesicles showed a greater significant difference between farms D and B ($p = 0.001$), with farms E, G, H, J, K and M most similar to D, and farms A, C and F most similar to B. The abundance of arbuscules showed a greater significant difference between farm E and farms B, G and J jointly ($p = 0.007$), with all remaining farms otherwise unrelated to one another. The presence of hyphae had a greater significant difference between farms D and B ($p = 0.001$), with farms C, K, M and H most similar to one another. No significant differences were found in the abundance of spores ($p = 0.761$).

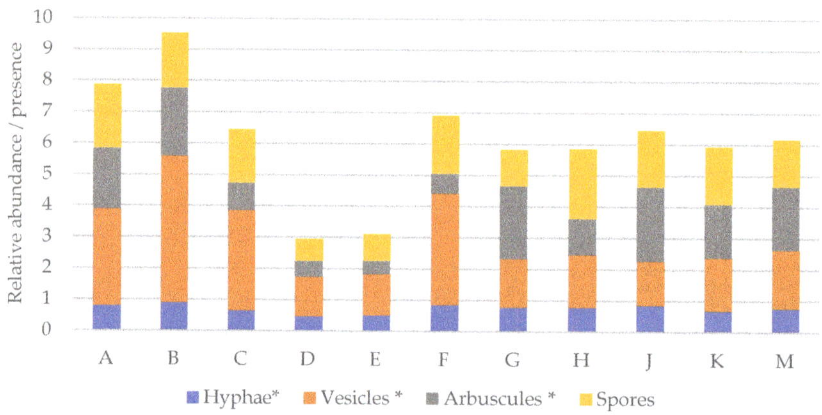

Figure 6. Farm B had a significantly higher abundance of arbuscules and vesicles and higher presence of hyphae compared to the other farms. Farm D had a significantly lower abundance of vesicles and lower presence of hyphae compared to the other farms. Farm E had the lowest abundance of arbuscules compared to the other farms.

It should be noted that farms C and F are organic, and the fields sampled on farms H and J were part of a crop rotation and, therefore, not permanent, despite being within the 5-year timespan stipulated in the selection process. However, these differences were not noticeable in comparison to the other farms (Figure 6).

Each farm showed considerable variation in the abundance and presence of the four AMF structures, and plant diversity. Here, the farms were ranked according to the highest abundance and presences of the four AMF structures (Figure 6) and are considered against each of the four agricultural interventions tested. It was observed that only the practice of grazing cattle and sheep together consistently appeared to correlate with the top-ranking farms (Table 3). No other treatments, which had previously yielded significant results, consistently appeared within these top performing farms.

Table 3. The farms which had the highest overall abundance or presence of the four AMF structures were ordered from highest to lowest (with farm B being the highest).

Overall Mycorrhizal Occurrence (by Farm)	Cattle and Sheep (Per Field)	Lime Applied (Per Farm)	Plant Diversity Ranked High (Per Field)	Actively Grazed (/Field/Temporal Period)
B	++	+	–	++++++
A	+	+	–	+++
F	++	+	++	+++
J	++	–	–	+++
G	++	–	–	+++++
H	++	–	–	++++
M	–	–	–	+++++

Table 3. *Cont.*

Overall Mycorrhizal Occurrence (by Farm)	Cattle and Sheep (Per Field)	Lime Applied (Per Farm)	Plant Diversity Ranked High (Per Field)	Actively Grazed (/Field/Temporal Period)
C	−	−	++	++++++
K	−	+	−	+++++
E	−	+	++	++++++
D	−	−	−	+++++

The two fields on each farm are marked with a (+) if they used both cows and sheep within their grazing rotation, or a (−) if they used either cows, or sheep. The farms which reported to use lime are marked with a (+) or a (−) if they did not use lime. The two fields on each farm are marked with a (+) if they ranked high on plant diversity or a (−) if they ranked either medium or low. The two fields are marked with a (+) if they had been actively grazed at each temporal period or (−) if they were not being actively grazed at each temporal period.

4. Discussion

This study investigated the effects of a selection of common agricultural management practices on the abundance of four structures of AMF. The results show that the most positive effect on AMF resulted from mixed grazing, while plant diversity had mixed results, and the application of lime, along with active grazing, negatively affected AMF. These variable results are discussed in more detail below.

4.1. The Effect of Mixing Livestock

The influence of mixing livestock types and the positive effect on the abundance of arbuscules and presence of hyphae is a novel outcome of this research as there is currently no other research into this particular phenomenon (Figure 1). Although research exists which investigates the influence of mixed grazing on other grassland functions, none refers specifically to the influence of mixed grazing on AMF. The results of this study, therefore, offer a new gateway for research to explore and examine the potential influence of mixed grazing on AMF.

Studies into the effects of mixed grazing or co-grazing on other aspects of grassland functions have found that combining cattle and sheep increases the soil organic content, reduces the bulk density and improves the species composition of the grassland [65–67]. Multiple studies conducted by Cuchillo-Hilario et al. have also shown that grazing sheep and cattle together alters the animals' grazing behaviours and, therefore, foraging selectivity and duration to the extent that plant diversity and botanical composition are also altered in a positive way [68–70]. A study by Zhang et al. (2022) also concluded that mixing livestock resulted in a higher turn-over of root growth and increased organic carbon soil content when compared to only one grazer type or no grazing at all [71]. Additionally, other influencing factors of mixing livestock may include the following: different grazing mechanisms such as ripping or biting, variable grazing duration and amounts consumed, body sizes and the associated weight impact on the soil, varying excretion composition and amount, and even potentially the sex of the animals as hormones within the excrements could affect the soil biome differently [67,72,73]. A recent meta-analysis of mixed grazing, conducted by Su et al. (2023) [65], highlighted the multi-dimensional benefits of mixing sheep and cattle, including effects on plant vegetation, both above and below ground, and the physical, chemical and biological effects on soil. However, with the exception of nematodes, the soil variables they investigated did not include any specific soil communities such as fungi or specifically AMF. As the role of hyphae and arbuscules are concerned predominantly with the transfer of nutrients between the fungi and the plant, the aforementioned beneficial effects exhibited within the soil by mixed grazing are, therefore, likely to be transferred to the AMF network also.

From the perspective of practical application, this initial analysis suggests mixing livestock types could be beneficial to AMF abundance in grassland soils. This study begins to fill this knowledge gap and highlights the need for further research to explore the value of mixed grazing alongside the existing and ongoing research which focuses on grazing intensity.

4.2. Active Grazing Impacts

The effect of grazing intensity on grasslands is one of the most widely studied areas in this field of research and the results from this study support previous findings which suggest grazing can negatively impact AMF (Figure 2) [43,45]. However, these results only indicated a reduction in the presence of hyphae, reflecting the findings of Faghihinia et al. (2020) [43], and support other suggestions [74] that the influence of grazing on AMF, whilst variable, may not necessarily be wholly detrimental.

In their study, Faghihinia et al. (2020) [43] found hyphal length was negatively correlated with increasing grazing intensity, but the number of root fragments colonised by hyphae and the amount of hyphal colonisation on those root fragments were not significantly affected. They also investigated arbuscule intensity, which did not express any significant correlation either. Earlier research by van der Heyde et al. (2017) [74] and Ren et al. (2017) [75] also showed hyphal length was negatively affected by grazing, but root colonisation itself was not. The results of this study appear to agree with those of Faghihinia et al. (2020) [43] and Heyde et al. (2017) [74].

Although it is unclear exactly why hyphal length is affected more severely than hyphal root colonisation, other studies have shown that carbon allocation to the plant roots decreases following the removal of above ground biomass by grazing, as available carbon is reserved for plant regrowth instead of being exchanged with the mycorrhizal associates [45,76]. A similar response also occurs with phosphorus, whereby it is absorbed by the plant for regrowth after vegetative removal, thereby altering the soil phosphorus availability [77]. Early research, supported by these later studies, also shows external hyphal growth is more sensitive to phosphorus availability than other nutrients, with phosphorus being a key trigger in hyphal growth [21,76]. This may go some way in explaining this phenomenon, but more research into the detailed mechanisms is needed. Additionally, the synergetic effects of grazing intensity and mixed grazing would be useful to explore as it is possible the impact of grazing intensity could be reduced or buffered by mixed stocking.

4.3. The Effects of Applying Lime

This research found that the application of lime negatively affected the abundance of both vesicles and root density, which appears to contradict the previous understanding of the benefits of liming (Figure 3).

Early research by Siqueira et al. (1984) [78] on the effect of soil acidity on mycorrhizal fungal colonisation suggests that reducing soil acidity improves colonisation, but it is now known that soil pH affects species of soil fungi differently, and this original assumption cannot be extrapolated across all AMF species [79]. There are currently no directly comparable data related specifically to the abundance of vesicles and the application of lime. However, research by Olsson et al. (2011) [80] identified phosphorus as a key element found within vesicles, along with calcium, sulphur and potassium. These particular nutrients become less available in soils with a pH of 6.5 or less, and therefore, the ability for AMF to provide a reserve of these nutrients for use by the host plants is highly advantageous in acidic soils. However, if lime is applied to readdress the pH balance, thus improving the availability of these nutrients, the need for vesicles is reduced as plants are able to receive them directly from the soil through their own roots or hyphal associations.

4.4. Plant Diversity

The relationship between plant diversity and AMF observed within this study appears to diverge from the published literature (Figure 4). Previous research has found that increased AMF correlates with increased plant diversity, but whilst both vesicle and arbuscule abundance occurred most frequently under medium plant diversity, arbuscule abundance significantly reduced under high plant diversity [81].

A study conducted by Horn et al. (2017) [82] found plant communities are not a strong predictor of AMF communities. Although their findings suggest plant community structure as a driver of AMF community structure, many compounding variables in the environment, such as soil requirements and temporal influences, meant the link was not necessarily universal across all plant and fungal interactions [82]. Research by Faghihinia et al. (2020) [42] also found plant diversity did not have a direct, linear effect on the presence or abundance of AMF, but was more variable and environmentally dependant. As plant diversity increases, the resource demands diversify, thus reducing inter-species competition and enabling the utilisation of different nutrients in different amounts. This trend has been observed in studies whereby increased plant species diversity improved soil resource use and correspondingly increased total plant biomass [83]. This theoretically leads to a reduced reliance on the AMF arbuscule exchanges sites within the root cells but an increase in storage vesicles, as excess nutrients are stored for later utilisation.

4.5. Implications and Limitations

The results of this study have highlighted further areas of potential research that could be applied to sustainable agricultural management practices related to both soil and grassland health within the livestock industry.

Considering the practical applications, utilising a mixed grazing approach could not only reduce reliance on artificial inputs and, therefore, reduce costs, but also offer farmers a more stable and resilient business model through diversification. Collaborations with other local farmers could also be encouraged through the sharing of livestock.

This study recommends more research is urgently required to better understand the effects of mixed grazing animals and AMF. The limitations of this study could be improved upon by increasing the sample size and the area of study to enable better extrapolation of the data, and by the addition of further parameters related to mixed grazing. This could include investigating the effectiveness of existing mixed grazing systems, such as 'co-grazing' or 'follow-on' grazing and the ability for mixed grazing to counteract the impact of grazing intensity.

5. Conclusions

This research aimed to investigate the impact of specific but widely used agricultural interventions on arbuscular mycorrhizal fungi in grazed grasslands.

The results of this research offer two novel insights: Mixed grazing promoted AMF, and the application of lime reduced AMF. The positive effect of mixed grazing is an important finding when considering the potentially negative effect of certain grazing intensities and, therefore, could have a significant impact on how livestock is managed in the future.

Grasslands play a key role within agriculture. As well as being an irreplaceable resource for the livestock industry, they provide invaluable habitats for wild species and multiple other ecosystem services. This research contributes to the knowledge required to inform best-practice grassland management, policy and governance and will help to inform the development and restoration of healthy, diverse and resilient grassland ecosystems.

Author Contributions: A.L.B. and P.L.S. designed the research. A.L.B. performed the research with guidance from P.L.S. and F.V.C., A.L.B. wrote the paper. P.L.S. and F.V.C. provided comments. All authors have read and agreed to the published version of the manuscript.

Funding: This research was performed as part of requirements for MSc at the RAU. No external funding was involved.

Institutional Review Board Statement: Not applicable.

Informed Consent Statement: Not applicable.

Data Availability Statement: Data is available from A.L.B.

Acknowledgments: Thanks to Hal Drysdale and the farmers from the Wye Soil Group who contributed their time and assistance and to Kathrin Heimbach for her assistance in the field.

Conflicts of Interest: No conflicts of interest exist.

References

1. Blair, J.; Nippert, J.; Briggs, J. Grassland Ecology. In *Ecology and the Environment*; Springer: Berlin/Heidelberg, Germany, 2014; pp. 389–423. [CrossRef]
2. Roberts, A.M. *Tamed: Ten Species That Changed Our World*; Windmill Books: London, UK, 2018.
3. Bengtsson, J.; Bullock, J.M.; Egoh, B.; Everson, C.; Everson, T.; O'Connor, T.; O'Farrell, P.J.; Smith, H.G.; Lindborg, R. Grasslands-more important for ecosystem services than you might think. *Ecosphere* **2019**, *10*, e02582. [CrossRef]
4. Bardgett, R.D.; Bullock, J.M.; Lavorel, S.; Manning, P.; Schaffner, U.; Ostle, N.; Chomel, M.; Durigan, G.; Fry, E.L.; Johnson, D.; et al. Combatting global grassland degradation. *Nat. Rev. Earth Environ.* **2021**, *2*, 720–735. [CrossRef]
5. Petermann, J.S.; Buzhdygan, O.Y. Grassland biodiversity. *Curr. Biol.* **2021**, *31*, R1195–R1201. [CrossRef]
6. Scurlock, J.M.O.; Hall, D.O. The global carbon sink: A grassland perspective. *Glob. Change Biol.* **1998**, *4*, 229–233. [CrossRef]
7. Jones, M.B.; Donnelly, A. Carbon sequestration in temperate grassland ecosystems and the influence of management, climate and elevated CO_2. *New Phytol.* **2004**, *164*, 423–439. [CrossRef]
8. Xu, S.; Eisenhauer, N.; Ferlian, O.; Zhang, J.; Zhou, G.; Lu, X.; Liu, C.; Zhang, D. Species richness promotes ecosystem carbon storage: Evidence from biodiversity-ecosystem functioning experiments. *Proc. R. Soc. B Biol. Sci.* **2020**, *287*, 20202063. [CrossRef]
9. Lemaire, G.; Hodgson, J.; Chabbi, A. *Grassland Productivity and Ecosystem Services*; CABI: Wallingford, UK, 2011.
10. O'Mara, F.P. The role of grasslands in food security and climate change. *Ann. Bot.* **2012**, *110*, 1263–1270. [CrossRef] [PubMed]
11. Eldridge, D.J.; Poore, A.G.B.; Ruiz-Colmenero, M.; Letnic, M.; Soliveres, S. Ecosystem structure, function, and composition in rangelands are negatively affected by livestock grazing. *Ecol. Appl.* **2016**, *26*, 1273–1283. [CrossRef]
12. Nordborg, M. *Holistic Management—A Critical Review of Allan Savory's Grazing Method*; Roos, E., Ed.; Centre for Organic Food & Farming: Umeå, Sweden, 2016.
13. Sirimarco, X.; Barral, M.P.; Villarino, S.H.; Laterra, P. Water regulation by grasslands: A global meta-analysis. *Ecohydrology* **2018**, *11*, e1934. [CrossRef]
14. Lai, L.; Kumar, S. A global meta-analysis of livestock grazing impacts on soil properties. *PLoS ONE* **2020**, *15*, e0236638. [CrossRef]
15. Kemp, D.R.; Michalk, D.L. Towards sustainable grassland and livestock management. *J. Agric. Sci.* **2007**, *145*, 543–564. [CrossRef]
16. Flack, S. *The Art and Science of Grazing: How Grass Farmers Can Create Sustainable Systems for Healthy Animals and Farm Ecosystems*; Chelsea Green Publishing: Chelsea, VT, USA, 2016.
17. Kapás, R.E.; Plue, J.; Kimberley, A.; Cousins, S.A.O. Grazing livestock increases both vegetation and seed bank diversity in remnant and restored grasslands. *J. Veg. Sci.* **2020**, *31*, 1053–1065. [CrossRef]
18. Voisin, R.; Horwitz, P.; Godrich, S.; Sambell, R.; Cullerton, K.; Devine, A. What goes in and what comes out: A scoping review of regenerative agricultural practices. *Agroecol. Sustain. Food Syst.* **2024**, *48*, 124–158. [CrossRef]
19. Lyons, K.G.; Török, P.; Hermann, J.-M.; Kiehl, K.; Kirmer, A.; Kollmann, J.; Overbeck, G.E.; Tischew, S.; Allen, E.B.; Bakker, J.D.; et al. Challenges and opportunities for grassland restoration: A global perspective of best practices in the era of climate change. *Glob. Ecol. Conserv.* **2023**, *46*, e02612. [CrossRef]
20. Smith, S.E.; Read, D.J. *Mycorrhizal Symbiosis*, 2nd ed.; Academic Press, Inc.: New York, NY, USA, 1997.
21. Lewis, J.D. Mycorrhizal Fungi, Evolution and Diversification of. *Encycl. Evol. Biol.* **2016**, *3*, 94–99. [CrossRef]
22. Malloch, D.W.; Pirozynski, K.A.; Raven, P.H. Ecological and evolutionary significance of mycorrhizal symbioses in vascular plants (A Review). *Proc. Natl. Acad. Sci. USA* **1980**, *77*, 2113–2118. [CrossRef]
23. Abbott, L.; Josland, S.; Lotinga, A.; Kruger, A.; Griffiths, L. (Eds.) *Encyclopedia of Organic Gardening—The Complete Guide to Natural & Chemical-Free Gardening*; Original Work Published 2001; The Henry Doubleday Research Association: London, UK, 2005; pp. 304–310.
24. Willis, A.; Rodrigues, B.F.; Harris, P.J.C. The Ecology of Arbuscular Mycorrhizal Fungi. *Crit. Rev. Plant Sci.* **2013**, *32*, 1–20. [CrossRef]
25. Chen, M.; Arato, M.; Borghi, L.; Nouri, E.; Reinhardt, D. Beneficial Services of Arbuscular Mycorrhizal Fungi—From Ecology to Application. *Front. Plant Sci.* **2018**, *9*, 408113. [CrossRef]
26. Bennett, A.E.; Groten, K. The Costs and Benefits of Plant–Arbuscular Mycorrhizal Fungal Interactions. *Annu. Rev. Plant Biol.* **2022**, *73*, 649–672. [CrossRef]
27. Tang, B.; Man, J.; Lehmann, A.; Rillig, M.C. Arbuscular mycorrhizal fungi benefit plants in response to major global change factors. *Ecol. Lett.* **2023**, *26*, 2087–2097. [CrossRef]
28. Bonfante, P.; Genre, A. Mechanisms underlying beneficial plant–fungus interactions in mycorrhizal symbiosis. *Nat. Commun.* **2010**, *1*, 48. [CrossRef] [PubMed]
29. Goss, M.J.; Carvalho, M.; Brito, I. *Chapter 3—The Roles of Arbuscular Mycorrhiza and Current Constraints to Their Intentional Use in Agriculture*; Goss, M.J., Carvalho, M., Brito, I., Eds.; ScienceDirect; Academic Press: New York, NY, USA, 2017.
30. Faghihinia, M.; Jansa, J.; Halverson, L.J.; Staddon, P.L. Hyphosphere microbiome of arbuscular mycorrhizal fungi: A realm of unknowns. *Biol. Fertil. Soils* **2022**, *59*, 17–34. [CrossRef]
31. Marschner, P. Rhizosphere Biology. In *Marschner's Mineral Nutrition of Higher Plants*; Academic Press: New York, NY, USA, 2012; pp. 369–388. [CrossRef]

32. Hart, M.; Klironomos, J. Colonization of roots by arbuscular mycorrhizal fungi using different sources of inoculum. *Mycorrhiza* **2002**, *12*, 181–184. [CrossRef] [PubMed]
33. Sullia, S.B. Use of Vesicular-Arbuscular Mycorrhiza (VAM) as Biofertilizer for Horticultural Plants in Developing Countries. In *Horticulture—New Technologies and Applications*; Springer: Berlin/Heidelberg, Germany, 1991; pp. 49–53. [CrossRef]
34. Duffy, E.M.; Cassells, A.C. Root Development | Mycorrhizae. In *Encyclopedia of Applied Plant Sciences*; Elsevier: Amsterdam, The Netherlands, 2003; pp. 1107–1115. [CrossRef]
35. Gilbert, M.; Nicolas, G.; Cinardi, G.; Van Boeckel, T.P.; Vanwambeke, S.O.; Wint, G.R.W.; Robinson, T.P. Global distribution data for cattle, buffaloes, horses, sheep, goats, pigs, chickens and ducks in 2010. *Sci. Data* **2018**, *5*, 180227. [CrossRef]
36. van Zanten, H.H.E.; Meerburg, B.G.; Bikker, P.; Herrero, M.; de Boer, I.J.M. Opinion paper: The role of livestock in a sustainable diet: A land-use perspective. *Animal* **2015**, *10*, 547–549. [CrossRef]
37. Török, P.; Penksza, K.; Tóth, E.; Kelemen, A.; Sonkoly, J.; Tóthmérész, B. Vegetation type and grazing intensity jointly shape grazing effects on grassland biodiversity. *Ecol. Evol.* **2018**, *8*, 10326–10335. [CrossRef]
38. Wilsey, B.J. *The Biology of Grasslands*; Oxford University Press: London, UK, 2018.
39. Gang, C.; Zhou, W.; Chen, Y.; Wang, Z.; Sun, Z.; Li, J.; Qi, J.; Odeh, I. Quantitative assessment of the contributions of climate change and human activities on global grassland degradation. *Environ. Earth Sci.* **2014**, *72*, 4273–4282. [CrossRef]
40. Faghihinia, M.; Zou, Y.; Chen, Z.; Bai, Y.; Li, W.; Marrs, R.; Staddon, P.L. Environmental drivers of grazing effects on arbuscular mycorrhizal fungi in grasslands. *Appl. Soil Ecol.* **2020**, *153*, 103591. [CrossRef]
41. Faghihinia, M.; Zou, Y.; Chen, Z.; Bai, Y.; Li, W.; Marrs, R.; Staddon, P.L. The response of grassland mycorrhizal fungal abundance to a range of long-term grazing intensities. *Rhizosphere* **2020**, *13*, 100178. [CrossRef]
42. Lugo, M.A.; Maza, M.E.G.; Cabello, M.N. Arbuscular Mycorrhizal Fungi in a Mountain Grassland II: Seasonal Variation of Colonization Studied, along with Its Relation to Grazing and Metabolic Host Type. *Mycologia* **2003**, *95*, 407. [CrossRef]
43. Yang, X.; Shen, Y.; Liu, N.; Wilson, G.W.T.; Cobb, A.B.; Zhang, Y. Defoliation and arbuscular mycorrhizal fungi shape plant communities in overgrazed semiarid grasslands. *Ecology* **2018**, *99*, 1847–1856. [CrossRef] [PubMed]
44. Faghihinia, M.; Zou, Y.; Bai, Y.; Dudáš, M.; Marrs, R.; Staddon, P.L. Grazing Intensity Rather than Host Plant's Palatability Shapes the Community of Arbuscular Mycorrhizal Fungi in a Steppe Grassland. *Microb. Ecol.* **2021**, *84*, 1062–1071. [CrossRef] [PubMed]
45. AHDB. Nutrient Management Guide (RB209) Section 1 Principles of Nutrient Management and Fertiliser Use 2. In *AHDB; RB209 Section 1 Principles of Nutrient Management and Fertiliser Use*; AHDB: New York, NY, USA, 2023.
46. Mandal, A.; Sarkar, B.; Mandal, S.; Vithanage, M.; Patra, A.K.; Manna, M.C. Impact of agrochemicals on soil health. In *Agrochemicals Detection, Treatment and Remediation*; Academic Press: New York, NY, USA, 2020; pp. 161–187. [CrossRef]
47. National Statistics. Welsh Agricultural Statistics. In *Welsh Government*; National Statistics: Wales, UK, 2018.
48. Statistics for Wales. Farming Facts and Figures, Wales 2022. In *Llywodraeth Cymru-Welsh Government*; Statistics for Wales: Wales, UK, 2022.
49. DEFRA. *An Official Statistics Publication: Defra Official Statistics Are Produced to the High Professional Standards Set Out in the Code of Practice for Official Statistics. Grazing and Cattle Housing (Section 1)*; DEFRA: London, UK, 2019.
50. Rollett, A.; Williams, J. *Assessment of Welsh Soil Issues in Context-Soil Policy Evidence Programme 2018–2019*; Soil Policy & Agricultural Land Use Planning Unit: UK, 2019.
51. Ashman, M.R.; Puri, G. *Essential Soil Science: A Clear and Concise Introduction to Soil Science*; Blackwell Science: Hoboken, NJ, USA, 2002.
52. Porter, W.M.; Robson, A.D.; Abbott, L.K. Field Survey of the Distribution of Vesicular-Arbuscular Mycorrhizal Fungi in Relation to Soil pH. *J. Appl. Ecol.* **1987**, *24*, 659–662. [CrossRef]
53. Davison, J.; Moora, M.; Semchenko, M.; Adenan, S.B.; Ahmed, T.; Akhmetzhanova, A.A.; Alatalo, J.M.; Al-Quraishy, S.; Andriyanova, E.; Anslan, S.; et al. Temperature and pH define the realised niche space of arbuscular mycorrhizal fungi. *New Phytol.* **2021**, *231*, 763–776. [CrossRef] [PubMed]
54. Schaub, S.; Finger, R.; Leiber, F.; Probst, S.; Kreuzer, M.; Weigelt, A.; Buchmann, N.; Scherer-Lorenzen, M. Plant diversity effects on forage quality, yield and revenues of semi-natural grasslands. *Nat. Commun.* **2020**, *11*, 768. [CrossRef]
55. AHDB. *Recommended Grass and Clover Lists for England and Wales 2023/24 Handbook*; AHDB: New York, NY, USA, 2023. Available online: https://projectblue.blob.core.windows.net/media/Default/Beef%20&%20Lamb/RGCL/RGCL%20Handbook%202023%20WEB.pdf (accessed on 17 August 2024).
56. Stiles, W. Can Increasing Plant Species Richness in Grassland Maintain Yield and Improve Soil Carbon Storage? *Farming Connect*, 23 January 2017.
57. Sinker, C. *A Lateral Key to Common Grasses*; Shropshire Conservation Trust: Shrewsbury, MA, USA, 1975.
58. Spohn, M.; Spohn, R. *Wild Flowers of Britain and Europe*; A. & C. Black: London, UK, 2008.
59. Emberson, C. Wildlife Trusts. In *Bloomsbury Concise Wild Flower Guide*; Bloomsbury Wildlife: London, UK, 2014.
60. Wu, Q.S.; Cao, M.Q.; Zou, Y.N. A Simple and Nontoxic Ink and Acetic Acid Staining Technique for Arbuscular Mycorrhizal Structures. *Adv. Mater. Res.* **2012**, *518–523*, 679–682. [CrossRef]
61. Penn State. Staining of Mycorrhizal Fungi (AMF) Colonized Roots. Plone Site. 2022. Available online: https://plantscience.psu.edu/research/labs/roots/methods/methods-info/staining-of-mycorrhizal-fungi (accessed on 17 August 2024).
62. Vierheilig, H.; Coughlan, A.P.; Wyss, U.; Piché, Y. Ink and Vinegar, a Simple Staining Technique for Arbuscular-Mycorrhizal Fungi. *Appl. Environ. Microbiol.* **1998**, *64*, 5004–5007. [CrossRef]

63. Dixon, A.P.; Faber-Langendoen, D.; Josse, C.; Morrison, J.; Loucks, C.J. Distribution mapping of world grassland types. *J. Biogeogr.* **2014**, *41*, 2003–2019. [CrossRef]
64. Walker, C.; Harper, C.J.; Brundrett, M.C.; Krings, M. Looking for Arbuscular Mycorrhizal Fungi in the Fossil Record. In *Transformative Paleobotany*; Academic Press: New York, NY, USA, 2018; pp. 481–517. [CrossRef]
65. Su, J.; Xu, F.; Zhang, Y. Grassland biodiversity and ecosystem functions benefit more from cattle than sheep in mixed grazing: A meta-analysis. *J. Environ. Manag.* **2023**, *337*, 117769. [CrossRef]
66. Abaye, A.O.; Allen, V.G.; Fontenot, J.P. Grazing Sheep and Cattle Together or Separately: Effect on Soils and Plants. *Agron. J.* **1997**, *89*, 380–386. [CrossRef]
67. Animut, G.; Goetsch, A.L. Co-grazing of sheep and goats: Benefits and constraints. *Small Rumin. Res.* **2008**, *77*, 127–145. [CrossRef]
68. Cuchillo Hilario, M.; Isselstein, J. Intake choices of cattle and sheep grazing alone or together on grass swards differing in plant species diversity. *Grassl. A Change World* **2010**, *15*, 922–1082.
69. Cuchillo Hilario, M.; Wrage-Mönnig, N.; Isselstein, J. Behavioral patterns of (co-)grazing cattle and sheep on swards differing in plant diversity. *Appl. Anim. Behav. Sci.* **2017**, *191*, 17–23. [CrossRef]
70. Cuchillo-Hilario, M.; Wrage-Mönnig, N.; Isselstein, J. Forage selectivity by cattle and sheep co-grazing swards differing in plant species diversity. *Grass Forage Sci.* **2017**, *73*, 320–329. [CrossRef]
71. Zhang, Y.; Wang, Z.; Liu, P.; Wang, C. Mixed cattle and sheep grazing reduces the root lifespan of the community in a desert steppe. *Ecol. Indic.* **2022**, *143*, 109422. [CrossRef]
72. Cowlishaw, S.J.; Alder, F.E. The grazing preferences of cattle and sheep. *J. Agric. Sci.* **1960**, *54*, 257–265. [CrossRef]
73. Forbes, T.D.A.; Hodgson, J. The reaction of grazing sheep and cattle to the presence of dung from the same or the other species. *Grass Forage Sci.* **1985**, *40*, 177–182. [CrossRef]
74. Heyde, M.; van der Bennett, J.A.; Pither, J.; Hart, M. Longterm effects of grazing on arbuscular mycorrhizal fungi. *Agric. Ecosyst. Environ.* **2017**, *243*, 27–33. [CrossRef]
75. Ren, H.; Gui, W.; Bai, Y.; Stein, C.; Rodrigues, J.L.M.; Wilson, G.W.T.; Cobb, A.B.; Zhang, Y.; Yang, G. Long-term effects of grazing and topography on extra-radical hyphae of arbuscular mycorrhizal fungi in semi-arid grasslands. *Mycorrhiza* **2017**, *28*, 117–127. [CrossRef]
76. Faghihinia, M.; Zou, Y.; Bai, Y.; Pourbakhtiar, A.; Marrs, R.; Staddon, P.L. Long-Term Grazing Intensity Impacts Belowground Carbon Allocation and Mycorrhizas Revealed by $13CO_2$ Pulse Labeling. *Rangel. Ecol. Manag.* **2023**, *86*, 64–72. [CrossRef]
77. Song, L.; Gong, J.; Zhang, Z.; Zhang, W.; Zhang, S.; Dong, J.; Dong, X.; Hu, Y.; Liu, Y. Changes in plant phosphorus demand and supply relationships in response to different grazing intensities affect the soil organic carbon stock of a temperate steppe. *Sci. Total Environ.* **2023**, *876*, 163225. [CrossRef] [PubMed]
78. Siqueira, J.O.; Hubbell, D.H.; Mahmud, A.W. Effect of liming on spore germination, germ tube growth and root colonization by vesicular-arbuscular mycorrhizal fungi. *Plant Soil* **1984**, *76*, 115–124. [CrossRef]
79. Yin, C.; Schlatter, D.C.; Kroese, D.R.; Paulitz, T.C.; Hagerty, C.H. Responses of Soil Fungal Communities to Lime Application in Wheat Fields in the Pacific Northwest. *Front. Microbiol.* **2021**, *12*, 576763. [CrossRef] [PubMed]
80. Olsson, P.A.; Hammer, E.C.; Pallon, J.; van Aarle, I.M.; Wallander, H. Elemental composition in vesicles of an arbuscular mycorrhizal fungus, as revealed by PIXE analysis. *Fungal Biol.* **2011**, *115*, 643–648. [CrossRef] [PubMed]
81. Bever, J.D.; Schultz, P.A.; Pringle, A.; Morton, J.B. Arbuscular Mycorrhizal Fungi: More Diverse than Meets the Eye, and the Ecological Tale of Why. *BioScience* **2001**, *51*, 923. [CrossRef]
82. Horn, S.S.; Hempel, S.; Verbruggen, E.; Rillig, M.C.; Caruso, T. Linking the community structure of arbuscular mycorrhizal fungi and plants: A story of interdependence? *Int. Soc. Microb. Ecol.* **2017**, *11*, 1400–1411. [CrossRef]
83. Tilman, D.; Lehman, C.L.; Thomson, K.T. Plant diversity and ecosystem productivity: Theoretical considerations. *Proc. Natl. Acad. Sci. USA* **1997**, *94*, 1857–1861. [CrossRef]

MDPI AG
Grosspeteranlage 5
4052 Basel
Switzerland
Tel.: +41 61 683 77 34

Soil Systems Editorial Office
E-mail: soilsystems@mdpi.com
www.mdpi.com/journal/soilsystems

www.ingramcontent.com/pod-product-compliance
Lightning Source LLC
Chambersburg PA
CBHW061842210926
51850CB00006B/948